Lecture Notes in Computer Science 3521

Commenced Publication in 1973
Founding and Former Series Editors:
Gerhard Goos, Juris Hartmanis, and Jan van Leeuwen

Nimrod Megiddo   Yinfeng Xu
Binhai Zhu (Eds.)

Algorithmic
Applications
in Management

First International Conference, AAIM 2005
Xian, China, June 22-25, 2005
Proceedings

 Springer

Volume Editors

Nimrod Megiddo
IBM Almaden Research Center
650 Harry Road, San Jose, CA 95120, USA
E-mail: megiddo@almaden.ibm.com

Yinfeng Xu
Xi'an Jiaotong University
School of Management
28 West Xianning Road, Xi'an, Shaanxi 710049, China
E-mail: yfxu@mail.xjtu.edu.cn

Binhai Zhu
Montana State University
Department of Computer Science
EPS 357, Bozeman, MT 59717-3880, USA
E-mail: bhz@cs.montana.edu

Library of Congress Control Number: 2005926828

CR Subject Classification (1998): F.2.1-2, E.1, G.1-3, J.1

ISSN   0302-9743
ISBN-10   3-540-26224-5 Springer Berlin Heidelberg New York
ISBN-13   978-3-540-26224-4 Springer Berlin Heidelberg New York

Springer is a part of Springer Science+Business Media

springeronline.com

© Springer-Verlag Berlin Heidelberg 2005
Printed in Germany

Typesetting: Camera-ready by author, data conversion by Scientific Publishing Services, Chennai, India
Printed on acid-free paper   SPIN: 11496199   06/3142   5 4 3 2 1 0

Preface

The papers in this volume were presented at the 1st International Conference on Algorithmic Applications in Management (AAIM 2005), held June 22–25, 2005 in Xi'an, China. The topics cover algorithmic applications in most management-related areas.

Submissions to the conference this year were conducted electronically. A total of 140 papers were submitted, of which 46 were accepted. The papers were evaluated by an international Program Committee consisting of Franz Aurenhammer, Sergey Bereg, Danny Z. Chen, Jian Chen, Zhixiang Chen, Edith Cohen, Xiaotie Deng, Michael Goldwasser, Jason Hartline, Wen-Lian Hsu, Haijun Huang, Minghui Jiang, Ellis Johnson, Naoki Katoh, Masakazu Kojima, Mohammad Mahdian, Nimrod Megiddo, Zhongping Qin, Panos Pardalos, Chung Keung Poon, Bruce Reed, Shouyang Wang, Peter Widmayer, Yinfeng Xu, Frances Yao, Yinyu Ye and Binhai Zhu.

The submitted papers were from Canada, Chile, China, Finland, Germany, Hong Kong, Iran, Israel, Japan, Korea, Malaysia, Mexico, Netherlands, Singapore, UK and USA. Each paper was evaluated by at least two Program Committee members, assisted in some cases by subreferees. In addition to selected papers, the conference also included two invited presentations, by Ellis Johnson and Yinyu Ye, and two invited papers, by Xujin Chen et al. and Siu-Wing Cheng et al.

We thank all the people who made this meeting possible: the authors for submitting papers, the Program Committee members and external referees (listed on the pages that follows) for their excellent work, and the two invited speakers. Finally, we thank NSF of China and the School of Management, Xi'an Jiaotong University for their support, and the local organizers and our colleagues for their assistance.

June 2005

Nimrod Megiddo
Yinfeng Xu
Binhai Zhu

Organization

Program Committee Chairs

Nimrod Megiddo, IBM Almaden Research Center, USA
Yinfeng Xu, Xi'an Jiaotong University, China
Binhai Zhu, Montana State University, USA

Program Committee Members

Franz Aurenhammer, TU Graz, Austria
Sergey Bereg, University of Texas-Dallas, USA
Danny Z. Chen, University of Notre Dame, USA
Jian Chen, Tsinghua University, China
Zhixiang Chen, University of Texas-Pan American, USA
Edith Cohen, AT&T Research, USA
Xiaotie Deng, City University of Hong Kong, China
Michael Goldwasser, Saint Louis University, USA
Jason Hartline, Microsoft Research, USA
Wen-Lian Hsu, Academia Sinica, Taiwan
Haijun Huang, NSF China, China
Minghui Jiang, Montana State University, USA
Ellis Johnson, Georgia Institute of Technology, USA
Naoki Katoh, Kyoto, Japan
Masakazu Kojima, Tokyo Institute of Technology, Japan
Mohammad Mahdian, Microsoft Research, USA
Panos M. Pardalos, University of Florida, USA
Chung Keung Poon, City University of Hong Kong, China
Zhongping Qin, Huazhong University of Science and Technology, China
Bruce Reed, McGill University, Canada
Shouyang Wang, Chinese Academy of Sciences, China
Peter Widmayer, ETH Zurich, Switzerland
Frances Yao, City University of Hong Kong, China
Yinyu Ye, Stanford University, USA

Organizing Committee

Wentian Cui, Xi'an Jiaotong University, China
Qingchuan Xu, Xi'an Jiaotong University, China
Yinfeng Xu, Xi'an Jiaotong University, China
Binhai Zhu, Montana State University, USA

Referees

Jian-Hung Chen	Guojun Li	Keun-Pin Wu
Ning Chen	Becky Jie Liu	Qingchuan Xu
Xi Chen	Kamil Sarac	Weijun Xu
Andy Chun	Atsushi Takizawa	Tommy Guomin Yang
Wentian Cui	Shin'ichi Tanigawa	Hong Zhu
Qizhi Fang	Peng-Jun Wan	
Yukinobu Hamuro	Duncan Wong	

Table of Contents

Invited Lecture

Contributed Papers

Robust Airline Fleet Assignment: Imposing Station Purity Using Station Decomposition

Ellis L. Johnson

Georgia Institute of Technology, Atlanta, Georgia, USA
ejohnson@isye.gatech.edu

Fleet assignment models are used by many airlines to assign aircraft to flights in a schedule to maximize profit [1]. A major airline reported that the use of the fleet assignment model increased annual profits by more than $100 million [3] a year over three years. The results of fleet assignment models affect subsequent planning, marketing and operational processes within the airline. Anticipating these processes and developing solutions favorable to them can further increase the benefits of fleet assignment models. We develop fleet assignment solutions that increase planning flexibility and reduce cost by imposing station purity, limiting the number of fleet types allowed to serve each airport in the schedule [4]. Imposing station purity on the fleet assignment model can limit aircraft dispersion in the network and make solutions more robust relative to crew planning, maintenance planning and operations.

Because imposition of station purity constraints can significantly increase computational difficulty, we develop a solution approach, station decomposition, which takes advantage of airline network structure. Station decomposition is an instance of Dantzig-Wolfe decomposition and uses a column generation approach to solving the fleet assignment problem. We further improve the performance of station decomposition by developing a primal-dual method that increases solution quality and reduces running times. This method can be applied generally within the Dantzig-Wolfe decomposition framework to speed convergence. It avoids "instability of the duals" and minimizes the "tailing" effect.

Station decomposition solutions can be highly fractional causing excessive running times in the branch-and-bound phase. We develop a "fix, price, and unfix" heuristic to efficiently find integer solutions to the fleet assignment problem.

Station purity can provide benefits to airlines by reducing planned crew costs, maintenance costs, and the impact of operational disruptions. We show that purity can provide compelling benefits (up to $29 million per year) to airlines based on reduced maintenance costs alone. Benefits associated with reduced crew costs are estimated at $100 million per year, giving $129 million per year increased profit. We would expect additional savings in operations.

References

1. Abara, J. (1989), "Applying Integer Linear Programming to the Fleet Assignment Problem", *Interfaces* 19 pp. 20-28.

N. Megiddo, Y. Xu, and B. Zhu (Eds.): AAIM 2005, LNCS 3521, pp. 1–2, 2005.

2   E.L. Johnson

2. Hane, C.A., C. Barnhart, E.L. Johnson, R.E. Marsten, G.L. Nemhauser, G. Sigismondi (1995), "The Fleet Assignment Problem: Solving a Large-scale Integer Program", *Mathematical Programming* 70 pp. 211-232.
3. http://www.informs.org/Press/EDELMAN02.html
4. Smith, B.C., E.L. Johnson (2005), "Robust Airline Fleet Assignment: Imposing Station Purity Using Station Decomposition", Submitted.

Computing the Arrow-Debreu Competitive Market Equilibrium and Its Extensions

Yinyu Ye

Department of Management Science and Engineering,
Stanford University, Stanford, CA 94305, USA
yinyu-ye@stanford.edu

We consider the Arrow-Debreu competitive market equilibrium problem which was first formulated by Leon Walras in 1874 [12]. In this problem every one in a population of n players has an initial endowment of a divisible good and a utility function for consuming all goods—own and others. Every player sells the entire initial endowment and then uses the revenue to buy a bundle of goods such that his or her utility function is maximized. Walras asked whether prices could be set for everyone's good such that this is possible. An answer was given by Arrow and Debreu in 1954 [1] who showed that such equilibrium would exist if the utility functions were concave. Their proof was non-constructive and did not offer any algorithm to find such equilibrium prices.

Fisher was the first to consider algorithm to compute equilibrium prices for a related and different model where players are divided into two catalogs: producer and consumer. Consumers have money to buy good and maximize their individual utility functions; producer sell their goods for money. The equilibrium prices is an assignment of prices to goods so as when every consumer buys an maximal bundle of goods then the market clears, meaning that all the money is spent and all goods are sold. Fisher's model is a special case of Walras' model when money is also considered a commodity so that Arrow and Debreu's result applies. Eisenberg and Gale [6, 8] gave a convex optimization setting to formulate Fisher's model with linear utility functions. They constructed an concave objective function that is maximized at the equilibrium. Thus, finding an equilibrium became solve a convex optimization problem, and it could be solved by using the Ellipsoid method in polynomial time. Here, polynomial time means that one can compute an ϵ approximate equilibrium in a number of arithmetic operations bounded by polynomial in n and $\log \frac{1}{\epsilon}$. Devanur et al. [5] recently developed a "combinatorial" algorithm for solving Fisher's model with linear utility functions. Both approaches, Eisenberg-Gale and Devanur et al., did not apply to the more general Walras model.

Solving the Arrow-Debreu problem was proved to be more difficult. Eaves [7] showed that the problem with linear utility can be formulated as a linear complementarity problem (e.g. Cottle et al. [4]) so that Lemke's algorithm could compute the equilibrium, if it existed, in a finite time. It was also proved there that the equilibrium solution values were rational as solutions to an n^2-dimension system of linear equations of the original rational inputs. More recently, however,

N. Megiddo, Y. Xu, and B. Zhu (Eds.): AAIM 2005, LNCS 3521, pp. 3–5, 2005.
© Springer-Verlag Berlin Heidelberg 2005

Jain [9] has showed that Walras's model can be also formulated as a convex optimization, more precisely, a convex inequality problem, so that the Ellipsoid method again can be used in solving it. Remarkably, it turned out that the very same formulation was developed by Nenakhov and Primak [10] more than twenty years earlier. They found out a clean set of posinomial inequalities to describe the problem which is necessary and sufficient. This set of inequalities can be logarithmically transferred into a set of convex inequality, which technique was used for geometric programming in early 60's.

The goal of this talk, based on the paper [13], is threefold. First, we develop a polynomial-time interior-point algorithm to solve Fisher's model with linear utility. The complexity bound, $O(n^4 \log \frac{1}{\epsilon})$, of the algorithm is significantly lower than either the Ellipsoid or "combinatorial" algorithm mentioned above. Secondly, we present an interior-point algorithm, which is not primal-dual, for solving the Arrow-Debreu competitive market equilibrium problem with linear utility. The algorithm has an efficient barrier function for every convex inequality where the self-condordant coefficient is at most 2. Thus, the number of arithmetic operations of the algorithm is again bounded by $O(n^4 \log \frac{1}{\epsilon})$, which is substantially lower than the one obtained by the ellipsoid method and in line with the best complexity bound for linear programming of the same size. Finally, we generalize these results to homothetic and/or quasi-concave utilities, and to the market equilibrium in the presence of economies of production and non-exogenous activities.

References

1. K. Arrow and G. Debreu. "Existence of a competitive equilibrium for a competitive economy." *Econometrica 22*, no. 3 (July 1954): 265-290.
2. W. C. Brainard and H. E. Scarf. "How to compute equilibrium prices in 1891." Cowles Foundation Discussion Paper 1270, 2000.
3. B. Codenotti and K. Varadarajan. "Efficient computation of equilibrium prices for market with Leontief utilities." Working Paper, Department of Computer Science, The University of Iowa, 2004, To appear in ICALP, 2004.
4. R. Cottle, J. S. Pang, and R. E. Stone. *The Linear Complementarity Problem*, chapter 5.9 : Interior–point methods, pages 461–475. Academic Press, Boston, 1992.
5. N. R. Devanur, C. H. Papadimitriou, A. Saberi, and V. V. Vazirani. "Market equilibrium via a primal-dual-type algorithm." In The 43rd Annual IEEE Symposium on Foundations of Computer Science, page 389-395, 2002; journal version on http://www.cc.gatech.edu/ saberi/ (2004).
6. E. Eisenberg and D. Gale. "Consensus of subjective probabilities: The pari-mutuel method." *Annals Of Mathematical Statistics*, 30:165–168, 1959.
7. B. C. Eaves. "A finite algorithm for the linear exchange model." *J. of Mathematical Economics*, 3:197-203, 1976.
8. D. Gale. *The Theory of Linear Economic Models*. McGraw Hill, N.Y., 1960.
9. K. Jain. "A polynomial time algorithm for computing the Arrow-Debreu market equilibrium for linear utilities." Discussion paper, Microsoft Lab, Seattle, WA, 2003.

10. E. Nenakhov and M. Primak. "About one algorithm for finding the solution of the arrow-debreu model." *Kibernetica*, (3):127–128, 1983.
11. Yu. E. Nesterov and A. S. Nemirovskii. *Interior Point Polynomial Methods in Convex Programming: Theory and Algorithms.* SIAM Publications. SIAM, Philadelphia, 1993.
12. L. Walras. *Elements of Pure Economics, or the Theory of Social Wealth* 1874 (1899, 4th ed.; 1926, rev ed., 1954, Engl. Transl.)
13. Y. Ye. *A Path to the Arrow-Debreu Competitive Matket Equilibrium.* Working paper, posted February/23/2004 on www.stanford.edu/~yyye, Stanford, CA 94305, USA.

Complexity of Minimal Tree Routing and Coloring*

Xujin Chen[a], Xiaodong Hu[a], and Xiaohua Jia[b]

[a] Institute of Applied Mathematics, Chinese Academy of Sciences,
P. O. Box 2734, Beijing 100080, China
[b] Department of Computer Science, City University of Hong Kong,
Kowloon, Hong Kong, SAR China
{xchen, xdhu}@amss.ac.cn, jia@cs.cityu.edu.hk

Abstract. Let G be a undirected connected graph. Given a set of g groups each being a subset of $V(G)$, tree routing and coloring is to produce g trees in G and assign a color to each of them in such a way that all vertices in every group are connected by one of produced trees and no two trees sharing a common edge are assigned the same color. In this paper we study how to find a tree routing and coloring that uses minimal number of colors, which finds an application of setting up multicast connections in optical networks. We first prove $\Omega(g^{1-\varepsilon})$-inapproximability of the problem even when G is a mesh, and then we propose some approximation algorithms with provable performance guarantees for general graphs and some special graphs as well.

1 Introduction

Let G be a finite and undirected graph with vertex-set $V(G)$ with $|V(G)| = n$ and edge-set $E(G)$ with $|E(G)| = m$, and $\boldsymbol{\Gamma} = \{\Gamma_1, \ldots, \Gamma_g\}$ be a set of g subsets of $V(G)$, where each Γ_i is called a *group*. A *tree over* Γ_i is a tree in G with $\Gamma_i \subseteq V(T_i)$ for each $1 \leq i \leq g$. A family $\mathcal{T} = \{T_1, \ldots, T_g\}$ of trees is said to be a *tree family over* $\boldsymbol{\Gamma} = \{\Gamma_1, \ldots, \Gamma_g\}$ if there is a permutation ρ on $\{1, 2, \ldots, g\}$ such that $T_{\rho(i)}$ is a tree over Γ_i for each $1 \leq i \leq g$. A *coloring* $\{(T_i, c_i) \mid i = 1, 2, \ldots, g\}$ of a tree family $\{T_1, \ldots, T_g\}$ colors T_i with color c_i such that $c_i \neq c_j$ whenever $E(T_i) \cap E(T_j) \neq \emptyset$.

In this paper we consider *Minimal Tree Routing and Coloring* (Min-TRC) problem: given an instance $(G, \boldsymbol{\Gamma})$, construct a tree family $\{T_1, \ldots, T_g\}$ over $\boldsymbol{\Gamma}$ such that there exist a coloring $\{(T_i, c_i) \mid i = 1, 2, \ldots, g\}$ with minimal number of distinct colors in $\{c_1, \ldots, c_g\}$. This problem models establishing multicast connection in optical networks [17].

This paper aims at giving the explicit inapproximability/approximability analysis for Min-TRC problem. We prove that Min-TRC problem is not ap-

---

* This work was supported in part by the NSF of China under Grant No. 70221001 and 60373012.

N. Megiddo, Y. Xu, and B. Zhu (Eds.): AAIM 2005, LNCS 3521, pp. 6–15, 2005.
© Springer-Verlag Berlin Heidelberg 2005

proximable within $g^{1-\varepsilon}$ and the MIN-TRC problem in meshes or tori is not approximable within $\frac{1}{2}g^{1-\varepsilon}$ for any $\varepsilon > 0$, unless $NP = ZPP$. The lower bounds are best possible with the exceptional case of meshes. We also propose a greedy algorithms with $O(\log g)$ approximation performance ratio for large g. Our proof of inapproximability uses reductions from the vertex coloring problem, and the derivation of approximability relies on some nice graph properties.

The remainder of this paper is organized as follows. In section 2, we introduce notations, definitions, and give some preliminary results on MIN-TRC problem. In section 3, we prove the inapproximability results for the MIN-TRC problem in trees, meshes and tori. In section 4, we first propose a greedy algorithm for MIN-TRC problem for general graphs, and then make special considerations for some widely used graphs. In section 5, we conclude the paper with some remarks.

2 Preliminaries

In this section, we first introduce some terminologies and then describe some related works. Let $\boldsymbol{\Gamma}$ be a set of groups in graph G and let $\mathcal{T}$ be a tree family over $\boldsymbol{\Gamma}$, the *maximum load* of G with respect to $\mathcal{T}$, denoted by $L_{\mathcal{T}}$, refers to the maximum number of trees in $\mathcal{T}$ that use any single edge in G. The *minimum maximum load* of G is defined as $L(G, \boldsymbol{\Gamma}) \equiv \min\{L_{\mathcal{T}} \mid \mathcal{T}$ is a tree family over $\boldsymbol{\Gamma}\}$. It is evident that the minimum number of colors necessary for MIN-TRC is at least $L(G, \boldsymbol{\Gamma})$.

Given a graph G and $v \in V(G)$, let $\delta(v)$ denote the set of edges in $E(G)$ incident with v. A k-*coloring* of G is a function $\phi : V(G) \to \{1, 2, \ldots, k\}$ such that each color class $\{v \mid \phi(v) = i$ and $v \in V(G)\}$ contains no two adjacent vertices of G for each $1 \le i \le k$. We say that G is k-*colorable* if it admits a k-coloring. The *chromatic number* $\chi(G)$ is the minimum value of k for which G is k-colorable. The *vertex coloring* problem is to find a $\chi(G)$-coloring of G.

Clearly, when a tree family $\mathcal{T} = \{T_1, \ldots, T_g\}$ in graph G is associated with its *intersection graph* $G_{\mathcal{T}}$ with vertex-set $V(G_{\mathcal{T}}) = \{v_1, \ldots, v_g\}$ and edge-set $E(G_{\mathcal{T}}) = \{(v_i, v_j) \mid T_i$ and T_j share at least one edge in $G\}$, a k-coloring of $G_{\mathcal{T}}$ gives rise to a coloring of $\mathcal{T}$ with no more than k colors, and vice versa.

Most of previous works on MIN-TRC problem mainly focused on the special case in which every group has only two members. In this case, a tree over a group is simply a path connecting the two members and the problem is known as *Minimal Path Routing and Coloring* (MIN-PRC) problem. This version has been extensively studied for several topologies including trees, rings, and meshes. In particular, MIN-PRC problem is NP-hard for all the three topologies [18, 5, 2], and is approximable within 4/3 for trees [13], within 2 for rings [16], and within poly($\log \log m$) for 2-dimensional meshes [15]; but it is polynomial-time solvable when the underlying graph is a chain [8] or a bounded-degree tree [14].

Some other related works on MIN-TRC problem associated with vertex coloring problem. It was shown in [1] that the vertex coloring problem on graph G cannot be approximate within $n^{\frac{1}{7}-\varepsilon}$ for any $\varepsilon > 0$ assuming $NP \ne P$. Since the faith in the hypothesis $NP \ne ZPP$ is almost as strong as $NP \ne P$, the following negative and positive results [3, 9] prove the intractability of approximation for vertex coloring problem, which will be used in our discussion.

Theorem 1. *The vertex coloring problem on graph of n vertices is not approximable within $n^{1-\varepsilon}$ for any $\varepsilon > 0$, unless $NP = ZPP$, and it is approximable within $O(n(\log \log n)^2/(\log n)^3)$.*

3   Inapproximability

In this section, we shall show that MIN-TRC problem is as hard as vertex coloring problem, and then deduce the inapproximability results directly from Theorem 1. We shall particularly consider the MIN-TRC problem for three special graph topologies including trees, meshes and tori, which are widely used in the applications.

Our first inapproximability result is about the MIN-TRC problem in star graphs. A *star graph* is a tree with at most one node of degree greater than one, which is called the *center* of the star graph.

Theorem 2. *The MIN-TRC problem in tree graphs is not approximable within $\max\{g^{1-\varepsilon}, m^{\frac{1}{2}-\varepsilon}\}$ for any $\varepsilon > 0$, unless $NP = ZPP$.*

Proof. Given a graph H with $V(H) = \{v_1, \ldots, v_n\}$ and $E(H) = \{e_1, \ldots, e_m\}$, the star graph G with $m + 1$ vertices and m edges is defined by $V(G) := \{c, d_1, \ldots, d_m\}$ and $E(G) := \{(c, d_i) \,|\, i = 1, 2, \ldots, m\}$, and $\Gamma := \{\Gamma_1, \ldots, \Gamma_g\}$ is defined by $\Gamma_i := \{c\} \cup \left(\cup_{e_j \in \delta(v_i)} d_j\right)$ for each $1 \le i \le g$, where $g = n$. Let $\mathcal{T} = \{T_1, \ldots, T_g\}$ be the tree family over Γ such that T_i is the unique tree in G over Γ_i for each $1 \le i \le g$. Then $G_{\mathcal{T}} = H$. Thus under approximation-ratio-preserving reduction, the MIN-TRC problem in star graphs is equivalent to the vertex coloring problem, and then the conclusion follows from Theorem 1.   □

We now consider the MIN-TRC problem in meshes and tori. We will only present the analysis for meshes since the same argument is applicable to tori. For graph H with $V(H) = \{v_1, \ldots, v_n\}$ and $E(H) = \{e_1, \ldots, e_m\}$, we define groups $\Gamma = \{\Gamma_1, \ldots, \Gamma_g\}$ with $g = n$ on a $5m \times 5m$ mesh G as follows: Label vertices in G as in the Cartesian plane with their corners located at $(0, 0)$, $(0, 5m - 1)$, $(5m - 1, 0)$, and $(5m - 1, 5m - 1)$, respectively. And then associate each edge e_j $(1 \le j \le m)$ in H with two vertex sets in G: $R_j = \{(\ell, 5j - k) \,|\, \ell = 0, 1, \ldots, 5m - 1; k = 1, 2, \ldots, 5\}$ and $S_j = \{(5j - k, \ell) \,|\, \ell = 0, 1, \ldots, 5m - 1; k = 1, 2, \ldots, 5\}$. Notice that R_j (resp. S_j) consists of vertices located on five consecutive rows (resp. columns) of G, and

$$\text{each of } \{R_1, \ldots, R_m\} \text{ and } \{S_1, \ldots, S_m\} \text{ is a (disjoint) partition of } V(G), \quad (1)$$

$$R_j \cap S_k \text{ induces a } 5 \times 5 \text{ submesh } G_{jk} \text{ of } G, \text{ for } 1 \le j, k \le m. \quad (2)$$

Now corresponding to each vertex v_i in G, set the i-th group as

$$\Gamma_i := \bigcup_{e_j \in \delta(v_i)} (R_j \cup S_j) \tag{3}$$

which is the union of $R_j \cup S_j$ for all e_j incident with v_i.

Lemma 1. *Given a graph H, there exist a mesh G, a set $\boldsymbol{\Gamma} = \{\Gamma_1, \ldots, \Gamma_g\}$ of g groups in G, and a tree family $\mathcal{T}$ over $\boldsymbol{\Gamma}$ such that*

(i) $G_{\mathcal{T}} = H$ and $L_{\mathcal{T}} = 2$; and
(ii) there does not exist three distinct integers $i, j, k \in \{1, 2, \ldots, g\}$ such that $(v_i, v_j) \in E(H)$ and T'_i, T'_j, T'_k are pairwise edge-disjoint, for any tree family $\mathcal{T}' = \{T'_1, \ldots, T'_g\}$ in G in which T'_i is a tree over Γ_i for each $1 \leq i \leq g$.

Proof. To justify (i), let us construct a tree family $\mathcal{T} = \{T_1, \ldots, T_g\}$ as follows: Each T_i is a tree obtained from its vertex-set $V(T_i) := \Gamma_i$ in two steps. In the first step, for every e_j incident with v_i, we add five rows each connecting all vertices in $\{(\ell, 5j - k) \,|\, \ell = 0, 1, \ldots, 5m - 1\}$ for each $1 \leq k \leq 5$. Then the horizontal edges on the five rows span R_j. Summing over all $e_j \in \delta(v_i)$, in total $5|\delta(v_i)|$ rows are added. In the second step, we use vertical edges with both ends in S_j for some $e_j \in \delta(v_i)$ to connect the $5|\delta(v_i)|$ rows and the rest vertices in Γ_i under the condition that the resulting graph is a tree. Though there are many possible T_i's, we can just pick any one of them.

By the construction, $L_{\mathcal{T}} = 2$ is obvious; it suffices to show that the intersection graph $G_{\mathcal{T}}$ of $\mathcal{T}$ is identical with H. Indeed, for every edge $(v_h, v_i) = e_j \in E(H)$, trees T_h and T_i in G share common edges on the rows that span R_j. On the other hand, for every pair of nonadjacent vertices v_h and v_i in $V(H)$, since $\delta(v_h) \cap \delta(v_i) = \emptyset$, we deduce from (1) that $R_j \cap R_k \neq \emptyset \neq S_j \cap S_k$ for all $e_j \in \delta(v_h), e_k \in \delta(v_i)$. Therefore, combining the definitions of Γ_h and Γ_i and the constructions of T_h and T_i, we can deduce from (3) that T_h and T_i shares neither a common horizontal edge nor a common vertical edge. In other words, T_h and T_i are edge-disjoint. Thus $G_{\mathcal{T}} = H$ as desired.

We now prove (ii) by contradiction argument. Suppose that $(v_i, v_j) = e_p \in E(H)$ and T'_i, T'_j, T'_k are pairwise edge-disjoint. Since $e_p \in \delta(v_i) \cap \delta(v_j)$, by (3), both T'_i and T'_j contain $(R_p \cup S_p) \subseteq \Gamma_i \cap \Gamma_j$. Take $e_q \in \delta(v_k)$. Obviously $e_p \neq e_q$. Recalling (2), we have a 5×5 submesh G_{pq} in G induced by $R_p \cap S_q$. Note that the 25 vertices of G_{pq} are all contained in $\Gamma_i \cap \Gamma_j \cap \Gamma_k \subseteq V(T'_i) \cap V(T'_j) \cap V(T'_k)$, and hence every vertex in G_{pq} is incident with three distinct edges one from each of T'_i, T'_j, T'_k. Consequently none of T'_h, T'_i, T'_j can have a branching vertex in G_{pq}, and each of the 9 internal vertices of G_{pq} is a leaf of at least two of T'_i, T'_j, T'_k. Therefore there are in total at least 18 different paths in $T'_i \cup T'_j \cup T'_k$ connecting these leaves to the boundary of G_{pq} because every of T'_i, T'_j, T'_k has vertices outside G_{pq}. Two of those paths must have a common edge in G_{pq} since it has only 16 boundary vertices. The two different paths are contained in exactly one tree in $\{T'_h, T'_i, T'_j\}$. It follows that this tree has a branching vertex in G_{pq}. The contradiction establishes (ii). $\qquad\square$

Theorem 3. *The* MIN-TRC *problem in 2-dimensional meshes (tori) is not approximable within* $\frac{1}{2}g^{1-\varepsilon}$ *for any* $\varepsilon > 0$, *unless* $NP = ZPP$.

Proof. The theorem will follow directly from Theorem 1 if we can prove that, given a $r(g)$-approximation algorithm A for the MIN-TRC problem on g groups in a mesh, there is an algorithm B that can properly color any connected graph H on g vertices with at most $2r(g)\chi(H)$ colors in polynomial time. We now design such an algorithm B in following.

Given graph H, applying the method described in Lemma 1 yields the mesh G, the set $\Gamma = \{\Gamma_1, \ldots, \Gamma_g\}$ of g groups in G, and the tree family $\mathcal{T} = \{T_1, \ldots, T_n\}$ over Γ. Then $G_{\mathcal{T}} = H$, and $c_{opt}(G, \Gamma)$ is no more than $\chi(G_{\mathcal{T}}) = \chi(H)$. Observe that $2c_A(G, \Gamma) \leq 2r(g)c_{opt}(G, \Gamma) \leq 2r(g)\chi(H)$. Our task now is to design a polynomial-time algorithm B which colors H with at most $2c_A(G, \Gamma)$ colors. This can be accomplished as follows: First run algorithm A on given instance of MIN-TRC problem (Γ, G). Then in polynomial time A outputs a solution $\{(T_1', c_1), \ldots, (T_g', c_g)\}$ in which T_i' is the tree in G over Γ_i with color c_i for each $1 \leq i \leq n$. Considering the intersection graph $G_{\mathcal{T}'}$ of $\mathcal{T}' = \{T_1', \ldots, T_g'\}$, we deduce from Lemma 1(ii) that

$$(v_i, v_k) \in E(G_{\mathcal{T}'}) \text{ or } (v_j, v_k) \in E(G_{\mathcal{T}'}) \text{ for any distinct } v_i, v_j, v_k$$
$$\text{such that } (v_i, v_j) \in E(H) \text{ and } (v_i, v_j) \notin E(G_{\mathcal{T}'}). \quad (4)$$

Note that $G_{\mathcal{T}'}$ has a $c_A(G, \Gamma)$-coloring $\phi' : V \rightarrow \{1, 2, \ldots, c_A(G, \Gamma)\}$ with $\phi'(v_i) = c_i$ for each $1 \leq i \leq g$. We now prove that graph H' with vertex-set $V(H') := V(H) = V(G_{\mathcal{T}'})$ and edge-set $E(H') = E(H) \cup E(G_{\mathcal{T}'})$ has a $2c_A(G, \Gamma)$-coloring ϕ. If ϕ' is a proper coloring of H', then we are done since we can simply set $\phi := \phi'$; else we can assume, without loss of generality, that edge $e_i = (a_i, b_i) \in E(H) \setminus E(G_{\mathcal{T}'})$ such that $\phi'(a_i) = \phi'(b_i)$ for $i = 1, 2, \ldots, \ell$ ($\ell \leq m$). By (4), every vertex in $V(G_{\mathcal{T}'}) \setminus \{a_i, b_i\}$ is adjacent to a_i or b_i in $G_{\mathcal{T}'}$ and hence assigned by ϕ' a color different from $\phi'(a_i) = \phi'(b_i)$. It follows that all vertices a_i and b_i are distinct for $i = 1, 2, \ldots, \ell$, where $\ell \leq c_A(G, \Gamma)$, and

$$\phi(v) := \begin{cases} c_A(G, \Gamma) + i, & v = a_i, \text{ for some } 1 \leq i \leq \ell; \\ \phi(v), & v \in V(H) \setminus \{a_1, \ldots, a_\ell\}. \end{cases}$$

defines a $2c_A(G, \Gamma)$-coloring of H' as claimed. Since H is a subgraph of H', ϕ is also a $2c_A(\Gamma, G)$-coloring of H. □

The comparison between the results on trees (Theorem 2) and meshes (Theorem 3) shows that the a few more choices for routing method might make it little bit easier to approximate MIN-TRC. Naturally, we seek for strategies that enable us to exploit this freedom in order to obtain a little better approximation in meshes than that in trees. Unfortunately, two popular strategies for the MIN-TRC problem in meshes, the *Shortest Path Tree* (SPT) strategy and the *Single Path* (SP) strategy fail to achieve this goal. Under SPT strategy, every group is connected by a tree, called *shortest path tree*, such that a distinguished group member, called *source*, is connected to every member in the group through a shortest path. Under SP strategy, every tree over a group is a (single) path that spans all group members.

Theorem 4. *Algorithm using either* SPT *strategy or* SP *strategy can not guarantee to produce a $g^{1-\varepsilon}$-approximation for* MIN-TRC *in meshes for any $\varepsilon > 0$, unless $NP = ZPP$.*

Proof. Note that MIN-TRC problem under either of the strategies is equivalent to vertex coloring problem via approximation-ratio-preserving reductions. To see the equivalence, it suffices to define a set $\varGamma = \{\varGamma_1, \ldots, \varGamma_g\}$ of g groups in a mesh G for any given graph H with $V(H) = \{v_1, \ldots, v_g\}$ and $E(H) = \{e_1, \ldots, e_m\}$ such that (a) H is the intersection graph of some tree family over $\varGamma$, and (b) the intersection graph of any tree family over $\varGamma$ contains H as a subgraph.

In the case of SPT routing, we consider the $p \times p$ mesh G with $p = \max\{m, g\}$ and its four corners labelled as $(0,0), (0, p-1), (p-1, 0), (p-1, p-1)$. We define groups by $\varGamma_i = \{(i-1, k) \mid k = 0, 1, \ldots, p-1\} \cup \{(h, j-1) \mid e_j \in \delta(v_i); h = 0, 1, \ldots, p-1\}$ for each $1 \leq i \leq n$. In addition, we set $s_i = (i-1, p-1)$ as the source of $\varGamma_i$. We then have the tree family $\mathcal{F} = \{F_1, \ldots, F_g\}$ in which each F_i is the shortest path tree over $\varGamma_i$ that has exactly $p-1$ vertical edges. Clearly, $G_{\mathcal{F}} = H$ and condition (a) is satisfied. Notice that any shortest path tree over $\varGamma_i$ contains the column spanning $\{(i-1, k) \mid k = 0, 1, \ldots, p-1\}$. Hence for any $(v_h, v_i) = e_j \in E(H)$ with $h < i$, since $\{(h-1, j-1), (h, j-1), \ldots, (i-1, j-1)\} \subseteq \varGamma_h \cap \varGamma_i$, it remains to verify that any shortest path tree over $\varGamma_h$ and any shortest path tree over $\varGamma_i$ share at least one common edge. Thus condition (b) holds.

In the case of SP routing, we consider $5m \times 5m$ mesh G with its corners labelled as $(0,0), (0, 5m-1), (5m-1, 0), (5m-1, 5m-1)$ and define $\varGamma_i := \{(0, 5j-k) \mid e_j \in \delta(v_i); k = 1, 2, \ldots, 5\}$ for each $1 \leq i \leq n$. Then we have the tree family $\mathcal{P} = \{P_1, \ldots, P_g\}$ in which P_i is the path in G with vertex-set $V(P_i) := \varGamma_i \cup \{(h, 5j-5), (h, 5j-1) \mid e_j \in \delta(v_i); h = 1, 2, \ldots, i\} \cup \{(i, 5j-k) \mid e_j \notin \delta(v_i); k = 1, 2, \ldots, 5\}$. Observe that $G_{\mathcal{P}} = H$. Moreover, for any $(v_h, v_i) = e_j \in E(H)$, any path over $\varGamma_h$ and any path over $\varGamma_i$ must share an edge incident with one of the five vertices $(0, 5j-5), (0, 5j-4), (0, 5j-3), (0, 5j-2), (0, 5j-1)$ on the boundary of G since $\varGamma_h \cap \varGamma_i$ contains the five vertices and at least three of the five vertices are each incident with two edges from T_h and from T_i. It follows that both conditions (a) and (b) are satisfied. □

4   Approximability

In this section, we first describe and analyze a greedy algorithm for MIN-TRC problem in general graphs, and then we investigate approximations for the MIN-TRC problem in some special graph topologies including trees, tori, and rings.

4.1   Greedy Algorithm

The main philosophy of our greedy strategy is to rout trees using as minimal number of edges as possible. This idea comes from an intuition: a tree of less edges potentially has more chances to share the same color with others, and therefore potentially reduces the number of colors needed. In order to carry out the greedy strategy, it is worth noting that finding a tree over a given group Γ of minimum number of edges is a special case of *Minimum Steiner Tree* (MST) problem, which is NP-hard [4] and admits a 2-approximation algorithm [11].

Using the greedy strategy to save colors, GREEDY_COLOR below always tries to assign one color to as many groups as possible by constructing trees of less edges; It does not introduce a new color unless it has to. To phrase it differently, using DIS_TREES as a subroutine, GREEDY_COLOR iteratively finds a large maximal set of edge-disjoint trees over some currently unrooted groups, and assigns them with a unique color.

ALGORITHM GREEDY_COLOR
Input: A set Γ of groups in graph G.
Output: Tree routing and coloring $\mathcal{C} = \{(T_i, c_i) \mid i = 1, 2, \ldots, g\}$.
1. $i \leftarrow 0$, $\mathcal{B}_0 \leftarrow \emptyset$, $\Gamma_0 \leftarrow \Gamma$
2. **While** $\Gamma_i \neq \emptyset$ **do begin**
3. Run DIS_TREES on (G, Γ_i) // to find a maximal family $\mathcal{T}_i$ of edge-disjoint trees.
4. $\Gamma_{i+1} \leftarrow \Gamma_i \setminus \{\Gamma \mid \Gamma \text{ over } T, T \in \mathcal{T}_i\}$,
5. $\mathcal{B}_{i+1} \leftarrow \mathcal{B}_i \cup \{(T, i+1) \mid T \in \mathcal{T}_i\}$ //assign $i+1$ to all trees in $\mathcal{T}_i$
6. $i \leftarrow i+1$
7. **End-while**
8. Output $C \leftarrow i$ and $\mathcal{C} \leftarrow \mathcal{B}_C$

PROCEDURE DIS_TREES
Input: A set Γ of groups in graph G.
Output: A tree family $\mathcal{T}$ of edge-disjoint trees over a subset of Γ.
1. $\mathcal{T} \leftarrow \emptyset$
2. **Repeat**
3. $\mathcal{S} \leftarrow \{\text{MST in } G \setminus \cup_{T \in \mathcal{T}} E(T) \text{ over } \Gamma \mid \Gamma \in \Gamma \setminus \{\Gamma \text{ over } T \mid T \in \mathcal{T}\}\}$
4. Take $T \in \mathcal{S}$ such that $|E(T)| = \min_{S \in \mathcal{S}} |E(S)|$
5. $\mathcal{T} \leftarrow \mathcal{T} \cup \{T\}$
6. **Until** $\mathcal{S} = \emptyset$
7. Output $\mathcal{T}$

Theorem 5. GREEDY_COLOR *for the* MIN-TRC *problem in case of* $g \geq \sqrt{2m} + 1$ *achieves the performance ratio* $(\log\lceil(1 - e^{-\frac{1}{\sqrt{2m+1}}})g\rceil + 1)/(1 - e^{-\frac{1}{\sqrt{2m+1}}})$.

Proof. Let $c^* = c_{opt}(G, \Gamma)$ denote the minimum number of colors needed for the instance of MIN-TRC problem (G, Γ). As a special case DIS_TREES might deal with, the maximum edge-disjoint path problem is shown [10] to be approximable within $\sqrt{m} + 1$ via greedy selection of shortest paths. In view of the 2-approximation taken in Step 3 of DIS_TREES, a slight modification of the proof in [10] can prove that DIS_TREES computes a $(\sqrt{2m} + 1)$-approximation of a tree family of edge-disjoint trees over a maximum subset of Γ. Hence the standard theorem in [21] guarantees that GREEDY_COLOR uses c^* colors to color at least $(1 - e^{-1/(\sqrt{2m}+1)})g$ trees. In other words, $\beta_1 := |\mathcal{B}_{c^*}| \geq (1 - e^{-1/(\sqrt{2m}+1)})g$. Subsequently, GREEDY_COLOR uses the next c^* colors to color at least $(1 - e^{-1/(\sqrt{2m}+1)})|\Gamma_{c^*}|$ remaining trees, i.e., $\beta_2 = |\mathcal{B}_{2c^*} - \mathcal{B}_{c^*}| \geq \lceil(1 - e^{-1/(\sqrt{2m}+1)})|\Gamma_{c^*}|\rceil = \lceil(1 - e^{-1/(\sqrt{2m}+1)})(g - |\mathcal{B}_{c^*}|)\rceil$. Continuing in this way, we get for $i = 0, 1, \ldots, \lfloor C/c^* \rfloor$,

$$\beta_{i+1} = |\mathcal{B}_{(i+1)c^*} - \mathcal{B}_{ic^*}| \geq \lceil(1 - e^{\frac{-1}{\sqrt{2m}+1}})|\Gamma_{ic^*}|\rceil = \lceil(1 - e^{\frac{-1}{\sqrt{2m}+1}})(g - |\mathcal{B}_{ic^*}|)\rceil, \quad (5)$$

where $\mathcal{B}_0 = \emptyset$. Notice that $|\mathcal{B}_{jc^*}| = \sum_{i=1}^{j}\beta_j, j = 1, 2, \ldots, \lfloor C/c^* \rfloor$, $\sum_{i=1}^{\lfloor \frac{C}{c^*} \rfloor}\beta_i \leq g$, and

$$\frac{C}{c^*} < \underbrace{\frac{1}{\beta_1} + \cdots + \frac{1}{\beta_1}}_{\beta_1} + \underbrace{\frac{1}{\beta_2} + \cdots + \frac{1}{\beta_2}}_{\beta_2} + \cdots + \underbrace{\frac{1}{\beta_{\lfloor\frac{C}{c^*}\rfloor}} + \cdots + \frac{1}{\beta_{\lfloor\frac{C}{c^*}\rfloor}}}_{\beta_{\lfloor\frac{C}{c^*}\rfloor}} + 1. \quad (6)$$

It can be seen from (5) that the right hand side of (6) attains its maximum when $\beta_1 = \lceil(1 - e^{-1/(\sqrt{2m}+1)})g\rceil, \beta_i = \lceil(1 - e^{-1/(\sqrt{2m}+1)})(g - \sum_{j=1}^{i-1}\beta_j)\rceil, i = 2, 3, \ldots, \lfloor C/c^* \rfloor$. Also note from (5) that the right hand side of (6) can have at most $1/(1 - e^{-1/(\sqrt{2m}+1)})$ terms 1. Now it follows from $\beta_1 \geq \beta_2 \geq \cdots \geq \beta_{\lfloor C/c^* \rfloor}$ that $\frac{C}{c^*} < \frac{g}{\beta_1}\left(\frac{1}{\beta_1} + \frac{1}{\beta_1-1} + \cdots + \frac{1}{2}\right) + \frac{1}{1-e^{-1/(\sqrt{2m}+1)}} < \frac{g}{\beta_1}\log\beta_1 + \frac{1}{1-e^{-1/(\sqrt{2m}+1)}} = (1 - e^{-1/(\sqrt{2m}+1)})^{-1}(\log\lceil(1 - e^{-1/(\sqrt{2m}+1)})g\rceil + 1)$. □

4.2   Special Graphs

To gain some insights into the impact of graph topology on the approximability of MIN-TRC problem, we present in this subsection approximation algorithms with guaranteed performance ratios for three special graphs: trees, tori, and rings. (Detailed analysis is omitted due to the space limitation.)

Recall that in trees, MIN-TRC problem is equivalent to tree coloring problem: color all trees in $\mathcal{T}$ with a minimum number of colors since the tree family $\mathcal{T}$ over a given set Γ of groups is unique. We refer to groups in Γ simply as trees in $\mathcal{T}$. Straightforwardly, the approximation algorithm for the vertex coloring problem [9] on $G_\mathcal{T}$ carries over to the MIN-TRC problem on $\mathcal{T}$, and gives the following immediate result matching the lower bounds given in Theorem 2.

Theorem 6. *The* MIN-TRC *problem in tree graphs is approximable within* $O(g(\log \log g)^2/(\log g)^3)$.

When the sizes of groups are smaller than a constant k [6], i.e., $\max_{\Gamma \in \Gamma}|\Gamma| \leq k$, we have a tree family $\mathcal{T} = \{T_1, \ldots, T_g\}$, called a *k-tree family* such that

$$\text{the maximum degree of every } T_i \text{ is no more than } k. \quad (7)$$

Notice that the MIN-TRC problem on a k-tree family in a tree graph is NP-hard even when $k = 2$ [18]. Fortunately, by the nice property of k-tree family and the acyclic structure of the underlying graph, the constant approximation ratio for MIN-TRC problem is now achievable by applying FIRST_FIT algorithm which consists of g steps. At the i-th step, FIRST_FIT assigns T_i the first available color, which is the smallest positive integer that has not been assigned to any trees in $\{T_1, \ldots, T_{i-1}\}$ sharing an edge with T_i.

Theorem 7. *The* MIN-TRC *problem is approximable within k for any given k-tree family in a tree graph.*

We now make a brief discussion on tori. Consider an instance of MIN-TRC problem (G, Γ) where G is a 2-dimensional torus. Since G is 4-edge connected, there are two edge-disjoint spanning trees S and T in G [12, 20]. It is easy to see that either $\{(T, i), (S, i) \mid 1 \le i \le g/2\}$ (when g is even) or $\{(S, i), (T, i) \mid 1 \le i \le (g-1)/2\} \cup \{(S, (g+1)/2)\}$ (when g is odd) is a solution to (G, Γ). The following theorem shows that the lower bound given in Theorem 3 for tori is tight.

Theorem 8. *The* MIN-TRC *problem in tori is approximable within $\lceil g/2 \rceil$.*

Finally, we consider the MIN-TRC problem in rings, where a tree is simply a path traversing all vertices in a group. In this case, tree routing can be done by algorithm in [7] so that the maximum load of the graph is within a ratio 1.8 to the optimal one; Tree coloring can be done by 2-approximation algorithm of circular arc coloring [19]. In such way we can obtain a 3.6-approximation for MIN-TRC.

Theorem 9. *The* MIN-TRC *problem in rings is approximable within 3.6.*

5   Conclusions

In this paper we have studied the hardness of approximating the tree routing and coloring for minimizing number of colors used. As our main contribution, we proved strong negative result on the possibility of finding efficiently good approximate solutions to the MIN-TRC problem even when the underlying topology is a mesh or a torus. The positive results presented include several approximation algorithms designed for general graphs and for some special graphs. The $\Omega(g^{1-\varepsilon})$-inapproximability proved seems a good reason to stop the efforts for seeking approximation on the general problem. As the future work, it would be interesting to see if the lower bound given in Theorem 3 for the MIN-TRC problem in meshes is tight.

References

1. M. Bellare, O. Goldreich, and M. Sudan, Free bits and non-approximability-towards tight results, *SIAM Journal on Computing*, **27** (1998), 804-915.
2. T. Erlebach and K. Jansen, The complexity of path coloring and call scheduling, *Theoretical Computer Science*, **255** (2001), 33-50,

3. U. Feige and J. Kilian, J, Zero knowledge and the chromatic number, *Journal of Computer and System Sciences*, **57** (1998), 187-199.
4. M. R. Garey and D. S. Johnson, *Computers and Intractability: A Guide to the Theory of NP-completeness*, W. H. Freeman and Company, 1979.
5. M. C. Golumbic and R. E. Jamison, The edge intersection graphs of paths in a tree, *Journal of Combinatorial Theory, Ser. B*, **38** (1985), 8-22.
6. J. Gu, X.-D. Hu, X.-H. Jia, and M.-H. Zhang, Routing algorithm for multicast under multi-tree model in optical networks, *Theoretical Computer Science*, **314** (2004), 293-301.
7. Q. Gu and Y. Wang, Efficient algorithm for embedding hypergraphs in a cycle, *Lecture Notes in Computer Science*, **2913** (2003), 85-94.
8. U. I. Gupta, D. T. Lee, and Y.-T. Leung, Efficient algorithms for interval graphs and circular-arc graphs, *Networks*, **12** (1982), 459-467.
9. M. M. Halldórsson, A still better performance guarantee for approximate graph coloring, *Information Processing Letters*, **45** (1993), 19-23.
10. S. G. Kollipoulos and C. Stein, Approximating disjoint-path problems using greedy algorithms and packing integer programs, *Integer Programming and Combinatorial Optimization*, Houston, TX, 1998.
11. L. Kou, G. Markowsky, and L. Berman, A fast algorithm for steiner trees, *Acta Informatica*, **15** (1981), 141-145.
12. C. St. J. A. Nash-Williams, Edge disjoint spanning trees of finite graphs, *Journal of London Mathematical Society*, **36** (1961), 445-450.
13. T. Nishizcki and K. Kashiwagi, On the 1.1 edge-coloring of multigraphs, *SIAM Journal on Discrete Mathematics*, **3** (1990), 391-410.
14. C. Nomikos, Path coloring in graphs, *Ph.D Thesis*, Department of Electrical and Computer Engineering, NTUA, 1997.
15. Y. Rabani, Path coloring on the mesh, *Proceedings of the 37th Annual Symposium Foundations of Computer Science*, 1996, 400-409.
16. P. Raghavan and E. Upfal, Efficient routing in all-optical networks, *Proceedings of the 26th Annual ACM Symposium on Theory of Computing*, 1994, 134-143.
17. L. H. Sahasrabuddhe and B. Mukherjee, Light-trees: optical multicasting for improved performance in wavelength-routed networks, *IEEE Communications Magazine*, **37** (2) (1999), 67-73.
18. R. Tarjan, Decomposition by clique separators, *Discrete Mathematics*, **55** (1985) 221-232.
19. A. Tucker, Coloring a family of circular arcs, *SIAM Journal on Applied Mathematics*, **29** (1975), 493-502.
20. W. T. Tutte, On the problem of decomposing a graph into n connected factors, *Journal of London Mathematical Society*, **36** (1961), 221-230.
21. P. J. Wan and L. Liu, Maximal throughput in wavelength-routed optical networks, *DIMACS Series in Discrete Mathematics and Theoretical Computer Science*, **46** (1998), 15-26.

Energy Efficient Broadcasting and Multicasting in Static Wireless Ad Hoc Networks*

Siu-Wing Cheng[1], Xiaohua Jia[2],
Frankie Hung[2], and Yajun Wang[1]

[1] Department of Computer Science, HKUST, Hong Kong
{scheng, yalding}@cs.ust.hk
[2] Department of Computer Science, CityU, Hong Kong
{jia, frankie}@cs.cityu.edu.hk

Abstract. In this paper, we present three energy efficient broadcast and multicast routing algorithms for wireless ad hoc networks. The first algorithm computes a broadcast tree whose energy consumption is within a factor $2 + 2\ln(n-1)$ of the optimal. The second algorithm computes a multicast tree whose energy consumption is within a constant factor of the optimal. Our third algorithm, for a multicast request with a given duration, computes an optimal multicast tree such that the minimal remaining energy of nodes is maximized after the multicast session. This algorithm helps to maximize the lifetime of the network.

1 Introduction

In wireless networks, mobile hosts are powered by batteries and it may be impossible to recharge or replace batteries during a mission. Therefore, the limited battery lifetime imposes a constraint on the network performance. Energy efficiency becomes an important issue in the design of applications in wireless ad hoc networks. Extensive research has been done on the energy conservation for such kind of networks. Some works addressed the issue of using minimum energy to achieve a required network connectivity (also referred as topology control) [4, 6, 11]. Some other works focused on the energy efficient routing [7, 8, 13]. In this paper, we focus on the issue of energy efficient broadcast/multicast in wireless ad hoc networks. Note that broadcast is a special case of multicast.

Wieselthier et al. [12] proposed several energy-efficient broadcast/multicast algorithms, namely the Broadcast Incremental Power (BIP), Multicast Incremental Power (MIP), MST (minimum spanning tree), and SPT (shortest-path tree) algorithms. Wan et al. [10] exploited the geometrical properties of the Euclidean plane to analyze the BIP, MST, and SPT algorithms. Specifically, Wan et al. proved that the approximation ratio of MST is between 6 and 12, the approximation ratio of BIP is between 13/3 and 12, and the approximation ratio

---

* Research of the first and fourth authors are supported in part by Research Grant Council (HKUST 6190/02E), Hong Kong, China.

N. Megiddo, Y. Xu, and B. Zhu (Eds.): AAIM 2005, LNCS 3521, pp. 16–25, 2005.

of SPT is at least $m/2$, where m is the number of receiving nodes. Calinescu et al. [1] studied the problems of symmetric connectivity, strong connectivity, and broadcasting with approximately minimum power. They obtained algorithms with approximation ratio $O(1 + \ln n)$ for all three problems. Cheng et al. [2] studied the problem of broadcasting in large ad hoc networks and proposed a method MLE (Minimum Longest Edge) based on MST. This algorithm provides a scheme to balance the energy consumption among all nodes.

We introduce three algorithms in this paper: one for broadcasting and the other two for multicasting. We assume that there are n stationary nodes.

- Our broadcasting algorithm computes an undirected spanning tree in $O(n^3)$ time. For any node v, by rooting this spanning tree at v, we obtain a broadcast tree with v as the source. We prove that the power consumption of our broadcast tree is within a factor $2 + 2\ln(n-1)$ of the optimal.
- Our first multicasting algorithm aims at finding a multicast tree that has the minimum energy consumption for the model in which any two nodes in the network can reach each other using d^α Watt of power, where d is the Euclidean distance between the two nodes. Our multicasting algorithm runs in $O(n^2)$ time. By making use of a result by Wan et al. [10], we show that the power consumption of our multicast tree is within a constant factor of the optimal.
- Our second multicasting algorithm aims to balance the energy consumption among the nodes. Assuming that the initial energy levels at the nodes and the duration of the multicast session are known, our algorithm returns an optimal multicast tree in $O(n^2)$ time. Our multicast tree is optimal in the sense that the minimum node energy at the end of the multicast session is maximized.

2   Approximate Minimum Power Broadcasting

Our algorithm works with a weighted undirected complete graph G on the n nodes. The weight of each edge $v_i v_j$ of G is the power needed to transmit data directly between v_i and v_j. Let d_{ij} denote the Euclidean distance between v_i and v_j. Let L and U be the common lower bound and upper bound on the power levels of the nodes. For any edge $v_i v_j$, we set its weight $w(v_i, v_j)$ as follows. If $d_{ij}^\alpha < L$, then $w(v_i, v_j) = L$. If $L \le d_{ij}^\alpha \le U$, then $w(v_i, v_j) = d_{ij}^\alpha$. Otherwise, $w(v_i, v_j) = \infty$. If v_i and v_j are obstructed from each other by some physical obstacles, we can model this by setting $w(v_i, v_j) = \infty$.

2.1   Approximation Algorithm

Our approximation algorithm consists of two steps. The first step is the construction of *stars* to cover all the nodes of G. The second step is to construct an approximate broadcast tree from the stars. Recall that if two nodes v_i and v_j in G are blocked by some obstacle, the weight $w(v_i, v_j)$ is set to ∞ to model this.

Then our algorithm will never select the edge $v_i v_j$ as the broadcast cost would be infinity otherwise. (Our algorithm can in fact handle more general weight models, but the presence of obstacles seem to be the most natural application.)

A subgraph of G is a star if there is one center and the other nodes are leaves directly connected to the center. We give an overview of the construction of stars before giving the details. The construction proceeds in rounds. Before the first round, each node of G exists as a trivial connected component. In general, in the kth round, we select a star to connect some of the connected components together. This continues until we obtain a single connected component at the end of the mth round.

We introduce a few definitions and then give the details of the construction.

- $\mathcal{S}_k$ is the set of connected components before the kth round.
- For each node v_r, $C_k(v_r)$ denotes the connected component in $\mathcal{S}_k$ that contains v_r.
- For any node v_r and any component $C \in \mathcal{S}_k$, the distance $d(v_r, C)$ between v_r and C is equal to $\min_{v_s \in C} w(v_r, v_s)$.
- For any two distinct nodes v_r, v_s, $\mathcal{N}_k(v_r, v_s)$ denotes the collection of connected components $\{C \in \mathcal{S}_k : d(v_r, C) \leq w(v_r, v_s)\}$. The weight of $\mathcal{N}_k(v_r, v_s)$ is $w(v_r, v_s)$.
- The (v_r, v_s)-*star* is centered at v_r and its set of leaves is $\{v_i \in G : w(v_r, v_i) \leq w(v_r, v_s)\}$. The weight of the (v_r, v_s)-star is $w(v_r, v_s)$.

Algorithm. *Construct_stars*

1. initialize $m = 1$ and $\mathcal{S}_1$ to be the set of nodes in G;
2. initialize E to be an empty set;
3. **repeat**
4.   pick the $\mathcal{N}_m(v_r, v_s)$ that minimizes the ratio $w(v_r, v_s)/(|\mathcal{N}_m(v_r, v_s)| - 1)$;
5. $E := (v_r, v_s)$-star $\cup\ E$;
6.   construct the set $\mathcal{S}_{m+1}$ of connected components induced by E;
7. $m := m + 1$;
8. **until** $\mathcal{S}_m$ contains exactly one connected component;

When we add a (v_r, v_s)-star in step 5 of the algorithm, if a (v_r, v_i)-star already exists for some node v_i, then v_s must be further away from v_r than v_i as v_s is considered later. In order words, we grow the (v_r, v_i)-star to the (v_r, v_s)-star and the (v_r, v_i)-star no longer exists by itself.

We analyze the total weight of the stars picked by *Construct_stars*. Our analysis is based on the charging argument used for deriving the approximation ratio of the greedy set cover algorithm [9]. We introduce some notations to ease the analysis. Let $\mathcal{N}_k(v_r, v_s)$ be the collection picked in the kth round of the algorithm. When we pick $\mathcal{N}_k(v_r, v_s)$, we distribute the weight $w(v_r, v_s)$ as charges equally among the components in $\mathcal{N}_k(v_r, v_s)$ that does not contain v_r. That is, for each component $C \in \mathcal{N}_k(v_r, v_s)$ that does not contain v_r, C receives a charge of $w(v_r, v_s)/(|\mathcal{N}_k(v_r, v_s)| - 1)$ which we denote by price(C). The component containing v_r receives a zero charge. Next, we put the positively charged components in all collections $\mathcal{N}_k(\cdot)$'s picked, $1 \le k \le m$, into an ordered list $\mathcal{L} = (C_1, C_2, \cdots)$. $\mathcal{L}$ is ordered as follows: if $C_i \in \mathcal{N}_k(\cdot)$ and $C_j \in \mathcal{N}_{k'}(\cdot)$ where $k < k'$, then $i < j$; otherwise, C_i and C_j belong to the same $\mathcal{N}_k(\cdot)$ and they are ordered arbitrarily.

The charge distribution scheme implies that the total weight of stars picked by *Construct_stars* is equal to the total charge $\sum_{C_i \in \mathcal{L}} \text{price}(C_i)$. Let N_a denote the collection of components picked in the ath round of the algorithm. Then exactly $|N_a| - 1$ of these components carry positive charges and are put into $\mathcal{L}$. The ordering scheme of components in $\mathcal{L}$ implies that if $C_{i'}$ is the component in N_{k-1} with the largest index in $\mathcal{L}$, then

$$i' = \sum_{a=1}^{k-1}(|N_a| - 1). \qquad (1)$$

Let OPT denote the power consumption of an optimal broadcast tree. The following lemma relates price(C_i) to OPT.

Lemma 1. *For each $C_i \in \mathcal{L}$, $i < n$ and* price(C_i) $\le \text{OPT}/(n - i)$.

Proof. Let T^* be an optimal broadcast tree (T^* is a directed tree). We use N_k to denote the collection picked in the kth round of the algorithm. Assume that C_i belongs to N_k. Recall that $\mathcal{S}_k$ is the set of all components at the beginning of the kth round. We select a subset $\mathcal{E}$ of edges in T^* that forms a directed spanning tree of the components in $\mathcal{S}_k$ (i.e., if we collapse each component in $\mathcal{S}_k$ into a vertex, $\mathcal{E}$ is a directed spanning tree of the vertices obtained.). For each node v, we use $\mathcal{E}_v$ to denote the outgoing edges of v in $\mathcal{E}$. We use comp(v) to denote the components that contain the destinations of edges in $\mathcal{E}_v$. Let V be the subset of nodes such that $\mathcal{E}_v$ is non-empty. For each $v \in V$, the heaviest edge in $\mathcal{E}_v$ defines the power for v (the sum of which is at most OPT). We distribute the power of each node $v \in V$ equally among the components in comp(v) as charges at them. Consequently, the minimum positive charge at a component is at most $\text{OPT}/(|\mathcal{S}_k| - 1)$. Assume that the component with the minimum positive charge belongs to comp(v). Let (v, v_c) be the heaviest edge in $\mathcal{E}_v$. So comp(v) is a subset of $\mathcal{N}_k(v, v_c)$. It follows that $w(v, v_c)/(|\mathcal{N}_k(v, v_c)| - 1) \le \text{OPT}/(|\mathcal{S}_k| - 1)$. The greedy nature of *Construct_stars* guarantees that

$$\text{price}(C_i) \le \frac{w(v, v_c)}{|\mathcal{N}_k(v, v_c)| - 1} \le \frac{\text{OPT}}{|\mathcal{S}_k| - 1}. \qquad (2)$$

In the ath round, we merge $|N_a|$ components into one. That is, the number of components drops by $|N_a| - 1$. Thus

$$n - \sum_{a=1}^{k-1}(|N_a| - 1) = |\mathcal{S}_k|. \qquad (3)$$

By (1), if $C_{i'}$ is the component in N_{k-1} with the largest index in $\mathcal{L}$, then $i' = \sum_{a=1}^{k-1}(|N_a| - 1)$. Substituting this into (3) yields $n - i' = |\mathcal{S}_k|$. As $i \ge i' + 1$, we have $n - i \leq |\mathcal{S}_k| - 1$. Substituting this into (2), we obtain price$(C_i) \leq$ OPT$/(n-i)$. Recall that $C_i \in N_k$. The number of connected components in N_k is at least $i - i' + 1$. The term 1 comes from the fact that one component in N_k is not charged and it does not appear in $\mathcal{L}$. So the total number of connected components reduces by at least $i - i'$ in the kth round. Thus, $|\mathcal{S}_k| - (i - i') = (n - i') - (i - i') \geq 1$ which implies that $i < n$.

Lemma 2. *The stars can be constructed in $O(n^3)$ time and the total weight of the stars is within a factor $1 + \ln(n-1)$ of the power consumption of any optimal broadcast tree.*

Proof. We first analyze the approximation ratio. The total weight of the stars is equal to $\sum_{C_i \in \mathcal{L}}$ price(C_i). By Lemma 1, $\sum_{C_i \in \mathcal{L}}$ price$(C_i) \leq$ OPT $\cdot (1 + 1/2 + \cdots + 1/(n-1)) \leq$ OPT $\cdot (1 + \ln(n - 1))$. Next, we derive the running time. Before the first round, for any node v_r, we sort the other nodes v_s in increasing distances from v_r. Denote this sorted list by $L(v_r)$. The sorting takes $O(n^2 \log n)$ total time. At the beginning of the kth round, we assume that each node has been labeled the connected component that it belongs to. For each node v_r, we construct the $\mathcal{N}_m(v_r, v_s)$'s by scanning $L(v_r)$ in $O(n)$ time. Summing over all choices of v_r, the total time is $O(n^2)$. After picking the greedy choice, we merge a few connected components together. So we need to relabel the nodes in the connected components merged. This takes $O(n)$ time. In all, one iteration of the repeat loop takes $O(n^2)$ time. As there are no more than $n - 1$ iterations, the total time needed is $O(n^3)$.

The union of stars is a subgraph of G that contain all nodes in G. We compute the minimum spanning tree T_G of the union of stars which serves as the undirected version of the broadcast tree. Whenever we want to broadcast from a node v_r, we root T_G at v_r to obtain a directed spanning tree and broadcast from v_r using this directed spanning tree.

Theorem 1. *For any source node v_r, rooting T_G at v_r yields a broadcast tree whose power consumption is within a factor $2 + 2\ln(n - 1)$ of the optimal.*

Proof. Since T_G is the minimum spanning tree of the union of stars, T_G is the union of some trimmed stars. In the rooted T_G, each trimmed star is either rooted at the center of the star or rooted at a leaf of the star. In the first case, the power consumption of sending data through the trimmed star is at most the weight of the star. In the second case, the power consumption is also twice the weight of the star. Hence, by Lemma 2, the power consumption of the rooted T_G is within a factor $2 + 2\ln(n - 1)$ of the optimal.

3   Approximate Minimum Power Multicasting

In this section, we present an algorithm to find a multicast tree that spans a set of nodes and the total power consumption in the tree is minimized. We consider the obstacle-free energy model where each node v_i can reach any other node v_j using power d_{ij}^α. We show that one can construct a multicast tree whose power consumption is within a constant factor of the optimal. We will make use of the following result due to Wan et al. [10] for this model.

Theorem 2. *Assume that any two nodes v_i and v_j can reach each other using power d_{ij}^α. The weight of the minimum spanning tree is within a factor c of the power consumption of the optimal broadcast tree, where c is a constant between 6 and 12.*

We remark that the lower bound of 6 in Theorem 2 is achieved by an example with seven points. It is not known what the lower bound is as n grows.

Let K be the set of nodes of G that belong to the multicast group. Given any multicast tree T for K, we use $cost(T)$ to denote its power consumption and $weight(T)$ to denote the total edge weight of T. The following lemma shows that the minimum Steiner tree in G interconnecting K is a constant factor approximation of the optimal multicast tree for K no matter which node in K is the source.

Lemma 3. *Let c be the constant in Theorem 2. Let T_K be the minimum Steiner tree in G interconnecting K. The total edge weight of T_K is within a factor c of the power consumption of the optimal multicast tree for K.*

Proof. Let T^* be the optimal multicast tree for K. Let M be the set of vertices in T^*. We use G_M to denote the subgraph of G on M. Since G is a complete graph, G_M is also a complete graph. We use mst_M to denote the minimum spanning tree of G_M. Since the edge weight is equal to the Euclidean distance raised to a power $\alpha \geq 2$, mst_M is the same as the Euclidean minimum spanning tree of M. Clearly, T^* is a broadcast tree for G_M. So by Theorem 2, we have $weight(mst_M) \leq c \cdot cost(T^*)$. Finally, since M contains K, mst_M is a Steiner tree in G interconnecting K. So $weight(T_K) \leq weight(mst_M) \leq c \cdot cost(T^*)$.

Since $cost(T_K) \leq weight(T_K)$, T_K could be used as an approximate multicast tree. Unfortunately, it is NP-hard to compute the minimum Steiner tree [3]. Nevertheless, by Lemma 3, a constant factor approximation of T_K suffices and several algorithms are known for doing this. For example, a 2-approximation T of T_K can be computed in $O(|K|n^2)$ time. It follows that $cost(T) \leq weight(T) \leq 2\,weight(T_K)$, which is within a factor $2c$ of the power consumption of the optimal multicast tree for K. We briefly sketch the ideas of the algorithm.

First, compute the shortest path distances in G among the vertices in K. This takes $O(|K|n^2)$ time. Second, construct a complete graph H on K such that the edge weight between two vertices v_i and v_j in H is equal to the shortest path distance between them. This takes $O(|K|^2)$ time. Third, compute the minimum spanning tree of H in $O(|K|^2)$ time. Each edge of this minimum spanning tree corresponds to some shortest path in G. The union of all such shortest paths is a spanning subgraph G' of G. We compute a minimum spanning tree of G' which is the desired 2-approximation of T_K.

Using better algorithmic techniques, Mehlhorn [5] improved the running time to $O(n^2)$. The factor $2c$ is very pessimistic. In practice, the minimum spanning tree is a much better approximation of the optimal broadcast tree than predicted by Theorem 2. Thus, we also expect that an approximate Steiner tree performs much better in practice. We summarize this section with the following theorem.

Theorem 3. *Assume that there are n nodes such that any two nodes v_i and v_j can reach each other using power d_{ij}^α. Let K be a subset of the n nodes. It takes $O(n^2)$ time to compute a multicast tree for K whose power consumption is within a factor $2c$ of the optimal, where c is a constant between 6 and 12.*

4   Maxmin Node Energy Multicasting

In this section, we present another multicast routing algorithm that aims to maximize the minimum remaining node energy at the end of a multicast session. The goal is to make the network survive longer since the network may become disconnected if some node is out of power. This objective was first introduced by Cheng et al. [2] and it is vastly different from minimizing the total power consumption.

We assume that the duration of the session is known and we denote it by t. We use $E(v_i)$ to denote the initial amount of energy at the node v_i. Our algorithm works with a weighted directed complete graph G on the n nodes. The weight of each arc (v_i, v_j) of G is the remaining energy at v_i if data is sent from v_i to v_j for t seconds. Let d_{ij} denote the Euclidean distance between v_i and v_j. We can accommodate individual lower and upper bounds L_i and U_i on the power level of each node v_i. For any arc (v_i, v_j), we set its weight $w(v_i, v_j)$ as follows. If $d_{ij}^\alpha < L_i$, then $w(v_i, v_j) = E(v_i) - L_i \cdot t$. If $L_i \leq d_{ij}^\alpha \leq U_i$, then $w(v_i, v_j) = E(v_i) - d_{ij}^\alpha$. Otherwise, $w(v_i, v_j) = -\infty$.

We first describe a solution which is easier to explain. Let v_0 be the source. First, we sort the edges in non-increasing order of their weights. Second, we delete all the edges from G and reintroduce them one by one in the sorted order. We stop as soon as the edges added so far contain a directed Steiner tree of the multicast group rooted at v_0. If (v_i, v_j) is the last directed edge added, $w(v_i, v_j)$ is the maxmin remaining node energy. Why is this strategy correct? Since the nodes in the multicast group cannot be connected by a directed Steiner tree rooted at v_0 before the introduction of (v_i, v_j), it is impossible that the optimal solution uses edges with weight larger than $w(v_i, v_j)$. Furthermore, if the optimal solution uses edges with weight less than $w(v_i, v_j)$, the minimum remaining node energy would be less than $w(v_i, v_j)$. There are $\Theta(n^2)$ edges, so the sorting takes $O(n^2 \log n)$ time. Verifying the existence of a directed Steiner tree can be done in $O(n^2)$ time: a directed Steiner tree exists if and only if a breadth-first-search (using the edges added so far) starting from v_0 can reach all nodes in the multicast group. Instead of examining the edges one at a time in order of non-increasing weights, we can speed it up by binary searching for the edge at which a directed Steiner tree of the multicast group starts to exist. The binary search takes $O(\log n)$ rounds and so the whole procedure takes $O(n^2 \log n)$ time.

In the following, we show that we can find the optimal solution by a single invocation of the Prim's algorithm which takes $O(n^2)$ time. This is faster than the previous procedure in theory. We also expect that the single invocation of the Prim's algorithm to be faster in practice because during the binary search, the repeated graph search incurs a significant overhead.

Let K be the set of nodes consisting of the source v_0 and the nodes that v_0 will multicast to. Our algorithm works by constructing a directed maximum spanning tree mst of G and pruning mst to a multicast tree M_K for K.

We initialize the tree mst to contain v_0 alone. Before mst includes all vertices in G, we select the arc (v_a, v_b) such that $w(v_a, v_b) = \max_{v_j \in mst, v_k \notin mst} w(v_j, v_k)$. Then we add (v_a, v_b) to mst and repeat. In the end, mst is a directed maximum spanning tree of G. Next, we prune mst as follows. If there is a leaf in mst that is not a node in K, we remove that leaf and repeat as long as such a leaf can be found. The final pruned tree obtained is M_K. The following result shows that M_K is the optimal multicast tree.

Theorem 4. *The multicast tree M_K can be constructed in $O(n^2)$ time. Moreover, for any multicast tree T for K with v_0 as the source, after the multicasting, the minimum node energy left in M_K is at least the minimum node energy left in T.*

Proof. The time needed to construct M_K follows from the previous discussion. Let T^* be an optimal multicast tree. We use $w(T^*)$ to denote $\min_{(v_i, v_j) \in T^*} w(v_i, v_j)$ and $w(M_K)$ to denote $\min_{(v_i, v_j) \in M_K} w(v_i, v_j)$. Assume to the contrary that $w(T^*) > w(M_K)$. Let (v_r, v_s) be the arc in M_K such that $w(v_r, v_s) = w(M_K)$. Consider the construction of mst. At the time when (v_r, v_s) is included by our algorithm, the nodes in G are partitioned into two subsets S_1 and S_2, where S_1 contains the growing mst and S_2 contains the nodes remaining to be connected. Note that $v_0, v_r \in S_1$, and $v_s \in S_2$. Moreover, since (v_r, v_s) belongs to M_K, some descendant of v_s in M_K must be a node in K. It follows that S_2 contains some node in K. Since T^* includes this node too, T^* contains some arc (v_i, v_j) from S_1 to S_2. We have $w(v_i, v_j) \geq w(T^*) > w(M_K) = w(v_r, v_s)$. This is a contradiction since our algorithm should have preferred (v_i, v_j) to (v_r, v_s).

In fact, M_K also maximizes the minimum remaining node energy in the entire network, instead of just the energy of the nodes of the multicast tree. The following corollary gives a precise statement.

Corollary 1. *Let T be any multicast tree for K with v_0 as the source. Let E_1 be the minimum node energy left in the network after multicasting using M_K. Let E_2 be the minimum node energy left in the network after multicasting using T. Then $E_1 \geq E_2$.*

Proof. Observe that E_1 is equal to the minimum of $\min_{v_i \in G} E(v_i)$ and the minimum of $w(v_i, v_j)$ over all (v_i, v_j) in M_K. Similarly, E_2 is the minimum of $\min_{v_i \in G} E(v_i)$ and $\min_{(v_i, v_j) \in T} w(v_i, v_j)$. Clearly, $E_1 \geq E_2$ by Theorem 4.

5   Conclusion

We have discussed energy efficient broadcasting and multicasting in in a more general nodal energy model, as well as in the obstacle-free environments. Three energy efficient broadcast and multicast algorithms have been proposed. The main results of this paper are summarized as:

1. The proposal of a minimum energy broadcast routing algorithm for wireless networks in which each node has a lower and an upper bound of energy levels and the communication among certain nodes may be obstructed. The algorithm has a guaranteed performance bound $2 + 2\ln(n - 1)$.
2. The proposal of a minimum energy multicast routing algorithm in obstacle-free environments. The algorithm has a constant performance bound.
3. The proposal of an optimal maxmin nodal energy multicast routing algorithm. The algorithm is optimal in the sense that the minimal remaining energy of nodes is maximized after a multicast session. Different from the previous work, this algorithm does not assume that all nodes have the same initial energy level.

References

1. G. Calinescu, S. Kapoor, A. Olshevsky, and A. Zelikovsky. Network lifetime and power assignment in ad-hoc wireless networks, *Proceedings of the 11th Annual European Symposium on Algorithms*, 2003, 114–126.
2. M. X. Cheng, J. Sun, M. Min, D.-Z. Du. Energy efficient Broadcast and Multicast Routing in Ad Hoc Wireless Networks. *Proceedings of the 22nd IEEE International Performance, Computing, and Communications Conference*, Pheonix, 2003, 87–94.
3. M.R. Garey and D.S. Johnson. *Computers and Intractability: a guide to the theory of NP-completeness*, W.H. Freeman, 1979.
4. V. Kawadia and P. R. Kumar. Power Control and Clustering in Ad Hoc Networks. *IEEE INFOCOM 2003*, 459–469.
5. K. Mehlhorn. A faster approximation algorithm for the Steiner problem in graphs. *Information Processing Letters*, vol. 27, 1988, 125–128.
6. R. Ramanathan and R. Rosales-Hain. Topology Control of Multihop Wireless Networks Using Transmit Power Adjustment. *IEEE INFOCOM 2000*, 404–413.
7. V. Rodoplu and T. Meng. Minimum Energy Mobile Wireless Networks. *IEEE Journal on Selected Areas in Communications*, Vol. 17, No. 8, Aug 1999, 1333–1344.
8. C. K. Toh. Maximum Battery Life Routing to Support Ubiquitous Mobile Computing in Wireless Ad Hoc Networks. *IEEE Communications Magazine*, 39 (2001), 138–147.

9. V.V. Vazirani. *Approximation Algorithms*. Springer, 2001.
10. P. J. Wan, G. Calinescu, X. Y. Li, and O. Frieder. Minimum-Energy Broadcast Routing in Static Ad Hoc Wireless Networks. *IEEE INFOCOM 2001*, 1162–1171.
11. R. Wattenhofer, L. Li, P. Bahl, and Y.-M. Wang. Distributed Topology Control for Power Efficient Operation in Multihop Wireless Ad Hoc Networks. *IEEE IN-FOCOM 2001*, 1388–1397.
12. J.E. Wieselthier, G. D. Nguyen, and A. Ephremides. On the Construction of Energy-Efficient Broadcast and Multicast Trees in Wireless Networks. *IEEE IN-FOCOM 2000*, 585–594.
13. G. Zussman and A. Segall. Energy Efficient Routing in Ad Hoc Disaster Recovery Networks. *IEEE INFOCOM 2003*, 682–691.

An Algorithm for Nonconvex Lower Semicontinuous Optimization Problems

Oscar Cornejo Z.

Facultad de Ingeniería, Universidad Católica de la Ssma. Concepción,
Casilla 297 - Concepción - Chile
ocornejo@ucsc.cl

Abstract. In this paper we study an algorithm to find critical points of a lower semicontinuous nonconvex function. We use the Moreau regularization for a special type of functions belonging to the class of prox-regular functions which have very interesting algorithmic properties. We show that it is possible to generate an algorithm in order to obtain a critical point using the theory developed for the composite functions and also the results for the solutions of nonsmooth vectorial equations. We prove the convergence of the algorithm and some estimations of the convergence speed.[1]

Keywords: Variational Analysis, Moreau Approximation, Proximal Point Algorithm, Prox-Regularity, Nonsmooth Equations.

1 Introduction

Martinet, Refs. [9, 10] proposed an iterative procedure based on the proximal point algorithm for solving the following problem

$$(P) \qquad \min\{f(x) : x \in \mathbb{R}^n\}$$

where f is a closed proper convex function. This iterative method generates a sequence $\{u_k\}$ defined as

$$\frac{u_k - u_{k-1}}{\lambda} \in -\partial f(u_k).$$

For this sequence both convergence and rate of convergence results were established by Rockafellar in Ref. [17]. In the non-convex case, some of these ideas have recently begun to be studied for some class of functions. For example, there is a very nice set of results developed by Poliquin and Rockafellar, Ref. [13] on the class of prox-regular functions. This type of functions has very interesting properties which are essential for our goals. We consider the Moreau envelopes

---

[1] I would like to thank to Prof. A. Jofré for advising this work. Partially supported by FONDAP-Matemáticas Aplicadas.

N. Megiddo, Y. Xu, and B. Zhu (Eds.): AAIM 2005, LNCS 3521, pp. 26–36, 2005.

$e_\lambda f$ which provide a sort of regularization of f and use the fact that $e_\lambda f(\cdot) + r \|\cdot\|^2$ is a convex function for some r and $e_\lambda f$ is a lower-C^2 function under appropriate mild conditions. We will start the paper introducing a modification in the original defi-nition of prox-regularity asking for a uniformity with respect to the "curvature" of the function. This class of functions will be called r-prox-regulars and we will show that when a function f in this class is also locally Lipschitz, f is lower-C^2 and hence a strongly amenable function, which will allow us to write f as a composite function. With this idea in mind we will use the results proved by Pang, Hang and Rangaraj Ref. [12] to obtain a critical point of $f = G \circ F$ where G is a convex continuous function and F is a C^2 mapping. On the other hand, without the locally lispchitzianity over f, we will use the Moreau regularization again, and we will show that $e_\lambda f$ is strongly amenable. Then, we will find critical points of $e_\lambda f$, that is, the necessary optimality condition involves $\nabla e_\lambda f(x) = 0$. In order to solve this equation, we will use some recent developments in the area of nonsmooth equations and generalized Newton methods. Hence, we will use some results proved by L. Qi Ref. [18] to obtain superlinear convergence for the subproblem: find $\hat{x}$ such that $\nabla e_\lambda f(\hat{x}) = 0$. Finally, by using the extension of Attouch Theorem proved by Poliquin, Ref. [14] about epi-convergence and convergence of subgradients we will conclude going with λ to zero, that $\hat{x}$ is a critical point of f.

2   Preliminaries

We start this section with a notion close to the definition of prox-regular function at a point $\bar{x}$ introduced by Poliquin and Rockafellar, Ref. [13] for a locally lower semicontinuous function at $\bar{x}$, that is, a function whose epigraph is locally closed around $(\bar{x}, f(\bar{x}))$. Indeed, we will work with a modification of the original definition making the behaviour of the "curvature coefficient r" explicit and uniform.

Definition 1. *Let $f : \mathbb{R}^n \to \mathbb{R}$ be a finite function and locally l.s.c. at $\bar{x}$. We say that f is r-prox-regular with $r > 0$ at $\bar{x}$ if for each $\bar{v} \in \partial f(\bar{x})$ there exists $\epsilon > 0$ such that*

$$f(x') \geq f(x) + \langle v, x' - x \rangle - \frac{r}{2} \|x' - x\|^2 ; \ \forall \ x' \in \mathbf{B}(\bar{x}, \epsilon) \tag{1}$$

when $v \in \partial f(x)$, $\|x - \bar{x}\| < \epsilon$, $\|v - \bar{v}\| < \epsilon$, $f(x) < f(\bar{x}) + \epsilon$

Remark 1. Note that (1) requires v to be a proximal subgradient at x: denoted $v \in \partial_p f(x)$

This class of functions is large enough for our purpose. From the "geometrical" point of view, this class contains any convex proper and l.s.c. function, a lower-C^2 function and a strongly amenable function such that it is possible to draw a parabola under their epigraphs (touching the graph) with a curvature bounded from below by r.

Remark 2. There are two main differences -both strongly required in this paper-
with respect to the prox-regularity definition given by Poliquin and Rockafellar.
First, they don't ask uniformity on the curvature r and second the notion of
prox-regularity is defined for a point $\bar{x}$ and a subgradient $\bar{v}$ of the function f.
Thus, the class of prox-regular functions is much larger than the r-prox-regular
one. However, most questions about the r-prox-regularity of f at $\bar{x}$ can be stated
for a particular vector $\bar{v} \in \partial f(\bar{x})$ (fixe $\bar{v}$ in the definition 2.1). In this case we
would say that f is r-prox-regular at $(\bar{x}, \bar{v})$. For example, for each $\bar{v}$ (fixed) this
notion would be conveniently normalized to the case where $\bar{x} = 0$ and $\bar{v} = 0$,
moreover with $f(0) = 0$.

By using this terminology we could establish the following lemma, however in
the following sections we will always work with the notion of r-prox-regular.

Lemma 1. *We say that f is r prox-regular at $(\bar{x}, \bar{v})$ if and only if $g(x) = f(x +$
$\bar{x}) - \langle \bar{v}, x \rangle$ is r prox-regular at $\bar{x} = 0$, for $\bar{v} = 0$.*

3   Main Results

Let $f \colon \mathbb{R}^n \to \mathbb{R}$ be a (nonconvex) lower semicontinuous function, (for short
l.s.c.). In what follows we will study the following problem, namely, *finding crit-*
ical points x^* of the problem,

$$\min_{x \in \mathbb{R}^n} f(x), \tag{2}$$

that is, such that $0 \in \partial f(x^*)$, where $\partial f(x)$ is the Mordukhovich subgradient set
already called subgradient set in this paper .
We consider the Moreau envelope functions $e_\lambda f$ for $\lambda > 0$ where

$$e_\lambda f(x) = \min_{x'} \{ f(x') + \frac{1}{2\lambda} \| x' - x \|^2 \}$$

We start giving the following result which corresponds to a ready modification
of Poliquin and Rockafellar's result (Theorem [5.2], see Ref. [13]) and which is
useful for our purpose.

Lemma 2. *Suppose that f is r-prox-regular at $\bar{x}$, and let $\lambda \in (0, 1/r)$ and $g(x) =$*
$f(x + \bar{x}) - \langle \bar{v}, x \rangle$, where $\bar{v} \in \partial f(\bar{x})$. Then on some neighborhood of the origin,
the function

$$e_\lambda g(x) + \frac{r}{2(1 - \lambda r)} |x|^2$$

is convex, where r is the constant that appears in the r-prox-regular definition [1].
Furthermore, if λ is sufficiently small, then on a neighborhood of the origin, $e_\lambda g$
is lower-C^2.

We know, Ref. [16], that if f is lower-C^1 on an open set $O \subset \mathbb{R}^n$, then there
is a local representation of f as a subsmooth (max of a C^1 family of functions

parameterized on a compact set) function over a functions around each point $x \in O$ and a similar property is true for lower-C^2 functions. Moreover, for any compact set $B \subset O$, a common representation is valid for all points x in some open set O' satisfying $B \subset O' \subset O$. In what follows we give a new proof for the lower-C^2 case by using only a local representation of these functions as a difference between a convex and a quadratic function. A rather complex proof of this property was given by Rockafellar and Wets in their book.

Proposition 1. *Let K be a compact convex subset of $\mathbb{R}^n$, and let $f \colon \mathbb{R}^n \to \mathbb{R}$ be a lower-C^2 function on a set K, then there exist $\bar{r} > 0$ such that the function $h(x) = f(x) + (\bar{r}/2)\|x\|^2$ is convex on K, where $\bar{r}$ can be computed using a particular finite covering of K.*

Proof. As f is a lower-C^2 function then, for all $x \in K$, there exists an open set O_x such that $f(x) + (r_x/2)\|x\|^2$ is a convex function on O_x. The collection of open set $\{O_x\}_{x \in K}$ cover K, so by compactness there exist points $x_i \in K$ for $i = 1, 2, \ldots, m$, such that $K \subset \cup_{i=1}^{m} O_{x_i}$ and $\bar{r} > 0$ are large enough such that $h(x) = f(x) + (\bar{r}/2)\|x\|^2$ is a convex function on K.

We now introduce the notion of r-prox regularity on a set still fixing the "curvature coefficient r" on the set. This class will be interesting from the algorithmic point of view.

Definition 2. *A function f is said to be r-prox-regular on set C if it is r-prox-regular for each $x \in C$.*

So, we can now give the following theorem showing the relation between r-prox-regularity, the notion of lower-C^2 and the convexity.

Theorem 1. *If f is r-prox-regular on an compact convex set C, then $f(x)+r|x|^2$ is a convex function on C.*

4   Algorithm I. Lipschitz Case

In this section we will introduce a new algorithm to find critical points of the problem (P). We use the ideas developed by Pan, Hang and Rangaraj, (see Ref. [12]) so as to obtain a globally convergent method and some estimations of convergence speed. We start with the following result about r-prox-regularity and composite functions.

Proposition 2. *If f is r-prox-regular and locally Lipschitz on a convex compact set K then f is lower-C^2 on K with a common representation for all $x \in K$. Thus, $f = G \circ F$ where G is a convex function defined by $G(y, x) = y + f(x) + \frac{r}{2}\|x\|^2$ and F is the C^2 function such that $F(x) = (-\frac{r}{2}\|x\|^2, x)$.*

This last Proposition is essential for us because it is known that any d verifying $G(F(x) + F'(x)d) < G(F(x))$ is a descent direction of the original function f at

x. Thus, we have an easy way to check descent directions. The other consequence is the fact that function f is locally Lipschitzian and directionally differentiable; moreover, f is a Bouligand differentiable function. Thus, the directional derivative of f at $x \in \mathbb{R}^n$ in the direction $d \in \mathbb{R}^n$ is $f'(x, d) = G'(F(x), \nabla F(x)d)$. All these properties will allow us to use over $e_\lambda f$ some known results about global convergence as we show later. Following the determination of an algorithm we first introduce the merit function (iteration function or casting function) ψ defined by

$$\psi(x, d) = G(F(x) + \nabla F(x)d) - G(F(x)).$$ (3)

This function is important for finding descent directions. Now, we recall the following abstract convergence result proved by Pan, Hang and Rangaraj, (Ref. [12]), of the model algorithm specialized to minimize a composite convex function which will be adapted to our goals.

Lemma 3. *Let $f = G \circ F$ where $G \colon \mathbb{R} \times \mathbb{R}^n \to \mathbb{R}$ is a convex function bounded from below and $F \colon \mathbb{R}^n \to \mathbb{R} \times \mathbb{R}^n$ is a continuously differentiable function. Let us consider the merit function ψ defined by (3) and a sequence $\{B_k\}$ of matrices satisfying $\exists\, \alpha \geq \beta > 0$ such that $\forall\, x \in \mathbb{R}^n\colon \beta\, x^t x \leq x^t B_k x \leq \alpha\, x^t x, \; \forall\, k$. Suppose that $\{x_k\}$ is a sequence generated as $x_{k+1} = x_k + t_k\, d_k; \; \forall\, k \geq 0$, where the direction d_k is computed solving the problem,*

$$(P_k) \quad \min\{\psi(x_k, d) + \frac{1}{2} d^t B_k\, d : d \in \mathbb{R}^n\},$$

and the step size t_k is calculated by Armijo's rule. Then, every accumulation point of $\{x_k\}$ is a critical point of the composite function f.

Remark 3. In the original version of the above lemma the accumulation point was actually a Dini stationary point but a Dini stationary point set is always a critical point. Therefore, the above lemma is a direct consequence.

Now, we can state the first algorithm of this paper which seeks critical points of a lipschitzian function.

Algorithm I

(1). Let $\rho, \; \sigma \in (0, 1)$ be given. Let $x_0 \in \mathbb{R}^n$ be arbitrary. Set $k = 0$
(2). Given x_k, compute a global optimal solution d_k of the problem

$$(P_k) \quad \min\{\psi(x_k, d) + \frac{1}{2} d^t B_k\, d : d \in \mathbb{R}^n\}$$

- Stop if the optimum objective value of problem (P_k) is zero; in this case, x_k is a desired critical point of f
- Otherwise, let m_k be the smallest nonnegative integer m such that

$$f(x_k + \rho^m\, d_k) - f(x_k) \leq -\frac{\sigma}{2} d_k^t B_k\, d_k$$

   Set $x_{k+1} = x_k + \rho^{m_k} d_k$
(3). Repeat the general step with $k + 1$ replacing k while x_{k+1} fails the stop criteria.

Remark 4. Problem P_k can be written as

$$\min\{f(x_k + d) + \frac{r}{2}|d|^2 - f(x_k) + \frac{1}{2}d^t B_k d : d \in \mathbb{R}^n\} \tag{4}$$

Remark 5. The line search step in the algorithm follows the usual Armijo rule; the integer m_k can be determined after a finite number of trials starting with $m = 0, 1, 2, \cdots,$

Remark 6. The sequence $\{x_k\}$ satisfies

$$f(x_{k+1}) \leq f(x_k) - \frac{\sigma}{2}\rho^{m_k}d_k^t B_k d_k < f(x_k),$$

thus, $\{f(x_k)\}$ is strictly decreasing. Moreover, if function f is bounded from below, then the sequence $\{f(x_k)\}$ converges and hence $\{f(x_{k+1}) - f(x_k)\} \to 0$.

In step 2) of Algorithm I, we require to solve problem 4 which can be written as (P_z) $\min\{z(d) : d \in \mathbb{R}^n\}$ where $z(d) = f(x_k + d) + \frac{r}{2}|d|^2 + \frac{1}{2}d^t B_k d$. This problem has obviously a nonempty optimal solution set denoted $S(P)$. One way to solve this problem is by using the proximal point algorithm. Martinet (Refs. [9, 10]) proposed an iterative procedure to solve (P): starting from $d_0 \in \mathbb{R}^n$ he defines the sequence d_k recursively such that

$$(Prox) \quad (d_k - d_{k-1})/\lambda \in -\partial z(d_k),$$

which is equivalent to

$$(P_k^z) \qquad d_k = \operatorname{argmin}\{z(d) + (1/2\lambda)|d - d_{k-1}|^2 : d \in \mathbb{R}^n\}$$

Inexact versions of $(Prox)$ are essential in order to produce implementable methods. Indeed, exact minimization in (P_k^z) can be replaced by

$$d_k \in \epsilon_k - \operatorname{argmin} \{z(d) + (1/2\lambda)|d - d_{k-1}|^2 : d \in \mathbb{R}^n\},$$

or equivalently,

$$(d_k - d_{k-1})/\lambda \in -\partial_{\epsilon_k} z(d_k).$$

This last problem can be solved via the "explicit" bundle algorithm described, and introduced by Ref. [3], which in our setting takes the following form:

Algorithm: Direction Search

(0). Select $\gamma \in \partial z(u)$, and let $y_0 = u - \lambda\gamma$; $\phi_0(y) = z(u) + \langle \gamma, y - u \rangle$

(1). If $z(y_j) - \phi_j(y_j) \leq \epsilon$ stop; otherwise, determine y_{j+1} as follows.

(2). Let $w_j = (u - y_j)/\lambda \in \partial\phi_j(y_j)$, and select $\gamma_j \in \partial z(y_j)$

(3). Take any function $\phi_{j+1} \leq z$ such that

$$\phi_{j+1}(y) \geq \max\{\phi_j(y_j) + \langle w_j, y - y_j \rangle; g(y_j) + \langle \gamma_j, y - y_j \rangle\}$$

(4). Take for y_{j+1} the solution of the minimization problem

$$\min\{\phi_{j+1}(y) + (1/2\lambda)|y - u|^2 : y \in \mathbb{R}^n\}$$

(5). Update $j \leftarrow j + 1$, and go to Step 1.

One of the main points of this algorithm is the fact that the computation of y_j in Step 4 is explicit. It can be proved that this algorithm stops after a finite number of steps, see Refs. [3, 2, 7]. Cominetti [4] estimates the number of steps to provide an answer. When the algorithm "stops" we have the solution point $y = y_j$ since

$$(u - y_j)/\lambda \in \partial\phi_j(y_j) \cap \partial_\epsilon z(y_j).$$

Now we can summarize all these ideas in a simple result of convergence for the proposed Algorithm I. *For this purpose, we assume that the sequence generated by the Algorithm I is contained in a known box C (convex and compact).*

Theorem 2. *Let f be r-prox-regular and locally lipschitzian function on the box C. Let $\{x_k\}$ be the sequence generated by Algorithm I, then every accumulation point of $\{x_k\}$ is a critical point of f.*

5   Algorithm II. Nonlipschitz Case

We know now the fact that when f is convex then $e_\lambda f$ is a C^{1+} convex function for every $\lambda > 0$. This fact has important consequence in variational analysis and optimization. The study of $e_\lambda f$ when f is not convex has received attention only recently. In Ref. [13], it was shown that $e_\lambda f$ is differentiable with $\nabla e_\lambda f(\bar{x}) = 0$, in fact of class C^{1+} with $\nabla e_\lambda f(x) = \lambda^{-1}[I - P_\lambda f](x)$ where

$$P_\lambda g(x) = \text{argmin}_w\{g(w) + \frac{1}{2\lambda}|w - x|^2\}$$

is the proximal mapping. We show in the next proposition that $e_\lambda f$ is a composite function. This simple result will be crucial for our goals. In what follows we assume that the solutions of (P) are contained in a known box (convex and compact) K.

Proposition 3. *Let f be an r-prox-regular function on K. Then there exist functions G and F such that $e_\lambda f = G \circ F$ over K, where G is a convex continuous function and F is a C^2 mapping.*

Now, we come back to problem (P), that is, how to find critical points of a nonlipschitzian function f. This problem obviously involves the necessary optimality condition $\nabla e_\lambda f(x) = \nabla G(F(x))\nabla F(x) = 0$. At this point, we use the developments in the area for solving nonsmooth equations by generalized

Newton methods in order to obtain a critical point of $\nabla_{e_\lambda} f$. This area has been focused on numerical solutions of a nonsmooth equation $T(x) = 0$, where the mapping $T: \mathbb{R}^n \to \mathbb{R}^n$ is assumed to be locally Lipschitzian. For such a function Rademacher's theorem implies that T is differentiable almost everywhere . We denote the set of points where T is differentiable by D_T. Let $\nabla T(x)$ be the $n \times m$ Jacobian matrix of partial derivatives whenever x is a point where the partial derivatives exist. This property was used by F. Clarke Ref. [8] to introduce for each $x \in \mathbb{R}^n$ the generalized subdfferential of T defined by $\partial_c T(x) = \mathrm{co}\{\lim_{x_i \to x} \nabla T(x_i) : x_i \in D_T\}$, which is a nonempty convex compact set. With this definition we recall the important notion for algorithmic purposes called the semismooth functions which were originally introduced by Mifflin, Ref. [11] for functions on $\mathbb{R}^n$. Convex functions, smooth functions and subsmooth functions are examples of semismooth functions. Moreover, the sums, differences, products, and composites of semismooth functions are semismooth. We now recall the notion of semismooth for such functions.

Definition 3. *We say that T is semismooth at x if T is lipschitzian near x and*
$$\lim_{\substack{V \in \partial_c T(x+th') \\ h' \to h,\, t \downarrow 0}} \{V h'\} \ \text{exists for any } h \in \mathbb{R}^n.$$

For a given $x \in \mathbb{R}^n$, the Clarke's generalized subdifferential $\partial_c T(x)$ is the convex hull of the following set called B-subdifferential:

$$\partial_B T(x) = \{ \lim_{x_i \to x} \nabla T(x_i) : x_i \in D_T \}$$

Now, we recall the notion of BD-regularity introduced by Qi, Ref.[18], which will be useful later.

Definition 4. *We say that T is BD-regular at x if all the elements in $\partial_B T(x)$ which themselves are $n \times n$ matrices, are nonsingular.*

As we mentioned before we seek as a first step zeroes of $\nabla_{e_\lambda} f$, i.e. we want to solve the system of nonsmooth equations: $H(x) = \nabla_{e_\lambda} f(x) = 0$. This equation will be solved by using an algorithm which generates a sequence $\{x_k\}$ such that $x_{k+1} = x_k - V_k^{-1} H(x_k)$, where $V_k \in \partial_B H(x_k)$. Thus, we propose the following algorithm to obtain a critical point of $\nabla_{e_\lambda} f$

Generic Algorithm II

(0). Let x_0 be arbitrary.
(1). For $k = 0, 1, 2, \cdots$, let d_k be a solution of the linear equation: $\nabla_{e_\lambda} f(x_k) + V_k \, d = 0$, where $V_k \in \partial_B H(x_k)$
(2). For $k = 0, 1, 2, \cdots$,, define $x_{k+1} = x_k + d_k$
(3). Test x_{k+1} for convergence. Repeat the general step with $k + 1$ replacing k if x_{k+1} fails the convergence test.

At this point we observe that as $e_\lambda f$ is C^{1+} thus, we can assume that $\nabla_{e_\lambda} f$ is *semismooth*. In what follows we use the notation $H(\cdot) := \nabla_{e_\lambda} f(\cdot)$ which is a locally lipschitzian function. Thus, the Clarkes generalized of $H(x)$ or the generalized Hessian of $e_\lambda f$ is given by $\partial H(x) = \mathrm{co}\{\lim \nabla H(x_j) : x_j \to x, \, x_j \in$

D_H}. Hence, by using the following result, developed by Qi, see Ref.[18], we will obtain superlinear convergence for our subproblem: find $\hat{x}_\lambda$ such that $0 = \nabla e_\lambda f(\hat{x}_\lambda)$.

Proposition 4. *Let x_λ^* be a solution of $H(x) = 0$ and assume that H is semismooth and BD-regular at x_λ^*. Then Algorithm II is well defined and convergent to x_λ^* superlinearly when the starting point of the algorithm is close enough to x_λ^*. Moreover, if $H(x_k) \neq 0$ for all k, then the norm of H decreases superlinearly.*

Remark 7. In Algorithm II it is possible to choose a step length, $t_k > 0$ such that $x_{k+1} = x_k + t_k\, d_k$ where the stepsize t_k can be determined by some line search procedures such as the Armijo rule.

In summary, we have a critical point x_λ^* of $\nabla e_\lambda f$. Moreover, the sequence generated converges Q-superlineary to x_λ^*. But, we want to find a critical point of our original problem, that is, $0 \in \partial f(x)$.

For this purpose, we will use an extension of Attouch's theorem. We recall that Attouch [1] showed that for convex functions the epi-convergence is the appropriate concept of convergence when we are interested in convergence of their subdifferentials. He showed that a sequence $\{f_n\}$ of lower semicontinuous proper convex functions epi-converge to f if and only if the sets $\{\text{gph } \partial f_n\}$ converge to gph ∂f and there exists $\{(x_n, u_n)\} \in$ gph f_n converging to $\{(x, u)\}$ with $u \in \partial f(x)$ and $f_n(x_n)$ converging to $f(x)$. The extension of Attouch's Theorem involves primal lower nice functions. These functions were first introduced by Poliquin in Ref. [15].

Definition 5. *A lower function $f\colon \mathbb{R}^n \to \mathbb{R} \cup \{\infty\}$ is said to be primal lower nice (p.l.n.) at $\bar{x}$ if there exist $c > 0$, $\epsilon > 0$, and $\rho > 0$ such that if $r \geq \rho$, $\|u\| \leq cr$, $\|x - \bar{x}\| < \epsilon$ and $u \in \partial_p f(x)$ then*

$$f(x') \geq f(x) + \langle u, x' - x \rangle - (r/2)\|x - x'\|^2 \quad \text{if } \|x - x'\| < \epsilon \tag{5}$$

In Ref. [15], Proposition [3.5] proved that for these functions the subgradients are actually proximal subgradients. Poliquin, Ref. [14] showed an equivalent characterization of primal lower nice functions: the subgradients of p.l.n. functions are "t-monotone" i.e. $\partial f + tI$ is a monotone set valued mapping. We know that $e_\lambda f$ is lower-C^2, therefore it is p.l.n. Now, in the following lemma, the extension of Attouch Theorem proved by Poliquin, (see Ref. [14]) is given.

Lemma 4. *Assume that $e_\lambda f$ is l.s.c. uniformly minorized by a quadratic and p.l.n. at $\bar{x}$ with constants c, ϵ, and $\rho > 0$ (see Definition 5). If $e_\lambda f$ epiconverge to f, denoted $e_\lambda f \xrightarrow{e} f$ on $\|x - \bar{x}\| < \epsilon$ then gph $\partial e_\lambda f \to$ gph ∂f (Painleve-Kuratowski sense) on $\|x - \bar{x}\| < \epsilon$ and there exist gph $\partial e_{\lambda_n} f \ni (x_n, u_n) \to (\tilde{x}, \tilde{u}) \in$ gph ∂f with $e_{\lambda_n} f(x_n) \to f(\tilde{x})$ and $\|\tilde{x} - \bar{x}\| < \epsilon$*

Now, we give the following result, which shows that we can compute by using Algorithm II a critical point of f

Theorem 3. *Let λ_k a sequence converging to zero. For each $\lambda_k > 0$, we denote $x^*_{\lambda_k}$ an accumulation point of the sequence generated by algorithm II. Assume that $x^*_{\lambda_k}$ or a subsequence converges to x^* when k goes to infinity, then $0 \in \partial f(x^*)$, that is, x^* is a critical point of f.*

Proof. As $e_{\lambda_k} f \xrightarrow{e} f$ when λ_k goes to zero, then by means of Lemma 4 we conclude that $\mathrm{gph}\, \nabla e_{\lambda_k} f \xrightarrow{k} \mathrm{gph}\, \partial f$. Moreover, we get $\mathrm{gph}\, \nabla e_{\lambda_k} f \ni (x^*_{\lambda_k}, 0) \to (x^*, 0) \in \mathrm{gph}\, \partial f$, that is, x^* is a critical point of f as required.

6   Conclusions

We have proposed two algorithms to solve an unconstrained nonsmooth optimization problem. In the first case, we have generated an algorithm converging to a critical point of f when this function is locally lipschitzian and r-prox-regular. This last class of functions has shown suitable properties from the algorithmic point of view. In the second case, we attack the problem when function f is only r-prox-regular (nonlipschitzian). Thus, by using the Moreau approximation and some Mifflin and Qi's ideas we have proved of the convergence of the generic Algorithm II proposed in this paper. With respect to speed of convergence, we can guarantee the superlinear convergence of the critical point given in Algorithm II. A remaining question is to study the speed of both algorithms I and II, which certainly will depend on the line search algorithms involved.

References

1. Attouch, H. [1984], Variational Convergence of Functions and Operators, Pitman, Londres.
2. Auslander, A. [1987], *Numerical Methods for Nondifferentiable Convex Optimization*, Mathematical Programming Studies, 30, 102-127.
3. Bahraoui, M. A. [1994], *Suites Diagonalement Stationnaires en Optimisation Convexe*, Thesis, Université de Montpellier.
4. Cominetti, R. [1997], *Coupling the Proximal Point Algorithm with Approximation Methods*, Journal of Optimization Theory and Applications, 95, 581-600.
5. Correa, R., A. Jofré and L. Thibault [1992], *Characterization of Lower Semicontinuous Convex Funtions*, Proceedings of the American Mathematical Society, 116, 6-72.
6. Correa R., A. Jofré and L. Thibault [1994], *Subdifferential Characterization of Convexity*, in Recents Advanced in Nonnsmooth Optimization , edited by D. Du, L. Qi and R. Womersley, pp. 18-23, World Scientific Publishing, Singapore.
7. Correa R. and C. Lemarechal [1993], *Convergence of Some Algorithms for Convex Minimization*, Mathematical Programming, 62, 261-275.
8. Clarke, F. H. [1983], Optimization and Nonsmooth Analysis, Wiley, New York.
9. Martinet, B. [1972], *Algorithmes pour la Résolution de Problémes d'Optimisation et de Minimax*, Thesis, Université de Grenoble.

10. Martinet, B. [1970], *Régularisation d'Inequations Variationnelles par Approxima-tions Successives*, Revue Française d'Informatique et Recherche Opérationnelle, 4, 154-159.
11. Mifflin, R. [1977], *Semismooth and Semiconvex Functions in Constrained Optimiza-tion*, SIAM Journal Control and Optimization, 15, 97-972.
12. Pang J. S., S.p. Hang and N. Rangaraj [1991], *Minimization of Locally Lipschitzian Functions*, SIAM Journal Optimization, 1, 57-82.
13. Poliquin, R. and T. Rockafellar [1996], *Prox-Regular Functions in Variational Anal-ysis*, Transactions of the American Mathematical Society, 348, 1805-138.
14. Poliquin, R. [1992], *An Extension of Attouch's Theorem and It's Application to Second Order Epi-Differentiation of Convexly Composite Functions*, Transactions of the American Mathematical Society, 332, 861-874.
15. Poliquin, R. [1991], *Integration of Subdifferentials of Nonconvex Functions*, Non-linear Analysis, Theory Methods and Applications, 17, 385-398.
16. Rockafellar, T. and R. Wets [1998], Variational Analysis, Springer.
17. Rockafellar, T. [1976], *Monotone Operators and the Proximal Point Algorithm*, SIAM Journal Control and Optimization, 14, 877-898.
18. Qi, L. [1993], *Convergence Analysis of some Algorithms for Solving Nonsmooth Equations*, Mathematics of Operations Research, 18, 227-244.

A Risk-Reward Competitive Analysis of the Bahncard Problem*

Lili Ding, Chunlin Xin, and Jian Chen

School of Management, Xi'an Jiaotong University,
Shaanxi, P.R. China
dinglili0220@sohu.com

Abstract. Competitive analysis for all investors in the Bahncard problem (a railway pass of the Deutsche Bundesbahn company) has received much attention in recent years. In contrast to this common approach, which selects the riskless outcome and achieves the optimal competitive ratio, this paper introduces a risk-reward competitive strategy to achieve flexibility. Namely, we extend the traditional competitive analysis to provide a framework in which the travellers can develop optimal trading strategies based on their risk tolerance and investing capability. We further present a surprisingly flexible competitive ratio of $r_A^* = 1 + \frac{1-\beta}{(2-\beta)t-(1-\beta)}$ for the Bahncard problem, where t is the risk tolerance and β is the percentage of discount with respect to this strategy. Then substituting $t = 1$ into the above equation, we obtain the $(2 - \beta)$-competitive ratio which is the best attainable result presented by Fleischer.

1   Introduction

An extensive study of the online problems began in the 1980s in the seminal work of Sleator and Tarjan [9] on list accessing and paging algorithms. Within the theoretical computer science community, the competitive ratio has become a standard approach for the analysis of the online problems. Nevertheless, one argument against the use of this approach is that the online players are inherently risk-averse as they are optimized with respect to the worst-case event sequences. A number of approaches have been developed in an attempt to remedy this situation. Raghavan [3] attempted to remedy this by proposing a competitive strategy against the statistical adversary whose request sequence was required to satisfy certain distributional requirements. Al-Binali [1] analyzed a financial game using the competitive analysis framework to include a flexible risk management mechanism. Our risk-reward competitive analysis blends the two approaches to allow the online players to benefit from their own capability in correctly forecasting the coming request sequences, but also allow them to control their risk of performing and then selects a set of near optimal algorithms. The property

---

* This research is supported by NSF of China under Grants 10371094 and 70471035.

N. Megiddo, Y. Xu, and B. Zhu (Eds.): AAIM 2005, LNCS 3521, pp. 37–45, 2005.

can be favorable for the online players who may prefer somewhat inferior but
guaranteed performance to better average performance.

In this paper we study the Bahncard problem originally proposed by Fleischer
[2]. We use the notation of $BP(C, \beta, T)$ to denote this problem where a Bahncard
costs C, reduces any regular ticket price p to βp, and is valid for time T. For
example, $BP(240\, DM, \frac{1}{2}, 1\, year)$ means that if the traveller spends $240\, DM$ for
a Bahncard, he is entitled to a percentage discount of 50% price reduction on
nearly all train tickets. Fleischer presented a $(2 - \beta)$-competitive solution using
the traditional deterministic online algorithm for $BP(C, \beta, T)$. However, another
key factor of $BP(C, \beta, T)$ (which is not considered in [2]) is the risk. Because in
most cases the investors do not seek to minimize risk, but to manage it.

Recognizing the need for risk management in $BP(C, \beta, T)$, we offer such a framework that generalizes competitive analysis and allows for flexible risk management.
The framework extends traditional competitive analysis by introducing two ingredients: risk and capability. This is because the online investor often owns such capability that can make a correct forecast about the future requirements. Therefore,
for $BP(C, \beta, T)$ the investor can estimate his own investing capability to choose a
maximum acceptable risk level t and a set of forecasts F, and then develop an algorithm that can maximize the reward should his forecast be correct. It is hoped
that by posing reasonable forecast the online investor can boost performance significantly as long as the input sequences conform to the forecasts.

This paper is organized as follows. Section 2 outlines the traditional competitive
analysis of $BP(C, \beta, T)$ by Fleischer. A detailed risk algorithm $\hat{A}$ and a proposed
competitive solution are given in section 3 while a flexible competitive ratio of $r_{\hat{A}}^* = 1 + \frac{1-\beta}{(2-\beta)t-(1-\beta)}$ is achieved. Section 4 presents some empirical results with respect
to the "Youth Discount Card of France Railway NO.CARTE 12-25". Finally, we
conclude with section 5, pointing out directions for future research.

2   The Bahncard Problem

For the problem of $BP(C, \beta, T)$, an investor has to decide whether to buy a Bahncard without any knowledge of the coming travel sequences $\delta_1, \delta_2, \ldots, \delta_n$. Here
each δ_i can be denoted by $\delta_i(t_i, p_i)$, which means the investor has to face the regular ticket price p_i at the travel time t_i. On the other hand, when purchasing a
Bahncard at the travel time t_i, the investor can obtain a reduced price βp_i during
the time interval $[t_i, t_i + T)$. Then the competitive analysis of online algorithms
often can help the online investor make decisions.

2.1   Competitive Analysis

Assume that an optimal offline investor knows exactly the future travel sequences,
and whether to buy Bahncards depends on the less cost incurred according to the
following rules:

1. the offline investor chooses a competitive algorithm with respect to the critical
cost $\frac{C}{1-\beta}$, which is the break-even point for any algorithms (see, for example, [2]).

2. during the time interval I, let $p_i^I(\delta) = \Sigma_{t_i \in I} p_i$ denote the all money, and $C_A^I(\delta) = \Sigma_{t_i \in I} C_A(\delta_i)$ be the total cost incurred by an algorithm A. If $p_i^I(\delta) \leq \frac{C}{1-\beta}$, the time interval I is called the cheap interval (OPT never buys a Bahncard), otherwise it is called the expensive interval.

3. for the online investor the chance of buying Bahncards by A is represented by $k_A(\delta) = (k_1, k_2, \ldots, k_m)$. That is, for the finite δ, an algorithm A buys Bahncards at times $0 \leq k_1 < \cdots < k_m$. Here epoch $[k_i, k_{i+1})$(except for, possibly, the first and last one) starts with an expensive phase $[k_i, k_i+T)$ and follows by a cheap phase $[k_i + T, k_{i+1})$. Then the total cost of A on δ is $C_A(\delta) = K_A(\delta) \cdot C + \Sigma_{i \geq 1} C_A(\delta_i)$, where $K_A(\delta)$ is the length m of the buying-chance.

For each request sequence, there is an optimal buying-chance $K_{OPT}(\delta)$ such that $C_{OPT}(\delta) = K_{OPT}(\delta) + min\Sigma_{i \geq 1} C(\delta_i)$. Fleischer [2] showed that given n travel requests, an optimal buying-chance of Bahncards and its minimal cost can be computed in $O(n)$ time. Therefore in this paper we also do not distinguish clearly between an algorithm A and the buying-chance $K_A(\delta)$. For simplicity, assume that $K_A(\delta) = 1$. Fleischer presented us a traditional competitive ratio of $BP(C, \beta, T)$ as follows:

Theorem 1. *No deterministic online algorithm for $BP(C, \beta, T)$ has the competitive ratio better than $(2 - \beta)$-competitive.*

Corollary 1. *If $p_i = p$ for $\forall i = 1, 2, \ldots, n$, $T \longrightarrow \infty$, the Bahncard problem is degenerated as the ski-rental problem.*

Proof. According to the preceding rules, the offline investor knows exactly the travelling itinerary and so will choose an online algorithm A of $min(np, C + \beta np)$. At the same time the offline investor can force any online investor to spend regular ticket price for $(n - 1)$ days by waiting until the online investor buys a Bahncard, and then decide not to travel again. Hence the online investor hopes to minimize the competitive ratio

$$r_A = \frac{C_A(P)}{C_{OPT}(p)} = \frac{C + (n-1)p + \beta p}{min(np, C + \beta np)}$$

subject to the constraints:

1. $p_i = p$ for $\forall i = 1, 2, \ldots, n$, $T \longrightarrow \infty$
2. $np \in [p, \frac{C}{1-\beta} - \varepsilon] \cup [\frac{C}{1-\beta} - \varepsilon, \infty)$ and ε is a small constant.

An algorithm A that is to buy tickets at the regular price until the total cost up to $\frac{C}{1-\beta} - \varepsilon$, and then decide to purchase a Bahncard can achieve the lower bound of

$$r^* = inf(r_A) = 2 - \beta - \frac{(1 - \beta)^2 p}{C} = 2 - \beta - \frac{(1 - \beta)^2}{n^*}$$

where $n^* = \frac{C}{p}$. If there is $\beta = 0$ such that

$$r^* = inf(r_A) = 2 - \frac{1}{n^*}.$$

$\square$

3 Competitive Analysis of the Bahncard Problem with a Risk-Reward Strategy

From the above analysis, the competitive algorithm A can overcome the weakness of the worst-case analysis. But the competitive ratio obtained by the traditional competitive analysis always enlarges the range for us to choose. In this section, we state a risk-reward strategy that extends the traditional competitive analysis by introducing two ingredients: risk and capability. We define the risk of an online algorithm to the ratio of the competitive ratio of the algorithm to the optimal competitive ratio, which is a smooth extension to coincide with the classical competitive analysis(see [1]). In fact the online investor pursues the richer reward with respect to his risk preference. Then we relate the reward to the investing capability which is defined as a subset of the correct forecasts.Therefore our reward is the ratio of the optimal competitive ratio to the restricted ratio should the forecasts come true.

3.1 Definition of a Risk-Reward Strategy

The basic definition of the risk-reward strategy is described as follows: when an online investor is risk-averse, he will use the traditional online algorithm A and achieve the optimal competitive ratio. If the online investor is a risk-seeker, our risk-reward strategy allows him to provide and benefit from his capability but also allows him to control his risk of performing with respect to the optimal offline algorithm using a risk algorithm $\hat{A}$. Because such online investor could always forecast exactly the coming requests, and then beat the optimal competitive ratio obtained by the classical competitive analysis.

First, the definition of the competitive ratio is presented as follows. For a request sequence $\delta\langle\delta_1, \delta_2, \ldots, \delta_n\rangle \in R^n$, the total cost of the online algorithm A is denoted by $C_A(\delta)$, and the optimal offline cost on δ is defined as $C_{OPT}(\delta) = min(cost_A(\delta))$. An algorithm A is $r-$ competitive if there exists a constant α such that

$$C_A(\delta) \leq r_A\, C_{opt}(\delta) + \alpha.$$

The optimal competitive ratio for the same problem is

$$r^* = inf_\delta(r_A) = inf_\delta \frac{C_A(\delta)}{C_{OPT}(\delta)} \qquad \forall\ \delta \in R^n.$$

Second, we link the risk tolerance t (where $t = 1$, the investor is risk-averse; $t > 1$, as the risk-seeker) with the competitive ratio r to denote the risk function $f(t, r)$. That is, the online investor could prefer to such risk tolerance t, with respect to the fluctuation of the competitive ratio of the algorithm to the optimal competitive ratio, such as

$$\frac{r_A - r^*}{r^*} = t.$$

Therefore, define the risk function of a risk algorithm $\hat{A}$ to be

$$f(t, r) = \{t, r \mid r_{\hat{A}} \leq tr^* \quad t \geq 1\}.$$

This risk function presents that the online investor will choose a risk algorithm $\hat{A}$ compared to the optimal online algorithm according to his own risk tolerance t.

Then we describe the reward function $R(t, r)$ on the basis of the risk preference over the optimal online algorithm. Facing with the coming request δ, the online investor with investing capability could always make a correct forecast $\bar{\delta}$, denoted by $F(\bar{\delta}) \in \delta$ ($F(\bar{\delta}) \notin \delta$ implies the false forecast). Hence, if there is $F(\bar{\delta}) \in \delta$, the online investor could achieve the better competitive ratio by a risk algorithm $\hat{A}$:

$$r_{\hat{A}}^* = inf_{F(\bar{\delta}) \in \delta}(r_{\hat{A}}) = \frac{C_{\hat{A}}(\bar{\delta})}{C_{OPT}(\bar{\delta})} \, .$$

Comparing $r_{\hat{A}}^*$ with r^*, we use the improvement of this risk algorithm $\hat{A}$ over the optimal online algorithm to measure the reward function $R(t, r)$:

$$R(t, r) = sup_{\hat{A} \in f(t, r)}\{t, r \mid \frac{r^* - r_{\hat{A}}^*}{r^* - 1} \quad F(\bar{\delta}) \in \delta\} \, .$$

Generally speaking, the bound of R can be presented as $R(t, r) \in [0, 1]$. This lower bound could be attained, when the online algorithm with the correct forecast exactly equals to the optimal offline algorithm, and then $\exists \hat{A}$, such that $r_{\hat{A}}^* \longrightarrow 1 \implies \exists \hat{A}$, such that $R = 0$. For the upper bound, when the online investor with such risk tolerance $t = 1$, note that the online risker algorithm $\hat{A}$ will always be the optimal online algorithm: $r_{\hat{A}}^* = r^* \implies R = 1$.

3.2   The Bahncard Problem with a Risk-Reward Strategy

We now analyze $BP(C, \beta, T)$ using this risk-reward strategy. Fleischer presented us an online algorithm A, in which the investor purchased the Bahncard at the time t_i for the total cost of the regular ticket price up to the critical cost $\frac{C}{1-\beta} - \varepsilon$, where ε is a constant, to achieve the optimal competitive ratio r^*. In contrast, we are concerned with the investing capability and risk preference of $BP(C, \beta, T)$. As an investing talent, the online investor has confidence in making a correct forecast on δ. Due to the preceding definition, denote the forecasts by $F(\bar{\delta}) = \{\bar{\delta}_1, \bar{\delta}_2 \ldots \bar{\delta}_n \mid \exists i$ such that $\sum \bar{\delta}_i \leq \frac{C}{1-\beta}$ or $\sum \bar{\delta}_i \leq \frac{C}{1-\beta}\}$, where $\frac{C}{1-\beta}$ is the break-even point. Suppose that the forecasts for the coming requests are correct, the online investor will buy a Bahncard at the early stage, guaranteeing the competitive ratio $r_{\hat{A}}$ less than $t\, r^*$. Within the limitation of $t\, r^*$, a risk algorithm $\hat{A}$ will compute a restricted optimal competitive ratio $r_{\hat{A}}^* = inf\, r_{\hat{A}}$. The analysis of the risk algorithm $\hat{A}$ is as follows:

Theorem 2. *With a correct forecast of* $F_1(\bar{\delta}) = \{\bar{\delta} \mid \sum \bar{\delta}_i \leq \frac{C}{1-\beta}\}$ *, the restricted optimal competitive ratio for* $BP(C, \beta, T)$ *is* $r_{\hat{A}}^* = 1$.

Proof. In this case, the best choice for the online investor is just what the offline investor does. They never purchase a Bahncard. Therefore, if $C_{\hat{A}}(\bar{\delta}) = C_{OPT}(\bar{\delta})$, then the restricted competitive ratio satisfies $r_{\hat{A}}^* = 1$ subject to $\{r_{\hat{A}}^* \mid r_{\hat{A}}^* \leq t\, (2 - \beta) \quad t \geq 1\}$.   □

Corollary 2. *With a correct forecast $F_1(\bar{\delta})$, the online investor could compute the maximal achievable reward of 1.*

Proof. From the above result of $r^*_{\hat{A}} = 1$, the reward is $R(t, r) = \frac{r^* - r^*_{\hat{A}}}{r^* - 1} = 1$. $\square$

For the online investor, the forecast of Theorem 2 means the higher risk the richest. However, there is a more complex situation with the second kind of forecast $F_2(\bar{\delta}) = \{\bar{\delta} \mid \sum \bar{\delta}_i \geq \frac{C}{1-\beta}\}$. Note that the online investor with a correct forecast 2 has to choose time t_i to purchase a Bahncard so as to keep the competitive ratio under the upper bound of $t r^*$.

Lemma 1. *With a correct forecast of $F_2(\bar{\delta}) = \{\bar{\delta} \mid \sum \bar{\delta}_i \geq \frac{C}{1-\beta}\}$, the optimal chance to buy a Bahncard is at the time of t_i, when the total regular ticket price amounts to $\sum p_i = \frac{C}{t(2-\beta)} - \varepsilon$.*

Proof. There are two cases:
Case 1: $\sum p_i \leq \frac{C}{1-\beta} - \varepsilon$ where ε is a small constant, then

$$r^*_{\hat{A}} = \frac{.C + \sum p_i + \beta\varepsilon}{\sum p_i + \varepsilon}.$$

For $r^*_{\hat{A}} \leq t r^*$ and $\varepsilon \longrightarrow 0$, note that

$$r^*_{\hat{A}} = \frac{C + \sum p_i + \beta\varepsilon}{\sum p_i + \varepsilon} \leq (2-\beta)t \Longrightarrow \sum p_i \geq \frac{C}{(2-\beta)t - 1}.$$

Case 2: $\sum p_i \geq \frac{C}{1-\beta} - \varepsilon$ where ε is a small constant, then

$$r^*_{\hat{A}} = \frac{C + \sum p_i + \beta\varepsilon}{C + \beta(\sum p_i + \varepsilon)}.$$

subject to the same conditions: $r^*_{\hat{A}} \leq t r^*$ and $\varepsilon \longrightarrow 0$, the restricted optimal competitive ratio is

$$r^*_{\hat{A}} = \frac{C + \sum p_i + \beta\varepsilon}{C + \beta(\sum p_i + \varepsilon)} \leq (2-\beta)t.$$

From the above inequality, the bound of the total cost is

$$\sum p_i \leq \frac{C(2-\beta)t - C}{1 - (2-\beta)t\beta} \qquad iff\ 1 - \frac{\sqrt{t^2 - t}}{t} \leq \beta \leq 1.$$

According to the Theorem 3, set $\sum p_i = \frac{C}{(2-\beta)t - 1}$, the restricted optimal competitive ratio with the risk-reward strategy can be obtained. $\square$

Through the risk-reward strategy, it can be shown that the restricted optimal competitive ratio is less than the optimal competitive ratio.

Theorem 3. *The restricted optimal competitive ratio with the risk-reward strategy for $BP(C, \beta, T)$ is denoted by $r_{\hat{A}}* = 1 + \frac{1-\beta}{(2-\beta)t - (1-\beta)}$.*

Proof. Should the $F_2(\bar{\delta}) = \{\bar{\delta} \mid \sum \bar{\delta}_i \geq \frac{C}{1-\beta}\}$ be correct, the online investor could have such competitive ratio function that $r_{\hat{A}}^* = f(\sum p_i) = \frac{C + \sum p_i + \beta\varepsilon}{C + \beta(\sum p_i + \varepsilon)}$. The monotonous increasing character of this function ensures that there exists a lower bound of $\sum p_i = \frac{C}{(2-\beta)t-1}$ such that $r_{\hat{A}}* = 1 + \frac{1-\beta}{(2-\beta)t-(1-\beta)}$. □

Corollary 3. *The relation of the optimal competitive ratio between the strategy in [2] and this risk-reward strategy for the Bahncard problem is that $\{r_{\hat{A}}^* \mid r_{\hat{A}}^* \in [1, r^*]\}$.*

Proof. Assume that β is a constant, then the restricted optimal competitive ratio fluctuates with the risk tolerance t. Hence, the online investors can estimate their own abilities to choose the different risk preference, such as $t = 1 \implies r_{\hat{A}}* = 1 + \frac{1-\beta}{(2-\beta)t-(1-\beta)} = 2 - \beta = r^*$. Nevertheless some investors would like benefit from accepting increased risk. Note that as $t \longrightarrow \infty$, the restricted optimal competitive ratio approaches to the minimum ratio of $r_{\hat{A}}^* = 1$. It is shown that the optimal competitive ratio in [2] is a special case of our results obtained by the risk-reward strategy. □

Corollary 4. *The reward of the risk algorithm $\hat{A}$ for the Bahncard problem with the correct forecast of $F_2(\bar{\delta})$ can achieve the maximum value: $R(t, r) \longrightarrow 1$.*

Proof. $R(t, r) = 1 - \frac{1}{(2-\beta)t-(1-\beta)} \approx 1$ as $t \longrightarrow \infty$. □

4   Some Empirical Results

In this section, we provide a example of "Youth Discount Card of France Railway NO.CARTE 12-25" to explore the relationship between the risk tolerance t and the reward R about the competitive ratio. Based on the actual data resource of the "Youth Discount Card of France Railway NO.CARTE 12-25", we present the table 1 to respect the various figures. In the table 1, we set C, β, and T, and compute the $r_{\hat{A}}^*$ and r^* according to the different tolerance t and the critical cost $\frac{C}{(2-\beta)t-1}$.

Table 1. Setting $C = 43.00EU$, $\beta = 25\%$, and $T = 1\,year$, the restricted competitive ratio $r_{\hat{A}}^*$ can be achieved according to the various risk tolerance t

t	$\frac{C}{(2-\beta)t-1}$	$r_{\hat{A}}^*$	r^*
1.00	57.33	1.7500	1.7500
1.10	47.51	1.6382	1.7500
1.20	39.10	1.5560	1.7500
1.25	36.21	1.5218	1.7500

With respect to the risk-reward strategy, the most important for an online investor is to hold the great capability of investing. Because he can carry on the correct forecast to pursue relatively less competitive ratio with such capability. From the above table 1, as the risk-averse, the online investor will wait until the total cost of regular ticket price amounts to 57.33. However, a risk-seeker would like to bear the risk loss(when the forecast is incorrect.)to purchase a Bahncard at the earlier time such as the critical cost satisfies $\frac{C}{(2-\beta)\,t-1} = 39.10$. Obviously, in this case the online investor has to face the opportunity loss of 18.23EU. Nevertheless, in return, when the forecast is correct, the less competitive ratio of 1.5560 can be obtained. Next, we present the risk-reward strategy to help the online investor judge his style according to his own investing capability. If $t = 1.20$, then the reward ratio $R = \frac{r_* - r_{\hat{A}}^*}{r_* - 1} = 25.80\%$. This means that the online investor bears the possible cost loss of 31.70%, but he could obtain the improvement of 25.80% about the reward.

Table 2. Setting $C = 43.00EU$, $\beta = 50\%$, and $T = 1\,year$, the restricted competitive ratio $r_{\hat{A}}^*$ can be achieved according to the various risk tolerance t

t	$\frac{C}{(2-\beta)\,t-1}$	$r_{\hat{A}}^*$	r^*
1.00	86.00	1.5000	1.5000
1.10	74.79	1.4651	1.5000
1.20	71.67	1.4546	1.5000
1.25	49.14	1.3637	1.5000

From the above analysis, we know that the risk tolerance t influences the restricted optimal competitive ratio $r_{\hat{A}}^*$. After we increase the percentage discount β up to 50%, a new status presents to us. In table 2, we find that at the same risk tolerance all restricted optimal competitive ratios are less than the ones in table 1. Therefore, the higher the discount ratio is, the more the online investor concerns with the risk-reward strategy.

5   Conclusion

In this paper, the investing capability and the risk tolerance are introduced into the online Bahncard problem so that the competitive analysis is more realistic. Compared with the results of Fleischer [2], a more flexible competitive ratio can be achieved. However, there are some interesting problems as follows.

⋆ In this paper, we assume that β is a constant. But the percentage discount β often takes on the different values at some stage which make the competitive analysis more complex.
⋆ For the risk-reward strategy, the risk tolerance t can be in the form of interval number, better simulating the behavior preference, such as $t \in [a,\ b]$.
⋆ The currency value over time and the transaction cost are of importance to the restricted optimal competitive ratio. In some cases, the competitive ratio may be seriously influenced by these factors.

References

1. B.al-Binali. A risk-reward framework for the competitive analysis of financial games. *Algorithmica*, **25**: 99-115,1999.
2. R.Fleischer. On the Bahncard problem. *COCOON'98*, LNCS 1449, 65-74,1998.
3. P.Raghavan. A Statistical Adversary for on-line Algorithms. *DIMACS Series in Discrete Mathematics*, **7** : 79-83,1992.
4. D.P.Helmbold, R.E.Schapire, Y.Singer and M.K.Warmuth .Online Portfolio Selection Using Multiplicative Updates. *Mathematical Finance*, **8** : 325-347,1998.
5. Borodin, R. El-Yaniv. *Online Computation and Competitive Analysis*, Cambridge University Press, London, U.K.,1998.
6. R.El-Yaniv, Kaniel and N.Linial. Competitive Optimal On-line Leasing.*Algorithmica*, **25**: 116-140,1999.
7. A. R. Karlin, C. Kenyon and D. Randall. Dynamic TCP Acknowledgement and Other Stories about $e/(e - 1)$, *Proc. STOC '01*, **25**: 502–509,2001.
8. R.El-Yaniv. Competitive Solution for online Financial Problem. *ACM Computing Surveys*, **25**(1): 28-70,1998.
9. D.D. Sleator, R.E.Tarjan. Amortized Efficiency of List Update and Paging Rules. *Communication of the ACM*, **28**,202-208,1985.
10. A.Chou, R.CooperstockJ , R.El-Yaniv, M.Klugerman and T.Leighton. The statistical adversary allows optimal money-making trading strategies. *In the Proceedings of the 6th Annual ACM-SIAM Symposium on Discrete Algorithms*,1995.

Competitive Strategies for On-line Production Order Disposal Problem*

Feifeng Zheng, Wenqiang Dai, Peng Xiao, and Yun Zhao

School of Management, Xi'an Jiaotong University,
Shaanxi, P.R. China
zhengff@mailst.xjtu.edu.cn

Abstract. In this paper we study the on-line production order disposal problem considering preemption and abortion penalty. We discuss the cases when orders have uniform and nonuniform lengths. For the case of uniform order length, the GR strategy is proved to be $2\rho + 2\sqrt{(1+\rho)^2 + \rho} + 3$ -competitive, where $\rho \geq 0$ is the coefficient of the punishment. For the case of nonuniform order lengths, GR is $2(\lambda + \lambda\rho) + 2\sqrt{(\lambda + \lambda\rho)^2 + \lambda\rho} + 1$ -competitive where λ is the ratio of length between the longest and shortest orders. When abortion penalty is not counted, the ER strategy is proposed and proved to be $e\lambda + e + 1$ -competitive, where $e \approx 2.718$. The result is much better than that of GR. We show that ER is not competitive when abortion penalty is counted.

1   Introduction

The production order disposal problem is one of the typical management problems in manufacturing industry. A manufacturer representing an organization or a company may receive and dispose many production orders during a time period and he can dispose at most one order at any time due to resource constraints. Under the on-line model, the manufacturer can not obtain the order information, such as processing time and profit, until the order arrives. So, he needs to dynamically adjust the production schedule for arrival orders to maximize his profit.

If we view the manufacturer as a single machine and orders as jobs, the on-line production order disposal problem can be transformed into the classical on-line single machine scheduling problem, which has been extensively studied. Liu and Cheng [1] considered a preemption-restart model with the objective being minimizing the maximum delay. They proved that if each job's release time is arbitrary then the problem is strongly NP-hard. Dauzere and Sevaux [2] studied the problem of minimizing the number of tardy jobs. They discussed the case of different release time and due dates and proposed a branch and bound scheme that produce good empirical results. Daniel Ng et al. [3] presented an $O(n^2 \log n)$ time algorithm for scheduling costs involving earliness/tardiness and

---

* This research is supported by NSF of China under Grants 10371094 and 70471035.

N. Megiddo, Y. Xu, and B. Zhu (Eds.): AAIM 2005, LNCS 3521, pp. 46–54, 2005.

the number of tardy jobs, where variable common due date and resource allocation were considered. In the above research it is assumed that all information is
known at the beginning, i.e., it is in an off-line manner. Anderson and Potts [4]
considered the on-line scheduling problem where preemption is forbidden and
the goal is to minimize the total weighted completion time. They presented a
2-competitive algorithm and proved that the ratio is tight. Chou [5] analyzed
the on-line problem of minimizing weighted completion time and presented an
on-line algorithm that has the asymptotic performance ratio equaling 1 as the
job number increases. Marek etc. [6] discussed the preemption-restart model to
maximize the number of satisfied job where jobs all have the same weight and
proposed a deterministic 3/2-competitive algorithm.

In this paper we discuss the preemption-restart model for the on-line production order disposal problem where each order has variable profit or weight
which can not be obtained until the order is finished. As the manufacturer can
not finish all the arrival orders due to the resource limitation, he must decide
whether to abort the current order and start another one for more profit. However, if he makes an abortion then there is an abortion penalty. His objective
is to maximize the net profit, which is the total profit of the completed orders
minus the total penalty of abortions. We consider both the cases when orders are
of uniform and non-uniform lengths. Section 3 discusses the former case and the
GR strategy is shown to be competitive. The latter case is discussed in section 4
where GR is proved to be competitive, and an improved strategy ER is proposed
when there is no abortion penalty. However, ER does not work when there is an
abortion penalty. Section 5 concludes this paper.

2   Model and Notations

Now we define the problem formally. Assume that a sequence of production orders, $S = (J_1, J_2, \dots)$, arrive one by one in the production order disposal system.
When order J_i arrives, its four parameters, a_i, l_i, d_i, w_i, become known on its arrival, where a_i denotes the arrival time, l_i the length of processing time, d_i the
deadline and w_i the profit of J_i respectively, but we do not know what is the
next order and when it arrives. For any strategy A, generally it can not start all
orders in S, so that some jobs of S are rearranged by their service start times
in $\sigma = (J'_1, J'_2, \dots)$, where $\sigma \subset S$ is produced by A and order J'_i is started by A
at time s'_i. The orders in σ are indexed such that $s'_i < s'_{i+1}$. If $s'_{i+1} < s'_i + l'_i$, we
say order J'_i is *aborted* by J'_{i+1}; otherwise J'_i is said to be *completed*. If an order
arrives in the system but is not served at once, it is called a *waiting order*. A
waiting order J_i will exit the system at time t if $t + l_i > d_i$. An aborted order
may be restarted from the beginning some time later provided it can still meet
its deadline. Thus, orders J'_i and J'_{i+k} in σ may be the same order in S if J'_i is
aborted and restarted by A at time t'_i and t'_{i+k} respectively; however, they are
regarded as different orders in σ. The objective of A is to maximize the net profit

within a time period. The penalty of an abortion is defined as follows. If order J'_j is aborted, the penalty equals $\rho w'_j$, where $\rho \geq 0$ is called the punishment coefficient.

Given an input I (a set of orders) and an on-line strategy A, denote by $S_A(I)$ and $S^*(I)$ the schedules produced by A and by an optimal off-line strategy on I respectively. Denote by $|S(I)|$ and $|P(I)|$ the total profit of completed orders and total penalty in the schedule $S(I)$, respectively. We use the competitive ratio analysis [7] to measure the quality of the schedules produced by A, then the competitive ratio of A is defined as $r_A = \sup_I \frac{|S^*(I)| - |P^*(I)|}{|S_A(I)| - |P_A(I)|}$, where I can be any order input sequence. $r_A \geq 1$ always holds.

3   Uniform Order Length: The GR Strategy

In this section we consider the case that each order has the same disposal time. Assume without loss of generality that $l_i = 1$ holds for any order J_i. In the following we give a proposition and then present the Greedy Strategy (abbr. GR).

Proposition 1. *For the case of uniform order length, any deterministic on-line strategy must use the abortion function to be competitive.*

Proof. Assume that a competitive strategy A does not abort any order it starts. We construct such an input sequence that at time $t = 0$, order J_1 with tight deadline and profit w_1 arrives. A has two choices. If it starts to serve J_1 at once, another order J_2 with tight deadline and profit $w_2 = \alpha w_1$ will arrive at time $t = 1/2$ and no more orders come later. The optimal off-line strategy, OPT, will select to serve J_2 and then have profit α times that of A. Since α can be arbitrarily large, A is not competitive in this case. If A does not start J_1 at time $t = 0$, OPT will serve it, and no more order comes later. Then A gains zero and can not be competitive either. Both of the two cases contradict the assumption. Hence, A must utilize the abortion function to be competitive.   □

Now we present GR as follows. Assume that when a new order $J_{j'}$ arrives at time t, GR is serving order J_j. GR will abort J_j to serve $J_{j'}$ if $w_{j'} \geq c w_j$, otherwise it continues serving J_j, where $c > 1 + \rho$ is some constant determined later. When GR finishes an order, it starts a waiting or new arriving order with the most amount of profit. If there is a tie, GR selects the one with the earliest deadline or arbitrary one if they are of the same deadlines. We can see that the abortion is triggered according to the relative profit of currently served order and new arriving order. The rest of this section is devoted to the proof of the following theorem.

Theorem 1. *GR is $2\rho + 2\sqrt{(1+\rho)^2 + \rho} + 3$ -competitive, where $\rho \geq 0$ is the punishment coefficient.*

Proof. Given any order disposal sequence produced by GR, σ, it can always be divided into such m sub-sequences, $\sigma = (\sigma_1, \sigma_2, \dots, \sigma_m)$, that in each sub-sequence $\sigma_i = (J_{i,0}, J_{i,1}, \dots, J_{i,k_i})$ $(1 \leq i \leq m)$, only the last order J_{i,k_i} is completed and all other orders are aborted by GR. If $k_i = 0$, the sub-sequence consists of only one order completed by GR. Since every order in σ is indexed by its service start time, if an order is aborted and restarted later, GR regards it as another order when it is restarted. So, we can regard that any aborted order will not be served again by GR in the whole σ. When σ_i is understood from the context, we will simply denote it by $\sigma_i = (J_0, J_1, \dots, J_k)$ for convenience.

We will first analyze an arbitrary sub-sequence σ_i and then extend the result to the whole sequence σ. Because OPT is the optimal off-line strategy, we suppose that it does not abort any order it starts for it will stay idle rather than start an order to be aborted later, i.e., OPT will not waste any order. So, OPT has no penalty and its net profit equal its profit, and if OPT starts an order, it means that OPT will complete that order. Denote by $|O(\sigma)|$, $|O(\sigma_i)|$ the profit OPT gains in σ and σ_i respectively. Especially, let $O(\sigma_i) = \{O_b(\sigma_i), O_r(\sigma_i)\}$, where $O_b(\sigma_i)$ includes orders started by OPT during GR is serving σ_i, and $O_r(\sigma_i)$ includes those of σ_i started by OPT after GR has finished σ_i. $O_r(\sigma_i)$ at most includes J_k, otherwise if OPT starts an order of σ_i other than J_k when GR has completed σ_i, GR can also serve that order for it has not been satisfied, contradicting that any aborted order will not be served again by GR in σ. Hence, we have $|O_r(\sigma_i)| \leq w_k$. In σ_i, denote by $|O(J_j)|$ $(0 \leq j \leq k)$ the profit of orders started by OPT during GR is serving J_j. Since $l_j = 1$, $O(J_j)$ includes at most one order and $|O(J_j)| < cw_j$ holds for the order does not abort J_j. Denote by $|G(\sigma)|$, $|G(\sigma_i)|$ the total profit GR gains and $|P_G(\sigma)|$, $|P_G(\sigma_i)|$ the total penalty GR receives in σ and σ_i respectively.

By the construction of σ_i, we have $w_j \leq \frac{1}{c}w_{j+1}$ and then $w_j \leq (\frac{1}{c})^{k-j}w_k$ for $0 \leq j < k$, and $|G(\sigma_i)| = w_k$. On the other hand, there are totally k abortions in σ_i, so

$$|P_G(\sigma_i)| = \sum_{j=0}^{k-1} \rho w_j \leq \sum_{j=0}^{k-1} \rho(\frac{1}{c})^{k-j}w_k = \frac{\rho}{c-1}[1 - (\frac{1}{c})^k]w_k < \frac{\rho}{c-1}w_k.$$

Thus, the net profit of GR in σ_i satisfies

$$|G(\sigma_i)| - |P_G(\sigma_i)| > (1 - \frac{\rho}{c-1})w_k. \tag{1}$$

For $O_b(\sigma_i)$, since $|O(J_j)| < cw_j \leq (\frac{1}{c})^{k-j-1}w_k$ holds for $0 \leq j \leq k$,

$$|O_b(\sigma_i)| = \sum_{j=0}^{k} |O(J_j)| < \sum_{j=0}^{k} (\frac{1}{c})^{k-j-1}w_k < \frac{c^2}{c-1}w_k. \tag{2}$$

We already know $|O_r(\sigma_i)| \leq w_k$. Combining inequalities (1,2), for σ_i

$$\frac{|O(\sigma_i)|}{|G(\sigma_i)| - |P_G(\sigma_i)|} = \frac{|O_b(\sigma_i)| + |O_r(\sigma_i)|}{|G(\sigma_i)| - |P_G(\sigma_i)|} < \frac{(\frac{c^2}{c-1} + 1)w_k}{(1 - \frac{\rho}{c-1})w_k} = \frac{c^2 + c - 1}{c - 1 - \rho}.$$

Hence, we have for σ that

$$\frac{|O(\sigma)|}{|G(\sigma)| - |P_{GR}(\sigma)|} = \frac{\sum_{i=1}^{m} (|O_b(\sigma_i)| + |O_r(\sigma_i)|)}{\sum_{i=1}^{m} (|G(\sigma_i)| - |P_G(\sigma_i)|)} < \frac{c^2 + c - 1}{c - 1 - \rho}.$$

Taking the derivative of $\frac{c^2+c-1}{c-1-\rho}$ with respect to c and equating it with zero, we obtain that $c^2 - 2c - 2\rho c - \rho = 0$. Solving the equation, $c = 1 + \rho + \sqrt{(1 + \rho)^2 + \rho}$, and then the minimum value of $\frac{c^2+c-1}{c-1-\rho}$ is $2\rho + 2\sqrt{(1 + \rho)^2 + \rho} + 3$, i.e., GR is $2\rho + 2\sqrt{(1 + \rho)^2 + \rho} + 3$ -competitive.   □

By Theorem 1, the competitive ratio of GR is a function of ρ. Taking $\rho = 0$, 0.1, 0.25, 0.5, 1.0, 2.0, 3.0, the competitive ratio equals 5.0, 5.5, 6.2, 7.3, 9.5, 13.6, 17.7, respectively. We can see that the competitive ratio increases about 4 times faster than ρ does, i.e., the competitive ratio varies in an approximately linear speed with respect to ρ.

4   Nonuniform Order Lengths: The ER Strategy

In this section we consider the case that orders have nonuniform disposal time. Let $\lambda = \lceil \frac{the\ longest\ order\ length}{the\ shortest\ order\ length} \rceil$. Without loss of generality, we assume that the size of the shortest order equals 1 and the longest order is λ long. Note that Proposition 1 still holds in this case.

For the performance of GR in this case, we have the following theorem.

Theorem 2. *GR is* $2(\lambda + \lambda\rho) + 2\sqrt{(\lambda + \lambda\rho)^2 + \lambda\rho} + 1$ *-competitive in the case of nonuniform order lengths.*

The proof is omitted here. Note that Since $l_j \leq \lambda$, $O(J_j)$ includes at most λ orders and then $|O(J_j)| < \lambda c w_j$ holds. The other proof is similar to that in the case of uniform order length. Especially, GR is $4\lambda + 1$ -competitive when $\rho = 0$.

In the following we propose the ER strategy and discuss its competitive performance without considering the abortion penalty. The main idea of ER is to increase its abortion ratio exponentially with respect to the processed percent of currently served order.

ER is described as follows. When ER is serving order J_j, the abortion ratio equals $c^{(t-s_j)/l_j}$ at time $t \in (s_j, s_j + l_j)$, where $c > 2$ is some constant determined later, that is, if a new arriving order $J_{j'}$ has profit at least $c^{(t-s_j)/l_j}$ times that of currently served order J_j, ER will abort J_j to serve $J_{j'}$. When ER finishes an order, it begins a waiting or new arriving order with the most amount of profit. If there is a tie, ER selects the one with the earliest deadline or the shortest length, and arbitrary one if they are of the same deadline or the same length. For ER, we have the following theorem.

Theorem 3. *In the case of non-uniform order lengths and no abortion penalty, ER is $e\lambda + e + 1$ -competitive, where $e \approx 2.718$.*

Proof. Similar to that in the case of uniform order length, given any sequence σ produced by ER, it can always be divided into such m sub-sequences, $\sigma = (\sigma_1, \sigma_2, \ldots, \sigma_m)$, that in each $\sigma_i = (J_{i,0}, J_{i,1}, \ldots, J_{i,k_i})$ $(1 \le i \le m)$, only order J_{i,k_i} is completed by ER and all other orders are aborted. As in the case of uniform order length, we denote σ_i by $\sigma_i = (J_0, J_1, \ldots, J_k)$ in the following discussion. $|E(\sigma)|$, $|E(\sigma_i)|$, $|P_E(\sigma)|$ and $|P_E(\sigma_i)|$ are similarly defined as those of GR, and $|O(\sigma)|$, $|O(\sigma_i)|$, $O_b(\sigma_i)$, $O_r(\sigma_i)$ and $O(J_j)$ $(0 \le j \le k)$ are the same as those in the case of uniform order length. Since $l_j \ge 1$ in the case of nonuniform order lengths, $O(J_j)$ may include several orders.

We will first analyze the special case that all orders started by ER have size λ, i.e., $l_j = \lambda$ $(0 \le j \le k)$, and prove that Theorem 3 holds in the special case. Then we show that compared with OPT, ER performs worst in the special case. Thus, Theorem 3 holds in all cases.

In the special case, we will bound $|O(J_j)|$, $|O(\sigma_i)|$ and $|O(\sigma)|$ step by step. First, we discuss $O(J_k)$, which includes at most λ orders for $l_k = \lambda$. If OPT starts an order at time $t \in (s_k, s_k + \lambda)$, the order must have profit strictly less than $c^{(t-s_k)/\lambda} w_k$, otherwise it will abort J_k, contradicting that J_k is completed by ER. Assume that the first and last orders in $O(J_k)$ start at time t_k^1 and t_k^λ respectively, then the last order has profit less than $c^{(t-s_k)/\lambda} w_k$. Since t_k^λ is earlier than the end of J_k, we have $t_k^\lambda - s_k < \lambda$ and then $c^{(t-s_k)/\lambda} w_k < c w_k$. Thus,

$$|O(J_k)| < w_k \int_{t_k^1}^{t_k^\lambda} c^{\frac{t-s_k}{\lambda}} dt + c^{\frac{t_k^\lambda - s_k}{\lambda}} w_k < w_k \int_{t_k^1}^{t_k^\lambda} c^{\frac{t-s_k}{\lambda}} dt + c w_k. \qquad (3)$$

For $O(J_{k-1})$, its last order must start not later than $t_k^1 - 1$, otherwise OPT can not finish the order. Define t_{k-1}^1 and t_{k-1}^λ similarly, then $t_k^1 - t_{k-1}^\lambda \ge 1$ holds. And the last order in $O(J_{k-1})$ has profit less than $c^{(t_{k-1}^\lambda - s_{k-1})/\lambda} w_{k-1}$. So,

$$|O(J_{k-1})| < w_{k-1} \int_{t_{k-1}^1}^{t_{k-1}^\lambda} c^{\frac{t-s_{k-1}}{\lambda}} dt + c^{\frac{t_{k-1}^\lambda - s_{k-1}}{\lambda}} w_{k-1}. \qquad (4)$$

Since J_k aborts J_{k-1} at time s_k, which is later than t_{k-1}^λ, we have $w_k \ge c^{\frac{s_k - s_{k-1}}{\lambda}} w_{k-1} > c^{\frac{t_{k-1}^\lambda - s_{k-1}}{\lambda}} w_{k-1}$. Together with $t_k^1 - t_{k-1}^\lambda \ge 1$,

$$c^{\frac{t_{k-1}^\lambda - s_{k-1}}{\lambda}} w_{k-1} < w_{k-1} \int_{t_{k-1}^\lambda}^{s_k} c^{\frac{t-s_{k-1}}{\lambda}} dt + w_k \int_{s_k}^{t_k^1} c^{\frac{t-s_k}{\lambda}} dt. \qquad (5)$$

Combining inequalities (4) and (5), $|O(J_{k-1})|$ is bounded as so.

$$|O(J_{k-1})| < w_{k-1} \int_{t_{k-1}^1}^{t_{k-1}^\lambda} c^{\frac{t-s_{k-1}}{\lambda}} dt + w_{k-1} \int_{t_{k-1}^\lambda}^{s_k} c^{\frac{t-s_{k-1}}{\lambda}} dt + w_k \int_{s_k}^{t_k^1} c^{\frac{t-s_k}{\lambda}} dt$$

$$= w_{k-1} \int_{t_{k-1}^1}^{s_k} c^{\frac{t-s_{k-1}}{\lambda}} dt + w_k \int_{s_k}^{t_k^1} c^{\frac{t-s_k}{\lambda}} dt. \qquad (6)$$

According to inequalities (3) and (6),

$$|O(J_{k-1})| + |O(J_k)|$$

$$< w_{k-1} \int_{t_{k-1}^1}^{s_k} c^{\frac{t-s_{k-1}}{\lambda}} dt + w_k \int_{s_k}^{t_k^1} c^{\frac{t-s_k}{\lambda}} dt + w_k \int_{t_k^1}^{t_k^\lambda} c^{\frac{t-s_k}{\lambda}} dt + cw_k$$

$$= w_{k-1} \int_{t_{k-1}^1}^{s_k} c^{\frac{t-s_{k-1}}{\lambda}} dt + w_k \int_{s_k}^{t_k^\lambda} c^{\frac{t-s_k}{\lambda}} dt + cw_k. \tag{7}$$

For $O(J_j)$ ($0 \leq j \leq k - 2$), t_j^1 and t_j^λ are similarly defined and the discussion is similar to that of $O(J_{k-1})$. So,

$$|O(J_j)| < w_j \int_{t_j^1}^{s_{j+1}} c^{\frac{t-s_j}{\lambda}} dt + w_{j+1} \int_{s_{j+1}}^{t_{j|1}^1} c^{\frac{t-s_{j+1}}{\lambda}} dt. \tag{8}$$

By inequalities (7) and (8), we can bound $|O(\sigma_i)|$ as follows.

$$|O_b(\sigma_i)| = \sum_{j=0}^{k} |O(J_j)|$$

$$< w_0 \int_{t_0^1}^{s_1} c^{\frac{t-s_0}{\lambda}} dt + \sum_{j=1}^{k-1} w_j \int_{s_j}^{s_{j+1}} c^{\frac{t-s_j}{\lambda}} dt + w_k \int_{s_k}^{t_k^\lambda} c^{\frac{t-s_k}{\lambda}} dt + cw_k. \tag{9}$$

To solve the above integral, we need a detailed analysis. Since J_{j+1} aborts J_j, we have $w_{j+1} \geq c^{(s_{j+1}-s_j)/\lambda} w_j$ for $1 \leq j \leq k - 1$. Thus,

$$w_j \int_{s_j}^{s_{j+1}} c^{\frac{t-s_j}{\lambda}} dt \leq \frac{w_{j+1}}{c^{\frac{s_{j+1}-s_j}{\lambda}}} \int_{s_j}^{s_{j+1}} c^{\frac{t-s_j}{\lambda}} dt = w_{j+1} \int_{s_j}^{s_{j+1}} c^{\frac{t-s_{j+1}}{\lambda}} dt. \tag{10}$$

Using inequality (10), together with $w_k \geq c^{(s_k-s_{k-1})/\lambda} w_{k-1} \geq c^{(s_k-s_j)/\lambda} w_j$ and $\int_{t_0^1}^{s_1} c^{\frac{t-s_0}{\lambda}} dt \leq \int_{s_0}^{s_1} c^{\frac{t-s_0}{\lambda}} dt$ for $s_0 \leq t_0^1$, we can solve inequality (9) as follows.

$$|O_b(\sigma_i)| < w_0 \int_{t_0^1}^{s_1} c^{\frac{t-s_0}{\lambda}} dt + w_k \int_{s_1}^{s_k} c^{\frac{t-s_k}{\lambda}} dt + w_k \int_{s_k}^{t_k^\lambda} c^{\frac{t-s_k}{\lambda}} dt + cw_k$$

$$\leq w_k \int_{s_0}^{t_k^\lambda} c^{\frac{t-s_k}{\lambda}} dt + cw_k$$

$$< (\frac{c\lambda}{\ln c} + c)w_k. \tag{11}$$

Note that the last inequality holds for $t_k^\lambda - s_k < \lambda$. For $O_r(\sigma_i)$, $|O_r(\sigma_i)| \leq w_k$ holds, the analysis is similar to that in the case of uniform order length. Thus,

$$|O(\sigma_i)| = |O_b(\sigma_i)| + |O_r(\sigma_i)| < (\frac{c\lambda}{\ln c} + c + 1)w_k.$$

For ER, we have $|E(\sigma_i)| = w_k$ due to the construction of σ_i. Hence, we have for the whole sequence σ that

$$\frac{|O(\sigma)|}{|E(\sigma)|} = \frac{\sum\limits_{i=1}^{m} |O(\sigma_i)|}{\sum\limits_{i=1}^{m} |E(\sigma_i)|} < \frac{\sum\limits_{i=1}^{m} (\frac{c\lambda}{\ln c} + c + 1) w_k}{\sum\limits_{i=1}^{m} w_k} = \frac{c\lambda}{\ln c} + c + 1. \qquad (12)$$

Taking $c = e$, ER is $e\lambda + e + 1$ -competitive in the special case.

To complete the theorem, it suffices to prove that compared with OPT, ER performs worst in the special case. First, we give a lemma.

Lemma 1. *Given any two exponential functions $f_1(x) = a^{\frac{x}{c_1}}$ and $f_2(x) = a^{\frac{x}{c_2}}$, where $c_2 > c_1 > 0$ and $a > 1$, $\int_{x_1}^{x_3} f_1(x)dx \leq \int_{x_2}^{x_4} f_2(x)dx$ holds for any $x_3, x_4 > x_1, x_2$ satisfying $f_1(x_1) = f_2(x_2)$ and $f_1(x_3) = f_2(x_4)$.*

Proof. Since $a^{\frac{x_1}{c_1}} = a^{\frac{x_2}{c_2}}$, $a^{\frac{x_3}{c_1}} = a^{\frac{x_4}{c_2}}$ and $x_3, x_4 > x_1, x_2$, we have

$$\int_{x_2}^{x_4} f_2(x)dx - \int_{x_1}^{x_3} f_1(x)dx = \frac{c_2}{\ln a}\left(a^{\frac{x_4}{c_2}} - a^{\frac{x_2}{c_2}}\right) - \frac{c_1}{\ln a}\left(a^{\frac{x_3}{c_1}} - a^{\frac{x_1}{c_1}}\right)$$

$$= \frac{c_2 - c_1}{\ln a}\left(a^{\frac{x_4}{c_2}} - a^{\frac{x_2}{c_2}}\right)$$

$$> 0. \qquad \square$$

For σ_i with $l_j = \lambda (0 \leq j \leq k)$, we construct another sub-sequence $\sigma_i' = (J_0', J_1', \dots, J_k')$, where $w_j' = w_j$ and $l_j' = l_j$ except that $l_p' < \lambda$ $(0 \leq p \leq k)$. According to Lemma 1, $|O(J_p')| < |O(J_p)|$ holds where $|O(J_p')|, |O(J_p)|$ are the profit of the orders started by OPT during ER is serving J_p', J_p respectively. Thus, $|O(\sigma_i')| < |O(\sigma_i)|$. Combining $|E(\sigma_i)| = |E(\sigma_i')| = w_k$, we have $|O(\sigma_i')| / |E(\sigma_i')| < |O(\sigma_i)| / |E(\sigma_i)|$. If there are several orders shorter than λ in σ_i', the analysis is similar and the result keeps the same. Combining inequality (12), $|O(\sigma)|/|E(\sigma)| < e\lambda + e + 1$ holds for all cases, that is, ER is $e\lambda + e + 1$ -competitive. $\qquad \square$

Combining Theorems 2 and 3, when there is no abortion penalty and $\lambda \geq 3$, ER is much better than GR. However, the following theorem shows that ER is very bad if there is an abortion penalty, no matter how small ρ is.

Theorem 4. *ER is not competitive if there is an abortion penalty.*

Proof. We construct such an input sequence that $\sigma = (J_0, J_1, \dots, J_q)$ where $a_q - a_0 = \lambda$, $a_{i+1} > a_i$, $l_i = \lambda$ and $w_{i+1} = e^{(a_{i+1} - a_i)/l_i} w_i$ $(i = 0, 1, \dots, q - 1)$. Then J_{i+1} aborts J_i at time a_{i+1} due to the construction of ER, and $|E(\sigma)| = w_q = e^{(a_q - a_0)/\lambda} w_0 = e w_0$. However, $|P_E(\sigma)| = \sum\limits_{j=0}^{q-1} \rho w_j > q\rho w_0$ for $w_j > w_0$.

Thus, $|E(\sigma)| - |P_E(\sigma)| < (e - q\rho) w_0$. For any $\rho > 0$, q can be large enough so that $e - q\rho \leq 0$ holds, which means ER obtains no positive net profit. However,

OPT can at least finish order J_q and obtain profit $e^{(a_q-a_o)/\lambda}w_0$. According to the definition of competitive ratio, ER is not competitive. □

Theorems 2 and 4 imply that GR is much better than ER when there is an abortion penalty.

5 Conclusion

In this paper, we study the on-line production order disposal problem with preemption-restart model. Both the cases that orders have uniform and nonuniform lengths are studied. For the former case, the GR strategy is shown to be competitive no matter whether there is an abortion penalty. For the latter case, we propose the ER strategy and prove that it performs much better than GR when there is no abortion penalty. But ER does not work at all when there is an abortion penalty.

References

1. L. Zhaohui, T.C. Edwin Cheng, Scheduling with job release dates, delivery times and preemption penalties. *Information Processing Letters*, 82(2):107-111, 2002
2. S. Dauz ere-Peres, M. Sevaux, An exact method to minimize the number of tardy jobs in single machine scheduling. *Journal of scheduling*, 7(6):405-420, 2004
3. C. T. Daniel Ng, T. C. Edwin Cheng, Mikhail Y. Kovalyov and Lam S. S., Single machine scheduling with a variable common due date and resource-dependent processing times. *Computers and Operations Research*, 30(8):1173-1185, 2003
4. E.J. Anderson, C.N. Potts, Online scheduling of a single machine to minimize total weighted completion time. *Mathematics of Operations Research*, 29(3):686-697, 2004
5. Chou C.F.M., Asymptotic performance ratio of an online algorithm for the single machine scheduling with release dates. *IEEE Transactions on Automatic Control*, 49(5):772-776, 2004
6. Marek Chrobak, Wojciech Jawor, Jiri Sgall, Toms Tichy, Online Scheduling of Equal-Length Jobs: Randomization and Restarts Help. ICALP 2004: 358-370.
7. Sleator D.D., Trjan R.E., Amortized Efficiency of List Update and Paging Rules. *Communications of the ACM*, 28(2):202-208, 1985

Automatic Timetabling Using Artificial Immune System

Yulan He, Siu Cheung Hui, and Edmund Ming-Kit Lai

School of Computer Engineering,
Nanyang Technological University,
Nanyang Aveue, Singapore 639798
{asylhe, asschui, asmklai}@ntu.edu.sg

Abstract. University timetabling problem is a very common and seemingly simple, but yet very difficult problem to solve in practice. While solution definitely exists (evidenced by the fact that we do hold classes), an automated optimal schedule is very difficult to derive at present. There were successful attempts to address this problem using heuristics search methods. However, until now, university timetabling is still largely done by hand, because a typical university setting requires numerous customized complicated constraints that are difficult to model or automate. In addition, there is a problem of certain constraints being inviolable, while others are merely desirable. This paper intends to address the university timetabling problem that is highly constrained using Artificial Immune System. Empirical study on course timetabling for the School of Computer Engineering (SCE), Nanyang Technological University (NTU), Singapore as well as the benchmark dataset provided by the Metaheuristic Network shows that our proposed approach gives better results than those obtained using the Genetic Algorithm (GA).

1 Introduction

Over four decades, timetabling problem has been a major attraction of scientists from various disciplines. A practical timetabling problem usually involves complex constraints and a large number of events and is considered as *NP-hard* [1]. Traditional methods used to solve the timetabling problem include graph coloring with constraint manipulation [2] and clustering algorithm [3]. The graph coloring algorithm represents the timetabling problem as graphs where events (courses/exams) are represented as vertices and conflicts between the events are represented by edges. The constraint manipulation is done by scheduling the nodes with maximal degree (large number of conflicts) early. The clustering algorithm splits a set of events into groups which satisfy hard constraints and then the groups are assigned to time periods to fulfil the soft constraints. Both methods can only produce sub-optimal solutions.

Over the last two decades a variety of meta-heuristic approaches such as simulated annealing [4], tabu search [5], genetic algorithms (GA) [1,?] and hybrid approaches [6] have been investigated for timetabling. In meta-heuristic the

N. Megiddo, Y. Xu, and B. Zhu (Eds.): AAIM 2005, LNCS 3521, pp. 55–65, 2005.

emphasis is on performing a deep exploration of the most promising regions of the solution space. Meta-heuristic methods begin with one or more initial solutions and employ search strategies to avoid local optima. These meta-heuristic algorithms can produce high quality solutions but often incur high computational cost. Moreover, the procedures are usually context dependent and require finely tuned parameters which may make their extension to other situations difficult [7].

In this paper, the Artificial Immune System (AIS), and in particular the Clonal Selection Algorithm (CLONALG), is investigated to solve the university timetabling problem. It has certain similarities to Genetic Algorithms (GA) since both of them could be characterized as evolutionary-like algorithms and are inspired by biological metaphors, though CLONALG is based on the biological metaphor of the immune system whilst GA is inspired by the Darwinian evolutionary theory. The preliminary experimental results indicate that CLONALG performs better than GA when tested on the university timetabling benchmark data.

The rest of the paper is organized as follows. Section 2 describes the university timetabling problem. Section 3 presents the background of artificial immune system and specifically the clonal selection algorithm. Applying the CLONALG algorithm on the university timetabling problem is discussed in section 4. Performance comparison of CLONALG and GA by applying these two algorithms on the benchmark data provided by the Metaheuristic Network [8] is discussed in section 5. Finally, section 6 concludes the paper.

2   University Timetabling Problem

A timetabling problem, which is a subset of scheduling problem, can be defined as a combinatorial optimization problem of assigning resources to events being placed in period of time, satisfying a set of predefined constraints. The combinatorial problems are typically classified as *NP-bard* as they take practically infinite time to find any optimal solution and are indeed computationally intractable.

This paper focuses on the university course timetabling problem. In any optimization problem, there are objectives, decisions to make, resources available and related constraints. In a course timetabling problem, resources available are faculty, students, subjects being offered, time periods and rooms. A solution must group these resources together to produce a timetable to adhere to certain conditions set by the timetabler managing the timetable.

A typical university timetable is characterized by the following elements:

- *Timeslot*, $T = \{t_0, t_1, \cdots, t_m\}$. It is defined as the time interval during which a lecture, a tutorial, or a laboratory takes place. Every day has 9 timeslots. Each timeslot has a default value of 1-hour duration and the default starting time is on half an hour boundary. Thus, the total number of timeslots in a 5-day week is 45.

- *Room, $R = \{r_0, r_1, \cdots, r_n\}$*. Each event must take place in a particular room. A room can be classified by their functions or properties. In general, there are three types of room: lecture theater, tutorial room, or laboratory.
- *Subject, $S = \{s_0, s_1, \cdots, s_p\}$*. A subject is described by a name, title, number of students enrolled.
- *Subject Grouping, $G = \{g_0, g_1, \cdots, g_u\}$*. For any set of subjects having students in common, they can be grouped together as one subject group. For example, every course year could have 3 different subject groups, the *main group*, in which students follow a normal path and take four years to get a Bachelor degree, the *accelerated group*, in which students take three and a half years to get the Bachelor degree, and the *take ahead* group, in which students are allowed to take subjects from higher course years.
- *Class groups, $C = \{c_1, c_2, \cdots, c_v\}$*. Students are divided into a set of class groups in order to be accommodated with a particular room's size. Hence, for a particular subject with n students, there will be (usually) 1 lecture group, $n/TutorialRoomCapacity$ tutorial groups, and $n/LaboratoryRoomCapacity$ laboratory groups.
- *Event, $E = \{e_1, e_2, \cdots, e_w\}$*. As mentioned above, a subject group contains a list of subjects, and each subject has i lectures, j class groups of tutorial and/or k laboratory groups. Each event must consist of the following five elements: the subject group, the subject, the class group, the allocated room, and the allocated timeslot.

Two types of constraints are defined for the timetabling problem, *hard constraints* and *soft constraints*. All feasible and meaningful solutions must satisfy hard constraints. However, some or all of the soft constraints may be violated provided that the penalty costs associated with them are kept to minimum.

Typical hard constraints are:

- *Room occupancy.* The first condition for an event to be liable for an allocation in room r_i and at time t_j is that no other events can happen in room r_i at time t_j.
- *Room type.* The type of the room must conform to the type of the event.
- *Room capacity.* The capacity of the room must be large enough to accommodate the event.
- *Conflict.* The conflict constraint only deals with events of the same subject group. Hence, to check the validity of an event e_k (to be allocated) in room r_i, and timeslot t_j, the system must check against any other events of the same subject group in the same timeslot t_j.
- *Lecture spread constraint.* It is only applicable to the lecture event. There can only be one lecture for one subject in a particular day and there must be no more than 3 lectures in total in the same day.

Typical soft constraints are:

- *Event spread constraint.* For each class group c_i, the system will check the number of lecture, tutorial, and also laboratory events from day 0 (Monday)

to day 4 (Friday). The penalty will be given if the total number of events
per day of each class group c_i is less than 2 or greater than 5.
 - *Noon punishment.* Events should be avoided to be scheduled during the lunch
   time between 12:30pm to 1:30pm.
 - *Consecutive event consideration.* It is preferable that the same type of events
   takes place in the same venue and are scheduled consecutively.
 - *Last timeslot punishment.* Events, if possible, should not happen in the last
   timeslot of the day.

3   Artificial Immune System

Artificial Immune Systems (AIS) are inspired from nature immune systems.
The powerful information processing capabilities of the immune system, such
as feature extraction, pattern recognition, learning, memory, and its distributive nature provide rich metaphors for its artificial counterpart [9]. Specifically,
three immunological principles are primarily used in a piecemeal in AIS methods.
These include the immune network theory, the mechanisms of negative selection,
and the clonal selection principles. In this paper, only clonal selection principles
adapted from [10] will be discussed.

3.1   Clonal Selection Algorithm (CLONALG)

The clonal selection principle is used as the fundamental basis for the development of the clonal selection algorithm (CLONALG) [10]. The algorithm is
presented in Figure 1, and works as follows. First, a initial set of population
P_r and an empty memory set M are generated (step 1). Then The *selection*
process selects the n best cells or antibodies to generate a new population P_n
based on the affinity measure (step 2). These n best individuals of the population are cloned (reproduced) by the *clonal* process, giving rise to a temporary
population of clones, C (step 3). The higher the affinity, the larger the number
of clones generated for each of the n selected antibodies. The affinity *maturation*
process then mutates the antibodies to create the population C^* (step 4). During mutation, it assigns a lower mutation rate for higher affinity antibodies than
low affinity antibodies. The idea is that the antibodies close to a local optimum
need only be fine-tuned, whereas antibodies far from an optimum should move
larger steps towards an optimum or other regions of the affinity landscape [11].
The *reselection* process reselects the improved antibodies from C^* to update the
memory set M (step 5). Finally, The *diversity introduction* process replaces d
low affinity antibodies with new ones N_d (step 6).

   The *selection* and *maturation* processes lead the population towards more
stimulated to antigens, while the *clonal* and *diversity introduction* processes help
to maintain the diversity of the population.

   Although CLONALG with population-based search is characterized as an
evolutionary-like algorithm, there are some important differences between CLON-
ALG and genetic algorithm (GA) [4]. Firstly, CONALG was developed with

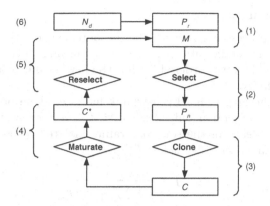

Fig. 1. Clonal Selection Algorithm (CLONALG)

inspiration from the immunological theory whereas GA was based on the Darwinian evolution. Secondly, in CLONALG, the solution is usually extracted from the memory pool constituted during the whole evolutionary process. In GA, the final solution is usually gathered from the population of the last generation of evolution.

4   Applying CLONALG to Timetabling

In order to apply CLONALG to the university timetabling problem, an antibody is represented as an *event* matrix $\mathbf{E}_{rooms \times timeslots}$. There are n number of rooms $\{r_0, r_1, \cdots, r_n\}$, m timeslots $\{t_0, t_1, \cdots, t_m\}$, and w events to be scheduled $\{e_0, e_1, \cdots, e_w\}$. $\mathbf{E}[r_i][t_j] = e_k$, means that an event e_k takes place in room r_i at time t_j.

4.1   Initialization

The CLONALG starts with a creation of an initial population which contains a set of feasible solutions, or antibodies, and is created randomly regardless to their affinity measurement value. Each antibody must follow the scheduling stages defined below. In the first stage, the most constraining events - events involved in largest number of constraints - are to be scheduled. The set of events requiring a special laboratory, or having to appear in the timetable in consecutive periods, or an event whose *room allocated, time allocated* attributes are set, have the higher weight and need to be scheduled first. Some events that have more relax constraints are to be scheduled next. For instance, the lecture events that require large-sized classrooms are randomly assigned to the corresponding rooms prior to those requiring small-sized ones. Apart from this, the rest of the events are randomly scheduled accordingly without violating any hard constraint.

4.2   Selection

The *selection* process begins with the measurement of the affinity of each antibody. These antibodies are then ordered according to the affinity calculated. The first antibody in the ordered list has the lowest affinity and the last one has the highest affinity.

Since the checking of hard constraints has been done beforehand, the affinity measurement function only deals with the soft constraints, and penalties will be given to each of the soft constraints violated respectively. The affinity measurement function is defined as

$$f = \frac{1}{1 + \sum_{i=1}^{k} w_i \cdot n_i} \tag{1}$$

where k is the total number of the soft constraints defined, n_i is the number of a certain kind of soft constraints within a particular antibody, w_i is its attached penalty or weight.

After the ordering of antibodies, n highest affinity antibodies are selected to produce a new population P_n. If we choose $n = N$, i.e., the number of highest affinity individuals equals to the number of candidates, each member of the population will constitute a potential candidate solution locally, characterizing a greedy search. In addition, if all the individuals are accounted locally, their clones will have the same size [1]. The value of the parameter n was determined empirically and will be elaborated in section 5.

4.3   Cloning

Antibodies in the population will be duplicated proportional to their affinity and enters the clone population C of size N_c, which is computed by equation 2

$$N_c = \sum_{i=1}^{n} \text{round}(\frac{\beta \cdot N}{i}) \tag{2}$$

where N_c is the total amount of clones generated, β is a multiplying factor, N is the total amount of antibodies and round($\cdot$) is the operator that rounds its argument towards the closest integer. Each term of this sum corresponds to the clone size of each selected antibody, e.g., for $N = 100$ and $\beta = 1$, the highest affinity antibody ($i = 1$) will produce 100 clones, while the second highest affinity antibody produces 50 clones, and so on.

4.4   Maturation

In the CLONALG algorithm, the mutation rate of a cell is inversely proportional to the affinity of the cell. It gives the chance for low affinity cells to "mutate"

---

[1] In order to maintain the best antibodies for each clone during evolution, it is possible to keep one original (parent) antibody for each clone unmutated during the maturation process.

more in order to improve its affinity. Since the mutations always result in better affinity antibodies, the immune system always climbs up the hill towards higher affinity antibody, leading to local optima.

A timetable **E** picked from the *selection* process will be mutated with mutation rate inversely proportional to its affinity measurement function. The mutation rate affects the maximum iteration of the hill climbing procedure. Hill climbing has two criteria to stop: maximum iteration or minimum soft constraints violated (which is set to 0).

At each iteration, the algorithm selects randomly a scheduled event to be moved. The selection of the move can be made either by randomly sampling a set of moves, or by exhaustively exploring the neighborhood looking for the best move (also known as Steepest Descent). In any way, the move should not lead to an infeasible timetable. The cost of applying the move to the chosen event will be calculated, which accounts for the number of violated soft constraints. If it gives a better value, or the same as before, the move will be accepted and applied to the timetable, moving an event from one room-timeslot to another room-timeslot.

4.5   Reselection and Diversity Introduction

There is a slight modification in the *reselection* step of the CLONALG algorithm (Figure 1), instead of choosing n highest affinity clones, the clones will be first compared with their parents. If one of the clones gives a better affinity than its parent, then the clone will be selected to enter the new population of P_n. If the parent is better, then the parent will be selected. Therefore, the n highest affinity antibodies, either the parents or the clones, will be selected to compose the new population of P_n, and d low affinity antibodies are to be replaced after every 5 generations by the *diversity introduction* process. This scheduling is supposed to leave a breathing time in order to allow achievement of the local optima, followed by replacement of the poorer individuals.

5   Experiments

The proposed approach has been developed to solve two problems, the timetabling problem for the School of Computer Engineering (SCE), Nanyang Technological University, and the university course timetabling benchmark problem. To cater for two different inputs and constraints defined by the two problems, two separate systems have been developed based on the CLONALG algorithm.

For the SCE timetabling problem, the system has been tested using a sample dataset of SCE year 1 main group, academic year 2002/2003, semester 1 and incorporating the constraints that have been discussed in section 2. Another system, which was built to solve the benchmark problem, has been tested using the small instances of the datasets which can be downloaded from http://iridia.ulb.ac.be/~msampels/ttmn.data/. The characteristics of both problems are given in Table 1. Entries represented as "-" means the corresponding characteristic is undefined for that particular problem.

Table 1. Characteristics of the timetabling problems

	SCE Problem	Benchmark Problem
Num events	116	100
Num rooms	8	5
Num features	-	5
Approx features per room	-	3
Percent feature use	-	70
Num students	450	80
Max events per student	30	20
Max students per event	200	20
Num subject groupings	5	-

CLONALG has several user-defined parameters and different settings of these parameters would affect the performance of the algorithm. Two parameters are discussed here, namely n, the number of antibodies to be selected for cloning, and β, parameter affecting the number of clones generated from the antibody population.

5.1   Sensitivity with Relation to n

CLONALG sensitivity with relation to n has been evaluated with β (of equation 2) being fixed to 1. Figure 2 presents the results on the maximum, minimum and mean obtained after running CLONALG 10 times with 50 and 100 generations for SCE timetabling and benchmark problem respectively. n takes the value ranging from 2 to 10. The figure shows that the average cost of calculating the antibody affinity in the population decreases as n increases. This behavior was expected, because the higher the value of n, the larger the number of antibodies to be mutated and eventually leads to the better solutions. On the other hand, n has a strong influence on the size of the antibody clone population as

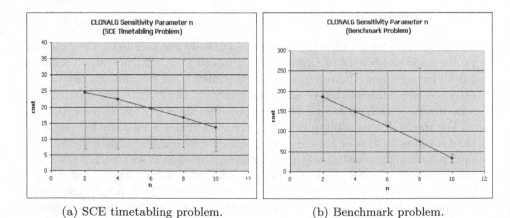

(a) SCE timetabling problem.   (b) Benchmark problem.

Fig. 2. CLONALG sensitivity with relation to n

Table 2. Computational time versus different settings of n

	Computational Time (sec)	
n	SCE Problem	Benchmark Problem
2	11.9	16.6
4	15.4	31.8
6	17.6	48.3
8	19.6	64.8
10	20.7	86.9

described in equation 2, and larger values of n imply a higher computational cost. The computational time for different settings of n is given in Table 2.

5.2 Sensitivity with Relation to β

To study the CLONALG sensitivity with relation to β, n was set to N (population size), and β took the following values, $\beta = \{0.1, 0.2, 0.4, 0.8, 1.5\}$. It can be observed from Figure 3 that the cost of calculating the antibody affinity in the population converges when β is set to 0.8 or higher. This is because the probability of getting better antibodies in a larger clonal population is higher. Nevertheless, the computational time increases linearly with β as can be seen from Table 3.

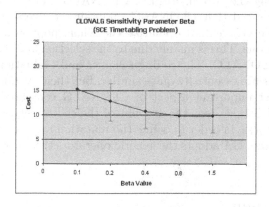

Fig. 3. CLONALG sensitivity parameter β

Table 3. Computational time versus different settings of β

β	0.1	0.2	0.4	0.8	1.5
Computational Time (sec)	13.6	18.7	29.9	50.3	88.1

5.3   Comparison of CLONALG and GA

The results presented in Table 4 are the best affinity antibodies out of 20 runs of CLONALG, with the number of generations set to 100, population size set to 10, best antibody set to 8, and the number of antibodies to be replaced at each generation set to 2.

The solution score is calculated based on the rules defined in [12] which essentially counts the number of soft constraint violations. The smaller the solution score, the better the algorithm is. It can be observed from Table 4 that CLONALG gives a better result than GA but with longer computational time.

Table 4. Results comparison of different dataset of Benchmark Problem

| | CLONALG | | GA | |
Data Set	Solution Best Score	Computation Time (sec)	Solution Best Score	Computation Time (sec)
small1.tim	19	62.9	36	12.2
small2.tim	24	62.6	37	12.8
small3.tim	20	63.1	38	11.1
small4.tim	14	81.3	51	18.8
small5.tim	12	61.2	31	12.7

By analyzing the two algorithms, CLONALG and GA, and the results obtained, it is noticed that CLONALG maintains a diverse set of local optimal solutions, while GA tends to polarize the whole population of individuals towards the best one. This is mainly due to the selection and reproduction schemes adopted by CLONALG. The coding schemes and evaluation functions of these two algorithms are essentially quite similar, but their evolutionary search is different. Another important aspect of CLONALG compared to GA is the fact that CLONALG takes into the account of the cell affinity, corresponding to an individual's fitness, in order to define the mutation rate applied to each member of the population. GA adopts the genetic operators that disregard the individual fitness.

6   Conclusion

This paper applies the Clonal Selection Algorithm (CLONALG) to solve the course timetabling problem for the School of Computer Engineering (SCE), Nanyang Technological University (NTU), Singapore as well as the university course timetabling problem provided by the Metaheuristics Network.

CLONALG, which is the algorithm based on the biological metaphor of the immune system, is proved to be effective to diverge the population. By comparing CLONALG with GA, it is observed that the main steps composing the GAs are embodied in CLONALG, allowing us to characterize it as an evolutionary-like algorithm. However, while GA uses a vocabulary borrowed from the natural genetics and are inspired in the Darwinian evolution, the CLONALG makes use immunological terminology to describe the Antigen-Antibody interactions and cellular evolution. CLONALG performs its search through the mechanisms of somatic mutation and receptor editing, balancing the exploitation of the best solutions with the exploration of the neighborhood.

For future work, we will test the system with SCE full data (four years courses) and add the lecturer constraints, etc. Some improvements on the heuristics used in producing the initial candidates could also be made in order to get better final solutions. In addition, implementation of other algorithms is essential for further studies, to compare the results of the two problems investigated here.

References

1. E. Yu and K.S. Sung. A genetic algorithm for a university weekly courses timetabling problem. *International Transactions in Operational Research*, 9(6):703–717, 2002.
2. E.K. Burke, D.G. Elliman, and R.F. Weare. A university timetabling system based on graph colouring and constraint manipulation. *Journal of Research on Computing in Education*, 27(1):1–18, 1994.
3. N. Balakrishnan, A. Lucena, and R. T. Wong. Scheduling examinations to reduce second-order conflicts. *Computers and Operational Research*, 19(5):353–361, 1992.
4. E.K. Burke, A. Eckersley, B. McCollum, S. Petrovic, and R. Qu. Using simulated annealing to study behaviour of various exam timetabling data sets. In *Proceedings of the Fifth Metaheuristics International Conference (MIC 2003)*, Kyoto, Japan, August 2003.
5. B. Jaumard, J.-F. Cordeau, and R. Morales. *Efficient Timetabling Solution with Tabu Search*. http://www.idsia.ch/Files/ttcomp2002/jaumard.pdf, 2003. Avaliable from Metaheuristics Network - Intenational Timetabling Competition.
6. M. Chiarandini, K. Socha, M. Birattari, and O. Rossi-Doria. An effective hybrid approach for the university course timetabling problem. *Journal of Scheduling*, 2003. To appear.
7. G. Laporte, M. Gendreau, J-Y.Potvin, and F. Semet. Classical and modern heuristics for the vehicle routing problem. *International Transactions in Operational Research*, 7:285–300, 2000.
8. *Metaheuristics Network*. http://www.metaheuristics.org/.
9. D. Dasgupta, Z. Ji, and F. Gonzlez. Artificial immune system (ais) research in the last five years. In *Proceedings of the International Conference on Evolutionary Computation Conference (CEC)*, Canbara, Australia, December 2003.
10. L.N. de Castro and F.J. Von Zuben. Learning and optimization using the clonal selection principle. *IEEE Transactions on Evolutionary Computation, Special Issue on Artificial Immune Systems*, 6(3):239–251, 2002.
11. L.N. de Castro and J.I. Timmis. Artificial immune system as a novel soft computing paradigm. *Soft Computing*, 7(8):526–544, 2003.
12. *University Course Timetabling Benchmark Solution Score Calculation*. http://www.idsia.ch/Files/ttcomp2002/IC_Problem/node2.html.

Improved Algorithms for Two Single Machine Scheduling Problems*

Yong He, Weiya Zhong, and Huikun Gu

Department of Mathematics, State Key Lab of CAD &CG,
Zhejiang University, Hangzhou 310027, P. R. China
mathhey@zju.edu.cn

Abstract. In this paper we investigate two single machine scheduling problems. The first problem addresses a class of the two-stage scheduling problem in which the first stage is job production and the second stage is job delivery. For the case that jobs are processed on a single machine and delivered by a single vehicle to one customer area, with the objective of minimizing the time when all jobs are completed and delivered to the customer area and the vehicle returns to the machine, an approximation algorithm with a worst-case ratio of 5/3 is known and no approximation can have a worst-case of 3/2 unless $P = NP$. We present an improved approximation algorithm with a worst-case ratio of 53/35, which only leaves a gap of 1/70. The second problem is a single machine scheduling problem subject to a period of maintenance. The objective is to minimize the total completion time. The best known approximation algorithm has a worst-case ratio of 20/17. We present a polynomial time approximation scheme.

1   Introduction

In this paper, we consider two single machine scheduling problems, which have strong background in supply chain management and manufacture management.

The first problem is a scheduling problem with job delivery coordination, which is first proposed by Chang and Lee [2], and can be described as follows: We are given n jobs $N = \{J_1, J_2, \cdots, J_n\}$ which must be first non-preemptively processed in a manufacturing system and then delivered to respective customers. Job J_j, $j = 1, 2, \cdots, n$, needs a processing time of p_j in the manufacturing system, and has a size s_j which represents the physical space J_i occupies when this job is loaded in the vehicle. One vehicle is available to deliver finished jobs in batches, and has a capacity z which means that finished jobs can be arranged to fit in the physical space provided by the vehicle as long as their total size does not exceed z. The vehicle is initially located at the manufacturing facility. All jobs delivered together in one shipment are defined as a delivery batch. A transportation time depending on customer area is associated with each delivery

---

* Research supported by the TRAPOYT of China and NSFC (10271110, 60021201).

N. Megiddo, Y. Xu, and B. Zhu (Eds.): AAIM 2005, LNCS 3521, pp. 66–76, 2005.

batch. Furthermore, we define a one customer area as a location where a group of customers are located in close proximity to each other. The goal is to find a schedule for processing jobs in manufacturing system and delivering finished jobs to the corresponding customers such that the time required for all jobs in N to be processed and delivered to the respective customer(s) is minimized. To evaluate this goal, we define the makespan of a schedule, denoted by $C_{\max}$, as *the time when the vehicle finishes delivering the last batch to the customer site(s) and returns to the machine(s)*. Then the problem is to find a schedule to minimize makespan.

As we know, coordination of activities among different stages in the supply chain has become one of the most important topic in production and operations management research in last decade. For the research on the coordination of production and delivery schedule, one may refer to [2]. Different from traditional scheduling problems which implicitly assume that there are infinitely many vehicles for delivering finished products to their destinations so that finished products can be transported to customers without delay, the above problem incorporates the delivery plan of a vehicle into a manufacturing system. It models a class of the two-stage scheduling problem in which the first stage is job production and the second stage is job delivery. The focus is on the study of the integration of production scheduling with delivery of finished products to customers, which measures the customer service level.

Three strongly NP-hard cases of the above problem are considered in [2]: For the case that the manufacturing system consists of a single machine and there is one customer area, Chang and Lee presented a polynomial time algorithm $H1$ with a worst-case ratio of $5/3$, while no polynomial time algorithm can have a worst-case ratio of smaller than $3/2$ unless $P = NP$. For the case that the manufacturing system consists of two identical machines and there is one customer area, they presented a polynomial time algorithm $H2$ with a worst-case ratio of 2. For the case that the manufacturing system consists of a single machine and there are two customer areas, they presented a polynomial time algorithm $H3$ with a worst-case ratio of 2, too.

In this paper, we revisit the first case of the above problem, which is denoted by $1 \rightarrow D, k = 1 | v = 1, c = z | C_{\max}$. Here "$1 \rightarrow D, k = 1$" means that jobs are first processed on a single machine and then delivered to customer(s) who are located in one area. "$v = 1, c = z$" means that there is only one vehicle with capacity z. We will present a modified algorithm $MH1$ with a worst-case ratio of $3/2 + 1/70 = 53/35$, which greatly improves the known upper bound of $5/3$ and is quite close to the lower bound of $3/2$.

The second considered problem in this paper is a single machine scheduling problem with a machine availability constraint, which can be described as follows: We are given n jobs $N = \{J_1, J_2, \cdots, J_n\}$ which must be processed on a single machine. Job J_j has a processing time p_j, $j = 1, 2, \cdots, n$. All jobs are available at time zero, whereas the machine has a maintenance period during the processing of jobs, i.e., the machine cannot process any job during the given time window $[R, R + L]$. Preemptions are not allowed. Hence a job that is preempted due to the maintenance must be restarted after the machine is repaired. The objective is to find a schedule such that the total completion time is minimized. This problem is denoted by $1, h_1 || \sum C_i$ [6].

Adiri et al. [1] and Lee and Liman [5] showed that the problem $1, h_1 || \sum C_i$ is NP-hard. They also studied algorithm SPT (*Shortest Processing Time*) as an approximation algorithm solving this problem. Lee and Liman [5] proved that the worst-case ratio of SPT is 9/7. Recently, Sadfi et al. [6] proposed a modified algorithm $MSPT$ with a worst-case ratio of 20/17. This algorithm is based on a post-optimization of the SPT algorithm by applying a 2-OPT procedure. In this paper, we will extend this idea to a general k, k-exchange procedure. Then we will propose a polynomial time approximation scheme ($PTAS$) based on this new procedure. Hence our result greatly improves the known result.

2   Problem $1 \rightarrow D, k = 1|v = 1, c = z|C_{\max}$

This section is devoted to the problem $1 \rightarrow D, k = 1|v = 1, c = z|C_{\max}$. Let P be the total processing time of all the jobs. Let t the one-way transformation time between the machine and the customer, therefore each delivery has the same transportation time $T = 2t$.

2.1   Preliminaries and Algorithm Description

Property 1. [2] *There exists an optimal schedule for the problem $1 \rightarrow D, k = 1|v = 1, c = z|C_{\max}$ that satisfies the following conditions:*

(1) *Jobs are processed on the machine without idle time.*
(2) *Jobs assigned to one batch are processed consecutively on the machine.*
(3) *Jobs assigned to one batch can be processed on the machine in any order.*
(4) *Batches are delivered in non-decreasing order of the total processing time of jobs in each batch.*

Therefore, only schedules satisfying the above properties are considered further. Also, batches will be indexed and delivered in non-decreasing order of the total processing time of the jobs in each batch.

Lemma 1. ([2]) *For any schedule satisfying Property 1, if $C_{\max} > P + T$, then $P_1 < T$ and $C_{\max} = P_1 + KT$, in which P_1 denotes the total processing time of the jobs in the first batch and K denotes the number of batches in the schedule.*

Algorithms FF (*First Fit*) and FFD (*First Fit Decreasing*) are two classical algorithms for the bin-packing problem. We will apply them as sub-procedures for solving our problem. Note that algorithms FF and FFD are based on the job sizes and the vehicle capacity z in this paper. For an instance I of the bin-packing problem, let $OPT(I)$, $FF(I)$, $FFD(I)$ be the numbers of used bins in an optimal solution, the solutions yielded by FF and FFD, respectively.

Lemma 2. (1)([7]) $FF(I) \leq \frac{7}{4}OPT(I)$; (2)([8]) $FFD(I) \leq \frac{11}{9}OPT(I) + 1$.

The following algorithm $H1$ was proposed in [2] for solving our problem.

Algorithm $H1$:

1. Assign jobs to batches by algorithm FFD. Let the total number of resulting batches be b_1.
2. Define P_k as the total processing time of the jobs in the k-th batch, $k = 1, 2, \cdots, b_1$. Reindex these batches such that $P_1 \leq P_2 \leq \cdots \leq P_{b_1}$, and denote the k-th batch as B_k.
3. Starting with B_1, assign jobs in B_k to the machine, for $k = 1, 2, \cdots, b_1$. Jobs within each batch can be sequenced in an arbitrary order.
4. Dispatch each finished but undelivered batch whenever the vehicle becomes available. If multiple batches have been completed when the vehicle becomes available, dispatch the batch with the smallest index.

It is clear that the time complexity of $H1$ is $O(n \log n)$. It is shown in [2] that the worst-case ratio of $H1$ is $5/3$. Furthermore, since the bin-packing problem is a special case of our problem, it is impossible to have a polynomial time approximation algorithm with a worst-case ratio of $3/2$ unless $P = NP$.

Our improved algorithm applies a fully polynomial time approximation scheme ($FPTAS$) of the knapsack problem. Recall that for any instance of the knapsack problem, we are given n items, each with a profit and a size, and a knapsack with limited capacity. We wish to put items into the knapsack such that the total size of the selected items is not greater than the knapsack capacity and the total profit of the selected items is maximized. For this NP-hard problem, among others, Lawler [4] proposed an $FPTAS$ with a time complexity of $O(n \log(\frac{1}{\epsilon}) + \frac{1}{\epsilon^4})$, where $1 - \epsilon$ is the worst-case ratio; and Kellerer and Pferschy [3] also proposed an $FPTAS$ with a time complexity of $O(n \min\{\log n, \log \frac{1}{\epsilon}\} + \frac{1}{\epsilon^2} \min\{n, \frac{1}{\epsilon} \log \frac{1}{\epsilon}\})$.

Now we are ready to present our improved algorithm.

Algorithm $MH1$:

1. Run algorithm $H1$. Let the obtained schedule be σ_1 with makespan C_1. If $b_1 \neq 3$, stop; Else, go to Step 2.
2. Construct an instance of the knapsack problem as follows: for each job J_j, $j = 1, 2, \cdots, n$, construct an item with profit p_j and size s_j, and let the knapsack capacity be z. Run any $FPTAS$ for the knapsack problem with $\epsilon = \frac{2}{35}$, and denote by N_1 the set of items put into the knapsack. Reindex all jobs such that N_1 is at the head.
3. Assign jobs to batches by algorithm FF. Let the total number of resulting batches be b_2.
4. Run Steps 2-4 of algorithm $H1$ except that denote by B'_k the k-th batch, and by P'_k the total processing times of B'_k, $k = 1, 2, \cdots, b_2$. Let the obtained schedule be σ_2 with makespan C_2.
5. Compare C_1 and C_2. Select the smaller one as output.

Remark 1. The jobs corresponding to the items in N_1 are assigned to the same batch by algorithm FF in Step 3 of algorithm $MH1$.

When analyzing our algorithm, we use the following:
$b_L^* =$ the number of batches if the jobs are assigned to batches by an optimal algorithm of the bin-packing problem.
$b^* =$ the number of batches in the optimal schedule for our problem.
$P^* =$ the optimal value of the instance of the knapsack problem constructed in Step 2.
$C^* =$ the optimal makespan for our problem.
$C_{MH1} =$ the makespan produced by $MH1$.
$y =$ the total processing time of jobs in the first batch in the optimal solution.

Lemma 3. *If there are only two batches in the optimal schedule, $P - y \leq P^*$.*

Proof. Since jobs in each batch constitutes a feasible solution for the instance of the knapsack problem, and $P - y$ is the total processing time of the second batch in the optimal schedule, we have $P - y \leq P^*$.   □

Lemma 4. $P_{b_2}' \geq \frac{33}{35}P^*$, *in which P_{b_2}' denotes the total processing time of jobs in the last batch of σ_2 (if exists).*

Proof. From Remark 1, we know that there exists $k, 1 \leq k \leq b_2$, such that $P_k' \geq \frac{33}{35}P^*$. Since $P_1' \leq P_2' \leq \cdots \leq P_{b_2}'$, $P_{b_2}' \geq P_k' \geq \frac{33}{35}P^*$.   □

2.2   Worst-case Analysis of Algorithm $MH1$

Lemma 5. *If $b_1 \neq 3$, $\frac{C_{MH1}}{C^*} < \frac{53}{35}$.*

Proof. In this case, algorithm $MH1$ is just $H1$, hence $C_{MH1} = C_1$. Chang and Lee [2] proved $\frac{C_1}{C^*} \leq \frac{5}{3}$. To obtain $\frac{C_1}{C^*} < \frac{53}{35}$, more careful analysis are necessary.
From Lemma 1, we have

$$C^* = \max\{y + b^*T, P + T\}. \tag{1}$$

If $C_1 = P + T$, we have $C_1 = C^*$ clearly, and we are done. Hence, we suppose that $C_1 > P + T$ in the following. Then by Lemma 1,

$$C_1 = P_1 + b_1 T, \quad \text{and} \quad P_1 < T. \tag{2}$$

It is obvious that $b^* \geq 2$. Otherwise, $b_1 = b^* = 1$ and $MH1$ yields an optimal solution. If $b_1 \leq b^*$, by (1) and (2), we have

$$\frac{C_1}{C^*} \leq \frac{P_1 + b_1 T}{y + b^*T} < \frac{P_1 + b^*T}{b^*T} = 1 + \frac{P_1}{T} \cdot \frac{1}{b^*} < 1 + \frac{1}{b^*} \leq \frac{3}{2} < \frac{53}{35}. \tag{3}$$

Hence, we only need to consider the case that $b_1 > b^*$.

Noting that the jobs are assigned to batches according to algorithm FFD in $H1$, by Lemma 2(2), we have

$$b_1 \leq \frac{11}{9}b_L^* + 1 \leq \frac{11}{9}b^* + 1. \tag{4}$$

If $b^* = 2$, then $b_1 \leq \frac{31}{9} < 4$. From $b_1 > b^*$ we know $b_1 = 3$, contradicting the Lemma's assumption $b_1 \neq 3$. Therefore, we suppose $b^* \geq 3$ in the following. To obtain the desired worst-case ratio, we distinguish two cases according to (1).

Case 1 $C^* = y + b^*T$. Then (1) implies $y + b^*T \geq P + T$, i.e., $P \leq y + (b^* - 1)T$. Recall that $P_1 \leq P_2 \leq \cdots \leq P_{b_1}$. We establish $P_1 \leq \frac{P}{b_1} \leq \frac{y + (b^* - 1)T}{b_1}$, and thus

$$\frac{C_1}{C^*} = \frac{P_1 + b_1 T}{y + b^*T} \leq \frac{\frac{y + (b^* - 1)T}{b_1} + b_1 T}{y + b^*T} = \frac{1}{b_1} \cdot \frac{y + b^*T + (b_1^2 - 1)T}{y + b^*T}$$

$$= \frac{1}{b_1} + \frac{1}{b_1} \cdot \frac{(b_1^2 - 1)T}{y + b^*T} < \frac{1}{b_1} + \frac{1}{b_1} \cdot \frac{b_1^2 - 1}{b^*}. \tag{5}$$

If $b^* = 3$, (4) states that $b_1 \leq \frac{42}{9} < 5$. Combing it with $b_1 > b^*$, we have $b_1 = 4$. Then from (5), it follows that $\frac{C_1}{C^*} < \frac{3}{2}$.

Similarly, if $b^* = 4$, then $b_1 = 5$ and thus $\frac{C_1}{C^*} < \frac{7}{5}$; if $b^* = 5$, then $b_1 = 6, 7$ and thus $\frac{C_1}{C^*} < \frac{4}{3}$ (for $b_1 = 6$) or $\frac{C_1}{C^*} < \frac{53}{35}$ (for $b_1 = 7$); if $b^* = 6$, then $b_1 = 7, 8$, and thus $\frac{C_1}{C^*} < \frac{9}{7}$ (for $b_1 = 7$) or $\frac{C_1}{C^*} < \frac{23}{16}$ (for $b_1 = 8$).

If $b^* \geq 7$, (4) implies that $b^* \geq \frac{9(b_1 - 1)}{11}$. Substituting it into (5), we obtain

$$\frac{C_1}{C^*} < \frac{1}{b_1} + \frac{1}{b_1} \cdot \frac{(b_1^2 - 1)}{\frac{9(b_1 - 1)}{11}} = \frac{11}{9} + \frac{20}{9b_1} < \frac{11}{9} + \frac{20}{9} \cdot \frac{1}{8} = \frac{3}{2}, \tag{6}$$

where the last inequality is from $b_1 > b^* \geq 7$.

Case 2 $C^* = P + T$. Then (1) implies $C^* = P + T \geq y + b^*T > b^*T$. Combining it with (4), we have $P > (b^* - 1)T \geq (\frac{9}{11}b_1 - \frac{20}{11})T$. As $P_1 \leq \frac{P}{b_1}$, we conclude that

$$\frac{C_1}{C^*} = \frac{P_1 + b_1 T}{P + T} \leq \frac{\frac{P}{b_1} + b_1 T}{P + T} = \frac{1}{b_1} \cdot \frac{P + T + (b_1^2 - 1)T}{P + T}$$

$$= \frac{1}{b_1} + \frac{b_1^2 - 1}{b_1} \cdot \frac{T}{P + T} < \frac{1}{b_1} + \frac{b_1^2 - 1}{b_1} \cdot \frac{T}{b^*T} = \frac{1}{b_1} + \frac{b_1^2 - 1}{b_1} \cdot \frac{1}{b^*}. \tag{7}$$

Note that (7) is the same as (5). Therefore, the same arguments as those in Case 1 can complete the proof. □

Lemma 6. If $b_1 = 3$, $\frac{C_{MH1}}{C^*} < \frac{53}{35}$.

Proof. Similarly, we can suppose that $b^* \geq 2$. If $C_1 = P + T$ or $C_2 = P + T$, then $\min\{C_1, C_2\} = C^*$ by (1). Hence we suppose that $C_1 > P + T$ and $C_2 > P + T$.

Then by Lemma 1, $P_1 < T$ and $P_1' < T$, where P_1 and P_1' are the total processing times of jobs in the first batches in σ_1 and σ_2, respectively.

If $b_1 \leq b^*$, we have shown in Lemma 5 that $\frac{C_1}{C^*} < \frac{3}{2}$ (see the proof of (3)). Hence, we suppose $b_1 > b^*$. Then $b^* = 2$. Hence, by Lemma 2(1), we have $b_2 \leq \frac{7}{4} * 2 = \frac{7}{2}$, that is, $b_2 \leq 3$. If further $b_2 \leq b^*$, by similar arguments to show (3) in the proof of Lemma 5, we can obtain $\frac{C_2}{C^*} \leq \frac{3}{2}$. Hence, we suppose $b_2 > b^*$. Combining it with $b_2 \leq 3$ and $b^* = 2$, we know that $b_2 = 3$.

$b^* = 2$ and (1) states that $C^* = \max\{y+2T, P+T\}$. Two cases are considered as follows.

Case 1 $C^* = y + 2T$. Then $y + 2T \geq P + T$, and thus $P \leq y + T$. From Lemmas 3 and 4, we get $P_3' \geq \frac{33}{35}P^* \geq \frac{33}{35}(P - y)$. Noting that $P_1' \leq P_2' \leq P_3'$, we obtain

$$P_1' \leq \frac{P - P_3'}{2} \leq \frac{P - \frac{33}{35}(P-y)}{2} = \frac{2P + 33y}{70} \leq \frac{2(y+T) + 33y}{70} = \frac{y}{2} + \frac{T}{35}. \quad (8)$$

Therefore,

$$\frac{C_2}{C^*} = \frac{P_1' + 3T}{y + 2T} \leq \frac{\frac{y}{2} + \frac{T}{35} + 3T}{y + 2T}$$
$$= \frac{\frac{1}{2}(y + 2T) + (2 + \frac{1}{35})T}{y + 2T} < \frac{1}{2} + \frac{(2 + \frac{1}{35})T}{2T} = \frac{53}{35}. \quad (9)$$

Case 2 $C^* = P + T$. Then $P + T \geq y + 2T$, and thus $P \geq y + T$. From Lemmas 3 and 4, we know $P_3' \geq \frac{33}{35}P^* \geq \frac{33}{35}(P - y) > \frac{33}{35}T$. By $P_1' \leq P_2' \leq P_3'$ and $P = \sum_{i=1}^3 P_i'$, we have $P_1' \leq \frac{1}{2}(P - P_3') < \frac{1}{2}(P - \frac{33}{35}T)$. Therefore,

$$\frac{C_2}{C^*} = \frac{P_1' + 3T}{P + T} < \frac{\frac{1}{2}(P - \frac{33}{35}T) + 3T}{P + T}$$
$$= \frac{\frac{1}{2}(P + T) + (\frac{5}{2} - \frac{33}{70})T}{P + T} < \frac{1}{2} + \frac{(\frac{5}{2} - \frac{33}{70})T}{2T} = \frac{53}{35}. \quad \square$$

Theorem 1. $\frac{C_{MH1}}{C^*} < \frac{53}{35}$.

Proof. This is a direct conclusion of Lemmas 5 and 6. $\square$

Since both FFD, FF and $H1$ run in time $O(n \log n)$, and the $FPTAS$ of Lawler or Kellerer and Pferschy for the knapsack problem runs in time $O(n)$ when we take $\epsilon = 2/35$. Hence the time complexity of $MH1$ is $O(n \log n)$, the same as that of $H1$.

3   Problem $1, h_1 || \sum C_i$

This section addresses the problem $1, h_1 || \sum C_i$. To present our improved algorithm, we will propose a local search procedure, called k, k-*exchange procedure*, which is an extension of 2-OPT procedure proposed in [6].

Using the same notations as those in [5] and [6], denote by S^* and S an optimal schedule and the schedule yielded by algorithm SPT, respectively. Denote by B the set of the jobs scheduled before the maintenance period in S, and by A the set of the remaining jobs scheduled after it. Denote by X the set consisting of the $|B|$ jobs scheduled first in S^*, and by Y the set of the remaining $|A|$ jobs scheduled last. Denote by δ^* and δ the idle times on the machine before the maintenance period, respectively in the schedules S^* and S.

Definition 1. *Let $\bar{a} \leq |A|$, $\bar{b} \leq |B|$ and k are positive integers satisfying $k \geq \bar{b} \geq \bar{a}$. An k, k-exchange procedure is an exchange of $\bar{a}$ jobs in A with $\bar{b}$ jobs of B in the schedule S, under the constraint that the total processing times of $\bar{b}$ jobs in B plus δ is no less than the total processing times of $\bar{a}$ jobs in A. After exchange, the jobs are reordered before and after the maintenance in non-decreasing order of their processing times.*

Obviously an k, k-exchange procedure is essentially a post-optimization of the SPT schedule using local search method. With this procedure, our improved algorithm, denoted by $SPTE$ can be formulated as follows.

Algorithm $SPTE$:

1. Process all the jobs according to the SPT rule.
2. For a given positive integer k, try all k, k-exchange procedures to generate new schedules.
3. Choose the best one from the schedules generated in Steps 1 and 2 as output.

Clearly, by setting $k = 1$, the above algorithm becomes $MSPT$, which was proposed in [6]. Hence $SPTE$ is a generalization of $MSPT$. Sadfi et al. showed that $MSPT$ has a worst-case ratio of 20/17. Hence we assume that $k \geq 2$ in the following. We will show that $SPTE$ is a $PTAS$. It shows that local search method is powerful for the considered problem.

Denote by S' the schedule yielded by algorithm $SPTE$. With straightforward notation, B' and A' represent the job partition is S'. Finally we denote by C_i, C_i' and C_i^* the completion times of job J_i in schedules S, S' and S^*, respectively.

The following Lemmas 7 and 8 are cited from in [6], and Lemma 9 is parallel to Lemma 4 of that paper which can be shown similarly.

Lemma 7. ([6]) $\delta \geq \delta^*$.

Lemma 8. ([6]) *If (at least) one of set X is scheduled after the maintenance period in the optimal solution, then*

$$\sum_{i=1}^{n} C_i' \leq \sum_{i=1}^{n} C_i^* + (|Y| - 2)(\delta - \delta^*). \tag{10}$$

Lemma 9. *If (at least) $k + 1$ jobs of the set B are scheduled after the maintenance period in the optimal solution, then*

$$\sum_{i=1}^{n} C_i^* \geq \left(\frac{|Y|(|Y| + 1)}{2} + k + 1 \right) (\delta - \delta^*). \tag{11}$$

Theorem 2. *For any given integer $k \geq 2$, algorithm SPTE has a worst-case ratio of at most $1 + \frac{2}{5+2\sqrt{2k+8}}$, and runs in $O(n^{2k+1})$. Therefore, SPTE is a PTAS for the problem $1, h_1 || \sum C_i$.*

Proof. We first prove (10) and (11). Two cases are considered as follows.

Case 1 No k, k-exchange procedure exists. Then the number of jobs from A should be no less than the number of B, and $S' = S$. If $B = X$, SPT schedule is optimal. If $B \neq X$, in order to process some job(s) from set A before the maintenance period, we have to remove at least $k + 1$ jobs from set B. Hence, if S is not optimal, at least one job from set X is processed after the maintenance period in S^*. Hence the condition of Lemma 8 is satisfied, and thus (10) is true. Furthermore, at least $k+1$ jobs from set B are processed after the maintenance period in S^*, implying that the condition of Lemma 9 is satisfied and thus (11) is true.

Case 2 k, k-exchange procedures generate new schedules. Without loss of generality, we suppose that S^* cannot be generated by these procedures.

Suppose that the optimal schedule S^* can be generated by exchanging b' jobs from set B with a' jobs from set A, where $|A| \geq a' \geq 1$ and $|B| \geq b'$. Since S^* cannot be generated by any k, k-exchange procedure, we have $b' > k$. It states that the condition of Lemma 9 is satisfied. Hence, (11) is true.

Since the processing time of any job from set A is no less than that of any job from set B, $b' \geq a'$. We distinguish two subcases according to this inequality.

Subcase 1 $b' > a'$. Since S^* can be generated by exchanging b' jobs from B with a' jobs from A, the number of jobs processed before the maintenance period in S^* must be $|B| - b' + a' < |B|$. Since $|X| = |B|$, we know that at least one job from X is scheduled after the maintenance period in S^*. (10) follows.

Subcase 2 $b' = a'$. Then $a' > k \geq 2$. For this subcase, the condition of Lemma 8 may not be satisfied, but we show that (10) is still true assuming all jobs in X are processed before the maintenance period.

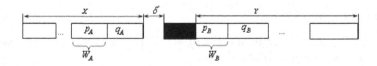

Fig. 1. Schedule S^*

Fig. 2. Schedule S''

Let q_A and q_B be the processing times of the biggest k jobs from A in X and the processing time of the biggest k jobs from B in Y, respectively. Let W_A and W_B be the sets of the other $a' - k$ jobs of $X \bigcap A$ and $Y \bigcap B$, respectively. Moreover, let p_A and p_B be the sums of their processing times. Note that $p_A \geq p_B$ by the construction of the schedule S. Now we construct a schedule S'' from S^* by exchanging the jobs of W_A and W_B and processing jobs in SPT order before and after the maintenance period. Let δ'' be the idle time on the machine before the maintenance period, X'' be the set consisting of the $|B|$ jobs scheduled first in S'', and Y'' be the set of the remaining $|A|$ jobs scheduled last. Let $\Delta = p_A - p_B \geq 0$ for short. Denote by C_i'' the completion times of job J_i in S''.

By comparing schedules S'' and S^* (see Fig. 1 and 2), we have $C_{[i]}^* \geq C_{[i]}''$, $1 \leq i \leq |B| - k$, and $C_{[i]}^* \geq C_{[i]}'' + \Delta$, $|B| - k + 1 \leq i \leq |B|$, where $C_{[i]}^*$ and $C_{[i]}''$ are the completion times of jobs at position i in S^* and S'', respectively. It follows that $\sum_{J_{[i]} \in X''} C_{[i]}'' \leq \sum_{J_{[i]} \in X} C_{[i]}^* - k\Delta$. On the other hand, $C_{[|B|+a']}'' - C_{[|B|+a']}^* = (R + L + q_B + p_A) - (R + L + p_B + q_B) = \Delta$. It implies that $C_{[i]}'' \leq C_{[i]}^* + \Delta$, $|B| + 1 \leq i \leq |B| + a' - 1$, and $C_{[i]}'' = C_{[i]}^* + \Delta$, $|B| + a' + 1 \leq i \leq n$. Summing these inequalities, and by $|Y| = n - |B|$, we obtain $\sum_{i=1}^n C_i'' \leq \sum_{i=1}^n C_i^* + (|Y| - k)\Delta \leq \sum_{i=1}^n C_i^* + (|Y| - 2)\Delta$. Since $p_A + q_A + \delta^* = p_B + q_A + \delta''$, we have $\Delta = \delta'' - \delta^* \leq \delta - \delta^*$ (due to $\delta'' \leq \delta$). Hence $\sum_{i=1}^n C_i'' \leq \sum_{i=1}^n C_i^* + (|Y| - 2)(\delta - \delta^*)$. Since $SPTE$ outputs the best schedule among all k, k-exchange procedures and S'' can be generated by an k, k-exchange procedure, we have $\sum_{i=1}^n C_i' \leq \sum_{i=1}^n C_i^* + (|Y| - 2)(\delta - \delta^*)$, which is just (10).

Now we are ready to get the worst-case ratio. By (10) and (11), we obtain

$$\frac{\sum_{i=1}^n C_i' - \sum_{i=1}^n C_i^*}{\sum_{i=1}^n C_i^*} \leq \frac{2(|Y| - 2)}{|Y|(|Y| + 1) + 2k + 2}.$$

Define $f(|Y|) = \frac{2(|Y|-2)}{|Y|(|Y|+1)+2k+2}$, $|Y| > 0$. By taking a derivation of $f(|Y|)$, we can see that it is increasing for $|Y| \leq 2 + \sqrt{2k+8}$ and is decreasing for $|Y| \geq 2 + \sqrt{2k+8}$ and hence reaches a maximum at $|Y| = 2 + \sqrt{2k+8}$ with $f(2 + \sqrt{2k+8}) = \frac{2}{5+2\sqrt{2k+8}}$. Therefore, the desired worst-case ratio of $SPTE$ follows.

It is obvious that there are at most $O(n^{2k})$ k, k-exchange procedures, and computing the objective function value of a schedule takes $O(n)$ time. Hence $SPTE$ runs in time $O(n^{2k+1})$, and is a $PTAS$ for $1, h_1 || \sum C_i$. $\square$

References

1. I. Adiri, J. Bruno, E. Frostig, A.H.G. Rinnooy Kan, Single machine flow-time schduling with a single breakdown, *Acta Informatica*, **26**, (1989), 679-696.
2. Y.C. Chang, C.Y. Lee, Machine scheduling with job delivery coordination, *European Journal of Operational Research*, **158**, (2004), 470–487.
3. H. Kellerer, U. Pferschy, A new fully polynomial approximation scheme for the knapsack problem, *Lecture Notes in Computer Science* **1444**, (1998), 123–134.

4. E. Lawler, Fast approximation algorithms for knapsack problems, *Mathematics of Operations Research*, **4**, (1979), 339–356.
5. C.Y. Lee, S.D. Liman, Single machine flow-time scheduling with scheduled main-teance, *Acta Informatica*, **29**, (1992), 375-382.
6. C. Sadfi, B. Penz, C. Rapine, J. Błazewicz, P. Formanowicz, An improved approximation algorithm for the single machine total completion time scheduling problem with availability constraints, *European Journal of Operational Research*, **161**, (2005), 3-10.
7. D. Simchi-Levi, New worst-case results for the bin packing problem, *Naval Research Logistics*, **41**, (1994), 579-585.
8. M. Yue, A simple proof of the inequality $FFD(L) \leq \frac{11}{9}OPT(L) + 1 \; \forall L$, for the FFD bin-packing algorithm, *Acta Math. Appl. Sinica*, **7**, (1991), 321-331.

N-Person Noncooperative Game with Infinite Strategic Space

Hui Yu[1], Jian Chen[1], and Caihong Sun[2]

[1] School of Economics and Management,
Tsinghua University, Beijing, China
yuhui@em.tsinghua.edu.cn
http://www.rccm.tsinghua.edu.cn
[2] School of Economy and Commerce,
Chongqing Technology and Business University,
Chongqing, China

Abstract. It is wel-known that Nash considered the n-person noncooperative game with each player having finite strategic set and proved the celebrated existence result of Nash equilibrium point (in mixed strategies). This paper investigates the n-person noncooperative game with each player having infinite strategic set. By considering these infinite strategic as complete metric spaces and based on a new finite equilibrium system, we obtain new existence result of Nash equilibrium. Then an algorithm is given to compute Nash equilibrium points and its convergence is proved.

1 Introduction

Nash equilibrium constitutes a central solution concept in game theory, which is a mathematical theory of socio-economic phenomena exhibiting interaction among decision-makers, called players, whose actions affect each other. In years past, economics has benefited greatly from the introduction of game-theoretic tools.

Classic n-person noncooperative game $\Gamma(\mathbb{N}, S, h)$(see [6]) in normal (strategic) form consists of the following:

(i) A set $\mathbb{N} = \{1, 2, \ldots, n\}$ of players.
(ii) For each player $i \in \mathbb{N}$, a finite set S^i of pure strategies. Let $S := S^1 \times S^2 \times \cdots \times S^n$ denote the set of n-tuples of pure strategies.
(iii) For each player $i \in \mathbb{N}$, a function $h^i : S \to \mathbb{R}$, called the payoff function. Let $h = (h^1, h^2, \ldots, h^n)$ denote the vector of payoff function.

A game is finite if the player set as well as the set of strategies available to each player is finite. So above game $\Gamma(\mathbb{N}, S, h)$ is a finite game. For this finite game, there maybe not solution for the game $\Gamma(\mathbb{N}, S, h)$(see [16]). So Nash(see, [11, 12]) considered its mixed extension of this finite game as following.

N. Megiddo, Y. Xu, and B. Zhu (Eds.): AAIM 2005, LNCS 3521, pp. 77–84, 2005.
© Springer-Verlag Berlin Heidelberg 2005

For the finite pure strategic set S^i, for the player $i \in \mathbb{N}$, his mixed strategic set is

$$\Delta(S^i) = \{x = (x(s))_{s \in S^i} : x(s) \geq 0 \text{ for all } s \in S^i \text{ and } \sum_{s \in S^i} x(s) = 1\}, \quad (1)$$

his expected payoff is

$$H^i(x) = \sum_{s \in S} x(s) h^i(s), \quad (2)$$

where $x(s) := \prod_{j \in \mathbb{N}} x^j(s^j)$ is the probability, under x, that the pure strategy n-tuple $s = (s^1, s^2, \dots, s^n)$ is played. Define payoff function $H^i : \Delta(S) \to \mathbb{R}$ for player i. Let $\Delta(S) = \Delta(S^1) \times \Delta(S^2) \times \cdots \times \Delta(S^n)$ and $H = H^1 \times H^2 \times \cdots \times H^n$. Since the mixed strategic set $\Delta(S)$ include not finite but infinite strategies, the finite strategic game is transformed to a n-person game with infinite strategic set. And since $\Delta(S)$ is compact and convex and payoff function H has especial convexity and continuity structure, Nash (see, [11, 12]) obtained his celebrated existence result of non-cooperative equilibrium using Kakutani's fixed pointed theorem (see [3])as follows.

Theorem 1 (Nash(see[11, 12])). *The mixed extension($\Gamma(\mathbb{N}, \Delta S, H)$) of the finite game has at least on strategic equilibrium.*

Nash's theorem has been generalized in many directions. One of most important generalization is following game $\Gamma(\mathbb{N}, X, \phi)$.

Theorem 2 (Debreu(see [2])). *Let, for each $i \in \mathbb{N}$, the strategic set X_i be nonempty, compact and convex subset of the Eulidean space E_i and $\phi : X \to R$ be a payoff function, which satisfies:*

(i) *For each $i \in \mathbb{N}$ and $x^i \in \prod_{j \in \mathbb{N} \setminus \{i\}} X_j$, the payoff function $y_i \to \phi(x^i, y_i)$ is convex;*
(ii) *For each $i \in \mathbb{N}$, the payoff function ϕ_i is continuous.*

Then there exists an $\bar{x} \in X$ such that, for each $i \in \mathbb{N}$

$$\phi_i(\bar{x}^i, y_i) \geq \phi_i(\bar{x}), \quad \forall y_i \in X_i$$

where $X = \prod_{i \in \mathbb{N}} X_i$ and $E = \prod_{i \in \mathbb{N}} E_i$.

Theorem 2 may be thought of as identifying conditions under which the strategy space X are like mixed strategy spaces $\Delta(S)$ for the finite games and the payoff function ϕ are like expected payoff H(see [9]). This result showed that when the player's (infinite) strategic set has compact and convex structure, under some conditions (continuity and convexity) of payoff function, the infinite (strategic) game has at least one Nash equilibrium.

The above is the classic transformation from finite (strategic) game to infinite (strategic) game. But we also know for infinite (strategic) sets, they also can be other topological structure such as completeness besides compactness and convexity. So it is a very natural question: is there a Nash equilibrium under complete strategic set for infinite strategic non-cooperative game? This paper gives a positive answer. Under some conditions, we build a new existence result of Nash equilibrium points by using finite equilibrium system theory(see [1]).

The task of detecting the Nash equilibria of n-person noncooperative game remains a challenging problem up-to-date. The most popular algorithms for calculating Nash equilibira are homotopy methods (see [4, 5, 7, 8, 10]). Other methods include neural network method (see [15]) and computational intelligence methods (see [13]). In this paper, we also give an algorithm method to compute the noncooperative equilibria in infinite (strategic) game and prove its convergence.

The paper is organized as follows. In Section 2, We give some notation and discuss the relationship between finite equilibrium system and n-person noncooperative game. Then the existence result of solution for the finite equilibrium system is obtained. In Section 3, equilibrium of n-person noncooperative game is given, an algorithm is proposed to compute this Nash equilibrium and its convergence is pointed out. In Section 4, some conclusions are given.

2   Finite Equilibrium System

This section gives an existence result of finite equilibrium system. We also let $\mathbb{N} = \{1, 2, \ldots, n\}$ be an index set, for each $i \in \mathbb{N}$, X_i be a subset of the Eulidean space E_i. Let $X = \prod_{i \in \mathbb{N}}$ and $E = \prod_{i \in \mathbb{N}} E_i$.

Definition 1. *The finite equilibrium system problem is finding of $\bar{x} \in X$ such that, for each $i \in \mathbb{N}$,*

$$f_i(\bar{x}, y_i) \geq 0, \quad y_i \in X_i. \tag{3}$$

where $f_i : X \times X_i \to \mathbb{R}$ be system payoff function.

Theorem 3. *For each $i \in \mathbb{N}$, let (E_i, d_i) be Eulidean space, where $d_i : X_i \times X_i \to \mathbb{R}^+$ is Eulidean distance function and for $x_i, y_i \in E_i$, $d_i(x_i, y_i) = ||x_i - y_i||$, the norm of $x_i - y_i$. Assume*

(i) *for each $i \in \mathbb{N}$, (X_i, d_i) be a complete metric subspace of E_i;*

(ii) *for each $i \in \mathbb{N}$, $f_i : X \times X_i \to \mathbb{R}$ is lower semicontinuous in the second argument and satisfies:*

$$f_i(x, x_i) = 0, \text{ for all } x \in X,$$

$$f_i(x, y_i) \leq f_i(x, z_i) + f_i(z, y_i), \quad \forall x, z \in X \text{ and } \forall y_i \in X_i;$$

(iii) *there is an $x(0) \in X$ such that, for every $i \in \mathbb{N}$, $\inf_{y_i \in X_i} f_i(x(0), y_i) > -\infty$;*

(iv) *for every $x \in X$, if for some $i \in \mathbb{N}$ with $\inf_{y_i \in X_i} f_i(x, y_i) < 0$, then there exists $z_i \in X_i$ with $z_i \neq x_i$ such that*

$$f_i(x, z_i) + d_i(x_i, z_i) \leq 0.$$

Then there exists $\bar{x} \in X$ such that, for each $i \in \mathbb{N}$,

$$f_i(\bar{x}, y_i) \geq 0, \quad \forall y_i \in X_i.$$

Proof. We prove this theorem by the following three steps.

(1) Construct inductively a sequence of point $x(j) \in X, j = 0, 1, 2 \ldots$,starting with $x(0)$ from (iii). Given $x(j), j = 0, 1, 2 \ldots$, we obtain $x(j+1)$ by the following way.
For each $i \in \mathbb{N}$, let the set

$$S_j^i = \{y_i \in X_i : f_i(x(j), y_i) + d_i(x(j)_i, y_i) \leq 0\}$$

$$r_j^i = \inf_{y_i \in S_j^i} f(x(j), y_i)$$

Let $S_j = \prod_{i \in \mathbb{N}} S_j^i$. It follows from (ii) that, for each $i \in \mathbb{N}$, $x(j)_i \in S_j^i$ and $r_j^i \leq 0$. Select $x(j+1) \in S_j$ such that, $\forall i \in \mathbb{N}$,

$$f_i(x(j), x(j+1)_i) \leq k r_j^i, \quad \text{where } k \in (0,1)\text{fixed}.$$

(2) In this step, we prove the sequence $x(j), j = 0, 1, 2, \ldots$ is convergence. From (ii) and $x(j+1)_i \in S_j^i$, we say that $S_{j+1}^i \subset S_j^i$, for every j and $i \in \mathbb{N}$. In fact, $\forall x_i \in S_{j+1}^i$, we have

$$f_i(x(j+1), x_i) + d_i(x(j+1)_i, x_i) \leq 0. \tag{4}$$

Since $x(j+1)_i \in S_j^i$, we also have

$$f_i(x(j), x(j+1)_i) + d_i(x(j)_i, x(j+1)_i) \leq 0. \tag{5}$$

From (6) and (5), and (ii), we have

$$f_i(x(j), x_i) + d_i(x(j)_i, x_i) \leq 0.$$

So for each $i \in \mathbb{N}$, $S_{j+1}^i \subset S_j^i$, further, $S_{j+1} \subset S_j$. From this we obtain, by virtue of (ii), for each $i \in \mathbb{N}$,

$$
\begin{aligned}
r_{j+1}^i &= \inf_{y_i \in S_{j+1}^i} f_i(x(j+1), y_i) \\
&\geq \inf_{y_i \in S_j^i} (f_i(x(j), y_i) - f_i(x(j), x(j+1)_i)) \\
&= r_j^i - f_i(x(j), x(j+1)_i) \\
&\geq (1-k) r_j^i.
\end{aligned}
$$

Therefore, $\forall i \in \mathbb{N}$, $0 \geq r_{j+1}^i \geq (1-k)^{j+1} r_0^i \to 0$. If $y_i \in S_{j+1}^i$ then

$$d_i(x(j+1)_i, y_i) \leq -f_i(x(j+1), y_i) \leq -r_{j+1}^i \to 0.$$

This implies that the diameter of the sets S_j^i tends to zero, for each $i \in \mathbb{N}$. Moreover for all $k \geq j$ one has $x(k)^i \in S_k^i \subset S_j^i$, hence $d_i(x(j)_i, x(k)_i) \leq -r_{j+1}^i$. It follows that for each $i \in \mathbb{N}$, $\{x(j)_i\}$ is Cauchy and converges to some $\bar{x}_i \in \cap_{j=0}^{\infty} S_j^i$. Since for each $i \in \mathbb{N}$, f_i is lower semicontinuous in the second argument, S_j^i is closed, for all j. Since the diameter of the set S_j^i tends to zero, it follows that $\cap_{j=0}^{\infty} S_j^i = \{\bar{x}_i\}$. So $\bar{x} = \prod_{i \in \mathbb{N}} \bar{x}_i = \prod_{i \in \mathbb{N}} \bar{x}_i \in \cap_{j=0}^{\infty} S_j$. So $\lim_{j \to \infty} x(j) = \bar{x}$.

(3) We obtain that $\bar{x}$ is finite equilibrium solution in this step. Assume that $\bar{x}$ does not meet the conclusion of the theorem, i.e. there is $i \in \mathbb{N}$ such that $f_i(\bar{x}, y_i) < 0$, for some $y_i \in X_i$. Hence, $\inf_{y_i \in X_i} f_i(\bar{x}, y_i) < 0$. Then by hypothesis, there exists $z_i \in X_i$ with $z_i \neq \bar{x}_i$ such that $f_i(\bar{x}, z_i) + d_i(\bar{x}_i, z_i) \leq 0$. since for this i, $\bar{x}_i \in \cap_{j=0}^{\infty} S_j^i$, i.e. for every j, $\bar{x}_i \in S_j^i$,

$$f_i(x(j), \bar{x}_i) + d_i(x(j)_i, \bar{x}_i) \leq 0.$$

Hence,

$$f_i(x(j), z_i) + d_i(x(j)_i, z_i) \leq 0$$

i.e.

$$z_i \in S_j^i \text{ for all } j$$

This implies

$$z_i \in \cap_{j=0}^{\infty} S_j^i, \forall j$$

Therefore, $x_i = \bar{x}_i$, this contradicts $x_i \neq \bar{x}_i$.   □

3   N-Person Noncooperative Game

In this section, we built an existence result of n-person noncooperative game with each player having infinite strategic set, which has complete structure, and give an algorithm to compute the corresponding Nash equilibrium points.

For n-person noncooperative game with infinite strategic set $\Gamma(\mathbb{N}, X, \phi)$, where $\mathbb{N} = \{1, 2, \cdots, n\}$ is player's set, for each $i \in \mathbb{N}$, X_i and $\phi_i : X \to R$ are strategic set and payoff function, respectively, its Nash equilibrium solution is finding of $\bar{x} \in X$ such that, for each $i \in \mathbb{N}$,

$$\phi_i(\bar{x}^i, y_i) \geq \phi_i(\bar{x}), \quad \forall y_i \in X_i,$$

where $X = \prod_{i \in \mathbb{N}} X_i$ and $x^i \in \prod_{j \in \mathbb{N} \setminus \{i\}} X_j$.

Theorem 4. *For each player $i \in \mathbb{N}$, let (X_i, d_i) be his strategic space, which is a complete metric subspace of Eulidean space (E_i, d_i) with $d_i : X_i \times X_i \to \mathbb{R}^+$ being Eulidean distance function and for $x_i, y_i \in E_i$, $d_i(x_i, y_i) = \|x_i - y_i\|$. His payoff function $\phi_i : X \to R$ satisfies following conditions:*

(i) *for every $x^i \in X^i$, $\phi_i(x^i, y_i)$ is lower semicontinuous on y_i and satisfies:*

$$\phi_i(x^i, y_i) \leq \phi_i(x^i, z_i) + \phi_i(z^i, y_i) - \phi_i(z), \quad \forall x, z \in X \text{ and } \forall y_i \in X_i. \quad (6)$$

(ii) *there is an $x(0) \in X$ such that $\forall i \in \mathbb{N}$, $\inf_{y_i \in X_i} \phi_i(x^i(0), y_i) > -\infty$.*

(iii) *for every $x \in X$, if for $i \in \mathbb{N}$ with $\inf_{y_i \in X_i} (\phi_i(x^i, y_i) - \phi_i(x)) < 0$, then there exists $z_i \in X_i$ with $z_i \neq x_i$ such that $\phi_i(x^i, z_i) - \phi_i(x) + d_i(x_i, z_i) \leq 0$.*

Then there exists $\bar{x} \in X$ such that, for each $i \in \mathbb{N}$,

$$\phi_i(\bar{x}^i, y_i) \geq \phi_i(\bar{x}), \quad \forall y_i \in X_i.$$

Proof. Just setting $f_i(x, y_i) = \phi_i(x^i, y_i) - \phi_i(x)$ and by using the theorem 3, we obtain that n-person noncooperative game has at least Nash equilibrium. □

Remark 1. Theorem 4 is different from Theroem 2 in the following two aspects.

(i) On the players' strategic sets, Theorem 4 need neither compactness nor convexity but completeness. Most general example of strategic set having complete space structure in economic literature is the set $[0, +\infty)$.

(ii) On the players' payoff function, Theorem 4 need neither continuity nor convexity but some special quantitative structure and lower semi-continuity. We should note that the condition (6) is satisfied by a type of payoff function

$$\phi(x^i, y_i) = l_i(x^i) + g_i(y_i), \quad \text{for any } (x^i, y_i) \in X$$

where $l_i : X \setminus X_i \to R$ and $g_i : X_i \to R$.

Example 1. Denote the set of players $\mathbb{N} = \{1, 2, \cdots, n\}$. For each player $i \in \mathbb{N}$, his strategic set is $X_i = [0, +\infty)$ and payoff is $\phi_i(x^i, y_i) = ||x^i|| + e^{y_i}$. It is easy to see that the strategic set is compact. So any existing Nash's equilibrium existence result (e.g. Theorem 2) cannot affirm the existence of equilibrium solution of this problem. But just using Theorem 4, there is at least one equilibrium solution. In fact, there is an only equilibrium solution, i.e., $(0, 0, \cdots, 0)$.

Next, we give an algorithm to compute the Nash equilibrium in this n-person noncooperative game with infinite strategic set.

Algorithm 1. Let n-person noncooperative game $\Gamma(\mathbb{N}, X, \phi)$ be given.

(0) Give $x(0)$, any small positive real number $\epsilon > 0$ and $k \in (0, 1)$.

(1) Solve the subproblem $r_j^i = \min_{y_i}(\phi(x^i(j), y_i) - \phi_i(x(j)))$ subject to

$$\phi_i(x(j)^i, y_i) - \phi_i(x(j)) + d_i(x(j)_i, y_i) \leq 0.$$

(2) Select $x(j+1)$. For any $i \in \mathbb{N}$, if $r_j^i = 0$, then $x(j+1)_i = x(j)_i$, else if $r_j^i < 0$ select $x(j+1)_i$ such that

$$\phi_i(x(j)^i, x(j+1)_i) - \phi_i(x(j)) \leq k \cdot r_j^i.$$

(3) If for some j and for all $i = 1, 2, \cdots, n$, $r_j^i > -\epsilon$, then stop, else if, turn to Step (1).

Theorem 5. *Under the hypotheses of the Theorem 4, the sequence of $x(j), j = 0, 1, 2, \cdots$ from Algorithm 1 is convergence.*

Proof. The convergence is guaranteed under the conditions of Theorem 4 by the step 2 in the proof of Theorem 3.

4   Conclusions

The method to prove the existence of at least a Nash equilibrium in n-person noncooperative game is using fixed point theorem. Nash gave his celebrated existence result of Nash equilibrium using famous Kakutani's fixed point theorem. In this paper, we obtain our result on the existence of Nash equilibrium by using contract fixed point theorem. This method is new and simultaneously gives an algorithm to compute this Nash equilibrium point.

Acknowledgements

The work was partially supported by China Postdoctoral Science Foundation (2004035049) and the National Natural Science Foundation of China (70321001,70401006).

References

1. Ansari, Q.H., Yao, J.C.: A fixed-point theorem and its application to the system of variational inequalities. Bulletion of the Australian Mathematical Society **59** (1999) 433–442
2. Debreu, G.: A social equilibrium existence theorem. Proceeding of National Academy of Sciences **38** (1952) 886–893
3. Glicksberg, I.L.: A further generalization of the Kakutani fixed-point theorem with application to Nash equilibrium points. Proceedings of the American Mathematical Society **3** (1952) 170–174.
4. Govindan, S., Wilson, R.: A global Newton method to compute Nash equilibria. Jouranl of Economic Theory **110** (2003) 65–86
5. Govindan, S.,Wilson,R.: Computing Nash equilibria by iterated polymatrix approximation. Journal of Economic Dynamics and Control **28** (2004) 1229–1241
6. Hart, S.: Games in extensive and strategic forms, R. Aumann, S. Hart (eds) Handbook of Game Theory, Horth-Holland **1** (2002) 19–40.
7. Herings, P.J.J., Peeters, R.J.A.P.: A differentiable homotopy to compute Nash equilibria of n-person games. Economic Theory **18** (2001) 159–185
8. Herings, P.J.J., Elzen, A.: Comptation of the Nash equilibrium selected by the tracing procedure in n-person games. Games and Economic Behavior **38** (2002) 89–117
9. Hillas, J., Kohlberg, E.: Foundations of strategic equilibrium. R. Aumann, S. Hart (eds) Handbook of Game Theory, Horth-Holland **3** (2002) 1597–1663
10. McKelvery,R.D., McLenna,A.: Computation of equilibira in finite games. In: Amman, H.M., Kendrick, D.A., Rust,J.(eds.) Handbook of computational economics, Amsterdam: Elservier **I** (1996) 87–142.
11. Nash, J.F.: Equilibrium point in n-person games. Proceedings of the National Academy of Sciences of the USA, **36** (1950) 48–49
12. Nash, J.F.: Non-cooperative games. Annals of Mathematics **54** (1951) 286–295
13. Pavildis, N.G., Parsopoulos, K.E., Vrahatis, M.N.: Computing Nash equilibria through computational intelligence methods. Journal of Computational and Applied Mathematics **175** (2005) 113-136

14. Prasad, K.: On the computability of Nash equilibria. Journal of Economic Dynamics and Control **21** (1997) 943–953
15. Sirakaya, S.: Genetic neural networks to approximate feedback Nash equilibria in dynamic games. Computers and Mathematics with Applications **46** (2003) 1493–1509
16. Sofronidis, N.E.: Undecidability of the existence of pure Nash equilibria. Economic Theory **23** (2004) 423–428

On the Online Dial-A-Ride Problem
with Time-Windows

Fanglei Yi and Lei Tian

School of Management, Xi'an Jiaotong University,
Xi'an, ShaanXi 710049, P. R. of China
{fangleiyi, ttianlei}@163.com

Abstract. In this paper the first results on the Online Dial-A-Ride Problem with Time-Windows (ODARPTW for short) are presented. Requests for rides appearing over time consist of two points in a metric space, a *source* and a *destination*. Servers transport objects of requests from sources to destinations. Each request specifies a deadline. If a request is not be served by its deadline, it will be called off. The goal is to plan the motion of servers in an online way so that the maximum number of requests is met by their deadlines. We perform competitive analysis of two deterministic strategies for the problem with a single server in two cases separately, where the server has unit capacity and where the server has infinite capacity. The competitive ratios of the strategies are obtained. We also prove a lower bound on the competitive ratio of any deterministic algorithm of $\frac{2-T}{2T}$ for a server with unit capacity and of $\frac{2-T}{2T}\lceil\frac{1}{T}\rceil$ for a server with infinite capacity, where T denotes the diameter of the metric space.

1 Introduction

In dial-a-ride problems (DARP for short) servers are traveling in some metric space to serve requests for rides. Each ride is characterized by two points in the metric space, a *source*, the starting point of the ride, and a *destination*, the ending point of the ride. The problem is to design routes for the servers through the metric space, such that all requested rides are made and some optimality criterion is met. A common characteristic of almost all the approaches to the study of the problem is the off-line point of view. The input is known completely beforehand. However, in many routing and scheduling applications the instance only becomes known in an online fashion. In other words, the input of the problem is communicated in successive steps.

In a natural setting of dial-a-ride problems requests for rides are presented over time while the servers are enroute serving other rides, making the problem an online optimization problem. Examples of such problems in practice are taxi and minibus services, courier services, and elevators.

In this paper we consider a class of variations of online DARP in which there is a time-window on each of the requests: a server moves from point to point in a metric space. Time is continuous and at any moment a request can arrive at a

N. Megiddo, Y. Xu, and B. Zhu (Eds.): AAIM 2005, LNCS 3521, pp. 85–94, 2005.

point in the space, requiring the server to carry the objects to the destination. Each request also specifies a deadline. If a request is to be served, the server must reach the point where the request originated during its time-window (the time between the request's arrival and its deadline). The goal of the algorithm is to serve as many incoming requests as possible by their deadlines. The online algorithm for online DARP neither has information about the release time of the last request nor has the total number of requests. It must determine the behavior of the server at a certain moment t of time as a function of all the requests released up to time t (and the current time t). In contrast, an off-line algorithm has information about all requests in the whole sequence already at time 0.

The ODARPTW and in general vehicle routing and scheduling problems have been widely studied for more than three decades (see[13] for a survey on the subject). We are inspired by recent exciting results in online routing and scheduling problems[4-6,10]. Most previous researches on online routing problems focused on the objectives of minimizing the makespan [4,5,7], the weighted sum of completion times [4,12], and the maximum/average flow time [11,13]. In the paper [6, 15, 16], results on the online k-taxi scheduling problem have been presented, in which a request consists of two points (a *source* and a *destination*) on a graph or in a metric space. Subsequently, a similar problem, online k-truck scheduling problem has been studied in [10]. Both of them (online k-taxi/truck scheduling) assumed that k servers (taxies or trucks) are all free when a new service request occurs, and the goal is to minimize the total distance traveled by servers. [4] studied the online DARP in which calls for rides come in while the server is traveling. The authors also considered two different cases, where the server has infinite capacity and where the server has finite capacity. Lipmann et al. [14] studied the online DARP under a restricted information model in which the information about the destination will be released (becomes known) while visiting the source. All of these previous work assumed that the requests could wait for any length of time until the server completed them. Results on online routing and scheduling problems, in which a certain job will be called off after waiting for a certain period of time, were presented in [8] firstly, studying the dynamic traveling repair problem, a degenerate form of ODRPTW presented in this paper. Subsequently, Krumke et al. [9] studied the similar problem in special case (such as in uniform metric space). We will pay our attention to the more general case that the requests for rides which have time-windows consist of two points (a *source* and a *destination*) in a general metric space. The goal is to serve as many requests as possible within their time-windows. We will give upper bounds for the competitive ratio of two algorithms. And several lower bounds for any deterministic algorithm will be shown in this paper.

2   The Model

Let $\mathcal{M} = (X, d)$ be a metric space with n points which is induced by an undirected unweighted graph $G = (V, E)$ with $V = X$, i.e., for each pair of points from the metric space $\mathcal{M}$ we have $d(x,y)$ that equals the shortest path length in G between vertices x and y. We also assume that $d(x,y) \leq d(x,z) + d(z,y)$ for all $x,y,z \in X$. An instance of the basic online DARP in the metric space $\mathcal{M}$ consists of a sequence $R = (r_1, r_2, \cdots, r_m)$ of *requests*. Each request is a triple $r_i = (t_i, a_i, b_i) \in \mathcal{R} \times X \times X$ with the following meaning: t_i, a real number, is the time that request r_i is released; $a_i \in X$ and $b_i \in X$ are the source and destination, respectively, between which the object corresponding to request r_i is to be transported. It is assumed that the sequence $R = (r_1, r_2, \cdots, r_m)$ of requests is given in order of non-decreasing release times, that is, $0 \leq t_1 \leq t_2 \leq \cdots \leq t_m$. A request is said to be *accepted request* if the corresponding object is picked up by the server at source, and a request is said to be *completed request* if the corresponding object is transported to the destination. We do not allow *preemption*: it is not allowed to drop an accepted request at any other place than its destination. This means, once a request is accepted, it will not be called off.

In this paper, we consider the following restrictions to ODARPTW: a) The speed of the server is constant 1. This means that, the time it takes to travel from one point to another is exactly the distance between the two points; b) The window sizes for all requests are uniform. We will normalize all values so that the window length is 1 for the remainder of this paper; c) The diameter of the metric space is bounded by a constant T, which is the maximum time required to travel between the two farthest points in the metric space.

We evaluate the quality of online algorithms by competitive analysis [1-3], which has become a standard yardstick to measure the performance. In competitive analysis, the performance of an online algorithm is compared to the performance of the optimal off-line algorithm, which knows about all future jobs. An algorithm A for online DARP with time-windows is called α-*competitive* if for any instance R the number of request completed by A is at least $1/\alpha$ times the number of request completed by an optimal off-line algorithm OPT.

3   Online Algorithms and Competitive Analysis

In this section we propose two algorithms, REPLAN and SMARTCHOICE, for ODARPTW, which are similar to BATCH algorithm and Double-Gain algorithm in [8]. And the performance guarantees of the two algorithms for the problem in a general metric space are shown in this section.

3.1   The REPLAN Algorithm (RE for short)

We will divide each step into *basic plan*. We shall think of basic plan as taking a certain interval of time to serve a certain subset of requests planed at the beginning of the interval. The length of time intervals is $1/2$. At the beginning of each interval, the server stops and replans: it computes an optimal schedule with $1/2$ length starting at the current position of the server, which can complete the maximum number of requests which arrived in the previous interval. Then it continues to use the new basic plan.

Theorem 1. *Provided that the server is unit-capacity $(z = 1)$, for any metric space with $T < 1/2$, RE is $4(\frac{1+2T}{1-2T})$-competitive.*

Proof. The server has capacity 1, i.e., it can carry at most 1 object at a time. Once the server has accepted a request, it is not allowed to accept another request until the *accepted request* is completed. Let R_i be the set of all requests arriving during interval i, and let $R_{i,j}$ be the subset of R_i that OPT completed during interval j. Obviously, $R_{i,j} = \emptyset$ for all $j \notin \{i, i+1, i+2, i+3\}$. Let $B_{OPT}(R)$ denote the number of requests completed by OPT on a request sequence R. Suppose the last request arrives in m^{th} interval, then

$$B_{OPT}(R) = \sum_{i=1}^{m}(|R_{i,i}| + |R_{i,i+1}| + |R_{i,i+2}| + |R_{i,i+3}|)$$

where $R_{i,i} \cup R_{i,i+1} \cup R_{i,i+2} \cup R_{i,i+3} \subseteq R_i$

When RE makes the basic plan at the beginning of interval i, it takes care of the set R_{i-1}, finds a new schedule and follows it. One of the options for RE is to pick the largest of $R_{i-1,i-1}, R_{i-1,i}, R_{i-1,i+1}$ and $R_{i-1,i+2}$. We distinguish between two cases depending on the state of OPT's server at the beginning of an interval.

The first case is that the sever of OPT is empty at the beginning of any interval. This means, each of $R_{i-1,i-1}, R_{i-1,i}, R_{i-1,i+1}$ and $R_{i-1,i+2}$ can be served (accepted and completed) by OPT during a single interval. That is to say, there is a tour of length at most $1/2$ that covers all of $R_{i-1,i-1}$. The same is true for $R_{i-1,i}, R_{i-1,i+1}$ and $R_{i-1,i+2}$. Let $R_{max} = \{R' : |R'| = max(|R_{i-1,i-1}|, |R_{i-1,i}|, |R_{i-1,i+1}|, |R_{i-1,i+2}|)\}$ be the set of requests which is the largest of $R_{i-1,i-1}, R_{i-1,i}, R_{i-1,i+1}$ and $R_{i-1,i+2}$. We denote by l_{max} the tour, which covers R_{max}. Now RE's server can take some time at most T to reach an optimal starting location on the route l_{max}, then it can take a remaining time of $1/2 - T$ to serve the most requests in R_{max} starting from the optimal location. Since it needs time $1/2$ to serve R_{max}, the server of RE will serve at least $(1 - 2T)|R_{max}|$ requests during the time of $1/2 - T$. Thus, the number of requests which RE's server can serve during interval i is at least $(1/2 - T)|R_{max}|$. Denoting the total number that RE can serve on the request sequence R by $B_{RE}(R)$, we have

$$B_{RE}(R) \geq (1 - 2T)\sum_{i=2}^{m+1} max(|R_{i-1,i-1}|, |R_{i-1,i}|, |R_{i-1,i+1}|, |R_{i-1,i+2}|)$$

$$\geq \left(\frac{1 - 2T}{4}\right)\sum_{i=1}^{m}(|R_{i,i}| + |R_{i,i+1}| + |R_{i-1,i+2}| + |R_{i-1,i+2}|)$$

$$= \frac{1 - 2T}{4}B_{OPT}(R)$$

The second case is that the OPT's server is currently carrying an object for a request r_e. Since T is the longest time for the server to travel between any two points, r_e can be completed within time T. This means that there is a tour of length $1/2 + T$, denoted by l_{max}, that can serve all the requests in R_{max}. So, RE's server can take time no more than T to pick its optimal starting location on the route l_{max}, and has a remaining time of $1/2 - T$ to serve requests from R_{max}. There will be at least $\frac{1-2T}{1+2T}|R_{max}|$ requests completed by RE's server during interval i. So we have

$$
\begin{aligned}
B_{RE}(R) &\geq \left(\frac{1-2T}{1+2T}\right) \sum_{i=2}^{m+1} max\big(|R_{i-1,i-1}|, |R_{i-1,i}|, |R_{i-1,i+1}|, |R_{i-1,i+2}|\big) \\
&\geq \frac{1}{4}\left(\frac{1-2T}{1+2T}\right) \sum_{i=1}^{m} \big(|R_{i,i}| + |R_{i,i+1}| + |R_{i-1,i+2}| + |R_{i-1,i+2}|\big) \\
&= \frac{1}{4}\left(\frac{1-2T}{1+2T}\right) B_{OPT}(R)
\end{aligned}
$$

This completes the proof the fact that RE is $4(\frac{1+2T}{1-2T})$ -*competitive* for unit-capacity in any $T < 1/2$ metric space. $\square$

3.2   The SMARTCHOICE Algorithm (SM for short)

The time is divided into intervals of length Δ. At the beginning of every interval (at time $i\Delta, i = 1, 2, \cdots$), SM will consider all yet unserved requests (including those that are currently carried by the server). We denote by $R^A(i\Delta)$ the set of all unserved requests at time $i\Delta$. SM then finds a new optimal route from its current position that can cover the maximum number of requests from $R^A(i\Delta)$ by their deadlines. Let $R^C(i\Delta)$ represents the set of outstanding requests that have yet to be served on SM's current route at time $i\Delta$, and $R^N(i\Delta)$ will denote the set of requests which are served on the new route for time $i\Delta$. At any beginning of intervals, SM's server will switch to the new route if the number of requests gained is at least λ times the number of requests lost by the switch. That is, $\big|R^N(i\Delta) - R^C(i\Delta)\big| \geq \lambda \big|R^C(i\Delta) - R^N(i\Delta)\big|$ for some $\lambda > 1$. All new requests that arrive during the interval are temporarily ignored until the next beginning of interval.

Lemma 1. *Let $\Delta = 1/2-T$ and R_{i-1} denotes the set of requests arriving during interval $[(i-1)\Delta, i\Delta)$. Provided that the server is unit-capacity ($z = 1$), during any interval of length Δ, the maximum number of requests that OPT's server can serve is at most $\lambda\big|R^C(i\Delta)\big|$.*

In the proof of lemma 1, we can also distinguish between two cases depending on the state of OPT's server at the beginning of an interval. The details of the proof are omitted for the page limitation.

Since the time-window of request is 1 and every request is completed within time T of its accepted, any request in R_{i-1} will be completed in the interval $[(i - 1)\Delta, i\Delta + 1 + T)$. According to lemma 1, during any interval of length Δ, the maximum number of requests that OPT's server can serve is at most

$\lambda \left| R^C(i\Delta) \right|$, the OPT's server can serve at most $\lambda \left\lceil \frac{\Delta+1+T}{\Delta} \right\rceil \left| R^C(i\Delta) \right|$ requests of R_{i-1} at any time. This means the total number of requests completed by OPT is at most

$$B_{OPT}(R) \leq \lambda \left\lceil \frac{\Delta+1+T}{\Delta} \right\rceil \sum_i \left| R^C(i\Delta) \right| \tag{1}$$

Now we will evaluate the number of requests that SM's server can complete. For each request that ever appears in SM's current route, consider a continuous interval of time in which it is in the current route. Each of intervals can be depicted as a line on a horizontal time axis with two endpoints. $\left| R^C(i\Delta) \right|$ is the number of lines that contain point $i\Delta$. Since every request expires or is completed within time $1 + T$ of its arrival, each line can cover at most $\left\lceil \frac{1+T}{\Delta} \right\rceil$ interval endings. This means that

$$\sum_i \left| R^C(i\Delta) \right| \leq \left\lceil \frac{1+T}{\Delta} \right\rceil I \tag{2}$$

where I is the total number of lines.

The number of the requests served by SM can be expressed by the inequation

$$B_{SM}(R) \geq I(\lambda - 1)/\lambda \tag{3}$$

The proof of inequation (3) is similar to paper [8].

Theorem 2. *Provided that the server is unit-capacity ($z = 1$), for any metric space with $T < 1/2$, SM is $8 \left\lceil \frac{3}{1-2T} \right\rceil \left\lceil \frac{1+T}{1-2T} \right\rceil$-competitive.*

Proof. According to inequation (2) and (3), we have

$$\sum_i \left| R^C(i\Delta) \right| \leq \left\lceil \frac{1+T}{\Delta} \right\rceil I \leq \frac{\lambda}{\lambda-1} \left\lceil \frac{1+T}{\Delta} \right\rceil B_{SM}(R)$$

Thus, from (1) we get

$$B_{OPT}(R) \leq \frac{\lambda^2}{\lambda-1} \left\lceil \frac{1+\Delta+T}{\Delta} \right\rceil \left\lceil \frac{1+T}{\Delta} \right\rceil B_{SM}(R)$$

Note that $\Delta = 1/2 - T$, we obtain

$$\frac{2\lambda^2}{\lambda-1} \left\lceil \frac{3}{1-2T} \right\rceil \left\lceil \frac{1+T}{1-2T} \right\rceil B_{SM}(R) \geq B_{OPT}(R)$$

The optimal choice for $\lambda > 1$ is 2, then $\frac{2\lambda^2}{\lambda-1} = 8$. This completes the proof. □

When we consider the case that the server has infinite capacity ($z = \infty$), we will obtain the following corollaries.

Corollary 1. *Provided that the server has infinite capacity ($z = \infty$), for any metric space with $T < 1/2$, RE is $\frac{3}{1-2T}$ -competitive.*

Corollary 2. *Provided that the server has infinite capacity $(z = \infty)$, for any metric space with $T < 1$, SM is $4\left\lceil\frac{2}{1-T}\right\rceil\left(\left\lceil\frac{2}{1-T}\right\rceil + 1\right)$-competitive.*

Since the server has infinite capacity, it can accept several service requests in succession, then complete them in the copestone. So no matter OPT or online algorithms, RE and SM, will firstly pick up objects from the requests' sources as many as possible until the last request disappears (expires or be served), then the server carries accepted objects to their destinations. The proofs of the two corollaries above are similar to the proof of theorem 1 and theorem 2, respectively.

4 Lower Bounds

In this section we derive lower bounds on the competitive ratio of any deterministic online algorithm for serving the requests in the versions of the problem. The results are obtained by considering the optimal algorithm as an adversary that specifies the request sequence in a way that the online algorithm performs badly.

Theorem 3. *Provided that the server has unit capacity $(z = 1)$, for any $T < 1/2$, there is a metric space with diameter T in which no deterministic online algorithm can obtain a competitive ratio less than $\frac{2-T}{2T}$.*

Proof. Let Q be a large enough integer, and k is an even number. Consider a metric space with Qk points that consists of Q groups which are denoted by g_q for $q = 1, 2, \cdots, Q$. Each group consists of two point sets, each with $k/2$ points. Let one set P_q^+ represent the set of pickup points and the other P_q^- be the set of delivery points in group g_q. Therefore, $\mid P_q^+ \mid = \mid P_q^- \mid = k/2$, $\mid P_q^+ \cup P_q^- \mid = k$ for any $q \in \{1, 2, \cdots, Q\}$. The distance between any two points is $1 - T$ units of length from the same group and $\frac{T}{2}$ units of length from different groups. Let $\tau = T$. At every time $t_i = i\tau (i = 0, 1, \ldots)$, the adversary will release the requests on the different points in the metric space depending on the state of the online algorithm's server at time t_i as follows:

1) If the online algorithm's server is working on one point in a certain group g_c or moving between two points which belong to the same group g_c at time t_i, then a request will be released on each point in set $P_{q'}^+$ for every $q' \in \{1, 2, \cdots, Q\}/\{c\}$, requiring the server to carry the objects to the corresponding destinations in set $P_{q'}^-$.

2) If the online algorithm's server is moving between two points which belong to two different groups, g_c and $g_{c'}$, then a request will be released on each point in set $P_{q'}^+$ for every $q' \in \{1, 2, \cdots, Q\}/\{c, c'\}$, requiring the server to carry the objects to the corresponding destinations in set $P_{q'}^-$. We notice that the points of the *source* and the *destination* of a certain request are in a same group.

The number of requests which the online algorithm's server can server that arrive at time $t_i(i = 0, 1, \ldots)$ is at most 1. The online server has to go to another group at some time after serving a request from the *current group* (where it is *currently* located) since no request will be released on any point in it. And the optimal time for the online server to leave for another group, denoted g_f, is at time $t_i + \varepsilon$, just after the adversary releases a request on each pickup point in every group (except for the *current group*), where $\varepsilon > 0$, an arbitrarily small number. When the online server arrives at one pickup point in group g_f at time $t_i + 1 - T + \varepsilon$, all the requests released *before* time t_i will be called off. There are only $k/2$ outstanding requests released at time t_i in group g_f. The online server will take $\frac{T}{2}$ units of time to server one request, ending at time $t_i + 1 - \frac{T}{2} + \varepsilon$. After that, if the online server tries to serve one more request at another point, all the outstanding requests will be called off when it arrives at the point since it has to spend at least $T/2$ units of time moving to the nearest pickup point from its current position. Thus, it will only be able to serve one of the requests that are released at time t_i. The total time that the online server needs to serve one request is $1 - T + \frac{T}{2} = 1 - \frac{T}{2}$.

On the other hand, the adversary will arrange to make its server stay at a group and serve the outstanding requests if the online server does not arrive at that group or is on the way to that group. Otherwise, the adversary's server will leave for a new group to continue to work. Note that it is always in the adversary's best interest to select the group that will be the last one which will be visited by the online server after the adversary's server leaves. So we can consider the adversary's server staying at one group all the while as long as the number of groups gets large enough. The adversary's server will arrange to arrive at a pickup point in the group at time t_i when the new requests are released. And it will continue to go to another pickup point in the same group for the next request just after the server completes one request. As a result, the adversary can serve one request within T units of time. Note that the online server needs at least $1 - \frac{T}{2}$ units of time to serve a request. So the competitive ratio is $\frac{2-T}{2T}$. Thus, we can say that no deterministic online algorithm can achieve a competitive ratio less than $\frac{2-T}{2T}$. The proof of theorem 3 is completed.

□

Theorem 4. *Provided that the server has infinite capacity ($z = \infty$), for any $T < 1/2$, there is a metric space with diameter T in which no deterministic online algorithm can obtain a competitive ratio less than $\frac{2-T}{2T} \lceil \frac{1}{T} \rceil$.*

Proof. The proof is similar to the proof of theorem 3. The server has infinite capacity, so when it arrives at a pickup point every time, it can pick up all the outstanding requests on that point. As long as the number of pickup points in a group is at least $\lceil \frac{2}{T} \rceil$, the adversary can serve $\lceil \frac{1}{T} \rceil$ requests within T units of time. Still, the online server will only be able to serve one request within $1 - \frac{T}{2}$ units of time with the same reasoning. So the competitive ratio is $\frac{2-T}{2T} \lceil \frac{1}{T} \rceil$. Therefore, the theorem holds.   □

5   Conclusions

Online dial-a-ride problems are occurring in a wide variety of practical settings and cover not only physical rides but also transportation means[1]. We consider a class of variations of this kind of problems, in which there is a uniform time-window on each of the requests. Two cases are considered separately, where $z = 1$ and where $z = \infty$. It will be interesting to extend the results to the case of non-uniform windows. Another interesting direction is to consider the case when the server has limited capacity $1 < z \leq C$, where C is a constant.

Acknowledgements

This work was partly supported by the NSF of China (NO.10371094,70471035).

References

1. Manasse M.S., McGeoch L.A., and Sleator D.D.: Competitive algorithms for server problems. Journal of Algorithms. **11**(2) (1990) 208–230
2. David S.B., Borodin A.: A new measure for the study of the on-line algorithm. Algorithmica. **11**(1) (1994) 73–91
3. Alon N., Karp R.M., Peleg D., et al.: A graph-theoretic game and its application to the k- server problem[J] . SIAM. J. Comput. **24**(1) (1995) 78–100
4. Feuerstein E. and Stougie L.: On-line single server dial-a-ride problems. Theoretical Computer Science. **268**(1) (2001) 91–105
5. Ascheuer N., Krumke S.O., and Rambau J.: Online dial-a-ride problems: Minimizing the completion time. Lecture Notes in Computer Science. (2000) 639–650
6. Xu Y.F., Wang K.L.: Scheduling for on-line taxi problem and competitive algorithms. Journal of Xi'an Jiao Tong University. **31**(1) (1997) 56–61
7. Ausiello G., Feuerstein E., Leonardi S., Stougie L., and Talamo M.: Algorithms for the on-line traveling salesman. Algorithmica. **29**(4) (2001) 560–581
8. Irani S., Lu X., and Regan A.: On-line algorithms for the dynamic traveling repair problem. In Proceedings of the 13th Annual ACM-SIAM Symposium on Discrete Algorithms. (2002) 517–524
9. Krumke S.O., Megow N., and Vredeveld T.: How to whack moles. Lecture Notes in Computer Science. **2909** (2004) 192–205
10. Ma W.M., XU Y.F., and Wang K.L.: On-line k-truck problem and its competitive algorithm. Journal of Global Optimization. **21**(1) (2001) 15–25
11. Hauptmeier D. Krumke S.O., and Rambau J.: The online dial-a-ride problem under reasonable load. Lecture Notes in Computer Science. (2000) 125–136
12. Krumke S.O., de Paepe W.E., Poensgen D., and Stougie L.: News from the online traveling repairman. Theoretical Computer Science. **295** (2003) 279–294
13. Krumke S.O., Laura L., Lipmann M., Marchetti-Spaccamela et al.: Non-abusiveness helps: An O(1)-competitive algorithm for minimizing the maximum flow time in the online traveling salesman problem. Lecture Notes in Computer Science. (2002) 200–214

14. Lipmann M.,Lu X.,de Paepe W.E. and Sitters R.A.: On-Line Dial-a-Ride Problems under a Restricted Information Model. Lecture Notes in Computer Science. **2461** (2002) 674–685
15. Xu, Y.F., Wang K.L. and Zhu B.: On the k-taxi problem. Journal of Information. **2** (1999) 429–434
16. Xu, Y.F., Wang K.L. and Ding J.H.: On-line k-taxi scheduling on a constrained graph and its competitive algorithm. Journal of System Engineering(P.R. China). **4** (1999)

Semidefinite Programming Based Approaches to Home-Away Assignment Problems in Sports Scheduling

Ayami Suzuka[1], Ryuhei Miyashiro[2],
Akiko Yoshise[3], and Tomomi Matsui[4]

[1] Graduate School of Systems and Information Engineering, University of Tsukuba,
Tsukuba, Ibaraki 305-8573, Japan
`asuzuka@sk.tsukuba.ac.jp`
[2] Institute of Symbiotic Science and Technology, Tokyo University of Agriculture and
Technology, Koganei, Tokyo 184-8588, Japan
`r-miya@cc.tuat.ac.jp`
[3] Graduate School of Systems and Information Engineering, University of Tsukuba,
Tsukuba, Ibaraki 305-8573, Japan
`yoshise@sk.tsukuba.ac.jp`
[4] Department of Mathematical Informatics,
Graduate School of Information Science and Technology,
The University of Tokyo, Bunkyo-ku, Tokyo 113-8656, Japan
`tomomi@misojiro.t.u-tokyo.ac.jp`

Abstract. For a given schedule of a round-robin tournament and a matrix of distances between homes of teams, an optimal home-away assignment problem is to find a home-away assignment that minimizes the total traveling distance. We propose a technique to transform the problem to MIN RES CUT. We apply Goemans and Williamson's 0.878-approximation algorithm for MAX RES CUT, which is based on a positive semidefinite programming relaxation, to the obtained MIN RES CUT instances. Computational experiments show that our approach quickly generates solutions of good approximation ratios.

Keywords: Sports timetabling; semidefinite programming; Goemans and Williamson's approximation algorithm.

1 Home-Away Assignment Problem

Recently, sports scheduling becomes one of the main topics in the area of scheduling (e.g., see *"Handbook of Scheduling"* Chapter 52 (Sports Scheduling) [2]). This paper deals with a home-away assignment problem that assigns home or away to each match of a (double) round-robin tournament so as to minimize the total traveling distance. We propose a technique to transform the problem to MIN RES CUT. We apply Goemans and Williamson's approximation algorithm for MAX RES CUT [5] and report the results of computational experiments.

N. Megiddo, Y. Xu, and B. Zhu (Eds.): AAIM 2005, LNCS 3521, pp. 95–103, 2005.

$T\backslash S$	1	2	3	4	5	6	7	8	9	10
1	3	3	4	4	6	2	5	2	6	5
2	5	5	6	3	3	1	4	1	4	6
3	1	1	5	2	2	4	6	6	5	4
4	6	6	1	1	5	3	2	5	2	3
5	2	2	3	6	4	6	1	4	3	1
6	4	4	2	5	1	5	3	3	1	2

$T\backslash S$	1	2	3	4	5	6	7	8	9	10
1	A	H	A	H	A	A	A	H	H	H
2	H	A	A	A	H	H	A	A	H	H
3	H	A	A	H	A	H	H	A	H	A
4	H	A	H	A	A	A	H	H	A	H
5	A	H	H	H	H	A	H	A	A	A
6	A	H	H	A	H	H	A	H	A	A

Fig. 1. A timetable and HA-assignment of six teams

In the following, we introduce a mathematical definition of the problem. Throughout this paper, we deal with a (double) round-robin tournament with the following properties:

- the number of teams (or players etc.) is $2n$, where $n \in \mathbb{N}$;
- the number of *slots*, i.e., the days when matches are held, is $2(2n-1)$;
- each team plays one match in each slot;
- each team has its home and each match is held at the home of one of the playing two teams;
- each team plays every other team twice;
- each team plays at the home of every other team exactly once.

Figure 1 is a schedule of a round-robin tournament, which is described as a pair of a timetable and a home-away assignment defined below.

A *timetable* is a matrix whose rows are indexed by a set of teams $T = \{1, 2, \ldots, 2n\}$ and columns are indexed by a set of slots $S = \{1, 2, \ldots, 4n-2\}$. Each entry of a timetable, say $\tau(t, s)$ $((t, s) \in T \times S)$, shows the opponent of team t in slot s. A timetable $\mathcal{T}$ should satisfy the following conditions:

- for each team $t \in T$, the t-th row of $\mathcal{T}$ contains each element of $T \setminus \{t\}$ exactly twice;
- for any $(t, s) \in T \times S$, $\tau(\tau(t, s), s) = t$.

For example, team 2 of Fig. 1 plays team 4 in slots 7 and 9, and the match in slot 7 is held at the home of team 4, while the other is held at the home of team 2.

A team is *at home* in slot s if the team plays a match at its home in s, otherwise said to be *at away* in s. A *home-away assignment* (HA-assignment for short) is a matrix whose rows are indexed by T and columns by S. Each entry of an HA-assignment, say $a_{t,s}$ $((t, s) \in T \times S)$, is either 'H' or 'A,' where 'H' means that in slot s team t is at home and 'A' is at away.

Given a timetable $\mathcal{T}$, an HA-assignment $\mathcal{A} = (a_{t,s})$ $((t, s) \in T \times S)$ is said to be *consistent* with $\mathcal{T}$ if the followings are satisfied: (C1) $\forall (t, s) \in T \times S$, $\{a_{t,s}, a_{\tau(t,s),s}\} = \{A, H\}$, and (C2) $\forall t \in T$, $[\tau(t, s) = \tau(t, s')$ and $s \neq s']$ implies $\{a_{t,s}, a_{t,s'}\} = \{A, H\}$ (Condition (C2) is assumed in an ordinary "double" round-robin tournament). A schedule of a round-robin tournament is described as a pair of a timetable and an HA-assignment consistent with the timetable.

A *distance matrix* $\mathcal{D}$ is a matrix with zero diagonals whose rows and columns are indexed by T such that the element $d(t, t')$ denotes the distance from the home of t to that of t'. We do not assume the symmetricity of $\mathcal{D}$ nor that the distance matrix satisfies triangle inequalities. Given a consistent pair of a timetable and an HA-assignment, the traveling distance of team t is the length of the route that starts from t's home, visits venues where matches are held in the order defined by the timetable and the HA-assignment, and returns to the home. The *total traveling distance* is the sum total of traveling distances of all the teams.

Given only a timetable of a round-robin tournament, one should decide a consistent HA-assignment to complete a schedule. In practical sports timetabling, the total traveling distance is required to be reduced [1, 11]. In this context, the home-away assignment problem is introduced as follows.

HA assignment Problem
Instance: a timetable $\mathcal{T}$ and a distance matrix $\mathcal{D}$.
Task: find an HA-assignment that is consistent with $\mathcal{T}$ and minimizes the total traveling distance.

We formulate the HA assignment problem as MIN RES CUT, and apply Goemans and Williamson's approximation algorithm [5], which is based on the semidefinite programming relaxation. Computational experiments show that our method quickly generates feasible solutions close to optimal.

The rest of this paper is organized as follows: Section 2 proposes formulations of the HA assignment problem as MIN RES CUT; Section 3 reports the results of computational experiments; Section 4 states conclusions.

The problem to find an HA-assignment that is consistent with a given timetable and minimizes the number of breaks (consecutive pairs of home-games) is called the break minimization problem. There are several previous results on this problem (see [10, 12, 3, 7] for example). In [7], Miyashiro and Matsui formulated the break minimization problem as MAX RES CUT and applied Goemans and Williamson's algorithm for MAX RES CUT. Our algorithm proposed in this paper is an extension of their procedure to HA assignment problems. However, we need a non-trivial technique, described in the next section, to extend their procedure to HA assignment problems.

2   Formulation as MIN RES CUT

We propose a formulation of the HA assignment problem as MIN RES CUT. First, we define the problem MIN RES CUT. Let $G = (V, E)$ be an undirected graph with a vertex set V and an edge set E. For any vertex subset $V' \subseteq V$, we define $\delta(V') = \{\{v_i, v_j\} : v_i, v_j \in V, \ v_i \notin V' \ni v_j\}$. The problem MIN RES CUT is defined as follows: given a graph $G = (V, E)$, a specified vertex $r \in V$, a weight function $w : E \longrightarrow \mathbb{R}$, and a set $E_{\text{cut}} \subseteq \{X \subseteq V : |X| = 2\}$, find a vertex subset V' that minimizes $\sum_{e \in \delta(V') \cap E} w(e)$ under the conditions that $r \notin V'$ and $E_{\text{cut}} \subseteq \delta(V')$ hold. Here we note that the condition $r \notin V'$ is redundunt for the definition of MIN RES CUT, because for any $V'' \subseteq V$, $\delta(V'') = \delta(V \setminus V'')$. The condition helps to formulate the HA assignment problem as MIN RES CUT. It is easy to show that MIN RES CUT is NP-hard even if $\forall e \in E$, $w(e) = 1$ holds.

The problem MAX RES CUT is the maximization version of MIN RES CUT, and Goemans and Williamson [5] proposed a 0.878-approximation algorithm for MAX RES CUT. Now we formulate the HA assignment problem as MIN RES CUT. Given a timetable $\mathcal{T} = (\tau(t,s))$ $((t,s) \in T \times S)$, let $G = (V, E)$ be an undirected graph with a vertex set V and an edge set E defined below. We introduce an artificial vertex r and define $V = \{v_{t,s} : (t,s) \in T \times S\} \cup \{r\}$, $E = \{\{v_{t,s-1}, v_{t,s}\} : t \in T, s \in S \setminus \{1\}\} \cup \{\{r, v_{t,s}\} : (t,s) \in T \times S\}$, and

$$E_{\mathrm{cut}} = \{\{v_{t,s}, v_{\tau(t,s),s}\} : (t,s) \in T \times S\}$$
$$\cup \{\{v_{t,s}, v_{t,s'}\} : t \in T, \; s, s' \in S, \; \tau(t,s) = \tau(t,s'), s \neq s'\}.$$

For a feasible solution V' of this MIN RES CUT instance, i.e., a vertex subset $V' \subseteq V$ satisfying $r \notin V'$ and $E_{\mathrm{cut}} \subseteq \delta(V')$, construct an HA-assignment $\mathcal{A} = (a_{t,s})$ $((t,s) \in T \times S)$ as follows: if $v_{t,s} \in V'$ then $a_{t,s} = \mathrm{A}$, else $a_{t,s} = \mathrm{H}$. This HA-assignment is consistent with $\mathcal{T}$ because (C1) each pair of vertices corresponding to a match is in E_{cut}, and (C2) for each team, every pair of vertices corresponding to matches with a common opponent is in E_{cut}. Obviously, for any consistent HA-assignment, there exists a unique corresponding feasible solution of the MIN RES CUT instance. Thus, there exists a bijection between the feasible set of MIN RES CUT and the set of consistent HA-assignments.

Next, we discuss the total traveling distance. In the following, we denote any singleton $\{v\}$ by v for simplicity. Given a pair of timetable $\mathcal{T}$ and an HA-assignment $\mathcal{A}$ consistent with $\mathcal{T}$, the traveling distance of team t between slots s and $s + 1$, denoted by $\ell(t,s)$, is defined as follows:

$$\ell(t,s) = \begin{cases} 0 & (\text{if } (a_{t,s}, a_{t,s+1}) = (\mathrm{H}, \mathrm{H})), \\ d(\tau(t,s), \tau(t,s+1)) & (\text{if } (a_{t,s}, a_{t,s+1}) = (\mathrm{A}, \mathrm{A})), \\ d(t, \tau(t,s+1)) & (\text{if } (a_{t,s}, a_{t,s+1}) = (\mathrm{H}, \mathrm{A})), \\ d(\tau(t,s), t) & (\text{if } (a_{t,s}, a_{t,s+1}) = (\mathrm{A}, \mathrm{H})). \end{cases}$$

In the following, we use the notations $t' = \tau(t,s)$ and $t'' = \tau(t, s+1)$ for simplicity. We show that the traveling distance $\ell(t,s)$ satisfy the following equations;

$$\begin{aligned} \ell(t,s) = \; & d(t',t'') \, |v_{t,s} \cap V'| \, |v_{t,s+1} \cap V'| \\ & + d(t,t'') \, (1 - |v_{t,s} \cap V'|) \, |v_{t,s+1} \cap V'| \\ & + d(t',t) \, |v_{t,s} \cap V'| \, (1 - |v_{t,s+1} \cap V'|) \\ = \; & d(t',t'') \frac{|\{v_{t,s}, r\} \cap \delta(V')| + |\{v_{t,s+1}, r\} \cap \delta(V')| - |\{v_{t,s}, v_{t,s+1}\} \cap \delta(V')|}{2} \\ & + d(t,t'') \frac{-|\{v_{t,s}, r\} \cap \delta(V')| + |\{v_{t,s+1}, r\} \cap \delta(V')| + |\{v_{t,s}, v_{t,s+1}\} \cap \delta(V')|}{2} \\ & + d(t',t) \frac{|\{v_{t,s}, r\} \cap \delta(V')| - |\{v_{t,s+1}, r\} \cap \delta(V')| + |\{v_{t,s}, v_{t,s+1}\} \cap \delta(V')|}{2} \\ = \; & \frac{d(t',t'') - d(t,t'') + d(t',t)}{2} |\{v_{t,s}, r\} \cap \delta(V')| \\ & + \frac{d(t',t'') + d(t,t'') - d(t',t)}{2} |\{v_{t,s+1}, r\} \cap \delta(V')| \\ & + \frac{-d(t',t'') + d(t,t'') + d(t',t)}{2} |\{v_{t,s}, v_{t,s+1}\} \cap \delta(V')|. \end{aligned}$$

The first equality is obvious, because $[a_{t,s} = A \iff |v_{t,s} \cap V'| = 1]$ and $[a_{t,s+1} = A \iff |v_{t,s+1} \cap V'| = 1]$. The second equality is obtained by applying the equations

$$|v_{t,s} \cap V'| = |\{v_{t,s}, r\} \cap \delta(V')|, \quad |v_{t,s+1} \cap V'| = |\{v_{t,s+1}, r\} \cap \delta(V')|, \quad (1)$$

and

$$|v_{t,s} \cap V'| \, |v_{t,s+1} \cap V'| $$
$$= \frac{|\{v_{t,s}, r\} \cap \delta(V')| + |\{v_{t,s+1}, r\} \cap \delta(V')| - |\{v_{t,s}, v_{t,s+1}\} \cap \delta(V')|}{2}. \quad (2)$$

Equations (1) and (2) are obtained from the properties that $r \notin V'$ and $\forall V' \subseteq V$, $|\delta(V') \cap \{\{r, v_{t,s}\}, \{r, v_{t,s+1}\}, \{v_{t,s}, v_{t,s+1}\}\}| \in \{0, 2\}$. The third equality is trivial. Here we note that, if we employ only Equations (1), $\ell(t, s)$ becomes a quadratic function of $|\{v_{t,s}, r\} \cap \delta(V')|$ and $|\{v_{t,s+1}, r\} \cap \delta(V')|$. Using Equation (2), we can transform the quadratic function to a linear function of $|\{v_{t,s}, r\} \cap \delta(V')|$, $|\{v_{t,s+1}, r\} \cap \delta(V')|$ and $|\{v_{t,s}, v_{t,s+1}\} \cap \delta(V')|$.

In a similar way, we can show that the traveling distance of team t before the first slot and after the last slot, denoted by $\ell(t, 0)$ and $\ell(t, 4n - 2)$ respectively, satisfy that

$$\ell(t, 0) = d(t, \tau(t, 1))|\{v_{t,1}, r\} \cap \delta(V')|,$$
$$\ell(t, 4n - 2) = d(\tau(t, 4n - 2), t)|\{v_{t,4n-2}, r\} \cap \delta(V')|.$$

From the above, the total traveling distance is represented by a linear function of variables $|e \cap \delta(V')|$ $(e \in E)$ as follows:

$$\sum_{t \in T} \sum_{s=0}^{4n-2} \ell(t,s) = \sum_{t \in T} \sum_{s=1}^{4n-3} \left(\begin{array}{l} \dfrac{d(t', t'') - d(t, t'') + d(t', t)}{2} |\{v_{t,s}, r\} \cap \delta(V')| \\[2mm] + \dfrac{d(t', t'') + d(t, t'') - d(t', t)}{2} |\{v_{t,s+1}, r\} \cap \delta(V')| \\[2mm] + \dfrac{-d(t', t'') + d(t, t'') + d(t', t)}{2} |\{v_{t,s}, v_{t,s+1}\} \cap \delta(V')| \end{array} \right)$$
$$+ \sum_{t \in T} d(t, \tau(t, 1))|\{v_{t,1}, r\} \cap \delta(V')|$$
$$+ \sum_{t \in T} d(\tau(t, 4n - 2), t)|\{v_{t,4n-2}, r\} \cap \delta(V')|.$$

Thus, by introducing an appropriate weight function $w : E \to \mathbb{R}_+$ (a precise description appears in Appendix), the total traveling distance satisfies that

$$\sum_{t \in T} \sum_{s=0}^{4n-2} \ell(t,s) = \sum_{e \in E} w(e)|e \cap \delta(V')| = \sum_{e \in E \cap \delta(V')} w(e)$$

and the objective function value of MIN RES CUT, with respect to $w(e)$, is equivalent to the total traveling distance. From the above, the HA assignment problem is formulated as MIN RES CUT.

Here we note that the break maximization problem, which maximizes the number of consecutive pairs of home-games, is a special case of our problem such that the distance between any pair of homes is equal to 1. It is shown in [8] that the break maximization problem is essentially equivalent to the break minimization problem, which is discussed in many papers [10, 12, 3, 7]. Thus, the break minimization problem is also a special case of the HA assignment problem discussed in this paper.

3   Computational Experiments

For MAX RES CUT, Goemans and Williamson [5] proposed a 0.878-randomized approximation algorithm using semidefinite programming. Here we apply Goemans and Williamson's algorithm to the proposed MIN RES CUT formulation of the HA assignment problem. In the following, we briefly explain the procedure. The algorithm consists of the following three steps.

1. Semidefinite Programming
   For a given instance of MIN RES CUT $(V, E, r, w, E_{\mathrm{cut}})$, let W be a matrix whose rows and columns are indexed by V such that $W_{ij} = W_{ji} = w(\{i, j\})$ if $\{i, j\} \in E$, otherwise $W_{ij} = W_{ji} = 0$. Then solve the following semidefinite programming problem:

$$\text{minimize } C \bullet X$$
$$\text{subject to } E_{ii,ii} \bullet X = 1 \quad (\forall i \in V),$$
$$E_{ij,ji} \bullet X = -2 \quad (\forall \{i, j\} \in E_{\mathrm{cut}}),$$
$$X \succeq O, \ X \text{ is symmetric}, \ X \in \mathbb{R}^{V \times V},$$

   where $C = (\mathrm{diag}(W\mathbf{1}) - W)/4$, $X \bullet Y = \sum_i \sum_j X_{ij} Y_{ij}$, $E_{ij,ji}$ is the matrix in which entries E_{ij} and E_{ji} are ones and every other entry is zero, and $X \succeq O$ means that X is positive semidefinite.

2. Cholesky Decomposition
   Decompose an (almost) optimal solution X_0 of the semidefinite programming problem in Step 1 into a matrix $\widehat{X}$ such that $X_0 = \widehat{X}^\top \widehat{X}$ (Cholesky decomposition).

3. Hyperplane Separation
   Generate a vector u at uniformly random on the surface of d-dimensional unit ball and put $V_1 = \{i \in V : u^\top \widehat{x}_i \geq 0\}$ where d is the number of rows of $\widehat{X}$ and $\widehat{x}_i$ is the column vector of $\widehat{X}$ index by $i \in V$. Output a vertex subset
   $$V' = \begin{cases} V_1 & (\text{if } r \notin V_1), \\ V \setminus V_1 & (\text{if } r \in V_1). \end{cases}$$

The above three steps terminate in polynomial time. Note that a practical procedure to obtain a good solution is to repeat Step 3 a number of times and output a solution with the best objective value.

Table 1. Results of computational experiments

(a)

#teams	ratio	SDP (s)		IP (s)	
	avg.	avg.	(s. d.)	avg.	(s. d.)
16	1.00119	81.8	(2.82)	13.2	(2.09)
18	1.11825	129.1	(23.99)	27.4	(8.87)
20	1.00119	233.1	(13.82)	61.1	(25.68)
22	1.00122	388.4	(15.10)	1550.3	(1124.54)
24	1.00478	617.7	(22.18)	68341.7	(124286.75)
26	—	989.4	(22.68)	—	—
30	—	2142.9	(104.27)	—	—

(b)

#teams	ratio	SDP (s)		IP (s)	
	avg.	avg.	(s. d.)	avg.	(s. d.)
16	1.00057	86.2	(2.82)	22.3	(7.71)
18	1.17323	123.9	(28.84)	61.1	(26.40)
20	1.00071	273.8	(19.26)	328.2	(419.23)
22	1.00099	393.0	(9.49)	1244.5	(748.60)
24	1.00121	664.1	(23.83)	13078.3	(19549.06)
26	—	1057.3	(30.40)	—	—
30	—	2226.3	(78.00)	—	—

#teams: the number of teams;
ratio: average of ratios of the optimal value and the objective function value of the
best solutions obtained by our procedure;
SDP : computational time for our procedure;
IP : computational time for integer programming;
avg.: average; s. d.: standard deviation.

Goemans and Williamson [5] showed that the maximization version of the
above algorithm finds a feasible solution of MAX RES CUT, and its expected
objective value is at least 0.87856 times the optimal value. In case of MIN RES
CUT, any non-trivial bound of approximation ratio of the above algorithm is
not known.

Finally, we report our computational results. Computational experiments
were performed as follows. Tables 1 (a) and (b) show the results when we generated 10 timetables for each size of $2n = 16, 18, 20, 22, 24, 26, 30$. We constructed
a timetable of "double" round robin tournament by concatenating two copies of
a timetable of "single" round robin tournament that is randomly created as the
method described in [3]. The results are shown in Table 1 (a). Table 1 (b) reports the results when each timetable is obtained by concatenating two mutually
different timetables of "single" round robin tournament. We used the distance
matrix obtained from TSP instance att48 from TSPLIB. We chose cities of
att48 with indices from 1 to $2n$.

For each instance, we applied Goemans and Williamson's algorithm and generated 10000 HA-assignments by executing Step 3 of the algorithm 10000 times. Finally, we output a solution with the best of generated 10000 solutions. In order to evaluate the quality of the best solutions, we solved the same instances with integer programming in a similar manner as Trick [12]. All computations were performed on Dell Dimension 8100 (CPU: Pentium4, 1.4GHz, RAM: 768MB, OS: Vine Linux 2.6) with SDPA 6.0 [13] for semidefinite programming problems and CPLEX 8.0 [6] for integer programming problems.

Table 1 shows the results of the experiments. In almost of all cases the average of approximation ratios is less than 1.18. We did not solved 26 and 30 teams instances with integer programming because it would not terminate within reasonable computational time. The computational time for our procedure is less than 670 seconds when $2n \leq 24$.

4   Conclusions

We proposed a formulation of HA assignment problems as MIN RES CUT problems, and performed computational experiments with Goemans and Williamson's algorithm for MAX RES CUT, based on semidefinite programming relaxation. Computational experiments showed that our approach is highly effective in terms of quality of solutions and computational speed, in particular, for a large instance.

References

1. Easton, K., Nemhauser, G. and Trick, M.: Solving the travelling tournament problem: a combined integer programming and constraint programming approach, In: Practice and Theory of Automated Timetabling IV (PATAT 2002), Lecture Notes in Computer Science, **2740** (2003) 100-109.
2. Easton, K., Nemhauser, G. and Trick, M.: Sports Scheduling, In: Handbook of Scheduling: Algorithms, Models, and Performance Analysis, Leung, J. Y-T. and Anderson, J. H. (Eds.), Chapman & Hall, 2004.
3. Elf, M., Jünger, M. and Rinaldi, G.: Minimizing breaks by maximizing cuts, Operations Research Letters, **31** (2003) 343–349.
4. Garey, M. R. and Johnson, D. S.: Computers and Intractability; a Guide to the Theory of NP-completeness, W. H. Freeman, New York, 1979.
5. Goemans, M. X. and Williamson, D. P.: Improved approximation algorithms for maximum cut and satisfiability problems using semidefinite programming, Journal of the ACM, **42** (1995) 1115–1145.
6. ILOG. ILOG CPLEX 8.0 User's Manual. ILOG, 2002.
7. Miyashiro, R. and Matsui, T.: Semidefinite programming based approaches to the break minimization problem, Computers and Operations Research (to appear).
8. Miyashiro, R. and Matsui, T.: A polynomial-time algorithm to find an equitable home–away assignment, Operations Research Letters, **33** (2005) 235–241.
9. Nemhauser, G. L. and Trick, M. A.: Scheduling a major college basketball conference, Operations Research, **46** (1998) 1–8.

10. Régin, J-C.: Minimization of the number of breaks in sports scheduling problems using constraint programming, In: Constraint Programming and Large Scale Discrete Optimization, DIMACS Series in Discrete Mathematics and Theoretical Computer Science, **57** (2001) 115–130.
11. Russel, R. A. and Leung, J. M. Y.: Devising a cost effective schedule for a baseball league, Operations Research, **42** (1994) 614-625.
12. Trick, M. A.: A schedule-then-break approach to sports timetabling, In: Practice and Theory of Automated Timetabling III (PATAT 2000), Lecture Notes in Computer Science, **2079** (2001) 242–253.
13. Yamashita, M., Fujisawa, K. and Kojima, M.: Implementation and evaluation of SDPA 6.0, Optimization Methods and Software **18** (2003) 491–505.

Appendix

We define a weight function $w : E \to \mathbb{R}_+$, which is discussed in Section 2, as follows:

$$
w(\{v_{t,s}, r\}) = \frac{d(\tau(t,s), \tau(t,s+1)) - d(t, \tau(t,s+1)) + d(\tau(t,s), t)}{2}
$$
$$
+ \frac{d(\tau(t,s-1), \tau(t,s)) + d(t, \tau(t,s)) - d(\tau(t,s-1), t)}{2}
$$
$$
(\forall t \in T, \forall s \in S \setminus \{1, 4n-2\}),
$$
$$
w(\{v_{t,1}, r\}) = d(t, \tau(t,1)) + \frac{d(\tau(t,1), \tau(t,2)) - d(t, \tau(t,2)) + d(\tau(t,1), t)}{2},
$$
$$
w(\{v_{t,4n-2}, r\}) = d(\tau(t,4n-2), t)
$$
$$
+ \frac{d(\tau(t,4n-3), \tau(t,4n-2)) + d(t, \tau(t,4n-2)) - d(\tau(t,4n-3), t)}{2},
$$
$$
w(\{v_{t,s}, v_{t,s+1}\}) = \frac{-d(\tau(t,s), \tau(t,s+1)) + d(t, \tau(t,s+1)) + d(\tau(t,s), t)}{2}
$$
$$
(\forall t \in T, \forall s \in S \setminus \{4n-2\}).
$$

Then, the total traveling distance satisfies that

$$
\sum_{t \in T} \sum_{s=0}^{4n-2} \ell(t,s) = \sum_{t \in T} \sum_{s=1}^{4n-3} \left(\begin{array}{l} \dfrac{d(t',t'') - d(t,t'') + d(t',t)}{2} |\{v_{t,s}, r\} \cap \delta(V')| \\ + \dfrac{d(t',t'') + d(t,t'') - d(t',t)}{2} |\{v_{t,s+1}, r\} \cap \delta(V')| \\ + \dfrac{-d(t',t'') + d(t,t'') + d(t',t)}{2} |\{v_{t,s}, v_{t,s+1}\} \cap \delta(V')| \end{array} \right)
$$
$$
+ \sum_{t \in T} d(t, \tau(t,1)) |\{v_{t,1}, r\} \cap \delta(V')| + \sum_{t \in T} d(\tau(t,4n-2), t) |\{v_{t,4n-2}, r\} \cap \delta(V')|
$$
$$
= \sum_{e \in E} w(e) |e \cap \delta(V')| = \sum_{e \in E \cap \delta(V')} w(e).
$$

Thus, the objective function value of MIN RES CUT with respect to $w(e)$ defined above is equivalent to the total traveling distance.

Coopetitive Game, Equilibrium and Their Applications*

Lihui Sun[1] and Xiaowen Xu[2]

[1] International Business College, Qingdao University,
Shandong, 266071, China
sun_lihui@sina.com
[2] School of Management,
Xi'an Jiaotong University, China

Abstract. Coopetition has become the current trend of economic activities. Coopetitive game is introduced through the comparison of the characteristics of noncooperative game and cooperative game. Furthermore, the coopetitive game is solved adopting one kind of Minimax theorem. Finally, the Cournot coopetition model is presented as an example, and the equilibrium is compared with Nash equilibrium.

1 Introduction

The current business environment, advances in information and communication technologies, and the resultant development of network and virtual organizations have led firms to cooperate and compete simultaneously. The term "co-opetition", coined by management professors Barry Nalebuff (Yale University) and Adam Brandenburger (Harvard University), refers to that phenomenon [1]. In the same year, Maria Bengtsson and Sören Kock entitled coopetition the phenomena including both cooperation and competition, and studied the cooperation and competition in business networks [2, 3]. In fact, cooperation and competition have been studied widely. According to the relationship of the aims in cooperation and competition theory, Deutsch divided the benefit body into three parts: cooperation, competition and independence. [4, 5] Claudia Loebbecke, Paul C.Van Fenema and Philip Powell paid much attention to the knowledge transfer under coopetition and presented the theory of interorganizational knowledge sharing during coopetition. [6, 7] Kjell Hausken studied cooperation and between-group competition and found that competition between groups in defection games might give rise to cooperation though the considerable cost of cooperation might be needed. [8] To Marc's theory, benefit body takes other's actions as positive exterior conditions in cooperation and in competition the other's actions are taken as negative exterior conditions [9].

---

* This research was supported by NSF of China(No. 10371094, No.70372050), foundations of Qingdao University and software science foundations of Qingdao city.

N. Megiddo, Y. Xu, and B. Zhu (Eds.): AAIM 2005, LNCS 3521, pp. 104–111, 2005.

In this paper, "coopetition" is defined as the phenomenon that differs from competition or cooperation, and stresses two faces of one relationship, cooperation and competition, in the same situation, in which competitors can strengthen their competitive advantages by cooperation. The "Coopetitive" is the adjective form of the coopetition. In the second section, coopetitive game is introduced and the comparison between noncooperative and cooperative game is studied. The coopetitive equilibrium is given by one kind of Minimax theorem in the third section. Economic examples and the comparisons between noncooperative and competitive game are made in the forth section. Conclusions can be made that the coopetitive game has a prodigious advantages both in modeling and algorithm.

2   Coopetitive Game and Coopetitive Equilibrium

2.1   Noncooperative Game and Cooperative Game and Their Comparison

Game theory can be classified into three types according to the interaction of the players: noncooperative game, cooperative game and coopetitive game. In noncooperative situation, players are self-concerned and each player makes decision by himself based on the strategy preferences. Each player maximizes his payoff against the others'. The equilibria can be obtained at the intersections of players' reaction functions and nearly all of them cannot obtain the satisfactory profits.

In cooperative situation, coalition without any conflict is supposed to construct through contract or nuisance suits commitment, etc. The coalition will maximizes its revenue and allocate it based on certain rules. Unfortunately, the coalitions is usually destroyed because of players' self-concerned actions or some details that are ignored in the cooperative process.

Coopetitive game is presented in this paper to avoid these conflicts. The self-concerned players can form coalition in competitive situation. At the same time, the coopetitive equilibria have advantages over those of noncooperative, and they are stable.

2.2   Coopetitive Game

Definition 1. *A Coopetition game $< N, (A_i), (uc_i) >$ includes:*

- *The set of players $1, 2, \cdots, I$.*
- *The pure strategy space A_i for each player i.*
- *The payoff coefficient functions $uc_i(a)$ for each player i.*

The payoff coefficient function is the standardization of the payoff function u_i, which gives player $i's$ Von Neumann-Morgenstern utility $u_i(a)$ for each profiles $a = (a_1, \cdots, a_I)$. The standardization of payoff function u_i is the ratio between the payoff that a player can get at a certain strategy profile and the highest payoff that he can gain. Therefore, the payoff coefficient function $u_i(a)$ denotes the satisfaction degree that player i can obtain under profile a.

Definition 2. *A subset B of the polytope A is called an action strategic extreme set, if $a, b \in A$ and $\lambda a + (1 - \lambda)b \in B$ for some $\lambda \in (0, 1)$ imply $a, b \in B$.*

For any $a \in A$, $M(a) = \{i' \in I | uc_i'(a) = \min_{i \in I} uc_i(a)\}$ is defined as the index set of a, which means the member of the players who obtain the lowest payoff coefficient function under profile a.

Definition 3. *A point a in A is called a critical strategy if there exists an extreme set B such that for $a \in A, b \in B$ and $M(a) \subseteq M(b)$ imply $M(a) = M(b)$. In the other words, a critical point is a point with maximum index set in certain extreme set.*

Definition 4. *Coopetitive equilibrium of game $< N, (A_i), (uc_i) >$ is some critical strategy $a^* \in A$, and $a^* = \arg\max_{a \in A} uc_{i \in M(a)}(a)$.*

The corresponding strategy profiles and the utility profiles under the equilibria are called equilibrium strategy profiles and equilibrium utility profiles respectively.

According to the definition, the coopetitive equilibrium can be obtained in this way: given the strategy profiles, each player finds the conservative (or minimum) payoff coefficients and selects a higher one among them. The coopetitive equilibrium is the one of the strategy profiles that much more players choose with higher satisfaction degrees, and is the counterbalance among players as well.

3   Minimax Theorem and Coopetitive Equilibrium

3.1   Minimax Theorem [10]

Let $\{G_i(x)\}_{i \in I}$ be a family of finitely many continuous concave functions on a polytope X and $I = \{1, \cdots, n\}$. Note that in general, $F(x) = \max_{i \in I} G_i(x)$ on X is not a concave function. However, its behavior is similar to a concave function.

A subset Y of the polytope X is called an extreme set of X if $x, y \in X$ and $\lambda x + (1 - \lambda)y \in X$ for some λ in the interval $(0, 1)$ imply $x, y \in X$. For example, every vertex is an extreme set and the set X, itself, is also an extreme set. For any $x \in X$, the index set of x is defined as $M(x) = \{i' \in I | G_i'(x) = \max_{i \in I} G_i(x)\}$. A point x in X is called a critical point if there exists an extreme set Y such that $x \in Y$ and that $y \in Y$ and $M(x) \subseteq M(y)$ imply $M(x) = M(y)$. In the other words, a critical point is a point with maximum $M(x)$ in some extreme set Y.

There is an intuitive interpretation for the critical points. Partition the polytope X into finitely many small regions $X_i' = \{x \in X | G_i'(x) = \max_{i \in I} G_i(x)\}$, every critical point is a "vertex" of some X_i'. Note that X_i' is not necessarily a polytope. If only one small region X_i' is considered, then we cannot say that the minimum value of $F(x)$ on X_i' takes place at "vertices" of X_i'. Thus, the following result is nontrivial.

Theorem 1. *Suppose that $F(x) = \max_{i \in I} G_i(x)$ where I is finite and $G_i(x)$ is a continuous, concave function. Then the minimum value of $F(x)$ for x over a polytope X is achieved at some critical points.*

The following corollaries can be obtained and the proofs are the same as that of theorem 1.

Corollary 1. *Suppose that $f(x) = \min_{i \in I} g_i(x)$ where I is finite and $g_i(x)$ is a continuous, convex function. Then the maximum value of $f(x)$ for x over a polytope X is achieved at some critical points.*

Corollary 2. *The critical strategy of the coopetition game $< N, (A_i), (uc_i) >$ is the critical point of the payoff coefficient function uc_i.*

3.2   Coopetitive Equilibrium

Based on the corollary 1 and corollary 2, the coopetitive equilibrium is one of the critical strategy profiles and the solution is to optimize:

$$\max_{a \in A} uc_{i \in M(a)}(a)$$

In this paper, we apply the following algorithm:

Step 1. Put into the set of players, the pure strategy space A_i and the payoff coefficient functions $uc_i(a)$ for each player i;

Step 2. Calculate the payoff coefficients of each player on every vertex;

Step 3. Let $i = 0, k = n - i$. When $k \neq 1$, let payoff coefficient functions be equal of any k players'; If there is no intersection of any k players' payoff coefficient functions, let $i = i + 1$ and repeat step 3; Otherwise, register the strategy profiles x, the utility profiles u, the sets M of players with higher payoff coefficients and their coefficients g_{max}, and the other's payoff coefficients g_{else} at any intersection, and go to step 4.
While $k = 1$, the coopetitive equilibria are the same as Nash equilibria.

Step 4. Maximize the payoff coefficients at the intersections of all the k payoff coefficient functions, and register the corresponding strategy profiles, which are the coopetitive equilibria.

4   Cournot Coopetition Model and Cournot Coopetitive Equilibrium

4.1   Cournot Coopetition Model

In Cournot Model, I oligarchs (firms) produce a homogeneous good. The strategies are quantities. All firms simultaneously choose their respective output lever

x_i from feasible sets $[0, \infty)$, They sell their outputs at the market-cleaning price $p(x)$, where $x = x_1 + x_2 + ... + x_I$. Firm i's cost of production is $C_i(x_i) = c_i x_i$, and firm i's total profit is $u_i(x_1, x_2, ..., x_I) = x_i p(x) - c_i(x_i)$. For linear demand $p(x) = max(0, a - x)$, the maximal profit that firm i can obtain in the monopoly market is $u_i^{max} = (a - c_i)^2/4$. Firm i's payoff coefficient function under strategy profile x is $uc_i(x) = u_i(x)/u_i^{max}$.

We call the constant a in the linear demand p potential demand. The coopetitive equilibria depend on the potential demands and the costs of firms.

Definition 5. *The demand-cost difference is defined as the difference between potential demand and the cost, say $(a - c_i)$; the cost-cost difference is defined as the cost difference of any two firms , say $(c_i - c_j)$.*

Lemma 1. *The demand-cost difference and the cost-cost difference determine the satisfaction degrees of the firms at critical points in coopetitive games.*

Proof. Suppose the costs of any two firms are c_i and c_j, $c_i > c_j$, $a - c_i = m(c_i - c_j)$, and then

$$g(x) = \frac{x_i(a - c_i - \sum_{k=1}^n x_k)}{(a - c_i)^2/4} = \frac{x_j(m(c_i - c_j) - \sum_{k=1}^n x_k)}{m^2(c_i - c_j)^2/4}$$

$\square$

The satisfaction degrees of the firm can be obtained given the demand-cost difference and the cost-cost difference.

Lemma 2. *Firms' satisfaction degrees at coopetitive equilibrium may differ from each other. The greater the demand-cost difference is or the smaller the cost-cost difference is, the closer the utilities of the firms are.*

Proof. Suppose the costs of any three firms are $c_i > c_j > c_k$, $a - c_i = m_1(c_j - c_k)$, $a - c_j = m_2(c_j - c_k)$ and $m_1 < m_2$.

Let $g^j = g^k$ at equilibria, and then

$$\frac{x_j(a - c_j - \sum_{l=1}^n x_l)}{m_2^2(c_j - c_k)^2/4} = \frac{x_k(a - c_k - \sum_{l=1}^n x_l)}{(m_2 + 1)^2(c_j - c_k)^2/4}$$

$$\frac{u^j(x)}{u^k(x)} = \frac{x_j(a - c_j - \sum_{l=1}^n x_l)}{x_k(a - c_k - \sum_{l=1}^n x_l)} = \frac{m_2^2}{(m_2 + 1)^2}$$

Let $g^i < g^j$, and then

$$\frac{u^i(x)}{u^j(x)} = \frac{x_i(a - c_i - \sum_{k=1}^n x_k)}{x_j(a - c_j - \sum_{k=1}^n x_k)} < \frac{(a - c_i)^2}{(a - c_j)^2} = \frac{m_1^2}{m_2^2}$$

$\square$

According to the above lemmas, theorem 2 can be obtained.

Theorem 2. *More players gain higher satisfaction degrees at coopetitive equilibrium. Moreover, the firms with lower cost can obtain a higher satisfaction degree, vice versa.*

4.2   Cournot Coopetitive equilibrium

4.2.1 Extreme Sets

According to the definition of extreme set and the feasible sets $0 \leq x_i \leq (a - c_i)$, all the n-dimension vectors $(x_1, x_2, \cdots, x_n)$, $x_i \in [a_i, b_i]$, $0 \leq a_i \leq x_i \leq b_i \leq (a - c_i)$ are extreme sets.

4.2.2 Critical Point

According to corollary 1, the critical points can be obtained at the vertices or at the interior critical points. From the feasible set $0 \leq x_i \leq (a - c_i)$ of firm i, the production of each firm is either zero or $a - c_i$, and there will be no profits for any firm and even internecine $(uc_i(x) \leq 0)$.

The interior critical points are the points whose index sets are not embraced by the other critical points and they can be obtained at the intersections of the payoff coefficient functions. The algorithm presented in section 3.2 is adopted to obtain all of the intersections approximately by simulation because of the difficulties in expressing the formula. For instance, for $a = 1.0, c_1 = 0.4, c_2 = 0.5, c_3 = 0.5, c_4 = 0.6$, we can divide the simulation span into 100 equal intervals, and set the iteration accuracy at 0.001. The iteration accuracy and the division step of the feasible sets can be changed for the different problems. There are many interior critical points whose index sets are $\{1, 2, 3\}, \{1, 2, 4\}, \{1, 3, 4\}$ and $\{2, 3, 4\}$.

4.2.3 The Coopetition Equilibrium

Comparing the utility coefficients at vertices and the interior critical points, the coopetitive equilibrium can be obtained. For the case of $a = 1.0, c_1 = 0.4, c_2 = 0.5, c_3 = 0.5, c_4 = 0.6$, the firms must choose the strategy profiles to maximize their satisfaction degrees. The maximum satisfaction degrees obtained at interior critical points, and the corresponding strategy profiles x, the utility profiles u, the sets M of players with higher satisfaction and their coefficients g_{max}, and the other's payoff coefficients g_{else} at any intersection are listed in Table 1. Which coalition can come into being among these coalitions? All of the firms will choose to participate the coalition in which they can obtain the highest satisfaction degrees. Coalition $\{1, 2, 3\}$ can give three members satisfaction degree 0.328 which is much higher than those of the other coalitions, therefore this coalition will come into being.

Table 1. Case 1: $a = 1.0, c_1 = 0.4, c_2 = 0.5, c_3 = 0.5, c_4 = 0.6$

M	x_1	x_2	x_3	x_4	g_{max}	g_{else}	u_1	u_2	u_3	u_4
$\{1,2,3\}^*$	0.09	0.09	0.09	0.002	0.328	0.064	0.030	0.021	0.021	0.0003
$\{1,2,4\}$	0.057	0.053	0.035	0.050	0.257	0.171	0.023	0.016	0.011	0.010
$\{1,3,4\}$	0.057	0.035	0.053	0.050	0.257	0.171	0.023	0.011	0.016	0.010
$\{2,3,4\}$	0.003	0.075	0.075	0.076	0.325	0.012	0.001	0.016	0.020	0.013

Several other examples are given in Table 2, 3 and 4 respectively. The strategies marked with asterisk denote the equilibrium strategy in the tables.

Remark 1. The above simulation results verify the theorem 2.

Table 2. Case 2: $a = 1.0, c_1 = 0.4, c_2 = 0.5, c_3 = 0.6, c_4 = 0.6$

M	x_1	x_2	x_3	x_4	g_{max}	u_1	u_2	u_3	u_4
{1,2,3,4}*	0.054	0.05	0.048	0.048	0.24	0.022	0.015	0.010	0.010

Table 3. Case 3: $a = 1.0, c_1 = 0.4, c_2 = 0.4, c_3 = 0.6, c_4 = 0.6$

M	x_1	x_2	x_3	x_4	g_{max}	g_{else}	u_1	u_2	u_3	u_4
{1,2,3}*	0.09	0.09	0.09	0.002	0.328	0.064	0.030	0.030	0.021	0.0003
{1,2,4}	0.087	0.087	0.005	0.106	0.305	0.017	0.027	0.027	0.001	0.012
{1,3,4}	0.078	0.009	0.075	0.078	0.312	0.036	0.028	0.003	0.020	0.013
{2,3,4}	0.009	0.078	0.075	0.078	0.312	0.036	0.0032	0.028	0.020	0.013

Table 4. Case 4: $a = 1.0, c_1 = 0.5, c_2 = 0.5, c_3 = 0.6, c_4 = 0.6$

M	x_1	x_2	x_3	x_4	g_{max}	u_1	u_2	u_3	u_4
{1,2,3,4}*	0.233	0.233	0.086	0.086	0.510	0.032	0.032	0.020	0.020
{1,2,3,4}	0.24	0.24	0.098	0.098	0.676	0.042	0.042	0.027	0.027
{1,2,3,4}	0.25	0.25	0.11	0.11	0.880	0.055	0.055	0.035	0.035

4.3   Comparison Between Coopetitive Equilibrium and Nash Equilibrium

The equilibrium strategies and equilibrium profits under Nash equilibrium in noncooperative game are as follows.

$$x_{Nash}^i = (a + \sum\nolimits_{j=1}^n c_j)/(n+1) - c_i \quad i = 1, 2, \cdots, n \tag{1}$$

$$u_{Nash}^i = \frac{1}{n+1}(a - c_i - \sum\nolimits_{j=1}^n x_j)(a + \sum\nolimits_{j=1}^n c_j - (n+1)c_i) \\ i = 1, 2, \cdots, n \tag{2}$$

Let $a=1.0$, $c_1=0.4$, $c_2=0.5$, $c_3=0.5$, $c_4=0.6$, the equilibrium strategy and equilibrium profit profiles are $x_{Nash} = (0.2, 0.1, 0.1, 0)$, $u_{Nash} = (0.04, 0.01, 0.01, 0)$ in noncooperative game. From Table 1, the coopetition equilibrium strategy is $x_{Cooptition} = (0.09, 0.09, 0.09, 0.002)$ and equilibrium profit profile is $u_{Coopetition} = (0.0295, 0.0205, 0.0205, 0.0131)$. By comparison, we can obtain that $\sum_{i=1}^n u_{Coopetition}^i > \sum_{i=1}^n u_{Nash}^i$, and $\sum_{i=1}^n x_{Nash}^i > \sum_{i=1}^n x_{Coopetition}^i$.

It is concluded that the coopetitive game has a prodigious advantage both in the modeling and algorithm. The coopetitive equilibrium can be obtained conveniently and convex or concave payoff coefficient functions are the only requirements.

5   Conclusions

In this paper, the advantages and disadvantages of noncooperative and cooperative game are compared and the coopetitive game is presented. The coopetitive equilibrium is defined and the algorithm is given by using one kind of Minimax theorem. This algorithm has great advantages and can solve games with irregular, non-differential concave or convex payoff functions.

The Cournot coopetitive model with linear demand function and asymmetric costs are studied as examples. Conclusions can be made that much more players can obtain higher satisfaction degrees at coopetitive equilibria, which are dependent on the costs and the potential demands. The comparison is made between noncooperative Nash equilibrium and coopetitive equilibrium. The algorithms for solving the coopetitive equilibria with non-linear demand function by minimax theorem need further study.

References

1. Nalebuff Barry, Brandenburger Adam. *Co-opetition*, Cambridge, MA: Harvard Business Press. 1996.
2. Maria Bengtsson, Sören Kock. Cooperation and Competition among Horizontal Actors in Business Networks. *Paper presented at the 6th Work-shop on Interorganizational Research*, Oslo, August 23–25, 1996.
3. Maria Bengtsson, Sören Kock. *"Coopetition" in Business Networks—to Cooperate and Compete Simultaneously*. Industrial Marketing Management. 2000; 29:411–426
4. Deutsch.M. *The relation of conflict*. New Haven, CT.: Yale University Press. 1973
5. Deutsch.M. "Fifty years of conflict", in Festinger. L.(ED).*Retrospection on social psychology*. New York: Oxford University Press.1980
6. Claudia Loebbecke, Paul C.van Fenema, Philip Powell. Knowledge Transfer Under Coopetition. *American Management System*, 1997.215-229
7. Loebbecke, C., and van Fenema, P. C. Towards a Theory of Interorganizational Knowledge Sharing during Coopetition. *Proceedings of European Conference on Information Systems, Aix-en-Provence*, France, 1998: 1632-1639.
8. Kjell Hausken,Stavanger. Cooperation and Between-group Competition[J]. *Journal of Economic Behavior & Organization*, 2000,42: 417–425
9. Marc W., Athony.z. Farming and cooperation in public games: an experiment with an interior solution. *Economic Letters*.1999,65:322-328
10. Du D.-Z, Hwang. An Approach for Proving Lower Bounds: Solution of Gilbert -Pollak's Conjecture on Steiner Ratio. 1990.*FOCS*. 76-85.

An Equilibrium Model in Urban Transit Riding and Fare Polices

Hai-Jun Huang[1], Qiong Tian[1], and Zi-You Gao[2]

[1] School of Economics and Management,
Beijing University of Aeronautics and Astronautics, Beijing 100083, China
hjhuang@mail.nsfc.gov.cn
[2] School of Traffic and Transportation, Beijing Jiaotong University,
Beijing 100044, China

Abstract. This paper deals with the riding behavior of commuters who take trains from a living place to a work place during morning rush hours. The total travel cost include early and late arrival penalties as well as carriage body capacity. An equivalent mathematical programming model which generates equilibrium riding behavior is presented. The number of actually chosen transit runs, passenger flow distributions, and fares resulted from three system configurations namely social optimum, monopoly by one company and duopoly competition, are investigated by numerical examples.

1 Introduction

One of the most insightful and tractable approaches used to study the trip-timing behavior is the bottleneck model proposed by Vickrey [14]. This model considers the commuting congestion on a highway with a single bottleneck between a residential area and a workplace. Each commuter is confronted with a trade-off between travel time, cost of queuing and schedule delay cost of early or late arrival at work. The travel cost experienced by a commuter is then determined by his or her departure time from home.

Using the deterministic queuing theory, Vickrey [14] first developed a departure time choice model which leads to a cost-equilibrium on all commuters. Later, this model has been extended by many others (see a reviewing paper, Arnott et al. [4]). Particularly, some researchers have successfully applied the bottleneck model to investigate the changes of individual's commuting behavior and the system's performances under various road-use pricing policies.

Glazer [7] and Cohen [6] investigated the welfare effects of road pricing on different commuting population. Braid [5] conducted the equilibrium analysis in the case of elastic demand. Arnott et al. [1, 2, 3, 4] made contributions in many aspects associated with bottleneck modeling, including the first-best and second-best tolling in networks with parallel routes. Verhoef [13] explored the economics of various road transportation regulations like road pricing. However, these studies are restricted to those transportation systems involving private cars (auto mode) only.

N. Megiddo, Y. Xu, and B. Zhu (Eds.): AAIM 2005, LNCS 3521, pp. 112–121, 2005.

Tabuchi [12] studied a competitive transportation system which contains transit and highway modes. He concluded that road pricing can be regarded as a measure for restraining auto use and providing revenue for mass transport improvement. Unlike the auto mode, the transit mode mainly depends on its fare level and service quality for attracting commuters. But Tabuchi [12] did not consider the body congestion in carriages and the time headways between trains or buses. It is obvious that these two factors affect people's travel behavior greatly. The body congestion also leads to the loss of independence and privacy (Horowitz and Sheth [8]). Recently, Huang [9, 10] extended Tabuchi's work through introducing the concept of body congestion and stressing the heterogeneity of commuters and the overall demand elasticity.

Progress has been made along the above studies in transit system modeling; nevertheless, a flaw still exists, i.e., the transit system's capacity is not considered, this means all transit commuters can arrive on time no matter how crowded the carriages may be and how long the time headway is. In a real transit system, the capacity of transporting passengers is reflected by the discrete time headways. It may be difficult for explicitly formulating a bottleneck model for transit systems, but there obviously exist a trade-off between the in-carriage congestion and the schedule delay caused by the time headways of dispatching transit runs. The purpose of this paper is to model this trade-off and investigate the system and user performances caused by different system configurations.

This paper is organized as follows. Section 2 formulates the equilibrium riding model through developing an equivalent mathematical program. The solution method of the problem is given in Section 3. Section 4 investigates three system operation configurations by comparing the system and user performances from numerical calculations. Section 5 concludes the paper.

2   Model Formulation

Consider a simplified corridor network in which a mass transit (e.g., subway) provides transportation service between H (a residential area) and W (a workplace). There are N identical commuters who must travel by transit mode from H to W, on every morning. To facilitate the presentation of the essential ideas of this paper, all commuters are assumed to be identical in perceiving the time value and schedule delay penalty. Let the moving time from H to W be a constant τ which covers the in-carriage time and access (egress) time from H to subway station (from subway station to W).

The total cost experienced by a commuter who travels from H to W can be formulated as $C = p + \alpha\tau + c_b + \delta$, where p is the transit fare, α is the unit cost of travel time, c_b is the body congestion cost occurring in carriages, and δ is the cost associated with schedule delay of early or late arrival at W. The body congestion cost is formulated by $c_b = \tau g(n)$, where $g(n)$ is a monotonically increasing function of the passenger flow n in the train unit, and $g(0) = 0$. When a commuter arrives at W early, let $\delta = \beta T_1$, where β is the unit cost of early arrival, T_1 equals the official work start time minus the arrival time; when he/she arrives late, let $\delta = \gamma T_2$, where γ is the unit cost of arrival late, T_2 equals the arrival time minus the official work start time.

Kraus and Yoshida [11] recently presented a railway commuting model in which the waiting time cost spent at stations is covered. Considering such a fact that in Beijing's subway system, commuters at each station always pack themselves into one coming train rather than wait for the next run, we hence set τ a constant.

Suppose that in the morning rush hours, all transit runs are equally dispatched from origin H to destination W with the headway t and move along the line by a constant speed. This means these trains can arrive at W by the fixed interval of t. We assume that every commuter always minimizes his/her own total cost when selecting a transit run. Because of the body congestion cost, some people prefer to take runs which can arrive at W earlier or later for preventing from high body congestion. In the equilibrium state, the N identical commuters are divided into l batches, each corresponding one transit run. Let $Z = \{a, \cdots, 2, 1, 0, -1, -2, \cdots, -b\}$ denote the set of dispatched runs, $i(\in Z)$ represent a specific run. That $i > 0$ means this train arrives at W early, $i < 0$ for a train arrival late, and $i = 0$ for a train arrival on time. In this study, overtaking between runs is not allowed, so there is only one train arrival on time. Let n_i be the number of passengers who ride on the ith run (note that the integer n_i is regarded as the traffic volume on train i and is hence treated as a real number in the following analyses). Their schedule delay cost is

$$\delta(i) = \begin{cases} i\beta t & i > 0 \\ 0 & i = 0 \\ -i\gamma t & i < 0 \end{cases} \tag{1}$$

The total travel cost paid by one commuter who rides on the ith run is

$$\delta(i) = \begin{cases} p + \alpha\tau + \tau g(n_i) + i\beta t & i > 0 \\ p + \alpha\tau + \tau g(n_i) & i = 0 \\ p + \alpha\tau + \tau g(n_i) - i\gamma t & i < 0 \end{cases} \tag{2}$$

An equilibrium state is reached when no commuter can reduce his/her total travel cost by unilaterally changing his/her transit run choice. This is called the user-equilibrium (UE) condition of transit riding behavior, mathematically stated below

$$\begin{cases} C(i) = C_0 & \text{if } n_i > 0 \\ C(i) \geq C_0 & \text{if } n_i = 0 \end{cases} \quad i \in Z \tag{3}$$

where C_0 is a constant representing the identical equilibrium cost.

We here show that with a given p and fixed τ, finding a passenger flow distribution $\{n_i \mid i \in Z\}$ satisfying the UE condition (24) is equivalent to solve the following mathematical program problem

$$\min L(\mathrm{n}) = \sum_{i \in Z} \left(\tau \int_0^{n_i} g(\omega) d\omega + n_i \delta(i) \right) \tag{4}$$

subject to the conservation condition of passenger flows and non-negative requirement

$$\sum_{i \in Z} n_i = N \tag{5}$$

$$n_i \geq 0 \qquad i \in Z \tag{6}$$

As in the classical user-equilibrium traffic assignment, the objective function (4) doesn't have any intuitive economic or behavioral interpretation. It is easy to show that the first-order conditions of the minimization problem (4)-(6) are

$$n_i \left(\tau g(n_i) + \delta(i) - \lambda \right) = 0 \qquad i \in Z \tag{7}$$

$$\tau g(n_i) + \delta(i) - \lambda \geq 0 \qquad i \in Z \tag{8}$$

$$\sum_{i \in Z} n_i = N \tag{9}$$

$$n_i \geq 0 \qquad i \in Z \tag{10}$$

where λ is the Lagrange multiplier associated with (5). Equation (7) shows that the travel cost of any run (excluding the pre-determined constant $p + \alpha \tau$ which doesn't affect the equilibrium riding behavior) with positive passenger flow, $i.e.$, $n_i > 0$, must equal a constant, λ, the Lagrange multiplier. This constant can be regarded as the minimum travel cost among all possible transit runs, as stated by (8) where the constant $p + \alpha \tau$ is not included. With this interpretation, it is clear that the first-order conditions are equivalent to the UE condition (24): any actually chosen transit run has the same and minimum travel cost.

The region of feasible solutions to mathematical program (4)-(6) is made by a linear equality and inequalities. This region is then a convex set. The objective function is strictly convex since all body congestion functions are monotonically increasing and the terms associated with schedule delay are linear, continuous. Hence, the problem (4)-(6) is a strictly convex mathematical program which has only one solution satisfying the UE condition.

3   Solution Method

We now further analyze the properties of the problem (4)-(6) and find a method to determine the number of transit runs that are actually chosen by passengers in equilibrium riding state. The number of actually chosen runs is the sum of all runs having positive passenger flows. This number can be obtained through solving the mathematical program (4)-(6) with a pre-determined, large enough set Z. Let $(n_i, i \in Z)$ be the solution and define $\theta_i = 1$ if $n_i > 0$ and $\theta_i = 0$ otherwise. Then, the number of actually chosen transit runs, denoted by l, is

$$l = \sum_{i \in Z} \theta_i \tag{11}$$

In addition, that l can be derived from the minimum travel cost λ defined in (7)-(8). It is known that in equilibrium, $\tau g(n_i) + \delta(i) - \lambda = 0$ for all $n_i > 0$. As $g(n_i) > 0$ for $n_i > 0$, then $\delta(i) < \lambda$. This states that the ith transit run doesn't need to be put into operation when $\delta(i) \geq \lambda$ holds. Referring to Equation (1), we first consider the case of arrival at W early. Let k and $k+1$ be two successive positive integers such that the inequality $k < \lambda/\beta t \leq k+1$ holds. We then have $i\beta t \geq \lambda$ for $i \geq k+1$ and $i\beta t < \lambda$ for $i \leq k$, which implies that the runs numbered by $(k, k-1, \cdots, 1, 0, \cdots)$ are chosen by passengers and the runs numbered by $(\cdots, k+1)$ are unnecessary. Hence, the number of actually chosen runs which arrive at W early is $[\lambda/\beta t] - 1 = k$, where $[x]$ is the smallest integer not less than x. Denote the numbers of transit runs which are actually chosen by passengers, and arrive at W early and late, by $\bar{a}(\leq a)$ and $\bar{b}(\leq b)$, respectively. We then have

$$\bar{a}(\leq a) = [\lambda/\beta t] - 1 \tag{12}$$

$$\bar{b}(\leq b) = [\lambda/\gamma t] - 1 \tag{13}$$

The total number of actually chosen transit runs in equilibrium is

$$l = \bar{a} + \bar{b} + 1 \tag{14}$$

From the condition, $\tau g(n_i) + \delta(i) - \lambda = 0$ for all $n_i > 0$, the passenger flow distribution is given by

$$n_i = \begin{cases} g^{-1}((\lambda - \delta(i))/\tau) & -\bar{b} \leq i \leq \bar{a} \\ 0 & -b \leq i < -\bar{b} \\ 0 & \bar{a} < i \leq a \end{cases} \tag{15}$$

Substituting (15) into (9), we have

$$\sum_{-\bar{b} \leq i \leq \bar{a}} g^{-1}\left(\frac{\lambda - \delta(i)}{\tau}\right) = N \tag{16}$$

Solving the above equation, we get the value of the Lagrange multiplier λ. In equilibrium, therefore, λ and l can be written as the functions of N and t, i.e.,

$$\lambda = U(N, t) \tag{17}$$

$$l = V(N, t) \tag{18}$$

These two functions cannot be expressed explicitly in most cases, except both the body congestion functions and the demand functions are linear. The final minimum travel cost for each commuter, including the constant $p + \alpha\tau$, is

$$C_0 = p + \alpha\tau + \lambda \tag{19}$$

4   Three System Configurations and Pricing Polices

Up to now, we have known that given a fare p and the time headway t, the number of actually used transit runs and passenger flow distribution can be endogenously determined by the equilibrium riding principle. Under different system configurations, the transit manager will set different service parameters (such as the fare and the headway) in order to achieve some targets. In this section, three types of system configurations, namely social optimum, monopoly and duopoly competition are investigated.

Let $D^{-1}(N)$ denote the marginal benefit from trip making or the inverse of demand function, $\mathrm{d}D^{-1}(N/\mathrm{d}N < 0)$. Let G be the fixed cost of initiating the transit system, F be the train-associated variable cost (e.g., the expenses for acquiring trains/carriages and employing drivers) and f be the passenger-associated variable cost (e.g., the electric power assumption of a transit run is dependent upon the loading passengers).

The system optimum configuration designed by government is in general to maximize the social welfare, i.e.,

$$\max_{N,t} SW = \int_0^N D^{-1}(\omega)\mathrm{d}\omega - [(\alpha\tau + \lambda + f)N + Fl + G] \tag{20}$$

where λ and l are determined by (16) and (17), respectively, representing the outputs of equilibrium riding behavior with given N and t. When N and t are solved from (20), one obtain the transit fare

$$p = D^{-1}(N) - (\alpha\tau + \lambda) \tag{21}$$

The monopoly configuration assumes that the transit system is operated by one company which aims at maximizing the company's net profit, i.e.

$$\max_{N,p,t} NF = pN - (fN + Fl + G) \tag{22}$$

subject to

$$D^{-1}(N) = \alpha\tau + \lambda + p \tag{23}$$

where λ and l are given by (16) and (17), respectively. Equation (26) states the equilibrium relationship between marginal trip benefit and individual's travel cost.

In a two-duopoly market, the transit line is operated by two independence companies which maximize their net profit simultaneously. The problem becomes

$$\begin{cases} \max_{N_1,p_1,t_1} NF_1 = p_1N_1 - (fN_1 + Fl_1 + G) \\ \max_{N_2,p_2,t_2} NF_2 = p_2N_2 - (fN_2 + Fl_2 + G) \end{cases} \tag{24}$$

subject to

$$D^{-1}(N) = \alpha\tau + \lambda_i + p_i \qquad i = 1, 2 \tag{25}$$

$$N = N_1 + N_2 \tag{26}$$

where λ_i and l_i, $i = 1, 2$, are given by (16) and (17), respectively.

For nonlinear body congestion function and demand function, it is hard to analytically investigate the solutions of the above three system configurations. In this paper, we conduct numerical experiments with following input data: (β,γ)=(10,50) (HK\$/hr), τ=1 hr, α=20 (HK\$/hr), F=1000 (HK\$/train), f=1 (HK\$/person), G=0 (HK\$). Clearly, all data are projected into one morning rush hour. The inverse of a linear demand function is $D^{-1}(N) = -\ln(N/N_0)/K$, where the potential demand is N_0=20,000 passengers. This function implies that for a larger K-value the demand is more sensitive to the marginal trip benefit and the final realized demand will be lower. The body congestion function is $g(n) = -1.0 \times \ln(1 - n/n_c))$, where the capacity of a transit run is $n_c = 500$ passengers per run.

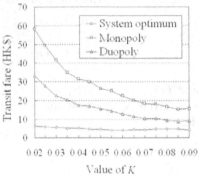

Fig. 1. Transit fares in three configs

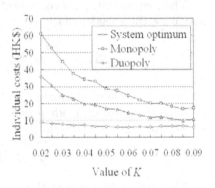

Fig. 2. Individual costs in three configs

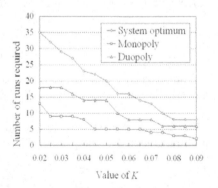

Fig. 3. Required transit runs

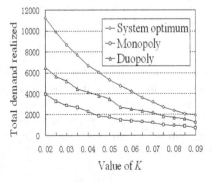

Fig. 4. Demands in three configs

Figs. 1-4 shows the fares, the individual travel costs, the number of actually used runs, and the realized travel demands by the three system configurations, respectively. It can be seen that the fare in system optimum is the lowest, then that in duopoly competition and that in monopoly is the highest; the same order is found for the individual travel costs; the number of actually chosen transit runs in system optimum is the most, then that in duopoly competition and that in monopoly is the fewest; the demand realized by system optimum is the most and that by monopoly is the fewest. When the value of K becomes larger, each of these four indexes comes down.

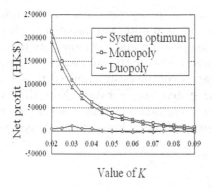
Fig. 5. Net profit in three configs

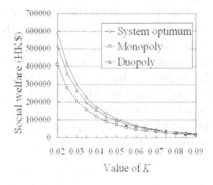
Fig. 6. Social welfare in three configs

Fig. 5 shows the net profit gained in the three system configurations. The system optimum configuration gives the lowest net profit, nearly K-value independent. The monopoly configuration results in more net profit for the company than the duopoly competition configuration though the differences are not significantly large. Fig. 6 shows the social welfare generated in the three configurations.

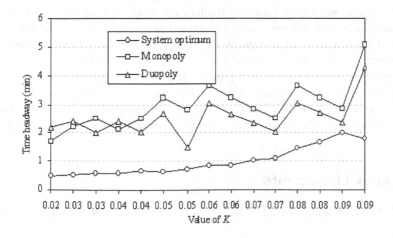
Fig. 7. Time headways in three configurations

It can be seen that, as expected, the system optimum configuration contributes the highest social welfare, then the duopoly competition, and the monopoly the lowest. The social welfare by the three configurations converge to the same value when the K-value approaches infinite.

The above numerical results show that the duopoly competition configuration can lead to lower bus fare, higher realized demand, less individual travel cost and more used runs than the monopoly configuration, although the two companies have the same variable costs. This should be contributed to the competition as stated in Microeconomics.

Finally, we check the time headways generated by the three system configurations, see Fig. 7. It can be seen that the system optimum dispatch the transit runs most frequently, then the duopoly competition and the monopoly in order. Oscillations can be observed in Fig. 7. This is because the number of transit runs is required to be integer and the headway is continuously adjusted to optimize the objective function designed for some system configuration. Note that some headways shown in Fig. 7 are less than 0.6 minutes, which may not match with reality since the input parameters of the example are selected without necessarily representing reasonable values.

5   Conclusions

In this paper we studied the equilibrium riding behavior for commuters who take trains from a living place to a work place during morning rush hours. With the aid of equilibrium riding model, in the elastic demand case we first formulated three system configurations and then compared their transit fares, numbers of actually used runs, demands, net profit and social welfare, by numerical experiments. We found that the monopoly configuration generates the highest transit fare and individual travel cost, then the duopoly competition and the system optimum configurations in order; the system optimum configuration needs the most transit runs and realizes the highest demands, then the duopoly competition and the monopoly configurations in order; the monopoly configuration produces more net profit for company but less social welfare than the duopoly competition configuration. The value of our study lies in the methodology adopted in this paper for analyzing different pricing mechanisms on the basis of equilibrium riding behavior. The approach presented in this paper can be extended to consider the continuously distributed value of time for dealing with riding behavior with heterogeneous commuters (Yang et al. [15]).

Acknowledgements

This research is supported by the National Natural Science Foundation of China.

References

1. Arnott, R., De Palma, A., Lindsey, R.: Schedule delay and departure time decisions with heterogeneous commuters. Transportation Research Record. **1197** (1988) 56-67

2. Arnott, R., De Palma, A., Lindsey, R.: Economics of a bottleneck. Journal of Urban Economics. **27** (1990) 111-130

3. Arnott, R., De Palma, A., Lindsey, R.: The welfare effects of congestion tolls with heterogeneous commuters. Journal of Transport Economics and Policy. **28** (1994) 139-161

4. Arnott, R., De Palma, A., Lindsey, R.: Recent developments in the bottleneck model. In: Button, K. J., Verhoef, E. (Ed.), Road Pricing, Traffic Congestion and the Environment: Issues of Efficiency and Social Feasibility. Aldershot, Edward Elgar. (1998) 79-110

5. Braid, R. M.: Uniform versus peak-load pricing of a bottleneck with elastic demand. Journal Urban Economics. **26** (1989) 320-327

6. Cohen, Y.: Commuter welfare under peak-period congestion tolls: Who gains and who loses? International Journal of Transport Economics. **14** (1987) 239-266

7. Glazer, A.: Congestion tolls and consumer welfare. Public Finance. **36** (1981) 77-83

8. Horowitz, A., Sheth, J.: Ride sharing to work: An attitudinal analysis. Transportation Research Record. **637** (1977) 1-8

9. Huang, H. J.: Fares and tolls in a competitive system with transit and highway: The case with two groups of commuters. Transportation Research E. **36** (2000) 267-284

10. Huang, H. J.: Pricing and logit-based mode choice models of a transit and highway system with elastic demand. European Journal of Operational Research. **140** (2002) 562-570

11. Kraus, M., Yoshida, Y.: The commuter's time-of use decision and optimal pricing and service in urban mass transit. Journal of Urban Economics. **51** (2002) 170-195

12. Tabuchi, T.: Bottleneck congestion and modal split. Journal of Urban Economics. **34** (1993) 414-431

13. Verhoef, E.: The Economics of Regulating Road Transportation. Edward Elgar, Brookfield. (1996)

14. Vickrey, W. S.: Congestion theory and transport investment. American Economic Review. **34** (1969) 414-431

15. Yang, H., Kong, H. Y., Meng, Q.: Value-of-time distributions and competitive bus services. Transportation Research E. **37** (2001) 411-424

Optimal Timing of Firms' R&D Investment Under Asymmetric Duopoly: A Real Options and Game-Theoretic Approach

Jianzu Wu and Huiyu Xuan

School of Management, Xi'an Jiaotong University,
P.O. Box 1605, 28 Xianning West Road,
Xi'an, Shaanxi 710049, China
wujianzu@hotmail.com, xuanhy@mail.xjtu.edu.cn

Abstract. In a real options and game-theoretic framework, this paper investigates the optimal timing of two asymmetric firms' R&D investment under uncertainty. There exist three types of equilibria that can occur in the choice of the R&D investment strategies, i.e. the preemptive, sequential and simultaneous equilibrium. The occurrence of a particular type of equilibrium is determined by the firms' relative payoff, which mainly depends on the level of operating cost asymmetry and first-mover advantage and the operating cost itself. We show that when the cost asymmetry and the first-mover advantage among firms are relatively small, two firms invest simultaneously. When the first-mover advantage is significant, the low-cost firm preempts the high-cost firm. In the situation where the asymmetry between firms becomes large enough, two firms invest sequentially. We also show that the lower the operating cost is, the more the incentive to become the leader has. So, with the operating cost decreasing, the possibility of simultaneous equilibrium and sequential equilibrium decreases, but that of preemptive equilibrium increases.

Keywords: Optimal timing, R&D Investment, Real Options, Game Theory, Asymmetric Duopoly.

1 Introduction

In the uncertain and competitive high-tech industry, one of the most important decisions of firms is when to invest in R&D, i.e. the R&D investment timing decision. A standard framework for the investment timing decision is the real options approach, which assumes that the opportunity to invest in R&D is analogous to an American call option on the investment project, and the timing of investment is economically equivalent to the optimal exercise decision for an option. In traditional real option modelling, the optimal exercise problem is always modelled as isolated optimal stopping problem without strategic interactions. According to real option theory, it is optimal to delay exercising the option to

N. Megiddo, Y. Xu, and B. Zhu (Eds.): AAIM 2005, LNCS 3521, pp. 122–131, 2005.

invest, even when it would be profitable to do so at once, in the hope of gaining a higher payoff in the future [1].

However, real options, unlike their financial counterparts, are rarely backed by legal contracts guaranteeing the holder's rights in precise terms. There are always other firms that also have access to the non-proprietary investment opportunity. When real options are held by a small number of competitive firms with an advantage to the first mover, each firm has incentive to invest early and its ability to delay is undermined by the fear of preemption. In order to deal with the tension between real options and strategic competition, this paper adapts the option game-theoretic approach that merges real options with game theory.

The study of option game theory started from the seminal paper of Smets [2], in which he studies the foreign direct investment decision in the duopoly market, considering both the exchange rate uncertainty and strategic interaction between two competitive firms. Dixit and Pindyck [3] summarize this model in their outstanding real option textbook. They assume that firms are not active in the market at original time and the firms as leader or follower is given exogenously. So there is only a preemptive equilibrium. Huisman and Kort [4] extend this model through introducing mix strategy equilibrium and assuming that firms are active in the market at original time, i.e. the profit cash flows are positive even they don't invest yet. Weeds [5] considers irreversible in competing R&D projects with uncertain returns under a winner-takes-all patent system. Huisman and Kort [6] study the optimal timing of technological innovation of a single firm in a duopoly framework. They examine the optimal technology investment decision of an individual firm, while taking into account the possible occurrence of better technologies in the future and competition of other firms.

While most of the above literature concentrate on symmetric oligopoly, Huisman [7] and Pawlina and Kort [8] analyze the situation where two firms' investment costs are asymmetric and one has cost advantage over another. In the real world of strategic investment, more or less the firms are not identical due various reasons. Thus the asymmetric duopoly model is more realistic compared with the symmetric cases and has more explanatory power of the real world.

The closest works to ours are the insightful paper by Pawlina and Kort [8] and book by Huisman [7] that address a similar question, but in their model, they assume the investment cost is asymmetric, while our paper focuses on operating cost asymmetry. In the all above literature, they assume that the operating cost of project are zero, i.e. once a firm successes in R&D, it markets the innovation at once and does not occur any further costs. This may be true for investment in natural resource and infrastructures, but for R&D investment, this is not true. In the previous, the lump-sum investment cost is usually much larger than its operating cost, so assumption of the operating cost is zero in the models doesn't put their conclusion into doubt. However, for the R&D investment, as the manufacturing-process innovation is more and more critical to product innovation [9], operating cost that mainly comes from production cost is not ignorant, even more important than investment cost.

The remainder of the paper is organized as follows. In Section 2, we present the basic model of two asymmetric competitive firms that face an exogenous stochastic market demand, and derive the optimal timing and strategies of firms as follower, leader and simultaneous investment respectively. In Section 3, we analyze different equilibria. Section 4 investigates the condition for equilibria. Concluding remarks are given in Section 5.

2   Model

Two risk-neutral firms, i and j, have the opportunity to invest in competing R&D projects. Project is directly competitive: the firms strive for the same patent and successful firm will get more market share and competitive advantage. The variable production cost, i.e. operating cost, of firm i is C_i. Two firms are asymmetric and one firm's operating cost is lower than the other's. Without loss of generality, we normalize C_1, which is the operating cost of the low-cost firm, to C, and C_2, which is the operating cost of the high-cost firm, is set equal to κC, where $\kappa \in [1, \infty)$ that denote the level of the operating cost asymmetry. Assuming there is not technological uncertainty, i.e. the R&D can success once investment is made, and the firms face only market demand uncertainty. The risk-free interest rate is r.

The firms face an inverse demand curve expressed by the market price of innovation $P(t)$ for firm i given by

$$P(t) = Y(t) D(N_i, N_j) \tag{1}$$

where $Y(t)$ is the stochastic demand shock following a geometric Brownian motion (GBM) with drift given by the following expression

$$dY(t) = \alpha Y(t) dt + \sigma Y(t) dz \tag{2}$$

where α is the drift parameter measuring the expected growth rate of $Y(t)$, σ is the instantaneous standard deviation or volatility parameter, and dz is the increment of a standard Wiener process. In the following pages, we use Y to denote $Y(t)$ without arising confuse.

$D(N_i, N_j)$ is a deterministic demand parameter for firm i, which depends on the status of firms i and j, N_k, for $k \in \{i, j\}$

$$N_k = \begin{cases} 0 & \text{if firm } k \text{ has not invested,} \\ 1 & \text{if firm } k \text{ has invested.} \end{cases} \tag{3}$$

The profit flow of firm i is

$$\pi(N_i, N_j) = YD(N_i, N_j) - N_i C_i \tag{4}$$

Because the irreversible investment in R&D increases the profit flow and the firm obtains higher profits if the competitor does not invest, the following restrictions on $\pi(N_i, N_j)$ are implied

$$\pi(1,0) > \pi(1,1) > \pi(0,0) > \pi(0,1) \tag{5}$$

Further we assume that there is the first-mover advantage to investment

$$\pi(1,0) - \pi(0,0) > \pi(1,1) - \pi(0,1) \tag{6}$$

Then we defined the level of the first-mover advantage is

$$\gamma \equiv \pi(1,0)/\pi(1,1) = [D(1,0) - C]/[D(1,1) - C] \tag{7}$$

Considering the decision of the competitor, there are three possibilities concerning the timing of firm i's investment. First, firm i may invest before its competitor (firm j) does and become the leader. Second, firm j may invest first and firm i becomes the follower. Finally, firm i and firm j may invest simultaneously, i.e. two firms invest at the same point in time.

In the next subsections, we establish the payoffs and optimal investment timing associated with the three situations described above. As in the standard approach used to solve dynamic games, we solve the problem backwards. First, we derive the optimal strategy of the follower under the strategies of the leader given. Then, we analyze the decision of the leader. Finally, the case of simultaneous investment is discussed.

2.1   Follower

Consider the optimal timing of firm i's investment when its competitor (firm j) has invested as the leader. According to the real option approach, this is an optimal stopping problem, i.e. there is optimal investment timing T_{iF}, when $t < T_{iF}$, waiting is optimal, and when $t \geq T_{iF}$, investment is optimal [3]. In other words, firm i will undertake the investment when profits are sufficiently large, i.e. when $Y(t)$ exceeds a certain threshold level denoted by Y_{iF}.

The payoff and the optimal investment timing of the firm can be calculated explicitly by applying the well-known standard dynamic programming methodology (see [3]). To save space, we provide the solution and refer the interested reader to [8] for further details. By solving the Bellman equation with corresponding value-matching and smooth-pasting, we arrive at the following expression for the payoff of firm i as the follower

$$F_i(Y) = \begin{cases} \frac{YD(0,1)}{r-\alpha} + \left(\frac{Y}{Y_{iF}}\right)^{\beta_1} \left(\frac{Y_{iF}[D(1,1) - D(0,1)]}{r-\alpha} - \frac{C_i}{r} - I\right), & Y \leq Y_{iF} \\ \frac{YD(1,1)}{r-\alpha} - \left(\frac{C_i}{r} + I\right), & Y \geq Y_{iF} \end{cases} \tag{8}$$

The interpretation of (8) is as follows. The first row is the present value of profits when the follower does not invest immediately. The first term is the payoff in case the follower does not invest forever, whereas the second term is the value of the option to invest. The second row is that the present value of cash flows resulting from immediate investment minus its costs. Where, Y_{iF} is the optimal investment threshold, and equals

$$Y_{iF} = \frac{\beta_1}{\beta_1 - 1} \frac{r - \alpha}{D(1,1) - D(0,1)} \left(\frac{C_i}{r} + I\right) \tag{9}$$

The corresponding optimal investment timing is

$$T_{iF} = \inf\left(t | Y \geq Y_{iF}\right) \tag{10}$$

2.2   Leader

Following a similar reasoning as in the previous subsection, we determine the payoff of firm i as the leader. After the firm j as a follower investment, i.e. $t \geq T_{jF}$, two firm's payoffs are equal as (8). When firm j is not investment, i.e. $t \leq T_{jF}$, the value function of firm i equals

$$\begin{aligned} L_i(Y) = &E\left\{\int_t^{T_{jF}} e^{-r(\tau-t)}\left[Y(\tau)D(1,0) - C_i\right]d\tau\right\} - I \\ &+ E\left\{\int_{T_{jF}}^{+\infty} e^{-r(\tau-T_{jF})}\left[Y(\tau)D(1,1) - C_i\right]d\tau\right\} \end{aligned} \tag{11}$$

where $T_{jF} = \inf\left(t|Y \geq Y_{jF}\right)$ is the optimal investment timing of the firm j as a follower, and Y_{jF} is corresponding investment threshold. The first two components of (11) correspond to the present value of the leader's profits realized until the moment of the follower's investment minus the leader's sunk cost. The second integral corresponds to the discounted perpetual stream of profits obtained after the investment of the follower. Considering the results of the follower problem, we can express the payoff of firm i as the leader in the following way

$$L_i(Y) = \begin{cases} \frac{YD(1,0)}{r-\alpha} + \left(\frac{Y}{Y_{jF}}\right)^{\beta_1} \frac{Y_{jF}[D(1,1)-D(1,0)]}{r-\alpha} - \left(\frac{C_i}{r} + I\right), Y \leq Y_{jF} \\ \frac{YD(1,1)}{r-\alpha} - \left(\frac{C_i}{r} + I\right), Y \geq Y_{jF} \end{cases} \tag{12}$$

The first row of (12) is the net present value of profits before the follower made the investment, i.e. the expected payoff in monopoly phase. The second row corresponds to the net present value of profits in a situation where it is optimal for the follower to invest immediately, i.e. the expected payoff in duopoly phase. The optimal investment threshold of firm i as the leader equals

$$Y_{iL} = \frac{\beta_1}{\beta_1 - 1} \frac{r - \alpha}{D(1,0) - D(0,0)} \left(\frac{C_i}{r} + I\right) \tag{13}$$

And the corresponding optimal investment timing is

$$T_{iL} = \inf\left(t|Y \geq Y_{iL}\right) \tag{14}$$

2.3   Simultaneous Investment

It is possible that the firms, despite the asymmetry in the operating cost, decide to invest simultaneously. The payoff of firm i investing at its optimal threshold simultaneously with firm j is

$$
S_i(Y) = \begin{cases} \frac{YD(0,0)}{r-\alpha} + \left(\frac{Y}{Y_{iS}}\right)^{\beta_1} \left(\frac{Y_{iS}[D(1,1)-D(0,0)]}{r-\alpha} - \frac{C_i}{r} - I\right), & Y \leq Y_{iS} \\ \frac{YD(1,1)}{r-\alpha} - \left(\frac{C_i}{r} + I\right), & Y \geq Y_{iS} \end{cases} \tag{15}
$$

Expression (15) is interpreted analogously to (8). And the corresponding optimal investment threshold and timing of two firms which invest simultaneously respectively is

$$
Y_{iS} = \frac{\beta_1}{\beta_1 - 1} \frac{r-\alpha}{D(1,1) - D(0,0)} \left(\frac{C_i}{r} + I\right) \tag{16}
$$

$$
T_{iS} = \inf\left(t | Y \geq Y_{iS}\right) \tag{17}
$$

3   Equilibria

There are three types of equilibria that can occur in the choice of strategies, i.e. the preemptive, sequential and simultaneous equilibrium. In this section we discuss the characteristics of each type of equilibrium and present the conditions under which each of them occurs.

3.1   Preemptive Equilibrium

The preemptive equilibrium occurs in the situation where both firms have an incentive to become the leader. When the cost advantage of the low-cost firm (firm 1) is relatively small, we has to take into account the fact that the high-cost firm (firm 2) will aim at preempting firm 1 as soon as a certain threshold is reached. This threshold, denoted by Y_{2P}, is the first hitting value of the process $Y(t)$ for which firm 2 is indifferent between being the leader and the follower. Formally, Y_{2P} is the smallest solution to $L_2(Y) - F_2(Y) = 0$.

When $Y_{2P} \geq Y_{1L}$, where Y_{1L} is the optimal investment threshold of firm 1 as the leader, the firm 2 has not opportunity to preempt because firm 1 will invest at Y_{1L} where firm 2's optimal strategy is waiting and becomes follower at Y_{2F}. When $Y_{2P} < Y_{1L}$, firm 1 uses the fact that firm 2 has no incentive to invest before Y_{2P} and preempts it by just an instant, i.e. firm 1 ϵ-preempts firm 2 at $Y_{2P} - \epsilon$, where ϵ is an infinitesimal. Therefore, firm 1 invests as soon as the process reaches the smaller of two values: Y_{2P} and Y_{1L}, that is, firm 1 invests at $\min\{Y_{2P}, Y_{1L}\}$ as leader and firm2 invests at Y_{2F} as follower.

3.2   Sequential Equilibrium

The sequential equilibrium occurs when firm 2 has no incentive to become the leader, i.e. when $F_2(Y) > L_2(Y)$ always holds. In this case, firm 1 simply maximizes the payoff of the monopoly investment opportunity and invests at the optimal timing, T_{1L}. As a result, firm 1 is able to invest at its unconditional threshold, Y_{1L}, and firm 2 invests at its follower threshold, Y_{2F}, sequentially.

3.3   Simultaneous Equilibrium

The simultaneous investment equilibrium occurs when there are no firms have incentive firstly to become a leader. In this case, the payoff of firm 1 associated with being the leader has to be lower than the payoff resulting from simultaneous investment at Y_{1S}, i.e. $L_1(min\{Y_{1L}, Y_{2P}\}) < S_1(Y_{1S})$. Otherwise, firm 1 will invest either at Y_{1L} or at Y_{2P} (depending on the level of cost asymmetry). Moreover, firm 2's follower threshold must be lower than Y_{1S}. In other words, firm 2 has to find it more profitable to respond to firm 1's investment at Y_{1S} immediately than to wait. Otherwise, firm 2 would wait and invest as the follower at Y_{2F}.

4   Equilibria Analysis

In this section, we mainly consider the impact of the level of operating cost asymmetry, κ, the level of first-mover advantage, γ, and the operating cost itself, C, on the equilibrium types following the method of Pawlina and Kort in [8]. However, it is noticeable that the first-mover advantage, as defined in (7), is related to the operating cost in our paper, which is different from [8] in which the first-mover advantage is independent of the investment cost.

First, we discuss the condition of the sequential equilibrium occurs. The sequential equilibrium occurs when firm 2 has no incentive to become the leader, i.e. its payoff as follower is always greater than its payoff as leader. Formally, we defined

$$\xi_2(Y) = L_2(Y) - F_2(Y) \tag{18}$$

Then, the condition of the sequential equilibrium occurs is that $\xi_2(Y) < 0$, for $Y \in [Y(0), Y_{2F}]$. Therefore, we are interested in finding a pair (Y^*, κ^*) that satisfies the following system of equations

$$\begin{cases} \xi_2(Y^*, \kappa^*) = 0 \\ \partial \xi_2(Y, \kappa^*)/\partial Y|_{Y=Y^*} = 0 \end{cases} \tag{19}$$

In other words, we are interested in a point (Y^*, κ^*) in which firm 2's leader function is tangent to the follower function. After substituting (8) and (12) into (18), and eliminating Y^*, we can obtain κ^*. As in other things equal, κ^* is the function of γ, i.e. $\kappa^* = f(\gamma)$, which separates the plate $\kappa - \gamma$ into two regions:

preemptive equilibrium and sequential equilibrium region (see Fig.1). For $\kappa < \kappa^*$, firm 1 needs to take into account possible preemption by firm2, whereas $\kappa \geq \kappa^*$ implies that two firms always invest sequentially at their optimal thresholds.

Next, we concentrate on determining the region in which the simultaneous equilibrium occurs. In order to do so, let us define

$$\xi_1(Y) = L_1(Y) - S_1(Y) \tag{20}$$

Firm 1 prefers simultaneous investment unless for some $Y(t)$ its leader payoff, $L_1(Y)$, exceeds the optimal simultaneous investment payoff, $S_1(Y)$. That is, simultaneous equilibrium occurs only if $\xi_1(Y) > 0$, for all $Y \in [Y_{1P}, Y_{2F}]$. Therefore, we are interested in finding a pair (Y^{**}, κ^{**}) that satisfies the following system of equations

$$\begin{cases} \xi_1(Y^{**}, \kappa^{**}) = 0 \\ \partial\xi_1(Y, \kappa^{**})/\partial Y|_{Y=Y^{**}} = 0 \end{cases} \tag{21}$$

In other words, we are interested in a point (Y^{**}, κ^{**}) in which firm 1's simultaneous investment function is tangent to its leader function. After substituting (12) and (15) into (20), and eliminating Y^{**}, we can obtain κ^{**}. In the same way as before, we know κ^{**} is the function of γ, i.e. $\kappa^{**} = f(\gamma)$, which separates the plate $\kappa - \gamma$ into two regions: simultaneous equilibrium and preemptive/sequential equilibrium region (see Fig.1). For $\kappa < \kappa^{**}$, the resulting equilibrium is of the simultaneous investment type, whereas for $\kappa \geq \kappa^{**}$ the sequential/preemptive investment equilibrium occurs. This implies that for a relatively high degree of asymmetry between firms, as in other things being equal, simultaneous investment is not optimal and either sequential or preemption equilibrium occurs.

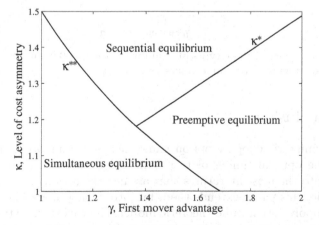

Fig. 1. The regions of three types of equilibria for $\alpha = 0.02$, $\sigma = 0.2$, $r = 0.05$, $D(0,1) = 0.25$, $D(0,0) = 0.5$, $D(1,1) = 1$, and $c = 0.25$

We present an illustration of when the resulting equilibria occur in a two-dimensional graph. In Fig.1 we depict the investment strategies as a function of the first-mover advantage, γ, and the investment cost asymmetry, κ. As it show, when the operating cost asymmetry is relatively small and there is no significant first-mover advantage, the firms invest simultaneously (a triangular area in the south-west). When the first-mover advantage becomes significant, firm 1 prefers being the leader to investing simultaneously. This results in the preemptive equilibrium (area in the south-east). Finally, if the asymmetry between firms is significant, the firms invest sequentially and firm 1 can act as a monopolist.

Finally, we consider the impact of the operating cost itself on the three types of equilibria. From (7), we know that the level of the first-mover advantage, γ, decreases with the operating cost, C. So the lower the operating cost is, the more the incentive to become the leader has. As Fig. 2 show, as in other things equal, with the operating cost decreasing, the possibility of simultaneous equilibrium and sequential equilibrium decreases, but that of preemptive equilibrium increases.

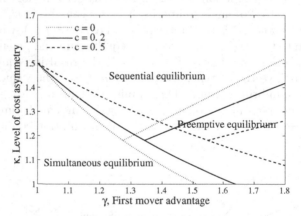

Fig. 2. The regions of three types of equilibria for $c = 0$, 0.2, and 0.5 respectively, and other parameters have the same values as in Fig.1

5   Conclusion

In an asymmetric duopoly option game-theoretic framework, this paper investigates the optimal timing of two firms' R&D investment under uncertainty. Contrary to the most literatures that assume the operating cost is zero and/or symmetric, this paper extends them by introducing operating cost asymmetry into a duopoly option game-theoretic model. There exist three types of equilibria that can occur in the R&D investment decision, i.e. the preemptive, sequential and simultaneous equilibrium. The occurrence of a particular type of equilibrium is determined by the firms' relative payoff, which mainly depends on the level of operating cost asymmetry and first-mover advantage and the operating cost itself. We show that when the cost asymmetry and the first-mover advantage among firms are relatively small, two firms invest simultaneously. When the first-mover advantage is significant, the low-cost firm preempts the high-cost firm. In the situation where the asymmetry between firms becomes large enough, two firms invest sequentially.

We also show that the lower the operating cost is, the more the incentive to become the leader has. This result has positive implication for Chinese firm who competes with its foreign rival in the same market. The Chinese firm has lower operating cost due to the relative cheap human resource, or at least its operating cost disadvantage is lower than its investment cost disadvantage in comparison with its foreign rival, who usual invests much more in R&D than Chinese firm does. So, it is possible for Chinese firms to compete with strong foreign multinational firms and get a leading position.

For sake of space we don't give our all numerical analysis results. The model could be extended in a number of ways. First, we can consider the situation where both the investment cost and operating cost are asymmetric. Second, we can study the impact of the volatility on equilibrium. Finally, we can analyze the situation where the information about cost is asymmetric, et al.

References

1. McDonald, R., Siegel, D. 1986. The Value of Waiting to Invest. Quarterly Journal of Economics, 101(4): 707-727.
2. Smets, F. 1991. Exporting versus FDI: The Effect of Uncertainty, Irreversibilities and Strategic Interactions, Working Paper: Yale University.
3. Dixit, A., Pindyck, R. 1994. Investment Under Uncertainty. Princeton: Princeton University Press.
4. Huisman, K., Kort, P. 1999. Effects of Strategic Interactions on the Option Value of Waiting, Working Paper: Tilburg University.
5. Weeds, H. 2002. Strategic delay in a real options model of R&D competition. Review of Economic Studies, 69(3): 729-747.
6. Huisman, K., Kort, P. 2003. Strategic investment in technological innovations. European Journal of Operational Research, 144(1): 209-223.
7. Huisman, K. 2001. Technology Investment: A Game Theoretic Real Options Approach. Boston: Kluwer Academic Publishers.
8. Pawlina, G., Kort, P. 2002. Real Options in an Asymmetric Duopoly: Who Benefits from Your Competitive Disadvantage?, Working Paper: Tilburg University.
9. Pisano, G., Wheelwright, S. 1995. The new logic of high-tech R&D. Harvard Business Review: 93-105.

Improvement of Genetic Algorithm and Its Application in Optimization of Fuzzy Traffic Control Algorithm

Jian Qiao, Huiyu Xuan, and Jinhu Jiang

School of Management, Xi'an Jiaotong University,
P. O. Box 1669, 28 N. West Xian Ning Road, Xi'an 710049, China
jian.qiao@163.com, {xuanhy,jiangjinhu}@mail.xjtu.edu.cn

Abstract. For the complex and time-varying traffic flow, single-strategy based fuzzy traffic control algorithms are not very ideal. In order to further improve the capacity of isolated intersection, we propose a multi-strategy fuzzy control algorithm to adapt to the variation of urban traffic flow, and then optimize its control rules and membership functions by using improved genetic algorithm. The simulation result shows that compared with traditional genetic algorithm, the efficiency of improved genetic algorithm is higher, and its performance is more stable. The multi-strategy fuzzy control model possesses the stronger self-adaptive competence and performance.

1 Introduction

For the stronger self-adaptability, fuzzy technology has been successfully applied in researching traffic control problem in recent years [1][2][3]. Nevertheless, aiming at the complex and time-varying traffic flow, if only one suit of control rules and membership functions is used in fuzzy control system, then satisfactory control effect can be acquired only under some special traffic conditions. In order to further improve fuzzy control system's self-adaptability and stability, a multi-strategy fuzzy control algorithm is proposed based on single-strategy fuzzy control algorithm which has been published in literature [4].

In fuzzy control system, control rules and membership functions are usually determined according to expert's experience. However, it will be very difficult to determine control rules and membership functions of multi-strategy fuzzy control algorithm in this way. Therefore, to design an optimization method that used to determine control rules and membership functions is very necessary. Genetic algorithm (GA) has been testified repeatedly being an efficient and stable global combined optimization technology. In this paper, a series of improvements on GA are carried out, and then the improved GA is used to optimize control rules and membership functions of multi-strategy fuzzy control algorithm. Simulation result shows that both in performance and efficiency, the improved GA is superior to the traditional GA, and the control effect of multi-strategy fuzzy control algorithm is better than that of single-strategy one.

N. Megiddo, Y. Xu, and B. Zhu (Eds.): AAIM 2005, LNCS 3521, pp. 132–141, 2005.

2   Multi-strategy Fuzzy Control Algorithm

2.1   Basic Structure of Fuzzy Control Model

Suppose loop detectors are embedded in stop-line and upstream-line of each entry lane of intersection respectively to detect arrival and go-away information of vehicles. The sequence number set of lane is $L = \{1, 2, ..., m\}$, the sequence number set of phase is $P = \{1, 2, ..., n\}$. Several concepts appeared in the following text has been defined in literature [4]. Subscript $i \in P$ denotes sequence number of phase, and subscript $j \in L$ denotes sequence number of lane in all variables. Let t denote time, its unit is second.

We have ever pointed out in literature [4], queue and average waiting time (AWT) are all the reflection of traffic demands of entry lanes, but anyone of them can not fully delegate the lane's traffic demands. Correlation and difference exist simultaneously between them, and anyone can not replace another one. They together reflect the actual traffic demand. For this reason, we propose a new concept—pass through need degree (PTND), and design two measurement indexes—actual pass through need degree (APTND) $n_{ij}(t)$ and righted pass through need degree (RPTND) $n_{ij}^r(t)$ to measure the actual traffic demand. The former is determined by actual queue $q_{ij}(t)$ and AWT $\bar{w}_{ij}(t)$, it is taken as the criterion of allocating the green time; and the later is determined by righted queue $q_{ij}^r(t)$ and $\bar{w}_{ij}(t)$, it is taken as the criterion of allocating the right-of-way. The reason of allocating right-of-way and green time respectively according to $n_{ij}^r(t)$ and $n_{ij}(t)$ is to ensure the rational allocation of them. $n_{ij}^r(t)$ is the embodiment of equity, and $n_{ij}(t)$ is the embodiment of efficiency.

Persons interested in detail of the control process see literature [4]. There are two suits of input/output variable combinations in the model. The first one's input variable combination is $(q_{ij}(t), \bar{w}_{ij}(t))$, its output variable is $n_{ij}(t)$; the second one's input variable combination is $(q_{ij}^r(t), \bar{w}_{ij}(t))$, its output variable is $n_{ij}^r(t)$. $n_{ij}(t)$ and $n_{ij}^r(t)$ is computed simultaneously, and the right-of-way and the green time are determined according to them respectively, so the model can be regarded as two parallel control systems. They share one suit of control rules and membership functions. In literature [4], there are $5 \times 5 = 25$ control rules in fuzzy control system. The number of control rules is one of key factors to influence the performance of control system. In order to further improve the system's real-time control competence, the model is modified as followings: the same three fuzzy subsets are defined for four input variables, and the same five fuzzy subsets are defined for two output variables. They are {S(Short), M(Medium), L(Long)} and {VL(Very Low), L(Low), M(Medium), H(High), VH(Very High)} respectively. Therefore, the number of operable control rules are $3 \times 3 = 9$. Mamdani's inference algorithm is used in fuzzy inference, and the center of gravity method is used to do defuzzification [5].

2.2   Multi-strategy Fuzzy Control

Although single-strategy fuzzy control technology possesses the competence of adapting to variation of the environment to a certain extent, when its control rules and membership functions are fixed, its self-adaptive competence is also limited in a special range, in other words, aiming at random, time-varying and non-linear traffic flow, if only one suit of control strategy (that means only one suit of control rules and membership functions) is used, no matter how to tune it, can we only acquire the most satisfactory control effect in a special traffic flow status, control effects of other cases may be not satisfactory comparatively. For this reason, it can be forecasted only if different control strategies is adopted in different traffic flow statuses, can we acquire the most satisfactory control effect in a larger range.

Although the amount of traffic flow statues are theoretically infinite, there does exist regular pattern, for example, in a working day, the density of traffic flow in the time of on and off duty is clearly large than that of other time; in daytime is clearly large than in nighttime, etc., therefore, although it is impossible to adopt different strategies in all different traffic flow statuses, can we select control strategies of key time points by using which the most satisfactory control effect can be acquired. The so-called key time point implies such a time point t, the density of traffic flow of partial or all directions exist extrema or larger changes in time range $[t - \Delta t, \ t + \Delta t]$. The more the key time points are set, the better the control effect can be acquired. That means we can select several key time points beforehand according to historical statistical traffic flow data of the intersection and take the combination of average arrival rate (AAR) of these points as key combinations, then optimize corresponding control rules and membership functions for acquiring the most satisfactory control effect. The optimal control effect can be approached ceaselessly by increasing the number of key combinations of average arrival rates (KCAARs) ceaselessly until the simulation result is satisfactory.

3   Improved Genetic Algorithm

Here, traditional GA is improved for optimizing fuzzy control rules and membership functions. Because the same fuzzy subsets of input/output variables, control rules and membership functions are shared by two groups of controls, so the following optimization method aims at them simultaneously.

3.1   Encoding of Control Rules

Let $\{q_i|q_i = i; i = 1, 2, 3\}$, $\{w_j|w_j = j; j = 1, 2, 3\}$ and $\{n_k|n_k = k; k = 1, 2, ..., 5\}$ denote queue, AWT and PTND respectively. Control rules are encoded with integer encoding method. A candidate solution is expressed in form of a chromosome, so a suit of control rules can be delegated by a chromosome, each gene on it delegates a control rule, and its value range is $\{n_k|n_k = k; k = 1, 2, ..., 5\}$. Suppose one gene of the chromosome is g_l, then its value range can also be denoted by set $\{g_l|g_l = n_k, l = (i - 1) \times 3 + j; n_k = k; i, j = 1, 2, 3; k = 1, 2, ..., 5\}$, the letter l denotes the location of gene on the chromosome, in other words, it is the sequence number of delegated control rule, the letter i and j denote the subscript of element queue and element AWT respectively, by which the control rule is constructed. For example, a control rule is "*IF q is M AND w is L, THEN n is H*", therefore, $i = 2$, $j = 3$, $l = (2-1) \times 3 + 3 = 6$, $n_k = k = 4$, that means this control rule is constructed by second element of queue set, third element of AWT set, and fourth element of PTND set, the location of gene on the chromosome is 6, so the sequence number of this gene is 6. Because the amount of control rules is 9, therefore the length of a chromosome is 9.

3.2   Encoding of Membership Functions

We take trapezoid as the basic geometric form of membership functions, therefore, its shape control parameters can be denoted by a quad $\langle a, b, c, d \rangle$ as showed in Fig 1. When $b = c$, the shape of membership function will transform into triangle, that means two geometric forms of membership functions are included in search space of candidate solutions, it is favorable for searching better candidate solution.

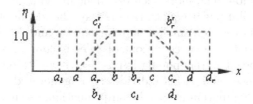

Fig. 1. Parameterized representation of membership function

The shape control parameters of membership functions of queue q, AWT w and PTND n can be denoted as followings respectively: $\langle a_i^q, b_i^q, c_i^q, d_i^q \rangle, i = 1, 2, 3$, $\langle a_i^w, b_i^w, c_i^w, d_i^w \rangle, i = 1, 2, 3$ and $\langle a_i^n, b_i^n, c_i^n, d_i^n \rangle, i = 1, 2, ..., 5$. Thus, there are 44 genes on a chromosome of membership functions, and it can be expressed as following:

$$a_1^q b_1^q c_1^q d_1^q ... a_3^q b_3^q c_3^q d_3^q a_1^w b_1^w c_1^w d_1^w ... a_3^w b_3^w c_3^w d_3^w a_1^n b_1^n c_1^n d_1^n ... a_5^n b_5^n c_5^n d_5^n. \tag{1}$$

The performance of traditional binary encoding method of GA is poor when applied to multidimensional and high-precision problems [6]. For this reason, we adopt the real value encoding method, in which a chromosome is constructed by a one-dimensional float-point array, and each gene is stored as an element of the array respectively. In this way, not only the search space is smaller and calculation precision is high enough, but also the size of search space is independent of calculation precision, the performance of algorithm can be improved effectively. Herrera et al. had ever given the following tuning domains of shape control parameters of membership functions [7]:

$$a \in [a_l, a_r] = [a - (b-a)/2, a + (b-a)/2], \tag{2}$$

$$b \in [b_l, b_r] = [b - (b - a)/2, b + (c - b)/2], \tag{3}$$

$$c \in [c_l, c_r] = [c - (c - b)/2, c + (d - c)/2], \tag{4}$$

$$d \in [d_l, d_r] = [d - (d - c)/2, d + (d - c)/2]. \tag{5}$$

Formula (3) and (4) indicate that the triangular membership functions can be formed only when $b = b_r$ and $c = c_l$ ($b_r = c_l$), so the search space is largely shrunk. In order to search better result, the tuning domains of b and c are extended in this paper as followings:

$$b \in [b_l, b'_r] = [b - (b - a)/2, c + (d - c)/2], \tag{6}$$

$$c \in [c'_l, c_r] = [b - (b - a)/2, c + (d - c)/2]. \tag{7}$$

3.3   Generation of Initial Population

Generally, the initial population of GA is generated randomly, so the individual's quality is difficultly ensured. If some excellent individuals can be selected into initial population, it equals that several evolutions have been finished before formal optimization. The more such excellent individuals are selected into, the more finished evolutions it equals. In this way, the ideal optimization result can also be acquired even in a smaller population size and smaller generations.

Because the solution space of control rules is comparatively smaller, so the first chromosome of initial population is set to be "123234345" according to expert's experience, while other individuals are generated randomly. The idea to generate initial population of membership functions is that suppose population size is s, n uniformly distributed points are selected respectively in domains (include two end points) that given by formula (2) to (5), so the number of these points is $n \times 44$. Partial selected combination results of these points and s randomly generated individuals are taken as candidates of initial population, and then s excellent individuals are selected as initial population according to everyone's fitness degree that is figured out by using simulation method. The size of candidates is too large if all combinations of $n \times 44$ points are taken as candidates, so n^2 representatives from them are selected as partial candidates, they can be expressed as followings:

$$
\begin{aligned}
&a^q_{1_i} b^q_{1_j} c^q_{1_{n-j}} d^q_{1_{n-i}} ... a^q_{3_i} b^q_{3_j} c^q_{3_{n-j}} d^q_{3_{n-i}} a^w_{1_i} b^w_{1_j} c^w_{1_{n-j}} d^w_{1_{n-i}} ... \\
&a^w_{3_i} b^w_{3_j} c^w_{3_{n-j}} d^w_{3_{n-i}} a^n_{1_i} b^n_{1_j} c^n_{1_{n-j}} d^n_{1_{n-i}} ... a^n_{5_i} b^n_{5_j} c^n_{5_{n-j}} d^n_{5_{n-i}}, \quad i, j = 0, 1, ..., n-1.
\end{aligned}
\tag{8}
$$

The mean of each symbol in formula (8) is same with that of formula (1). The size of candidates is $n^2 + s$. The bigger the value of n is or the more the candidates are selected from all combinations of $n \times 44$ points, the better the individual's quality is and the more the excellent individual's quantities are in initial population, while at the same time, the longer the filtering process time of initial population is, so n should be appropriate.

For only a small number of chromosomes of initial population are generated according to expert's experience, most of them are randomly generated, hence, not only the GA's original property and practicality may be ensured, but also its performance is improved.

3.4   Crossover and Mutation Operation

Crossover and mutation operation of control rules chromosomes are similar to that of simple GA that binary encoding method is used. In theory, the cases that the values of the first, forth, and seventh genes of a chromosome are 4 or 5, 5 and 5 respectively are irrational, because when a lane's queue and/or AWT are very short, its PTND is impossibly very large. Plenty of simulation experiments also support this conclusion. Once this case occurred in the processes of crossover and mutation operation, we will replace such chromosomes with the best one.

In order to saving the computation cost and enhancing the performance, max-min-arithmetical crossover algorithm proposed by Herrera et al. [7] is improved and then used for crossover operation of membership functions chromosomes. Suppose $C_v^t = (c_{v1}^t, ..., c_{vk}^t, ..., c_{vK}^t)$ and $C_w^t = (c_{w1}^t, ..., c_{wk}^t, ..., c_{wK}^t)$ are two selected chromosomes in tth generation of chromosomes, they will generate two offspring after crossover operation between them, the general items of them are:

$$c_{vk}^{t+1} = \begin{cases} ac_{wk}^t + (1-a)c_{vk}^t, & 0.00 \le p_v \le 0.25, \\ ac_{vk}^t + (1-a)c_{wk}^t, & 0.25 < p_v \le 0.50, \\ \min\{c_{wk}^t, c_{vk}^t\}, & 0.50 < p_v \le 0.75, \\ \max\{c_{wk}^t, c_{vk}^t\}, & 0.75 < p_v \le 1.00. \end{cases} \tag{9}$$

$$c_{wk}^{t+1} = \begin{cases} ac_{wk}^t + (1-a)c_{vk}^t, & 0.00 \le p_w \le 0.25, \\ ac_{vk}^t + (1-a)c_{wk}^t, & 0.25 < p_w \le 0.50, \\ \min\{c_{wk}^t, c_{vk}^t\}, & 0.50 < p_w \le 0.75, \\ \max\{c_{wk}^t, c_{vk}^t\}, & 0.75 < p_w \le 1.00. \end{cases} \tag{10}$$

Where, p_v and p_w are random numbers in the range of $[0, 1]$.

Non-uniform mutation algorithm proposed by Michalewicz [6] is used for mutation operation. Suppose $C_v^t = (c_{v1}, ..., c_{vk}, ..., c_{vK})$ is a chromosome of tth generation of chromosomes, and the element $c_{vk} \in [c_k^l, c_k^r]$ is selected to carry out mutation operation, the result is $C_v^{t+1} = (c_{v1}, ..., c_{vk}', ..., c_{vK})$, where,

$$c_{vk}' = \begin{cases} c_{vk} + \Delta(t, c_{vk}^r - c_{vk}), & b = 0, \\ c_{vk} + \Delta(t, c_{vk} - c_{vk}^l), & b = 1. \end{cases} \tag{11}$$

b is a random number of 0 or 1. The function $\Delta(t, y)$ is calculated as followings:

$$\Delta(t, y) = y(1 - r^{(1-t/T)^h}). \tag{12}$$

Where r is a random number in the range of $[0, 1]$, T is the maximum number of generations and h is a given constant, $\Delta(t, y)$ will return a value in the range of $[0, y]$, so the probability of $\Delta(t, y)$ approaches to 0 when t increases. This kind of property causes the operator to make a uniform search in initial space when t is smaller but very locally at later stages.

3.5   Amendment of Shape Parameters of Membership Function

Because the domain of b is same with that of c, so the irrational result $b > c$ may occur in the process of randomly generating chromosome and successive genetic operations. In this case, the result is amended as followings:

$$b' = \begin{cases} \min\{b,c\}, \ p = 0, \\ (b+c)/2, \ p = 1. \end{cases} \tag{13}$$

$$c' = \begin{cases} \max\{b,c\}, \ p = 0, \\ (b+c)/2, \ \ p = 1. \end{cases} \tag{14}$$

Where, p is a random number of 0 or 1. The reason of designing such an amendment method is to increase the probability of forming the triangular membership functions.

3.6   Evolution Strategy and Improvement of Computation Efficiency

We will adopt simulation method to compute average delay, and then take its reciprocal as the fitness degree of chromosomes. The optimization process of chromosomes includes two stages. In first stage, control rules chromosomes are optimized based on membership functions chromosomes that have been generated according to expert's experience, and then membership functions chromosomes are optimized in second stage according to control rules chromosomes that have been optimized in first stage.

For the characteristic of probability , there always be some chromosomes don't participate in crossover and mutation operation in the process of genetic operation. In order to reduce the computation costs, in the process of selecting operation, only the fitness degrees of new generated chromosomes will be computed, such computations will not be repeated for those old ones. In this way, computation time can be effectively saved. The less the crossover and mutation probabilities are, the more computation time is saved.

Because most of the improvements are not problem-oriented, so the improved GA is all-purpose to a great extent.

4   Simulation Result

All our simulation experiments are carried out by using a personal computer, its main hardware configurations are Intel Pentium III processor and 512MB memory, and operation system is Windows Server 2003. The parameters of simulation environment are as followings. The maximal queue length of each lane is 20. The green interval has three timing parameters, namely, lost time, minimum duration and maximum duration. Minimum duration is 12s, maximum duration is 60s, and lost time is 3s. Amber period is 3s, and green time is extended by 3s each time when needed. The arrival rate of upstream vehicles submits to Poisson distribution, left-turn rate and right-turn rate of each approach is 20%. The

saturation flow rates of through, left-turn and right-turn of each approach are
1v/s, 0.8v/s and 1v/s respectively. Phase arrangement is showed in Fig. 2.

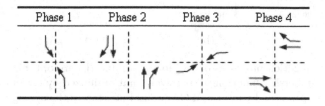

Fig. 2. Phase arrangement

The fixed-cycle control strategy is that the green time of through and right-
turn are all 60s, and that of left-turn is 20s. The vehicle-actuated control strategy
is that minimum green time is 12s. When there are vehicles arrive at within the
last one second, green time of current green phase is extended by 3s, otherwise,
switch right-of-way to the next phase. Maximum green time is 60s. KCAARs
of all directions listed in table 1 are adopted in all simulation experiments. The
simulation is carried out for 10 times in each KCAAR, each time length is 7200s,
and the average result of them is taken as final result.

Table 1. KCAARs of all directions

No. of KCAAR		1	2	3	4	5	6	7	8	9	10
AAR	East/West	400	400	400	400	800	800	800	1200	1200	1600
(veh./h)	North/South	400	800	1200	1600	800	1200	1600	1200	1600	1600

For the comparison, traditional GA and improved GA are respectively used to
optimize the single-strategy (SS) and multi-strategy (MS) fuzzy control model in
this paper. Here, the traditional GA's crossover operation is replaced with that of
improved GA's, otherwise, the computation time is too long to endure. Their parameters are set as followings: population size=30, generation size=30, crossover
probability=0.85, mutation probability=0.005, crossover parameter $a = 0.35$,
mutation parameter $h = 0.5$.

The optimization time of SS model that optimized by traditional and improved GA are 2.50h and 1.99h respectively, the latter improves by 20.4%. In
the optimization of MS model, the costs of time are 2.36h and 2.15h respectively, the latter improves by 8.9%. Fig. 3 shows the optimization processes of
SS Model, from which we can find that the performance of improved GA is obviously more excellent than that of traditional GA, especially in initial stages.
The optimization processes of MS model are also similar with that of SS model.

The simulation results of average delay of three fuzzy control models that
optimized by improved GA compared with two traditional methods are showed in

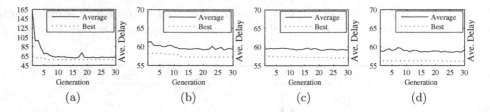

(a)   (b)   (c)   (d)

Fig. 3. Optimization processes of SS Model. (a) and (b) are optimization processes of rules and membership functions respectively that optimized by traditional GA. (c) and (d) are that optimized by improved GA. Solid line denotes the average value of average delays of population. Dotted line denotes the average delay of the best chromosome individual

Fig. 4. The average delay of fixed-cycle control is the longest, and when volumes of traffic flows of partial or all directions are larger, its average delay increases rapidly, this indicates that its performance worsens remarkably. The curve of vehicle-actuated control is comparatively lower and smoother. The curves of fuzzy control models are not only very smooth but also always in lowest location. It indicates their performances are not only best but also very stable.

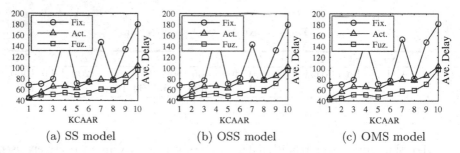

(a) SS model   (b) OSS model   (c) OMS model

Fig. 4. Average delay of different fuzzy control models and traditional control methods

The average delay improvements of three fuzzy control models compared with two traditional methods are listed in table 2. The average results show that SS model is superior to traditional methods. The performance of optimized single-strategy (OSS) model is superior to SS model, and optimized multi-strategy (OMS) model is superior to OSS model. Comparing the simulation results of 10 groups of KCAARs one by one, we can find that the performances of OSS model are almost all superior to SS model, and the most of OMS model is superior to OSS model. The cause of this result is that the optimization of OSS model seeks for optimal average result, so it is difficult to avoid some non-ideal results. Whereas the optimization of OMS model seeks for optimal individual results by applying different control strategies to different individuals, and therefore its average result and individual result are all more satisfactory. That means OMS model is superior to OSS model on the whole. Summarizing the aforementioned, after optimized by our improved GA, the OMS model has stronger self adaptive competence and stability, the average delay can be decreased and the capacity of intersection be improved effectively by using this model.

Table 2. Improvement(%) of average delay

No. of KCAAR		1	2	3	4	5	6	7	8	9	10	Ave.
SS	with Fix.	34.92	28.96	35.56	66.64	29.02	27.63	58.77	24.19	45.33	46.80	39.78
model	with Act.	1.89	9.33	22.40	19.16	19.37	27.12	22.68	22.86	14.14	7.82	16.68
OSS	with Fix.	35.27	31.97	33.25	67.74	32.10	34.69	58.89	25.67	45.85	46.79	41.22
model	with Act.	2.70	16.21	22.58	20.67	23.39	28.23	24.64	24.16	16.32	6.76	18.57
OMS	with Fix.	37.83	36.10	34.53	65.85	33.27	30.83	62.02	26.63	51.89	46.23	42.51
model	with Act.	6.85	21.47	22.68	22.41	23.43	27.32	26.76	23.96	18.26	5.42	19.85

5 Conclusion

Aiming at complex and time-varying urban traffic flow, only when control strategy can changes with it, can the most satisfactory control effect be acquired. For this reason, we propose a multi-strategy fuzzy traffic control model, and optimize control rules and membership functions by using our improved GA. The simulation result shows that improved GA is more efficient and stable than traditional GA, and compared with traditional control methods, OMS model possesses the stronger self-adaptive competence and stability. How to apply them in coordination control of multi-intersection will be researched in the future.

References

1. Pappis C.P., Mamdani E.H.: A Fuzzy Logic Controller for a Traffic Junction. IEEE Trans on Systems, Man, and Cybernetics, 1977, SMC-7(10): 707–717
2. Chou Chih-Hsun, Teng Jen-Chao: A fuzzy controller for traffic junction signals. Information Sciences, 2002, 143: 73–97
3. Niittymaki J., Pursula M.: Signal Control Using fuzzy logic. Fuzzy Sets and Systems, 2000, 116: 11–22
4. QIAO Jian, XUAN Hui-yu: A Need Degree-based Isolated Intersection Fuzzy Control Algorithm. Systems Engineering, 2004, 22(10): 59–64
5. Pedrycz W.: Fuzzy control and Fuzzy Systems, second extended edition. New York: Wiley, 1993
6. Michalewicz, Z.: Genetic Algorithms + Data Structures = Evolution Programs. 3rd edn. Springer-Verlag, Berlin Heidelberg New York (1996)
7. Herrera F., Lozano M., Verdegay J.L.: Tuning fuzzy logic controllers by genetic algorithms. International Journal of Approximate Reasoning, 1995, 12: 299–315

Facility Location in a Global View

Wenqiang Dai, Peng Xiao, and Ke Jiang

School of Management, Xi'an Jiaotong University,
Xi'an, 710049, P.R. China
wqdai@mail.xjtu.edu.cn

Abstract. Facility Location Problems have always been studied with
the assumption that the environment in the network is static and does
not change over time. In practice, however, the environment is usually
dynamic and we must consider the facility location in a global view.
In this paper, we impose the following additional constraints on input
facilities: the total number of facilities to be placed is not known in
advance and a facility cannot be removed once it is placed. We solve this
problem by presenting an algorithm to find a facility permutation such
that any prefix of the permutation of facilities is near-optimal over any
other facility subset.

1 Introduction

Variants of the facility location problem (FLP) have been studied extensively
in operation research and management science literatures [14, 3, 1]. The model
in typical theoretical work has addressed situations in which we want to locate
facility in a network and optimize an objective function in the static environment.
In practice, however, the environment is dynamic in many cases. For example, in
a commercial network, we do not know the exact number of facility in advance,
our business plan is to start with one facility, and then to gradually add a new
facility but never to remove a previously established facility. The same is true
for locating the public facility in community network, such as schools, hospitals,
etc.

Under the above considerations, the facilities must be constructed in the
global view. That is, we locate facilities in a way that their cost is optimal or
near-optimal over any other non-empty facility subset and, when the number of
needed facilities increases, no former facility can be removed. It is worth noting
that a typical facility location problem with these two constraints can be viewed
as the facility permutation problem whose goal is to specify the facility order that
minimizes the maximum ratio between the cost of any prefix of the permutation
and that of an any non-empty subset of all facilities.

Of the facility location models, our problem most closely resembles Undesirable
Facility Location Problem [13], the Online Median Problem [11] and the Incremental
Facility Location Problem [12]. All of these problems, as well as our problem, used
exactly the same assumption that the number of facilities is not known in advance

N. Megiddo, Y. Xu, and B. Zhu (Eds.): AAIM 2005, LNCS 3521, pp. 142–150, 2005.

and once the facility is placed, it can't be removed, but their objective formulations are different with ours. See the section 2 for some more discussions.

In this paper, we solve the facility location problem with above considerations by defining a model of locating facilities to optimize the location cost over any other cost of non-empty subset of all facilities, which we call strength facility location problem (SFLP). Although NP-hardness of SFLP has not been proven yet, finding an optimal solution for the problem seems to be difficult, in view of the NP-completeness of the relaxation version: classical Uncapacitated Facility Location Problem. This paper is concerned with approximation algorithm for this problem. Given a minimization problem, an algorithm is said to be a (polynomial) $r - approximation$ algorithm, if for any instance of the problem, the algorithm runs in polynomial time and produces a solution that has a cost at most $r \geq 1$ times the minimal cost, where r is called the *approximation ratio* of the algorithm.

Though not stated specifically, Mettu and Plaxton [11] first presented a 3-approximation algorithm for SFLP. But their solution does not generate the permutation of facilities. Based on their algorithm, this paper presents a group of simple and deterministic approximation algorithms for constructing some solutions of SFLP which is similar to, in simplicity and efficiency, the standard heuristics for facility location. Each approximation algorithm and the result are only linked to a single parameter, and for the different parameter, the solutions obtained by each algorithm are *nested*, namely, one of solution is the subset of another one, thus we obtain a permutation of all facilities by varying the parameter.

The rest of this paper is organized as follows. Section 2 specifies the problem and describes related works. Section 3 presents the approximation algorithm and proves the approximation ratio. The final section, section 4, concludes the paper and describes future research.

2   Problem Description and Related Work

This section describes the formulation of SFLP and gives an overview of related work. We first discuss the basic facility location model, the metric uncapacitated facility location problem (UFLP), in which we are given a graph with nonnegative edge costs. As motivation [7], the nodes can be thought of as customers, the facilities as service centers, and the distance between a customer and a service center as the cost of serving the customer by that center. Furthermore, each customer (node) has a weight, corresponding to the amount of requests. So, the cost of serving a customer becomes the weight of the customer node times its distances from the closest service center. In addition, each node is assigned a constructive cost to represent the cost of building a service center at that node. The total cost we wish to minimize is total constructive cost of chosen facilities plus the cost of serving all of the customer requests by the chosen facilities. The objective of the UFLP is to choose k nodes (as the facilities) so as to minimize the cost, where k is a given facility number. Compared with this, the goal of

SFLP is to minimize the cost over the cost of any number of facilities. More precisely, the problem formulation of SFLP is as follows:

Problem Formulation: Fixed a set of points U, a distance function $d : U \times U \longrightarrow R^+$, a nonnegative weight function $w : U \longrightarrow R$ and nonnegative construtive function $f : U \longrightarrow R$. We assume that the distance that d is a metric, that is, d is nonnegative, symmetric, satisfies the triangle inequality, and $d(x, y) = 0$ iff $x = y$. We define the distance of a point x to a point set S is $d(x, S) = \min_{y \in S} d(x, y)$ and $|S|$ is the number of points in S. Let $n = |U|$ denote the number of total customers, and for any subset $S \subseteq U$, let its cost be $cost(S) = \sum_{x \in S} f(x) + \sum_{y \in U} d(y, S) w(y)$. In contrast to UFLP whose objective is to find a subset $S \subseteq U$ such that it meet $\min_{|S|=k} cost(S)$, where k is the given positive integer, $1 \le k \le n$, the objective of SFLP is to give a subset $S \subseteq U$ such that it meet that $\min_{1 \le k \le n} \min_{|S|=k} cost(S)$.

Related Work: Facility location has been the subject of a great deal of previous work[5, 15, 9, 8, 2, 6], and here we just describe some typical theoretical analysis. This problem is MAX-$\mathcal{SNP}$ Hard and the first constant approximation algorithm was given by Shmoys et.al [15]; the approximation ratio was later improved to 1.728 by Charikar and Gula[5] and to 1.528 by Sviridenko [16]. Now the best approximation ratio is 1.52 given by Mahdian, Ye and Zhang [10], and the negative result is that no polynomial-time algorithm can achieve an approximation ratio less than 1.463 unless **NP**⊆DTIME$[n^{O(\log \log n)}]$ [8]. The methods they have used are based on, such as linear Programming rounding (e.g. [15]), local search (e.g. [5]), and the primal-dual method (e.g. [9]) etc. The reader is referred to Mahdian et.al [10] for a detail discussion.

Current et.al [4] firstly considered the facility location problem when the number of facilities is uncertain. Unlike ours, they used the criteria of the minimization of expected opportunity loss and the minimization of maximum regret to make decision about what's the number of facilities and where we locate those. Their solution, unfortunately, does not also generate a permutation of facilities and thus does not meet the demand that no point can be removed when the number of facilities increases.

Recently, the Undesirable Facility Location Problem [13], the Online Median Problem [11] and the Incremental Facility Location Problem [12] most resemble our problem, but their objective formulations are different with ours. The Undesirable Facility Location Problem seeks a solution to maximize the minimum distance between facilities and the minimum distance between facilities and existing non-obnoxious facilities [13]. In the Online Median Problem, the goal is to determinate a permutation of facilities that minimizes the maximum ratio between the service cost of any prefix of the permutation and that of an optimal *offline same-size* configurations [11], and this problem uses the formulation $cost(S) = \sum_{y \in U} d(y, S) w(y)$ to compute the cost. And for the Incremental Facility Location Problem [12], the cost formulation is $cost(S) = |S| \sum_{x \in S} f(x) + \sum_{y \in U} d(y, S) w(y)$ and the goal is also minimal over the *same-size* configurations.

3   Algorithm and Approximation Ratio

3.1   Greedy Selection of Mettu and Plaxton [11]

For the new requirement of our problem, we first present the algorithm of Mettu and Plaxton [11], and will need to build upon it. Roundly speaking, their algorithm is to compute the "value" of each ball about every node in the metric space to start with, then sort them by increasing order, then greedily pick up the point if they separate sufficient large, and so on until all n points are examined. These points are taken as service facilities. If the distance function is a metric and the separation distance is more twice longer than the maximal radius of ball, the cost of resulting location is within a factor three of optimal. Implicitly, their idea came from the work of Jain and Vazirani [9].

The following definition were used in [11], but we rewrite them by our notation for easy use later.

Definition 1. *A ball A is a pair (x, r_x), where $x \in U$ is the center and r_x is the radius of the ball, which is a nonnegative real.*

Definition 2. *Given a ball $A = (x, r_x)$, we let **Points**(A_x) denote the set $\{y \in U | d(x, y) \leq r_x\}$ and always directly use A_x instead of $Point(A_x)$. For example, we write "$a \in A_x$" and "$A_x \cup A_y$" instead of "$a \in Points(A_x)$" and "$Points(A_x) \sqcup Points(A_y)$", respectively.*

Definition 3. *The **value** of a ball $A = (x, r_x)$, denoted value(A_x), is defined by*

$$value(A_x) = \sum\nolimits_{y \in A_x} (r_x - d(x, y)) \cdot w(y)$$

Definition 4. *For any ball $A = (x, r_x)$ and any nonnegative real c, we define **cA** as the ball (x, cr_x).*

3.2   Constructing a Permutation

We will stick with the same computing of "value" of each point in the metric space, but we will no longer greedily select the facility by the former separate distance. Note that the longer the separate distance between two facilities is, the less the number of obtained facilities is. Intuitively, we may only need the distance to separate longer than twice and thus it will enable us to generate less facilities. Thus if the obtained facilities are nested each other and the near-optimal property is still held for fixed constant distance, the permutation of facilities, which will meet the constraints of our problem, will be obtained. Following we will give a positive answer for these considerations.

For easy reference, and to illustrate our notation, we write integrally the algorithm. In the following algorithm and the later analysis, we always assume that the weight of each point is larger than zero for the sake of convenience.

Input: (U, d), f and w;

Output: A non-empty subset $Z_n \subseteq U$.

Algorithm:

Step 1. For each point x, determine an associated ball $A = (x, r_x)$ such that $value(A_x) = f(x)$.

Step 2. Sorting r_x for all $x \in U$ increasingly, denoting the index of x after sorting is $\phi(x)$.

Step 3. Let $B_i = (x_i, r_{x_i})$ denote the ball $A = (x, r_x)$ such that $\phi(x) = i$, $0 \le i < n$. Let $Z_0 = \emptyset$.

Step 4. For $i = 0$ to $n - 1$: If $Z_i \cap (1 + \alpha)B_i = \emptyset$ then let $Z_{i+1} = Z_i \cup \{x_i\}$; otherwise, let $Z_{i+1} = Z_i$.

Where $\alpha \ge 1$ is a parameter which need to be inputed.

Throughout the rest of the paper, we always let Z denote the result of our algorithm.

Remark 1. We have $x_0 \in Z$, so $Z \ne \emptyset$.

Remark 2. Note that if $\phi(x) = i$, we have $value(B_i) = value(A_{x_i}) = f(x_i)$.

Remark 3. Following the same manner as analysis given in [11], it can be easily obtained that the total time complexity of above algorithm is $O(n^2)$ time for any fix constant $\alpha \ge 1$.

3.3   Performance Guarantee

We now show a strong guarantee for the facility location induced by our algorithm with any fixed input $\alpha \ge 1$.

Lemma 1. *For any point $x \in U$, there exists a point $y \in Z$ such that $\phi(y) \le \phi(x)$ and $d(x, y) \le (1 + \alpha)r_x$.*

Proof. If $x \in Z$, we can choose $y = x$ and this lemma is proven. Following we assume $x \notin Z$.

The proof is by contradiction. Assume that $\forall y \in Z$ with $\phi(y) \le \phi(x)$, we have $d(x, y) > (1 + \alpha)r_x$, that is, $Z_{\phi(x)} \cap (1 + \alpha)B_{\phi(x)} = \emptyset$. Thus we have x belongs to Z according to step 4 and this is contradicted with $x \notin Z$.   □

Lemma 2. *Let $x, y \in Z$ and $x \ne y$, then $d(x, y) > (1 + \alpha)max\{r_x, r_y\}$.*

Proof. Without loss of generality we assume $\phi(y) < \phi(x)$, thus we have $r_y \le r_x$ and $Z_{\phi(y)+1} \subseteq Z_{\phi(x)}$ and so $y \in Z_{\phi(x)}$. On the other hand, due to $x \in Z$ we must get $Z_{\phi(x)} \cap (1 + \alpha)B_{\phi(x)} = \emptyset$. Thus we obtain $d(x, y) > (1 + \alpha)r_x = (1 + \alpha)max\{r_x, r_y\}$.   □

For any point x and any non-empty subset $Y \subseteq U$, let

$$charge(x, Y) = d(x, Y) + \sum\nolimits_{y \in Y} max\{0, r_y - d(x,y)\}$$

Lemma 3 ([11]). *For any non-empty subset $Y \subseteq U$, we have*

$$cost(Y) = \sum\nolimits_{x \in U} charge(x, Y) \cdot w(x)$$

Lemma 4. *Let $x \in U$ be a point, let Y be a non-empty subset of U, and let y belong to Y. If $d(x, y) = d(x, Y)$ then $charge(x, Y) \geq max\{r_y, d(x, y)\}$*

Proof. If $x \notin A_y$, then $d(x, y) > r_y$, we have $charge(x, Y) \geq d(x, y) > r_y$. Otherwise, we have $d(x, y) \leq r_y$, then we have $charge(x, Y) \geq d(x, y) + (r_y - d(x, y)) = r_y \geq d(x, y)$. The lemma is proven. $\square$

Lemma 5. *Let $x \in U$ and $z \in Z$. If $x \in A_z$, then $charge(x, Z) \leq r_z$.*

Proof. Firstly assuming $x \in Z$, by $z \in Z$ and Lemma 2, we have $d(x, z) \geq (1 + \alpha)r_z$, which is contradicted with $x \in A_z$, so $x \notin Z$.

Now we prove $\forall y \in Z, y \neq z, d(y, x) > r_y$. Assuming this is not true, that is, $\exists y^*$ such that $y* \neq z$ and $d(y^*, x) \leq r_{y^*}$, so

$$(1 + \alpha)max\{r_{y^*}, r_z\} \leq d(y^*, z) \leq d(y^*, x) + d(x, z) \leq r_{y^*} + r_z$$

this is contradicted by $\alpha \geq 1$.

According to discussion above, we have that

$$charge(x, Z) = d(x, Z) + (r_z - d(z, x)) + \sum_{y \in Z, y \neq z} max\{0, r_y - d(y, x)\}$$
$$= d(x, Z) + (r_z - d(x, z)) \leq d(x, z) + (r_z - d(x, z)) \leq r_z$$

where $d(x, z) \leq r_z$ by x belongs to A_z and the third inequality by $z \in Z$. $\square$

Lemma 6. *Let $x \in U$ and $z \in Z$. If $x \notin A_z$, then $charge(x, Z) \leq d(x, z)$.*

Proof. If $\exists y \in Z$ such that $x \in A_y$, that is, $d(x, y) \leq r_y$. By Lemma 2 and Lemma 5, we have $d(y, z) \geq (1 + \alpha)max\{r_y, r_z\}$ and $charge(x, Z) \leq r_y$, respectively. Then we have $d(x, z) \geq d(y, z) - d(x, y) > (1 + \alpha)r_y - r_y = \alpha r_y \geq \alpha \cdot charge(x, Z) \geq charge(x, Z)$, that is, $charge(x, Z) \leq d(x, z)$.

Otherwise, if $\forall y \in Z$, we have $x \notin A_y$, by the definition of $charge(x, Z)$ and $z \in Z$ we have $charge(x, Z) = d(x, Z) \leq d(x, z)$. $\square$

Lemma 7. *For any point $x \in U$ and non-empty subset Y, $charge(x, Z) \leq (2 + \alpha)charge(x, Y)$.*

Proof. Let y be some point in Y such that $d(x, y) = d(x, Y)$. By Lemma 1, there exists a point $z \in Z$ such that $\phi(z) \leq \phi(y)$ and $d(y, z) \leq (1 + \alpha)r_y$.

Now if $x \in A_z$, then $charge(x, Z) \leq r_z$ by Lemma 5. Thus by $\phi(z) \leq \phi(y)$, we have $r_z \leq r_y$, and since Lemma 4 implies $charge(x, Y) \geq r_y$, we obtain $charge(x, Z) \leq charge(x, Y)$.

However if $x \notin A_z$, then $charge(x, Z) \leq d(x, z)$ by Lemma 6. By triangular inequality we have $charge(x, Z) \leq d(x, y) + d(y, z) \leq [d(x, y) + (1 + \alpha)r_y]$. Moreover, by Lemma 4, we have $charge(x, Y) \geq max\{r_y, d(x, y)\}$. Then we have

$$\frac{charge(x, Z)}{charge(x, Y)} \leq \frac{d(x, y) + (1 + \alpha)r_y}{max\{r_y, d(x, y)\}} \leq 2 + \alpha$$

and the lemma is proven.   □

With above lemma 3 and lemma 7, we can then clinch the following main result.

Theorem 1. *For any non-empty subset Y of U, we have*

$$cost(Z) \leq (2 + \alpha)cost(Y)$$

that is, the approximation ratio of our algorithm is $(2 + \alpha)$.

The best approximation ratio of our algorithm, which is 3 for $\alpha = 1$, is equal to the approximation ratio of Mettu and Plaxton. Though our best ratio is not less than theirs, our algorithm can present a permutation of facilities with the property that no facility can be removed as the number of facilities increases by varying the parameter α, and it can be easily used in practice since practical manager can choose the proper number of facilities by the proper α. Moreover, the result of our algorithm presents an online fashion for the different α and the different number of facilities. This property can give the practical manager much room to freely add a new facility or to delete a facility from the exist facilities.

4   Summary and Future Work

For more practical considerations, this paper presents two primary constraints on input facilities: the total number of facilities to be placed is not known in advance and a facility cannot be removed once it is placed. We gave a new variant of classic facility location problem, the strength facility location problem, and presented a group of constant-factor approximation algorithms. These algorithms, which all take quadratic time in the worst case, produce the solutions such that each result is the subset of another one. Thus the permutation of facilities obtained by all of the results present the property that no point is removed when the number of facility increases.

There are still various open problems for the future research. For example, it is interesting to design and analyze some algorithms to improve the approximation ratio. Actually, according to the proof of Theorem 1, the condition $\alpha \geq 1$ is the main bottleneck for improving the approximation ratio. So how to relax this condition may be one of the possible directions.

Acknowledgements

This research is supported by NSF of China under Grants 10371094 and 70471035. The authors would like to thank the anonymous referees for their valuable comments.

References

1. P. K. Agarwal and M. Sharir, Efficient Algorithms for Geometric Optimization. *ACM Computing Surveys*, 30(4), pp. 412-458, 1998.
2. F.Chudak. Improved algorithms for uncapacitated facility location problem. *Proc. 5th Conference on Integer Programming and Combinatorial Optimization*, LNCS 1412, pp.180-194,1998.
3. J. Current, M. Daskin and D. Schilling, Discrete network location models. In Z. Drezner and H. W. Hammacher, editors, *Facility Location: Applications and Theory*, Springer, pp.81-118, 2002.
4. J. Current, S. Ratick and C.Revelle, Dynamic facility location when the total number of facilities is uncertain: a decision analysis approach, *European Journal of Operation Research*, 110(3), pp. 597-609, 1998.
5. M. Charikar and S. Guha. Improved combinatorial algorithms for facility location and k−median problems, *Proc. FOCS'99*, pp.378-388, 1999.
6. F. Chudak and D. B. Shmoys. Improved approximation algorithms for capacitated facility location problem. *Proc. 10th ACM-SIAM Symposium on Discrete Algorithms*, 1999.
7. R. Fleischer, M. J. Golin and Z. Yan, Online Maintenance of k−medians and k−covers on a line, *Proc. 9th Scandinavian Workshop on Algorithm Theory*, LNCS 3111, pp.102-113, 2004.
8. S. Guha and S. Khuller. Greedy strikes back: Improved facility location algorithms, *Proc. of the 9th ACM-SIAM Symposium on Discrete Algorithms*,1998, Also in *Journal of Algorithms*, 31, pp. 228-248,1999.
9. K. Jain and V. Vazirani. Primal-dual approximation algorithms for metric facility location and k−median problems. *Proc. FOCS'99*, 1999.
10. M. Mahdian, Y. Ye and J. Zhang, Improved approximation algorithms for metric facility location problems, *Proc. of the 5th International Workshop on Approximation Algorithms for Combinatorial Optimization (APPROX'02)*, pp.229-242, 2002.
11. R. R. Mettu and C. G. Plaxton. The online median problem. *Proc. FOCS'00*, pp. 339-348, 2000.
12. C.G.Plaxton. Approximation algorithms for hierarchical location problems. *Proc. STOC'03*, pp. 40-49,2003.
13. Z. Qin, Y. Xu and B. Zhu, On some optimization problems in obnoxious facility location, *Proc. COCOON'00*, LNCS 1858, pp.320-329, 2000.

14. D. B. Shmoys, Approximation algorithms for facility location problems. In K. Jansen and S. Khuller, editors, *Approximation Algorithms for Combinatorial Optimization*, LNCS 1913, pp.27-33, 2000
15. D. B. Shmoys, E. Tardos and K. Aardal. Approximation algorithms for facility location problems, *Proc. STOC'97*, pp. 265-274,1997
16. M. Sviridenko. An 1.528-approximation algorithm for the metric uncapacitated facility location problem. *Proc. of the 9th Conference on Integer Programming and Combinatorial Optimization*, 2002.

Existence and Uniqueness of Strong Solutions for Stochastic Age-Dependent Population

Qimin Zhang[1,2] and Congzhao Han[1]

[1] School of Electronic and Information Engineering Xi'an Jiaotong University,
Xi'an Shaanxi 710049, P. R. China
zhangqimin64@sina.com
[2] School of Mathematics and Computer Science Ningxia University,
Yinchuan Ningxia 750021, P. R. China

Abstract. In this paper, we introduce a class of stochastic age-dependent population dynamic system. Applying the theory of stochastic functional differential equation, using Gronwall's lemma and Barkholder-Davis-Gundy's lemma, Existence and uniqueness of strong solution are proved for a class of stochastic age-dependent population dynamic system on Hilbert space. In particular, as a direct consequence our main results extend some of those from ordinary age-dependent population dynamic system.

1   Introduction

Consider the following Age-dependent Population dynamic system:

$$
\begin{cases}
\frac{\partial P}{\partial t} + \frac{\partial P}{\partial a} = -\mu(t,a)P + f(t,p), & \text{in } Q = (0,A) \times (0,T), \\
P(0,a) = P_0(a), & \text{in } [0,A], \\
P(t,0) = \int_0^A \beta(t,a)P(t,a)da, & \text{in } [0,T],
\end{cases}
\tag{1}
$$

where $P(t,a)$ is the density of individuals of age a at the time t. $\beta(t,a)$ is the fertility rate of females of age a at time t. $\mu(t,a)$ is the mortality rate of age a at time t. $f(t,P)$ denotes affects external environment for population system. using the Banach fixed point theorem and operator semi-group theory, respectively; the existence, uniqueness and stability of solutions equation (1) were discussed in [1-2]. Existence and Uniqueness of solutions for non-Stochastic Population system has been studied by many mathematicians[3-4]. However, given that Population system are often subject to environmental noise[5-6], it is important to discover whether the presence of such noise affects this result. Recently, stochastic population system has been studied by several authors . For instance, Mao studied environmental Brownian noise suppresses explosions in population dynamics[7]. In [8], under condition $f(t,P)$ with random migration perturbations, the existence and uniqueness of solution are proved. Suppose that the $f(t,P)$ is stochastically perturbed, with

$$
f(t,P) \to f(t,P) + g(t,P)\dot{\omega}(t)
$$

N. Megiddo, Y. Xu, and B. Zhu (Eds.): AAIM 2005, LNCS 3521, pp. 151–161, 2005.
© Springer-Verlag Berlin Heidelberg 2005

where $\dot{w}(t)$ is white noise represents the intensity of noise. Then this environmentally perturbed system may be described by the *Itô* equation

$$\begin{cases} d_t P = -\frac{\partial P}{\partial a}dt - \mu(t,a)Pdt + f(t,P)dt + g(t,P)d\omega_t, & \text{in } Q = (0,A) \times (0,T), \\ P(0,a) = P_0(a), & \text{in } [0,A], \\ P(t,0) = \int_0^A \beta(t,a)P(t,a)da, & \text{in } [0,T] \end{cases} \quad (2)$$

where $d_t p$ is the differential of P relative to t, i.e., $d_t P = \frac{\partial P}{\partial t}dt$.

In this paper, by virtue of a direct approach quite different from those mentioned above, we shall discussion the existence and uniqueness of strong solution for stochastic age-dependent population equation (2).

2   Preliminaries

Let $V = H^1([0,A]) \equiv \{\varphi | \varphi \in L^2([0,A]), \frac{\partial \varphi}{\partial x_i} \in L^2([0,A]),$ where $\frac{\partial \varphi}{\partial x_i}$ is generalized partial derivatives$\}$, V is a Sobolev space. The norm in V is defined as follows

$$\|\varphi\|_V = (\|\varphi\|_{L^2([0,A])}^2 + \sum_{i=1}^N \|\frac{\partial \varphi}{\partial x_i}\|_{L^2([0,A])}^2)^{\frac{1}{2}}.$$

$H = L^2([0,A])$. V' is the dual space of V. We denote by $\|\cdot\|$, $|\cdot|$ and $\|\cdot\|_*$ the norms in V, H and V' respectively; by $\langle \cdot, \cdot \rangle$ the duality product between V, V', and by $(\cdot, \cdot)$ the scalar product in H.

Let ω be a Wiener Process defined on a certain complete probability space $(\Omega, \mathcal{F}, \mathcal{P})$ and take in the separable Hilbert space K, with incremental covariance operator W. Let $(\mathcal{F}_t)_{t \geq 0}$ be the σ-algebras generated by $\{\omega_s, 0 \leq s \leq t\}$, then ω_t is a martingale relative to $(\mathcal{F}_t)_{t \geq 0}$ and we have the following representation of ω_t:

$$\omega_t = \sum_{i=1}^\infty \beta_i(t)e_i,$$

where $\{e_i\}_{i \geq 1}$ is an orthonormal set of eigenvectors of W, $\beta_i(t)$ are mutually independent real Wiener processes with incremental covariance $\lambda_i > 0$, $We_i = \lambda_i e_i$ and $trW = \sum_{i=1}^\infty \lambda_i$ (tr denotes the trace of an operator). Let an operator $B \in \mathcal{L}(K,H)$ be the space of all bounded linear operators from K into H ,we denote by $\|B\|_2$ its denotes the Hilbert-Schmidt norm, i.e.

$$\|B\|_2^2 = tr(BWB^T).$$

In this paper, ω_t taking its values in real space R, so that $W = 1$.

Let $\mu(t,a)$, $\beta(t,a)$ are nonnegative measurable, and

$$\begin{cases} 0 \leq \mu_0 \leq \mu(t,a) < \infty, & (t,a) \in Q, \\ 0 \leq \beta(t,a) \leq \bar{\beta} < \infty, & (t,a) \in Q. \end{cases}$$

Let operator $A : P \rightarrow -\frac{\partial P(t,a)}{\partial a}$, and $p \geq 2$. Assume the following hypotheses:
(a.1) $\exists \alpha > 0, \nu \in R$ such that;

$$-2\langle A(t, P), P \rangle + A\bar{\beta}^2 |P|^2 + \nu \geq \alpha \|P\|^p, \quad \forall P \in V, \quad a.e.t.;$$

Let $f(t, \cdot) : L^2_H \rightarrow H$ be a family of nonlinear operators defined a.e.t., and satisfy
(b.1) $f(t, 0) = 0$;
(b.2) $\exists k_1 > 0$ such that

$$|f(t, y) - f(t, x)| \leq k_1 \|y - x\|_C, \quad \forall x, y \in C, a.e.t;$$

(b.3) $t \in (0, T) \rightarrow f(t, x) \in H$ is Lebesgue-measurable $\forall x \in L^2_H$.
And let $g(t, \cdot) : L^2_H \rightarrow \mathcal{L}(K, H)$, the family of nonlinear operator defined a.e.t. and satisfy
(c.1) $g(t, 0) = 0$
(c.2) there exists $k_2 > 0$ such that

$$\|g(t, y) - g(t, x)\|_2 \leq k_2 \|y - x\|_C, \quad \forall x, y \in C, a.e.t,$$

(c.3)$g(t, x) \in \pounds(K, H)$ is Lebesgue-measurable $\forall x \in L^2_H$.
The objective in this paper is that under the conditions described above, we hopefully find a unique process $P_t \in I^p(0, T; V) \bigcap L^2(\Omega; C(0, T; H))$ such that

$$\begin{cases} P_t = P_0 - \int_0^t \frac{\partial P_s}{\partial a} ds - \int_0^t \mu(s, a) P_s ds \\ \quad + \int_0^t f(s, P_s) ds + \int_0^t g(s, P_s) d\omega_s, \forall t \in [0, T], \\ P(t, 0) = \int_0^A \beta(t, a) P_s da, \qquad \forall t \in [0, T], \end{cases} \tag{3}$$

where $P_t = P(t, a)$, $P_0 = P(0, a)$.

Definition 1. *let $(\Omega, \mathcal{F}, \{\mathcal{F}_t\}, P)$ be the stochastic basis and ω_t a Wiener process with covariance operator W. Suppose that P_0 is a random variable such that $E|P_0|^2 < \infty$. A stochastic process P_t is said to be a strong solution on Ω to the SDE (3) for $t \in [0, T]$ if the following conditions are satisfied:*

(a) P_t is a $\mathcal{F}_t$ -measurable random variable;
(b) $P_t \in I^p(0, T; V) \bigcap L^2(\Omega; C(0, T; H))$, $p > 1$, $T > 0$, where $I^p(0, T; V)$ denotes the space of all V-valued processes $(P_t)_{t \in [0, T]}$ (we will write P_t for short) measurable (from $[0, T] \times \Omega$ into V), and satisfying

$$E \int_0^T \|P_t\|^p dt < \infty.$$

Here $C(0, T; H)$ denotes the space of all continuous functions from $[0, T]$ to H;
(c) Eq.(3) is satisfied for every $t \in [0, T]$ with probability one.
    If T is replaced by ∞, P_t is called a global strong solution of (3).

3   Existence and Uniqueness of Solutions

3.1   Uniqueness of Solutions

Now we shall prove that there exists at most one solution of (3). This result will be deduced mainly from *Itô's* formula.

Theorem 1. *Assume the preceding hypotheses hold. Then, there exists at most one solution of (3) in $I^p(0, T; V) \bigcap L^2(\Omega; C(0, T; H))$.*

Proof. Suppose that $P_{1t}, P_{2t} \in I^p(0, T; V) \bigcap L^2(\Omega; C(0, T; H))$ are two solutions of (3). Then, applying *Itô's* formula to $|P_{1t} - P_{2t}|^2$, we obtain

$$
\begin{aligned}
&|P_{1t} - P_{2t}|^2 \\
&= 2 \int_0^t \langle -\frac{\partial P_{1s}}{\partial a} + \frac{\partial P_{2s}}{\partial a} - \mu(s, a)(P_{1s} - P_{2s}), P_{1s} - P_{2s} \rangle ds \\
&\quad + 2 \int_0^t (f(s, P_{1s}) - f(s, P_{2s}), P_{1s} - P_{2s}) ds \\
&\quad + 2 \int_0^t (P_{1s} - P_{2s}, (g(s, P_{1s}) - g(s, P_{2s})) d\omega_s) \\
&\quad + \int_0^t tr(g(s, P_{1s}) - g(s, P_{2s}) W (g(s, x_{1s}) - g(s, P_{2s}))^* ds \\
&\leq -2 \int_0^t \langle \frac{\partial (P_{1s} - P_{2s})}{\partial a}, P_{1s} - P_{2s} \rangle ds - 2\mu_0 \int_0^t (P_{1s} - P_{2s}, P_{1s} - P_{2s}) ds \\
&\quad + 2 \int_0^t (f(s, P_{1s}) - f(s, P_{2s}), P_{1s} - P_{2s}) ds + \int_0^t \|g(s, P_{1s}) - g(s, P_{2s})\|_2^2 ds \\
&\quad + 2 \int_0^t (P_{1s} - P_{2s}, (g(s, P_{1s}) - g(s, P_{2s})) d\omega_s).
\end{aligned}
$$

Since

$$
-\langle \frac{\partial (P_{1s} - P_{2s})}{\partial a}, P_{1s} - P_{2s} \rangle \leq \frac{1}{2} A \bar{\beta}^2 |P_{1s} - P_{2s}|^2.
$$

Therefore, we get that

$$
\begin{aligned}
&|P_{1t} - P_{2t}|^2 \\
&\leq A \bar{\beta}^2 \int_0^t |P_{1s} - P_{2s}|^2 ds + 2 \int_0^t |P_{1s} - P_{2s}| |f(s, P_{1s}) - f(s, P_{2s})| ds \\
&\quad - 2\mu_0 \int_0^t |x_s - y_s|^2 ds + \int_0^t \|g(s, P_{1s}) - g(s, P_{2s})\|_2^2 ds \\
&\quad + 2 \int_0^t (P_{1s} - P_{2s}, (g(s, P_{1s}) - g(s, P_{2s}) d\omega_s).
\end{aligned}
$$

Now, it follows from (b.2) and (c.2)that for any $t \in [0, T]$

$$E \sup_{0 \le s \le t} |P_{1s} - P_{2s}|^2$$

$$\le (|A\bar{\beta}^2 - 2\mu_0| + 1) \int_0^t E|P_{1s} - P_{2s}|^2 ds \tag{4}$$

$$+ (k_1^2 + k_2^2) \int_0^t E\|P_{1s} - P_{2s}\|_C^2 ds$$

$$+ 2E \sup_{0 \le s \le t} \int_0^s (P_{1r} - P_{2r}, (g(r, P_{1r}) - g(r, P_{2r}))d\omega_r).$$

However, by Burkholder-Davis-Gundy's inequality, we have

$$E[\sup_{0 \le s \le t} \int_0^s (P_{1r} - P_{2r}, (g(r, P_{1r}) - g(r, P_{2r}))d\omega_r)$$

$$\le 3E[\sup_{0 \le s \le t} |P_{1s} - P_{2s}|[\int_0^t \|g(s, P_{1s}) - g(s, P_{2s})\|_2^2 ds)^{1/2}]$$

$$\le \tfrac{1}{4}E[\sup_{0 \le s \le t} |P_{1s} - P_{2s}|^2 + K \int_0^t \|g(s, P_{1s}) - g(s, P_{2s})\|_2^2 ds \tag{5}$$

$$\le \tfrac{1}{4}E[\sup_{0 \le s \le t} |P_{1s} - P_{2s}|^2 + K \cdot k_2^2 \int_0^t E\|P_{1s} - P_{2s}\|_C^2 ds,$$

for some positive constant $K > 0$. On the other hand, we get

$$\int_0^t E\|P_{1s} - P_{2s}\|_C^2 ds \le \int_0^t E \sup_{0 \le r \le s} |P_{1r} - P_{2r}|^2 ds.$$

Thus, it follows from (4) and (5)

$$E \sup_{0 \le s \le t} |P_{1s} - P_{2s}|^2$$

$$\le 2(|A\bar{\beta}^2 - 2\mu_0| + 1 + k_1^2 + k_2^2 + 2Kk_2^2) \int_0^t E \sup_{0 \le r \le s} |P_{1r} - P_{2r}|^2 ds, \quad \forall t \in [0, T].$$

Now, Gronwall's lemma obviously implies uniqueness.   □

3.2   Existence of Strong Solutions

First of all,we state a theorem on existence and uniqueness of solutions of stochastic population system. Next, by means of this result we will prove the desired existence of solution of (3).

Theorem 2. *Assume the preceding hypotheses and $A\bar{\beta}^2 = 0$ holds. Then , there exist a unique process $P_t \in I^p(0, T; V) \bigcap L^2(\Omega; C(0, T; H))$ such that*

$$P_t = P_0 + \int_0^t [A(s, P_s) + f_1(s)]ds + M(t), P - a.s., \quad \forall t \in [0, T],$$

where $f_1 \in I^2(0,T;H)$, $P_0 \in L^2(\Omega, \mathcal{F}_0, P; H)$ and M_t is an H-valued continuous, square integrable $\mathcal{F}_t$-martingale. In addition, the following energy equality also holds:

$$|P_t|^2 = |P_0|^2 + 2\int_0^t \langle A(s, P_s), P_s\rangle ds + 2\int_0^t (f_1(s), P_s)ds$$
$$+2\int_0^t (P_t, dM_s) + tr\langle\langle M\rangle\rangle_t, \quad P - a.s., \quad \forall t \in [0,T],$$

where $\langle\langle M\rangle\rangle_t$ denotes the quadratic variation of M_t.

Proof. see Métivier and Pellaumail[9].   □

For the existence, we consider the equations

$$P_t^1 = P_0 + \int_0^t [-\frac{\partial P^1}{\partial a} - \frac{A\bar{\beta}^2}{2}P_s^1]ds, \quad t \in [0,T], \tag{6}$$

$$P^1(t,0) = \int_0^A \beta(t,a)P_t^1 da, \quad t \in [0.T], \tag{7}$$

$$P_t^{n+1} = P_0 + \int_0^t [-\frac{\partial P_s^{n+1}}{\partial a} - \frac{A\bar{\beta}^2}{2}P_s^{n+1}]ds + \int_0^t \frac{A\bar{\beta}^2}{2}P_s^n ds - \int_0^t \mu(s,a)P_s^n ds$$

$$+ \int_0^t f(s, P_s^n)ds + \int_0^t g(s, P_s^n)d\omega_s, \quad t \in [0,T], \quad \forall n \geq 1 \tag{8}$$

$$P^{n+1}(t,0) = \int_0^A \beta(t,a)P_t^{n+1} da, \quad t \in [0.T], \quad \forall n \geq 1. \tag{9}$$

Now, we want to prove that the sequence $\{P_t^n\}$ is convergent to a process P_t in $I^p(0,T;V) \cap L^2(\Omega; C(0,T;H))$, which will be the solution of (3). We shall first prove the following lemmas.

Lemma 1. $\{P_t^n\}$ is a Cauchy sequence in $L^2(\Omega; C(0,T;H))$.

Proof. For $n > 1$ and the process $P_t^{n+1} - P_t^n$, it following from Itô's formula

$$|P_t^{n+1} - P_t^n|^2$$

$$= 2\int_0^t \langle -\frac{\partial P_s^{n+1}}{\partial a} + \frac{\partial P_s^n}{\partial a}, P_s^{n+1} - P_s^n\rangle ds$$

$$-2\int_0^t (\mu(s,a)(P_s^n - P_s^{n-1}), P_s^{n+1} - P_s^n)ds - A\bar{\beta}^2\int_0^t |P_s^{n+1} - P_s^n|^2 ds$$

$$+A\bar{\beta}^2\int_0^t (P_s^{n+1} - P_s^n, P_s^n - P_s^{n-1})ds \tag{10}$$

$$+2\int_0^t (f(P_s^n) - f(P_s^{n-1}), P_s^{n+1} - P_s^n)ds$$

$$+2\int_0^t (P_s^{n+1} - P_s^n, (g(P_s^n) - g(P_s^{n-1}))d\omega_s) + \int_0^t \|g(P_s^n) - g(P_s^{n-1})\|_2^2 ds$$

where, by definition, $P_t^n := P^n(t, a), f(P_t^n) := f(t, P_t^n)$ and $g(P_t^n) := g(t, P_t^n)$. It is easy to deduce

$$|P_t^{n+1} - P_t^n|^2$$

$$\leq |A\bar{\beta}^2 - 2\mu_0| \int_0^t |P_s^{n+1} - P_s^n||P_s^n - P_s^{n-1}|ds$$

$$+2| \int_0^t (P_s^{n+1} - P^n, (g(P_s^n) - g(P_s^{n-1}))d\omega_s)| \quad (11)$$

$$+2 \int_0^t |f(P_s^n) - f(P_s^{n-1})||P_s^{n+1} - P_s^n|ds$$

$$+ \int_0^t \|g(P_s^n) - g(P_s^{n-1})\|_2^2 ds.$$

Consequently, (11) yields

$$E[\sup_{0 \leq \theta \leq t} |P_\theta^{n+1} - P_\theta^n|^2]$$

$$\leq |A\bar{\beta}^2 - 2\mu_0|E \int_0^t |P_s^{n+1} - P_s^n||P_s^n - P_s^{n-1}|ds$$

$$+2E[\sup_{0 \leq \theta \leq t} | \int_0^\theta (P_s^{n+1} - P_s^n, (y(P_s^n) - g(P_s^{n-1}))d\omega_s)|] \quad (12)$$

$$+2E \int_0^t |f(P_s^n) - f(P_s^{n-1})||P_s^{n+1} - P_s^n|ds$$

$$+E \int_0^t \|g(P_s^n) - g(P_s^{n-1})\|_2^2 ds.$$

Now, we estimate the terms on the right-hand side of (12)

$$|A\bar{\beta}^2 - 2\mu_0|E \int_0^t |P_s^{n+1} - P_s^n||P_s^n - P_s^{n-1}|ds$$

$$\leq \frac{1}{4}E[\sup_{0 \leq \theta \leq t} |P_\theta^{n+1} - P_\theta^n|^2] \quad (13)$$

$$+(A\bar{\beta}^2 - 2\mu_0)^2 T \int_0^t E[\sup_{0 \leq \theta \leq s} |P_\theta^n - P_\theta^{n-1}|^2]ds.$$

On the other hand, we can get from (c.2)

$$E \int_0^t \|g(P_s^n) - g(P_s^{n-1})\|_2^2 ds \leq k_2^2 E \int_0^t \sup_{0 \leq r \leq s} |P_r^n - P_r^{n-1}|^2 ds. \quad (14)$$

In a similar manner, from (b.2) we can obtain

$$2E \int_0^t |f(P_s^n) - f(P_s^{n-1})||P_s^{n+1} - P_s^n|ds$$
$$\leq \frac{1}{4}E[\sup_{0 \leq r \leq t} |P_r^{n+1} - P_r^n|] + 4k_1^2 T \int_0^t E[\sup_{0 \leq r \leq s} |P_r^n - P_r^{n-1}|^2]ds. \quad (15)$$

Now, Burkholder-Davis-Gundy's inequality implies

$$2E[\sup_{0\leq r\leq t} |\int_0^r (P_s^{n+1} - P_s^n, (g(P_s^n) - g(P_s^{n-1}))d\omega_s|]$$

$$\leq 6E[(\sup_{0\leq r\leq t} |P_r^{n+1} - P_r^n|^2) \int_0^t \|g(P_s^n) - g(P_s^{n-1})\|_2^2 ds]^{\frac{1}{2}} \qquad (16)$$

$$\leq \tfrac{1}{4}E\{\sup_{0\leq s\leq t} |P_r^{n+1} - P_r^n|^2 + 72k_2^2 \int_0^t E[\sup_{0\leq r\leq s} |P_r^n - P_r^{n-1}|^2]ds.$$

If we set

$$\varphi^n(t) = E[\sup_{0\leq\theta\leq t} |P_\theta^{n+1} - P_\theta^n|^2|]. \qquad (17)$$

then from (13)-(16), it could be deduced that there exists a positive constant $c > 0$ such that

$$\varphi^n(t) \leq \frac{3}{4}\varphi^n(t) + c\int_0^t \varphi^{n-1}(s)ds, \qquad (18)$$

and consequently there exists $k > 0$ such that

$$\varphi^n(t) \leq k\int_0^t \varphi^{n-1}(s)ds. \qquad (19)$$

By iteration from (19), we get

$$\varphi^n(t) \leq \frac{k^{n-1}T^{n-1}}{(n-1)!}\varphi^1(T), \quad \forall n > 1, \ \forall t \in [0,T]. \qquad (20)$$

Therefore,

$$E[\sup_{0\leq\theta\leq T} |P_\theta^{n+1} - P_\theta^n|^2] \leq \frac{k^{n-1}T^{n-1}}{(n-1)!}\varphi^1(T), \quad \forall n > 1. \qquad (21)$$

Obviously, (21) implies that $\{P_t^n\}$ is a Cauchy sequence $L^2(\Omega; C(0,T; H))$.   □

Lemma 2. *The sequence $\{P_t^n\}$ is bounded in $I^p(0,T; V)$.*

Proof. Indeed, applying *Itô's* formula to $|P_t^n|^2$ with $n \geq 2$ immediately yields

$$E|P^n(T)|^2$$

$$= 2E\int_0^T \langle -\frac{\partial P_s^n}{\partial a}, P_s^n\rangle ds - 2\int_0^T (\mu(s,a)P_s^n, P_s^n)ds$$

$$-A\bar\beta^2 E\int_0^T |P_s^n|^2 ds + E|P_0|^2 + 2E\int_0^T (f(P_s^{n-1}), P_s^n)ds \qquad (22)$$

$$+A\bar\beta^2 E\int_0^T (P_s^n, P_s^{n-1})ds + E\int_0^T \|g(P_s^{n-1})\|_2^2 ds.$$

Therefore,

$$2E\int_0^T \langle \frac{\partial P_s^n}{\partial a}, P_s^n \rangle ds + A\bar{\beta}^2 E \int_0^T |P_s^n|^2 ds$$

$$\leq E|P_0|^2 + 2E\int_0^T |f(P_s^{n-1})||P_s^n| ds \tag{23}$$

$$+A\bar{\beta}^2 E\int_0^T |P_s^n||P_s^{n-1}| ds + E\int_0^T \|g(P_s^{n-1})\|_2^2 ds.$$

Since $\{P^n\}$ is convergent in $L^2(\Omega; C(0, T; H))$, it will be bounded in this space. Now, it is not difficult to check that there exists a positive constant $k' > 0$ such that the right-hand side of (23) is bounded by this constant. We will estimate one of those terms. Firstly, we observe that

$$2E\int_0^T |f(P_s^{n-1}||P_s^n)| ds$$

$$\leq k_1 E \int_0^T [\|P_s^{n-1}\|_C^2 + |P_s^n|^2] ds$$

$$\leq Tk_1 E(\sup_{0\leq\theta\leq T} |P_\theta^{n-1}|^2) + k_1 T E(\sup_{0\leq\theta\leq T} |P_\theta^n|^2)$$

$$= Tk_1 \|P_t^{n-1}\|_{L^2(\Omega;C(0,T;H))} + k_1 T \|P_t^n\|_{L^2(\Omega;C(0,T;H))},$$

which, in addition to (23) and (a.1), leads to the following inequalities:

$$\alpha \int_0^T E\|P_s^n\|^p ds \leq 2E\int_0^T \langle \frac{\partial P_s^n}{\partial a}, P_s^n \rangle ds + A\bar{\beta}^2 E \int_0^T |P_s^n|^2 ds + \nu T \leq k',$$

and Lemma 2 is proved.   □

Lemma 3. *The limit of the sequence $\{P_t^n\}$ is a solution to (3).*

Proof. Firstly, we observe that Lemma 1 implies that there exists $P_t \in L^2(\Omega; C(0, T; H))$ such that $P_t^n \to P_t$ in $L^2(\Omega; C(0, T; H))$. Since (b.2) and (c.2) hold, we have $f(P_t^n) \to f(P_t)$(in $L^2(\Omega; L^\infty(0, T; H))$), and $g(P_t^n) \to g(P_t)$(in $L^2(\Omega; L^\infty; \mathcal{L}(K, H)))$).

Let $DP_t = \frac{\partial P_t}{\partial t} + \frac{\partial P_t}{\partial a}$. So

$$DP^n(t, a) = -\frac{A\bar{\beta}^2}{2} P_t^n dt - \mu(t, a) P_t^{n-1} dt$$

$$+\frac{A\bar{\beta}^2}{2} P_t^{n-1} dt + f(t, P_t^{n-1}) dt + g(t, P_t^{n-1}) d\omega_t.$$

By preceding analysis, we easily obtain that

$$\|DP^n\|_{V'} \leq M \leq \infty.$$

On the other hand, by virtue of lemma 2 $\{P_t^n\}$ has a subsequence which is weakly convergent in $I^p(0,T;V)$. But, since $P_t^n \to P_t$ in $L^2(\Omega;C(0,T;H))$, we can assure that $P_t^n \to P_t$ weakly in $I^p(0,T;V)$ (in the sequel, we will denote this by $P_t^n \rightharpoonup P_t$ in $I^p(0,T;V)$). In conclusion, we have proved:

$$P^n \to P \quad in \quad L^2(\Omega;C(0,T;H)), \tag{24}$$

$$f(P_t^n) \to f(P_t) \quad in \quad L^2(\Omega;L^\infty(0,T;H)), \tag{25}$$

$$g(P_t^n) \to g(P_t) \quad in \quad L^2(\Omega;L^\infty(0,T;\mathcal{L}(K,H))), \tag{26}$$

$$P_t^n \rightharpoonup P_t \quad in \quad I^p(0,T;V), \tag{27}$$

$$DP^n \rightharpoonup h \quad in \quad L^{p'}(\Omega \times (0,T);V').$$

Since the differential operator is continuous, so $DP = h$. The proof of Lemma 3 is now complete.   □

By Theorem 1 and Lemma 3, we can obtain the following Theorem

Theorem 3. *Assume (a.1),(b.1)-(b.3),(c.1)-(c.3) hold. Then, for each $P_0 \in H \bigcap V$, there exists a unique solution of the problem (3) in $I^p(0,T;V) \cap L^2(\Omega; C(0,T;H))$.*

Acknowledgements

The research was supported by Ningxia Natural Science Foundation(No.G002) (China); also was supported by Ningxia Higher School Science and Technique Research Foundation.

References

1. Jian, S., Jingyuan, Y.: Control Theory of Population. Science Press, Beijing(1985)
2. Jianzhong, S., Zongben, X.: The well posedness of the time-dependent population dynamics system and the optimal control on the fertility rate. Syst. Science Math. **21** (2001) 274–282
3. Anita, S.: Existence and uniqueness for the population dynamics with nonlinear diffusion. Diff. Integral Eqs. **4** (1991)835–850
4. Anita, S.: Analysis and control of age-dependent population dynamics. Kluwer Academic publishers, Netherlands(2000)
5. Kifer,Y.: Principal eigenvalues,topological pressure, and stochastic stability of equilibrium states. Israel J. Math. **70** (1990) 1–47
6. Ramanan, K., Zeitouni, O.: The quasi-stationary distribution for small random perturbations of certain one-dimensional maps. Stoc. Process. Appl. **86** (1999)25–51
7. Xuerong, M.: Environmental Brownian noise suppresses explosions in population dynamics. Stoc. Process. Appl. **97** (2002) 95–110

8. Guowei, Z., Shurong, H.: The existence and uniqueness of random periodic solutions for a class of nonlinear population dynamics with random periodic migration perturbations. J. Biomath. **17** (2002)60–63
9. Métivier, M., Pellaumail, J.: Stochastic Integration. Academic Press, New York (1980)

A PTAS for Scheduling on Agreeable Unrelated Parallel Batch Processing Machines with Dynamic Job Arrivals

Yuzhong Zhang, Zhigang Cao*, and Qingguo Bai

College of Operations Research and Management Science,
Qufu Normal University, Rizhao, Shandong, China
cullencao@eyou.com

Abstract. We consider the scheduling problem $R_m|r_j, B|C_{max}$ under the assumption of *agreement*, i.e., $p_{ij_1} \geq p_{ij_2}$ for some i implies $p_{ij_1} \geq p_{ij_2}$ for all $1 \leq i \leq m$, where p_{ij_1} and p_{ij_2} denote the processing times on machine M_i of jobs J_{j_1} and J_{j_2}, respectively. For the special case when the number of distinct release times t is constant and all processing times and release times integral, we propose a pseudo-polynomial time algorithm by approach of dynamic programming. Without the integral restriction, an FPTAS is provided. And for the general case with arbitrary t, we establish a PTAS.

1   Introduction

The batch scheduling problem has its deep root in the real world and has attracted a lot of attention recently. In the industry of semiconductor manufacturing , the last stage is the final testing (called the burn-in operation). In this stage, chips are loaded onto boards which are then placed in an oven and exposed to high temperature. Each chip has a pre-specified minimum burn-in time, and the burn-in oven has a limited capacity B. Up to B chips (which is called a batch) can be baked in an oven simultaneously, and the baking process is not allowed to be preempted, that is, once the processing of a batch is started, the oven is occupied until the process is completed. To ensure that no defective chips will pass to the customer, the processing time of a batch is that of the longest one among these chips. As the baking process in burn-in operations can be long compared to other testing operations(e.g.,120 hours as opposed to 4-5 hours for other operations), an effective algorithm for batching and scheduling is highly non-trivial. There has been a lot of work on different variants of batch scheduling. To have a better view, see Brucker's paper [2].

According to Graham et. al. [7], a scheduling problem can be denoted by a 3-tuple $\alpha|\ \beta|\ \gamma$, where α denotes the machine (i.e. oven here)environment, β the additional constraints on the jobs (i.e. chips here), and γ the objective function.

---

* Supported by NSF China 10171054.

N. Megiddo, Y. Xu, and B. Zhu (Eds.): AAIM 2005, LNCS 3521, pp. 162–171, 2005.

The model we consider in this paper is $R_m|r_j, B|C_{max}$, where R_m denotes unrelated parallel machines, implying that there are a constant number m of parallel machines, and for each job, the processing times on different machines may be different and unrelated. In this paper, we assume that $p_{ij_1} \geq p_{ij_2}$ for some i implies $p_{ij_1} \geq p_{ij_2}$ for all $1 \leq i \leq m$, where p_{ij_1} and p_{ij_2} denote the processing times on machine M_i of jobs J_{j_1} and J_{j_2}, respectively. Thus we can re-index all the jobs such that $p_{i1} \geq p_{i2} \geq \ldots \geq p_{in}$ holds for all $1 \leq i \leq m$, where m is the number of machines and n that of jobs. This assumption will be later referred to as *agreement*. In addition, for each job J_j, it is associated with a release time r_j, before which it can't be processed. And our objective is to minimize the makespan C_{max}, i.e., the time when all jobs are finished.

2 Previous Related Work and Our Contributions

As to the single machine variant, Ikura and Gimple [9] considered the $1|r_j, B|C_{max}$ problem when the jobs have equal processing times. They showed that this special case can be solved in $O(n)$ time using a simple strategy which they call First-Only-Empty algorithm. Lee and Uzsoy [12] gave an $O(n^2)$ time algorithm for the unbounded special case (i.e. $B \geq n$), and extended it to polynomial time algorithms for other special cases. When there are only two distinct release times, they proposed an algorithm running in $O(nB^2 P_{sum} P_{max})$ time , where P_{max} and P_{sum} are the maximum and total processing time, respectively. For the general $1|r_j, B|C_{max}$ problem, however, they only proposed a number of heuristics. Brucker et. al.[2] showed that the general problem is strongly NP-hard even if $B = 2$.

Finally, the problem was solved by Xiaotie Deng, C.K. Poon and Yuzhong Zhang[6]. They obtained a PTAS when there are arbitrary release times. Given the strongly NP-hardness, it is the best possible.

In their paper, they also considered the on-line version. They proved that there can't exist an on-line algorithm with competitive ratio less than $1 + \delta$, where $\delta = (\sqrt{5} - 1)/2$ is the golden ratio. And for the unbounded special case, they provided an algorithm with competitive ratio $1 + \delta$, and thus it is the most competitive. For the bounded variant(i.e. $B < n$),Guochuan Zhang, Xiaoqiang Cai and C.K. Wong [16]solved the special case with two release times, they also generalized their algorithm to the general case and showed a worst-case ratio of 2. Later, with the restriction of identical processing times, Guochuan Zhang et. al. studied the parallel machine variant and gave beautiful results[17]. Still for the on-line version, C.K. Poon and Wenci Yu get some new results recently.[14][15]

To the best of our knowledge, the only paper that considered the off-line parallel variant is by Shuguang Li, Guojun Li and Shaoqiang Zhang[11]. They deal with the case of identical parallel machines.

In this paper, we present a PTAS for the more general model $R_m|r_j, B|C_{max}$ under the assumption of agreement, where the batch size B is fixed, and which will be later referred to as $\overline{R}_m|r_j, B|C_{max}$. Given the strongly NP-hardness, it is the best possible in the sense of worst-case performance. Our result is the generalization of that of Xiaotie Deng et. al., and the approaches relied on are the same: dynamic programming and scaling-and-rounding.

The rest of the paper is organized as follows. In Subsection 3.1, necessary preliminary knowledge is given. In Subsection 3.2, we discuss the special case when there are constant release times and all the release times and processing times are integral. We propose an optimal algorithm with pseudo-polynomial running time. Without the restriction of integral inputs, we provide an FPTAS in Subsection 3.3. And in Subsection 3.4, for the more general case when there are arbitrary distinct release times, we establish a PTAS. Finally, in Section 4, we draw a conclusion of this paper and direct the further research in this area.

3   A PTAS for $\overline{R}_m|r_j, B|C_{max}$

3.1   Preliminaries

An algorithm $\mathcal{A}$ is a $(1+\varepsilon)$−approximation algorithm for a minimization problem if it produces a solution which is at most $(1+\varepsilon)$ times the optimal one. A family of algorithms $\{\mathcal{A}_\varepsilon\}_\varepsilon$ is called a PTAS (*Polynomial Time Approximation Scheme*)if, for every $\varepsilon > 0$, the algorithm $\mathcal{A}_\varepsilon$ is a $(1+\varepsilon)$−approximation algorithm running in time polynomial in the input size when ε is treated as constant. It is also an FPTAS (*Fully Polynomial Time Approximation Scheme*) if the running time is also polynomial in $\frac{1}{\varepsilon}$. For an NP-hard problem in the strong sense, it is impossible to obtain an FPTAS or a pseudo-polynomial time algorithm.

Throughout this paper, we will denote by $\lfloor x \rfloor$ the largest integer less than or equal to x.

Before describing our algorithms, we first explain the FBLPT (*Full Batch Largest Processing Time*) algorithm of Bartholdi, which computes the optimal solution for $1|B|C_{max}$.

Algorithm FBLPT
step 1. Index all the jobs such that $p_1 \geq p_2 \geq \ldots \geq p_n$.
step 2. From the first job onwards, group the adjacent jobs together into a batch such that all the batches are full except possibly the last one.
step 3. Schedule the batches in any arbitrary order.

From now on, a schedule is said to follow the FBLPT rule if it groups the jobs according to the FBLPT algorithm and schedules them in non-increasing order of the processing time.

3.2   A Special Case with Constant Distinct Release Times and Integral Inputs

We are given a set of jobs $\mathcal{J} = \{J_1, \ldots, J_n\}$ and a set of machines $\mathcal{M} = \{M_1, \ldots, M_m\}$ and t distinct release times $r_1 < r_2 < \ldots < r_t$, where each job J_j is associated with a release time $r_j' \in \{r_1, r_2, \ldots, r_t\}$ and processing times $p_{1j}, \ldots, p_{mj}$.

Under the assumption of agreement, we can assume that all the jobs have been re -indexed such that $p_{i1} \geq p_{i2} \geq \ldots \geq p_{in}$ holds for all $1 \leq i \leq m$. It's valuable to remark that this re-indexing is quite critical, which will be noticed later, thus the agreement assumption is indispensable and we can not extend our result to the general unrelated parallel machines.

Suppose π is a schedule for the problem, i.e., π assigns each job to some machine, batches the jobs on the same machine and determines the start time for each batch. Since a batch is always started at the time when all jobs in it have arrived or when its previous batch is completed, whichever comes later, a schedule is determined by the batching and processing order of all batches.

For any feasible schedule π, we can partition $\mathcal{J}$ into mt disjoint subsets $\mathcal{J}_{11}, \ldots .\mathcal{J}_{mt}$, where $\mathcal{J}_{ik}(\pi)$ is defined as the set of jobs that are scheduled on machine M_i and start at or after time r_k but strictly before r_{k+1}, where $1 \leq i \leq m, 1 \leq k \leq t$ and $r_{t+1} = \infty$. A key observation is that we can locally rearrange the schedule of batch $\mathcal{J}_{ik}(\pi)$, without increasing its makespan, so that each schedule follows the FBLPT rule.

Lemma 1. *For any feasible schedule π with makespan C_{max}, there exists a schedule π' with makespan $C'_{max} \leq C_{max}$ such that $\mathcal{J}_{ik}(\pi) = \mathcal{J}_{ik}(\pi')$ and the schedule for $\mathcal{J}_{ik}(\pi')$ follows the FBLPT rule for any $1 \leq i \leq m, 1 \leq k \leq t$.* □

Lemma 1 is the immediate result of the optimality of algorithm FBLPT for $1|B|C_{max}$.

By this lemma, the original scheduling problem boils down to that of partitioning $\mathcal{J}$ into a sequence of mt disjoint subsets $\mathcal{J}_{11}, \ldots, \mathcal{J}_{mt}$. Once this partition is fixed, we can assume that each subset $\mathcal{J}_{ik}$ will be processed by machine M_i and scheduled in the time interval $[r_k, r_{k+1})$ according to the FBLPT rule. Moreover, the first batch in $\mathcal{J}_{ik}$ will start at time r_k or when all jobs in $\mathcal{J}_{i(k-1)}$ have been completed, whichever comes later. In a word, the partitioning determines the batch scheduling completely. Therefore, an immediate idea is to check all the $(mt)^n$ possibilities. However, it is not practical at all, as the running time is exponential in n. To further reduce the running time, a dynamic programming will be relied on.

First of all, let's introduce the *state matrix*. Suppose that π is a feasible schedule and $\mathcal{J}_{11}(\pi), \ldots, \mathcal{J}_{mt}(\pi)$ the corresponding partition. The *state matrix* of π is defined as $\mathbf{S}(\pi) = (\mathbf{B}, \mathbf{C}, \mathbf{N})$, where $\mathbf{B} = (b_{ik})_{m \times t}, \mathbf{C} = (c_{ik})_{m \times t}, \mathbf{N} = (n_{ik})_{m \times t}$ are all mt-dimensional matrices. b_{ik}, c_{ik} and n_{ik} are defined as follows.

If $\mathcal{J}_{ik}(\pi)$ is not empty, we define b_{ik} as its delay time, i.e., the start time of its first batch minus r_k. $r_k > 0$ happens when the last batch of $\mathcal{J}_{i(k-1)}(\pi)$ is finished after r_k. Since there may be at most one batch in $\mathcal{J}_{i(k-1)}(\pi)$ that is finished after r_k, the maximum delay is at most P_{max}, where $P_{max} = max\{p_{ij} : 1 \leq i \leq m, 1 \leq j \leq n\}$. Thus $0 \leq b_{ik} \leq P_{max}$. Obviously, for any sensible schedule π, b_{i1} is zero for all $1 \leq i \leq m$. we also define the processing time of $\mathcal{J}_{ik}(\pi)$, denoted by c_{ik} as the time needed to complete all the jobs in $\mathcal{J}_{ik}(\pi)$. Thus $0 \leq c_{ik} \leq P_{sum}$, where $P_{sum} = max\{p_{i1} + p_{i2} + \ldots p_{in} : 1 \leq i \leq m\}$. n_{ik} is defined as the *size* of $\mathcal{J}_{ik}(\pi)$, i.e., the number of jobs in the last batch of $\mathcal{J}_{ik}(\pi)$. If the last batch is full, we define n_{ik} as zero. Thus n_{ik} ranges from 0 to B-1. And for empty $\mathcal{J}_{ik}(\pi)$ we define all the three variables as zero. It is easy to see that the state matrix is completely determined by a feasible schedule, and furthermore, it determines the objective value. As a matter of fact, $C_{max} = max\{r_t + b_{it} + c_{it} : 1 \leq i \leq m\}$. It is not that easy, however, to derive the schedule from a feasible state.

It is valuable to notice that, under the integral assumption, given an instance with arbitrarily large number of jobs, the number of possible states are much smaller than that of partitioning. In fact, there are no more than $(BP_{max}P_{sum})^n$ possible feasible states. Our basic idea is to check all the $(BP_{max}P_{sum})^n$ possibilities, find an optimal feasible state and derive the corresponding schedule. Now, we define L as the set of all possible states. $L = \{\mathbf{S} = (\mathbf{B}, \mathbf{C}, \mathbf{N})| \ for \ all \ 1 \leq i \leq m, 1 \leq k \leq t : 0 \leq b_{ik} \leq P_{max}, 0 \leq c_{ik} \leq P_{sum}, 0 \leq n_{ik} \leq B - 1; for \ all \ 1 \leq i \leq m, 1 \leq k \leq t - 1 : r_k + b_{ik} + c_{ik} \leq r_{k+1} + b_{i(k+1)}\}$, where $r_k + b_{ik} + c_{ik} \leq r_{k+1} + h_{i(k+1)}$ means that the last batch of $\mathcal{J}_{ik}(\pi)$ should be finished before or at the time when the first batch of $\mathcal{J}_{i(k+1)}(\pi)$ is started. We index all the members of L arbitrarily, say, $L = \{\mathbf{L}_1, \ldots, \mathbf{L}_q\}$. Here, q is at most $(BP_{max}P_{sum})^n$.

Next, we will establish the recursive relation between adjacent states for the dynamic programming.

Suppose $\mathbf{S} = (\mathbf{B}, \mathbf{C}, \mathbf{N})$ is a feasible state determined by a schedule for jobs $J_1, \ldots, J_h$. And suppose that $J_h \in \mathcal{J}_{ik}$. We define :

$$u(h, i, k, \mathbf{B}, \mathbf{C}, \mathbf{N}) = \begin{cases} (\mathbf{B}, \mathbf{C}_{ik}, \mathbf{N}_{ik}) & if \ n_{ik} = 1 \\ (\mathbf{B}, \ \mathbf{C}, \ \mathbf{N}_{ik}) & if \ n_{ik} = 0 \ or \ n_{ik} > 1 \end{cases}$$

where

$$\mathbf{C}_{ik} = \begin{pmatrix} c_{11}, & \cdots\cdots & , c_{1t} \\ \cdots\cdots & & \\ c_{i1}, \cdots, c_{ik} - p_{ih}, \cdots, c_{it} \\ \cdots\cdots & & \\ c_{m1}, & \cdots\cdots & , c_{mt} \end{pmatrix} \quad \mathbf{N}_{ik} = \begin{pmatrix} n_{11}, & \cdots\cdots & , n_{1t} \\ \cdots\cdots & & \\ n_{i1}, \cdots, (n_{ik} - 1) mod \ B, \cdots, n_{it} \\ \cdots\cdots & & \\ n_{m1}, & \cdots\cdots & , n_{mt} \end{pmatrix}$$

as one of the possible and feasible preceding states which are determined by schedules for jobs $J_1, \ldots J_{h-1}$.

We give some necessary remarks for this equation. In the case $n_{ik} = 1$, remember $J_h \in \mathcal{J}_{ik}$, it means that J_h itself forms the last batch of $\mathcal{J}_{ik}$. Compared with its preceding state, c_{ik} is increased by p_{ih} and n_{ik} by 1; In the case $n_{ik} = 0$ or $n_{ik} > 1$, J_h is inserted into the last batch of the preceding schedule. From the agreement assumption and the index of the jobs, the completion time c_{ik} is the same as the preceding one. And in both subcases, the *size* of $\mathcal{J}_{ik}$ equals $(n_{ik} - 1) mod \ B$.

It is essential to observe that, in any recursive equation, $\mathbf{B}$ remains the same. Therefore, in the first step of our algorithm, we should set the original values to range all the possibilities.

Now, we are about to describe the dynamic programming. In the algorithm, h records different stages(In stage h, the first h jobs $J_1, \ldots, J_h$ are scheduled.). $g(h, \mathbf{S})$ records one of the possible positions of J_h in state $\mathbf{S}$, i.e., $g(h, \mathbf{S}) = (i, k)$

means $J_h \in \mathcal{J}_{ik}$. $\mathbf{S}_h$ and $F(h)$ record the optimal state and optimal objective value in stage h, respectively.

Algorithm OMM (*Optimal Minimum Makespan*)

step 1. Let $S(1) = \{\mathbf{S} = (\mathbf{B}, \mathbf{C}, \mathbf{N}) \in L|$ for some $(i, k), 1 \leq i \leq m, i_1 \leq k \leq t : c_{ik} = p_{i1}, n_{ik} = 1,$ and other values in $\mathbf{C}, \mathbf{N}$ are all zero.$\}$ For each $\mathbf{S} \in S(1)$, say $n_{ik} = 1$, set $g(1, \mathbf{S}) := (i, k)$ and let $f(\mathbf{S}) = max\{r_t + b_{it} + c_{it} : 1 \leq i \leq m\}$; Set $\mathbf{S}_1 := argmin\{f(\mathbf{S}) : \mathbf{S} \in S(1)\}$, $F(1) := f(\mathbf{S}_1)$, $h := 2$;

step 2. If $h = n + 1$:

Case 1. $n = 1$: output $F(1)$ and $g(1, \mathbf{S}_1)$; Stop

Case 2. $n > 1$: set

$$\mathbf{S}_{n-1} := u(n, \mathbf{S}_n, g(n, \mathbf{S}_n));$$

$$\mathbf{S}_{n-2} := u(n - 1, \mathbf{S}_{n-1}, g(n - 1, \mathbf{S}_{n-1}));$$

$$\cdots$$

$$\mathbf{S}_1 := u(2, \mathbf{S}_2, g(2, \mathbf{S}_2)).$$

Output $F(n), g(n, \mathbf{S}_n), g(n - 1, \mathbf{S}_{n-1}), \cdots, g(1, \mathbf{S}_1)$; Stop

step 3. Set $S(h) := \emptyset$, $x := 1$. **step 4.** If there exists $(i, k), 1 \leq i \leq m$, $i_h \leq k < t$ such that $u(h, i, k, \mathbf{L}_x) \in S(h - 1)$, (where $u(h, i, k, \mathbf{L}_x)$ is defined the same as before), set $S(h) := S(h) \bigcup \{\mathbf{L}_x\}$, $g(h, \mathbf{S}) := (i, k)$

step 5. If $x < q$, set $x := x + 1$ and goto **step 4**.

step 6. For each $\mathbf{S} = (\mathbf{B}, \mathbf{C}, \mathbf{N}) \in S(h)$, let $f(\mathbf{S}) = max\{r_t + b_{it} + c_{it} : 1 \leq i \leq m\}$ and set $\mathbf{S}_h := argmin\{f(\mathbf{S}) : \mathbf{S} \in S(h)\}$, $F(h) := f(\mathbf{S}_h)$, $h := h + 1$, goto **step 2**.

It is easy to show that the running time of algorithm OMM is $O(nmt \cdot (BP_{sum}P_{max})^{mt})$. For constant m and t, it is a pseudo-polynomial in n. Hence the special case of $\overline{R}_m|r_j, B|C_{max}$ is not strongly NP-hard.

3.3   A special Case with Constant Distinct Release Times but Arbitrary Inputs

Algorithm OMM solves the $\overline{R}_m|r_j, B|C_{max}$ problem optimally when the input values are integral. We now extend this algorithm to an FPTAS for the same problem without the integral restriction.

Algorithm AMM(ε) (*Approximate Minimum Makespan*)

step 1. Let $M_0 = max\{r_{max}, P_{max}\}$ and $M = \varepsilon M_0/(2n)$

step 2. Construct a rounded down instance $(\tilde{\mathcal{J}}, B, m)$ such that each job J_j in $\tilde{\mathcal{J}}$ has release time $\tilde{r}_j = M\lfloor r_j/M \rfloor$ and processing times $\tilde{p}_{1j} = M\lfloor p_{1j}/M \rfloor, \ldots, \tilde{p}_{mj} = M\lfloor p_{mj}/M \rfloor$

step 3. Take M as a unit in the rounded down instance, i.e., scale it, we still denote the newly scaled instance by $(\tilde{\mathcal{J}}, B, m)$. Hence $(\tilde{\mathcal{J}}, B, m)$ satisfies the integral restrictions in Subsection 3.2. Call algorithm OMM to find an optimal schedule $\tilde{\pi}$ for $(\tilde{\mathcal{J}}, B, m)$.

step 4. Derive the corresponding original schedule π and output it as the solution to the original instance.

It is crucial to remark that the rounded-down instance still meets the demand of agreement, and thus we can call algorithm OMM to solve it. The same remark is also necessary for the rounding in the next subsection and we omit it.

Due to rounding, the start time of each batch in π may be later than that of the corresponding batch in $\tilde{\pi}$, and its processing time may also be longer, but at most longer by M (in the original scale). And furthermore, the start time of each batch in π is at most later than that of the corresponding batch in $\tilde{\pi}$ by $(n' + r')M$, where n' is the number of batches processed on the same machine and before this batch, and r' the number of distinct release times earlier than its start time. Therefore:

$$C_{max} \leq \tilde{C}_{max} + (n' + r')M \leq \tilde{C}_{max} + 2nM$$

where C_{max} is the makespan of π and $\tilde{C}_{max}$ that of $\tilde{\pi}$. Note that the release time and processing times of each job in $(\mathcal{J}, B, m)$ may be larger than that of the corresponding job in $(\tilde{\mathcal{J}}, B, m)$, and $\tilde{\pi}$ is the optimal schedule of $(\tilde{\mathcal{J}}, B, m)$, we have $\tilde{C}_{max} \leq C^*_{max}$, where C^*_{max} is the optimal makespan of $(\mathcal{J}, B, m)$. Thus:

$$\frac{C_{max}}{C^*_{max}} \leq \frac{\tilde{C}_{max} + 2nM}{C^*_{max}} \leq 1 + \frac{2nM}{C^*_{max}} = 1 + \frac{2n \cdot \varepsilon M_0/(2n)}{C^*_{max}} \leq 1 + \varepsilon$$

Now, let's calculate the running time. As the main time of algorithm $\mathrm{AMM}(\varepsilon)$ is in step 3, i.e., calling OMM to solve $(\tilde{\mathcal{J}}, B, m)$, its running time is:

$$O\left(nmt\left(B\lfloor P_{max}/M\rfloor\lfloor P_{sum}/M\rfloor\right)\right)$$

$$\leq O\left(nmt\left(\frac{BP_{max}P_{sum}}{M^2}\right)^{mt}\right)$$

$$\leq O\left(n^{mt+1}mtB^{mt}\left(\frac{BnP^2_{max}}{M^2}\right)^{mt}\right)$$

$$= O\left(n^{mt+1}mtB^{mt}\left(\frac{P_{max}}{M}\right)^{2mt}\right)$$

$$\leq O\left(n^{mt+1}mtB^{mt}\left(\frac{2n}{\varepsilon}\right)^{2mt}\right)$$

$$= O\left(n^{3mt+1}mt(4B)^{mt}\left(\frac{1}{\varepsilon}\right)^{2mt}\right)$$

It is polynomial both in n and $\frac{1}{\varepsilon}$ for constant m and t. According to the above analysis, we have :

Theorem 1. *The family of algorithms $\{AMM(\varepsilon)\}_\varepsilon$ is an FPTAS for $\overline{R}_m|r_j, B|C_{max}$ with the restriction of constant distinct release times.*

Next, we will extend algorithm AMM(ε) to a PTAS for problem $\overline{R}_m|r_j, B|C_{max}$ in which the number of distinct release times, t, is non-constant. Our basic idea is to round all the release times such that the number of distinct release times can be reduced to a constant (unrelated with n but related with given ε), and the according error is in our control.

3.4   A PTAS for $\overline{R}_m|r_j, B|C_{max}$

Algorithm AMM$_2(\varepsilon)$

step 1. Let $K = (\varepsilon/2)r_{max}$

step 2. Construct a rounded down instance $(\tilde{\mathcal{J}}, B, m)$. Each job J_j in $\tilde{\mathcal{J}}$ is associated with a release time $\tilde{r}_j = K\lfloor r_j/K \rfloor$ and the same processing times as the corresponding job in $\mathcal{J}$.

step 3. Call AMM($\varepsilon/2$) to find an $(1+\varepsilon/2)-$approximate schedule $\tilde{\pi}$ for the rounded down instance.

step 4. Output π as the approximate solution for the original instance, where the batching and sequencing in π are the same as that in $\tilde{\pi}$.

Theorem 2. *The family of algorithms* $\{AMM_2(\varepsilon)\}_\varepsilon$ *is a PTAS for agreeable* $R_m|r_j, B|C_{max}$

Proof. First of all, it is easy to see that there are at most $\lfloor 2/\varepsilon \rfloor + 1$ distinct release times in the rounded down instance, and thus the running time of AMM$_2(\varepsilon)$ is:

$$O\left(n^{3m((\lfloor \frac{2}{\varepsilon} \rfloor+1)+1)} m \left(\lfloor \frac{2}{\varepsilon} \rfloor + 1 \right) (4B)^{m(\frac{2}{\varepsilon}+1)} \left(\frac{1}{\varepsilon} \right)^{2m(\frac{2}{\varepsilon}+1)} \right)$$

$$\leq O\left(n^{\frac{6m}{\varepsilon}+3m+1} m \left(\frac{2}{\varepsilon} + 1 \right) (4B)^{(\frac{2}{\varepsilon}+1)m} \left(\frac{1}{\varepsilon} \right)^{2m(\frac{2}{\varepsilon}+1)} \right)$$

which is a polynomial in n (but exponential in $\frac{1}{\varepsilon}$).

As for the accuracy, denote by $\tilde{C}_{max}, \tilde{C}_{max}^*, C_{max}, C_{max}^*$ the makespan obtained in $\tilde{\pi}$, the optimal makespan of $(\tilde{\mathcal{J}}, B, m)$, makespan determined by π and the optimal makespan of $(\mathcal{J}, B, m)$, respectively. According to Theorem 3.3, $\tilde{C}_{max} \leq (1 + \frac{\varepsilon}{2})C_{max}$. And due to the rounding, $\tilde{C}_{max}^* \leq C_{max}^*$, $C_{max} \leq \tilde{C}_{max} + K$. Therefore,

$$C_{max} \leq \left(1 + \frac{\varepsilon}{2} \right) \tilde{C}_{max}^* + K$$

$$\leq \left(1 + \frac{\varepsilon}{2} \right) \tilde{C}_{max}^* + \left(\frac{\varepsilon}{2} \right) r_{max}$$

$$\leq \left(1 + \frac{\varepsilon}{2} \right) \tilde{C}_{max}^* + \frac{\varepsilon}{2} C_{max}^*$$

$$= (1 + \varepsilon) C_{max}^*$$

Hence the theorem. $\square$

4   Conclusion and Remarks

In this paper, we extend the result of Xiaotie Deng et. al. to a constant number of agreeable unrelated parallel machines. We give a PTAS for the $R_m|r_j, B|C_{max}$ problem under the assumption of agreement. Given the strongly NP-hardness, it is the best possible. For the more general problem with non-constant machines and/or without the agreement assumption, further discussion is still needed.

References

1. J.H. Ahmadi, R.H. Ahmadi, S. Dasu, and C.S. Tang: Batching and scheduling jobs on batch and discrete processors, Operations Research, **40**(1992) 750-763
2. P. Brucker, A. Gladky, H. Hoogeveen, M.Y. Kovalyow, C.N. Poots,T. Tautenhahn,and S.L. van de Velde: Scheduling a batching machine. Journal of Scheduling. **1**(1998), 31-54
3. J.J. Bartholdi: unpublished manuscript, 1988
4. V. Chandru, C.Y. Lee and R.Uzsoy: Minimizing total completion time on batch processing machine with job families. Operations Research Letters. **13**(1993a), 61-65
5. Xiaotie Deng, Haodi Feng, Pixing Zhang, Yuzhong Zhang, H. Zhu: Minimizing mean completion time in batch processing system. Algorithmica. **38**(4)(2004), 513-528
6. Xiaotie Deng, C.K. Poon and Yuzhong Zhang: Approximation Algorithms in batch scheduling. Journal of Combinational Optimization.**7**(2003), 247-2
7. R.L. Graham, E.L. Lawler, J.K. Lenstra and A.H.G. Rinnooy Kan: Optimization and approximation in deterministic sequencing and scheduling. Annals of Discrete Mathematics. **5**(1979), 287-326
8. C.R. Glassey and W.W. Weng: Dynamic batching heuristics for simultaneous processing. IEEE Transactions on Semiconductor Manufacturing.(1991) 77-82 Rochester, 1992
9. Y. Ikura and M. Gimple: Scheduling algorithm for a single batch processing machine. Operations Research Letters. **5**(1986), 61-65
10. C.L. Li and C.Y. Lee: Scheduling with agreeable release times and due dates on a batch processing machine. European Journal of Operational Research. **96**(1997), 564-569
11. S.Li, G. Li and S. Zhang: Minimizing Makespan with Release Times on Identical Parallel Batching Machines. *Discrete Applied Mathematics*,**1**, 2005
12. C.Y. Lee and R. Uzsoy: Minimizing makespan on a single batch processing machine with dynamic job arrivals. Technical report, Department of Industrial and System Engineering, University of Florida, January 1996
13. C.Y. Lee, R. Uzsoy, and L.A. Martin Vega: Efficient algorithms for scheduling semiconductor burn-in operations. Operations Research. **40**(1992), 764-775
14. C.K. Poon and Wenci Yu: A Flexible On-line Scheduling Algorithm for Batch Machine with Infinite Capacity. Annals of Operations Research, **133**(2005), 175-181

15. C.K. Poon and Wenci Yu: On-line Scheduling Algorithms for a Batch Machine with Finite Capacity. To appear in Journal of Combinatorial optimization, 2005
16. Guochuan Zhang, Xiaoqiang Cai and C.K.Wong: On-line algorithms for minimizing makespan on batch processing machines. Naval Research Logistics. 48(2001), 241-258
17. Guochuan Zhang, Xiaoqiang Cai and C.K. Wong: Optimal on-line algorithms for scheduling on parallel batch processing machines. IIE Transactions. 35(2003), 175-181

Linear Time Algorithms for Parallel Machine Scheduling*

Zhiyi Tan[1] and Yong He[1]

1. Department of Mathematics,
State Key Lab of CAD & CG Zhejiang University,
Hangzhou 310027, P.R. China
{tanzy, mathhey}@zju.edu.cn

Abstract. This paper addresses linear time algorithms for parallel machine scheduling problems. We introduce a kind of threshold algorithms and discuss their main features. Three linear time threshold algorithm classes DT, PT and DTm are studied thoroughly. For all classes, we study their best possible algorithms among each class. We also present their application to several scheduling problems. The new algorithms are better than classical algorithms in time complexity and/or worst-case ratio. Computer-aided proof method is used in the proof of main results, which greatly simplifies the proof and decreases case by case analysis.

1 Introduction

In the scheduling theory, the classical parallel machine scheduling problem $Pm \,||C_{\max}$ is of great importance, which can be described as follows. We are given n independent jobs $\mathcal{J} = \{p_1, p_2, \ldots, p_n\}$ which must be non-preemptively processed on m identical machines $\mathcal{M} = \{M_1, M_2, \ldots, M_m\}$. We identify the jobs with their processing times. Jobs and machines are available at time zero. The objective is to minimize the maximum machine load (makespan), where the *load* of a machine is the total processing time of the jobs assigned to it. In his pioneering work, Graham [3], [4] proposed algorithms LS and LPT based on primal greedy idea and proved that their worst-case ratios are $2-1/m$ and $4/3-1/(3m)$, respectively. Here the *worst-case ratio* of an algorithm A is defined as the smallest number c such that $C^A \leq cC^{OPT}$ for all instances, where C^A and C^{OPT} denote the makespan produced by A and the optimal makespan, respectively. Coffman et al. [2] devised an algorithm $MULTIFIT$ based on dual greedy idea, which has a worst-case ratio $13/11$. Hochbaum and Shmoys [8] generalized the idea to obtain a polynomial time approximation scheme ($PTAS$). Since then, dual and primal greedy ideas are two main tools in algorithm design.

Besides worst-case ratio, time complexity is also an important criterion of an approximation algorithm. Note that both LPT and $MULTIFIT$ run in time $O(n \log n)$, while that of $PTAS$ increases exponentially when the worst-case

* Supported by NSFC (10301028, 10271110, 60021201) and TRAPOYT of China.

N. Megiddo, Y. Xu, and B. Zhu (Eds.): AAIM 2005, LNCS 3521, pp. 172–182, 2005.

ratio tends to 1, which makes it only theoretical significance. Although LS runs in linear time, its worst-case ratio is not appealing. Hence it is an open question how to design an approximation algorithm with a smaller worst-case ratio and lower time complexity [9]. As an attempt, He et al. [5] proposed a linear time algorithm for $P2||C_{\max}$ with a worst-case ratio 12/11, which is based on dual greedy idea. Noting that the worst-case ratios of LPT and $MULTIFIT$ are 7/6 and 9/8 for $P2||C_{\max}$, the new algorithm beats them with respect to both two criteria: worst-case ratio and time complexity.

In this paper, we will further consider the design and analysis of linear time algorithms. We will generalize the method in [5] to obtain a *Dual Threshold Algorithm Class* (*DT* for short), and propose a new *Primal Threshold Algorithm Class* (*PT* for short) based on primal greedy idea. Moreover, in order to get the worst-case ratios, we will use computer-aided proof method which sounds a new technique to analyze deterministic algorithms for solving parallel machine scheduling. Dozens of linear programs are solved by computer as a substitution for complicated case by case analysis of approximation algorithms. Hence this method greatly simplifies the proof and decreases case by case analysis.

2   Main Idea and Results

In [5], He et al. designed a procedure with parameters for solving $P2||C_{\max}$, which is based on dual greedy idea. By a combination of three such procedures with different parameter values, a linear time algorithm with a worst-case ratio 12/11 was obtained. Let $T = \sum_{j=1}^{n} p_j$ and $\epsilon > 0$, then the main idea of this procedure is as follows. According to the value of T and expected worst-case ratio $1 + \epsilon$, appropriately set an upper threshold and a lower threshold, and then by properly assigning jobs the algorithm tries to make one machine load between the upper and lower thresholds, resulting in the worst-case ratio of $1 + \epsilon$. Even there is no way to realize it, the algorithm can also guarantee the expected worst-case ratio by dual greedy assignment rule. We call this procedure *Dual Threshold Algorithm*. Formally, the algorithm can be formulated as follows, where thresholds are $(1 + \epsilon)T/2$ and $(1 - \epsilon)T/2$, respectively.

Dual Threshold Algorithm $DT(\epsilon)$
While there exists at least one unassigned job, and p is the first such job, **we do**:

1. (**Normal Stopping Rule, NSR**) If there exists i, $i = 1, 2$, such that the new load of M_i will be in $[(1 - \epsilon)T/2, (1 + \epsilon)T/2]$ by assigning job p to M_i, then assign p to M_i, and all remaining jobs to another machine. Stop.
2. (**Abnormal Stopping Rule, ASR**) If no machine will have a new load at most $(1 + \epsilon)T/2$ by assigning job p to it, then assign p to the machine with smaller current load and all remaining jobs to another machine. Stop.
3. (**Dual Assignment Rule, DAR**) If the new load of the machine with larger load will not greater than $(1 + \epsilon)T/2$ by assigning p to it, then assign p to this machine. Otherwise, assign p to the machine with smaller load.

It is clear that $DT(\epsilon)$ defines an approximation algorithm class DT when ϵ varies from 0 to 1, called *Dual Threshold Algorithm Class*. We say that ϵ_0 is the *tight bound* of the algorithm class DT if the worst-case ratio of $DT(\epsilon_0)$ is $1 + \epsilon_0$, and there does not exist $\epsilon < \epsilon_0$ such that the worst-case ratio of $DT(\epsilon)$ is $1 + \epsilon$. Tight bound exposes the best possible worst-case performance that a threshold algorithm class can be achieved (i.e., the best possible algorithm among the class), which is the most important measure for the class.

Generally, a scheduling problem is called *offline*, if full information on jobs is available before scheduling. By contrast, if jobs arrive one by one, and we are required to schedule jobs irrevocably on the machines as soon as they are given, without any knowledge about jobs that follow later on, this problem is called *online*. Moreover, a scheduling problem is called *semi-online* if some partial additional information about jobs is available in advance, and we cannot rearrange any job which has been assigned to machines [10], [7]. Semi-online is neither offline nor online, but somehow in between. With these definitions, we know that algorithms LPT and $MULTIFIT$ are designed only for offline problems, while LS is for online problem. Although LS can be applied to solve offline and semi-online problems, its performance is not satisfactory. However, noting that $DT(\epsilon)$ only uses the information of T and runs in time $O(n)$, it can be applied to solve not only offline problems in linear time, but semi-online problems with better performance than that of LS as well. This may be the first feature of class DT. For example, it is shown in [10] that $DT(1/3)$ is the best possible algorithm for the semi-online problem $P2|sum|C_{\max}$, where sum means that the value of T is known in advance. Here *best possible* means that no other semi-online algorithm can be better with respect to worst-case ratio. Moreover, the algorithm $DT(1/9)$ with a preprocess is the best possible algorithm for the semi-online problem $P2|sum, decr|C_{\max}$ [11], where $decr$ means that jobs arrive in decreasing order of their processing times; and a variant of $DT(1/5)$ is the best possible algorithm for the semi-online problem $P2|sum, max|C_{\max}$ [11], where max means that the largest job processing time is known in advance.

The second feature of Dual Threshold Algorithm is that it can be modified to solve other problems without difficulty. For example, it was extended to solve the offline scheduling problem with machine available times $P2$, $r_i||C_{\max}$ [5] and the classical offline uniform machine scheduling problem $Q2||C_{\max}$ [1]. Linear time algorithms with small worst-case ratios were obtained. Moreover, in Section 4, we will deploy the basic idea of DT to obtain DTm for solving semi-online problem of $Pm||C_{\max}$. The worst-case ratio of $DTm(\frac{m-1}{m+1})$ for the problem $Pm|sum|C_{\max}$ is $2 - 2/(m+1)$, which is smaller than that of an algorithm in [6]. We will also show that the worst-case ratio of $DT3(2/13)$ for the offline problem $P3||C_{\max}$ is $15/13$, and the time complexity is $O(n)$. Hence $DT3(2/13)$ beats LPT with respect to both two criteria, and has the same worst-case ratio and lower time complexity compared to $MULTIFIT$.

Last, the performance of Dual Threshold Algorithm becomes better if one more information is used, besides the total processing times. To explain it, we introduce a new concept. For given $\epsilon > 0$, if all instances of a scheduling problem on m machines satisfy

$$p_1 \geq p_2 \geq \cdots \geq p_{\lfloor (m-1)/\epsilon \rfloor} \text{ and } p_j \leq p_{\lfloor (m-1)/\epsilon \rfloor}, \ \lfloor (m-1)/\epsilon \rfloor < j \leq n. \quad (1)$$

we say that the problem satisfies *Limited Decreasing Condition* (*LDC* for short). Obviously, for any instance of an offline problem, we can renumber all jobs in time $O(n/\epsilon)$ such that it satisfies (1). Moreover, semi-online problems with decreasing job processing times, such as $P|decr|C_{\max}$, also satisfy *LDC*. In this paper, we will show the tight bound of a threshold algorithm class with *LDC* is significantly smaller than that of a threshold algorithm class without *LDC*. Here an algorithm class (or an algorithm) is called *with (without) LDC* if jobs are (not) sorted to satisfy *LDC* before applying any one of the algorithm class (or the algorithm). Note that the time complexities of the algorithms with and without *LDC* are $O(n/\epsilon)$ and $O(n)$, respectively.

In Section 3, we will develop a new *Primal Threshold Algorithm Class PT* for $P2||C_{\max}$, which has similar properties as *DT*. It differs from *DT* only in the assignment rule, i.e. it assigns jobs by primal greedy method instead of dual greedy method. We will prove that the tight bounds are 4/3 and 8/7 for *DT* without and with *LDC*, and 3/2 and 9/8 for *PT* without and with *LDC*, respectively. Here the definition of tight bound of *PT* is the same as that of *DT*. From results we conclude that both *DT* and *PT* have advantage.

3   *DT* and *PT*

We begin with the description of the new class *PT* for solving $P2||C_{\max}$.

Primal Threshold Algorithm $PT(\epsilon)$
While there exists at least one unassigned job, and p is the first such job, **we do:**

1. (**NSR**) The same as $DT(\epsilon)$
2. (**ASR**) The same as $DT(\epsilon)$
3. (**Primal Assignment Rule, PAR**) Assign p to the machine with smaller current load.

As ϵ can vary from 0 to 1, $PT(\epsilon)$ also defines an algorithm class *PT* with time complexity $O(n)$.

Let $AT \in \{DT, PT\}$ be an algorithm class. If $AT(\epsilon)$ terminates by NSR, we have $C^{AT(\epsilon)}/C^{OPT} \leq 1 + \epsilon$. Hence we suppose in the following that $AT(\epsilon)$ terminates by ASR when analyzing its worst-case ratio. Then there exists a job p_t such that the loads of both machines are less than $(1 - \epsilon)T/2$ before $AT(\epsilon)$ assigning p_t, whereas assigning p_t to any machine would make its new load greater than $(1 + \epsilon)T/2$. We call such a job p_t as *critical job*. Obviously, $p_t > \epsilon T$ and p_t determines the algorithm makespan, that is, its completion time is $C^{AT(\epsilon)}$. Moreover, if the problem satisfies *LDC*, then $t < \lceil 1/\epsilon \rceil$ since $T \geq \sum_{i=1}^{t} p_i \geq t p_t > t \epsilon T$. This conclusion will make us only consider all possible assignments of small number of jobs when proving the worst-case ratio.

Theorem 1 gives a property of the threshold algorithm classes.

Theorem 1. *If the worst-case ratio of $AT(\epsilon_0)$ is $1 + \epsilon_0$, then for any ϵ, $\epsilon_0 < \epsilon < 1$, the worst-case ratio of $AT(\epsilon)$ is $1 + \epsilon$.*

Proof. We prove $C^{AT(\epsilon)}/C^{OPT} \leq 1 + \epsilon$ by contradiction. Assume that there exists an ϵ, $\epsilon_0 < \epsilon < 1$ such that $C^{AT(\epsilon)}/C^{OPT} > 1 + \epsilon$. Then algorithm $AT(\epsilon)$ terminates by ASR and the critical job p_t must exist. Since $[(1 - \epsilon_0)T/2, (1 + \epsilon_0)T/2] \subset [(1 - \epsilon)T/2, (1 + \epsilon)T/2]$, and the fact that assigning $p_1, p_2, \cdots, p_{t-1}$ by $AT(\epsilon)$ does not enter Steps 1 and 2 of $AT(\epsilon)$, we know that assigning $p_1, p_2, \cdots, p_{t-1}$ by $AT(\epsilon_0)$ does not enter Steps 1 and 2 of $AT(\epsilon_0)$. That is to say, the assignments of $p_1, p_2, \cdots, p_{t-1}$ in $AT(\epsilon)$ and $AT(\epsilon_0)$ are identical. Moreover, the assignment of the critical job p_t is identical, too. Hence $C^{AT(\epsilon_0)} = C^{AT(\epsilon)} > (1 + \epsilon)C^{OPT} > (1 + \epsilon_0)C^{OPT}$, a contradiction.

The following instance shows the worst-case ratio of $AT(\epsilon)$ cannot be smaller: $p_1 = 1 - \epsilon$, $p_2 = \epsilon$, $p_3 = \cdots = p_{M+2} = 1/M$, where $M > 1/\epsilon$ is an integer. □

Theorem 2. *The tight bounds of the algorithm class DT without and with LDC for $P2||C_{\max}$ are 4/3 and 8/7, respectively.*

Proof. We first consider DT without LDC. Note that $DT(1/3)$ is just the algorithm for solving the semi-online problem $P2|sum|C_{\max}$ [10]. It has already been proved that $C^{DT(1/3)}/C^{OPT} \leq 4/3$. We next prove that there does not exist $\epsilon < 1/3$ such that the worst-case ratio of $DT(\epsilon)$ is $1 + \epsilon$. To see it, for any $0 < \epsilon < 1/3$, consider the instance $p_1 = p_2 = 1/6$, $p_3 = p_4 = 1/3$. We have $C^{DT(\epsilon)} = 2/3 = 4C^{OPT}/3 > (1 + \epsilon)C^{OPT}$.

Next we consider DT with LDC. We first prove that the worst-case ratio of $DT(1/7)$ with LDC is 8/7. Note that at this time LDC turns into $p_1 \geq p_2 \geq \cdots \geq p_6$ and $p_i \leq p_6$ for $i > 6$. By normalizing all job processing times we assume $T = 14$. Since we only consider the case that $DT(1/7)$ terminates by ASR, $p_t > 2$ and $t < 7$. We distinguish several cases according to the value of t.

Case 1 $t = 6$. W.l.o.g., we assume p_1 is assigned to M_1. Then p_2 is also assigned to M_1. Otherwise, we have $p_1 + p_2 > 8$ and $T > p_1 + p_2 + \sum_{j=3}^{6} p_j > 8 + 4 \cdot 2 > 14$, which contradicts $T = 14$. Further, p_3 must be assigned to M_2. Otherwise, $p_1 + p_2 + p_3 \leq 6$. It implies that $p_4 + p_5 + p_6 \leq p_1 + p_2 + p_3 \leq 6$, contradicting the fact that p_6 is a critical job. By similar arguments, we can claim that p_4 and p_5 are both assigned to M_2. Since p_6 is the critical job, we know $C^{DT(1/7)} \leq \min\{p_1 + p_2 + p_6, \sum_{j=3}^{6} p_j\}$.

Consider the assignment of the first 6 jobs in an optimal schedule. If there exists a machine processing at least 4 jobs of them, we have $C^{OPT} \geq \sum_{j=3}^{6} p_j \geq C^{DT(1/7)}$. If each machine processes exactly 3 jobs of them, we have $C^{OPT} \geq p_1 + p_5 + p_6$. Suppose $C^{DT(1/7)}/C^{OPT} > 8/7$, then $p_1 + p_2 + p_6 \geq C^{DT(1/7)} > \frac{8}{7}C^{OPT} \geq \frac{8}{7}(p_1 + p_5 + p_6)$. Simplifying this inequality, we obtain $6p_2 > 8p_5 + p_6 > 8 \cdot 2 + 2$, i.e., $p_2 > 3$. Hence $T \geq p_1 + p_2 + \sum_{j=3}^{6} p_j > 2 \cdot 3 + 4 \cdot 2 = 14$, a contradiction.

Case 2 $t \leq 5$. First let $t = 5$. Assume p_1 is assigned to M_1. If p_2 is assigned to M_2, by algorithm rule, we have $p_1 + p_2 > 8$, and thus $T > p_1 + p_2 + \sum_{j=3}^{5} p_j > 14$.

Therefore, p_2 is assigned to M_1. Then p_3 and p_4 are assigned to M_2 (since $p_1 + p_2 + p_3 \geq p_1 + p_2 + p_4 > 6$, they are greater than 8 to avoid NSR). Hence $C^{DT(1/7)} \leq p_3 + p_4 + p_5 \leq C^{OPT}$. For $t \leq 4$, the result can be proved similarly, and omitted here due to length limitation.

The following instance shows that there does not exist $\epsilon < 1/7$ such that the worst-case ratio of $DT(\epsilon)$ with LDC is $1 + \epsilon$, i.e., the tight bound of the class DT with LDC cannot be smaller than $8/7$: $p_1 = p_2 = 3$, $p_3 = \cdots = p_6 = 2$.   □

The following Theorem 3 can be shown similarly.

Theorem 3. *The tight bound of PT without LDC for* $P2||C_{\max}$ *is* $3/2$.

Theorem 4. *The tight bound of PT with LDC for* $P2||C_{\max}$ *is* $9/8$.

From the above proofs, we know the larger the value of t is, the more complicated the proof is, since the possible situations of job assignments in an algorithm and the optimal schedule become more and more. Hence if an algorithm has a smaller worst-case ratio, it is a complex work to complete its proof. In the following, we propose to prove Theorem 4 by a computer-aided proof method, which lets us leave elaborate, but complicated and troublesome computation and analysis to computer.

Roughly speaking, in order to prove the worst-case ratio of an algorithm A is not greater than c, we show the optimal objective value of a series of linear programs with the objective of maximizing $C^A - cC^{OPT}$ is at most 0. The constraints of a linear program include the estimate of C^A, C^{OPT}, inequalities describing possible assignments of algorithm and the optimal solution, and ordering of the processing times of jobs, etc. The variables include C^A, C^{OPT} and the job processing times appearing in the constraints. Since linear programs can be easily solved by mathematical softwares such as Mathematica, LINDO, etc, we do not care about the numbers of the constraints and variables in our proof.

Proof of Theorem 4. We first prove $C^{PT(1/8)}/C^{OPT} \leq 9/8$. By normalization, we assume $T = 16$. Remember that the upper and lower thresholds are 9 and 7 respectively. Similarly, since we suppose the algorithm $PT(1/8)$ terminates by ASR, $p_t > 2$ and $t < 8$.

Case 1 $t = 7$. Consider the assignment of the first 6 jobs in $PT(1/8)$ schedule. Then each machine must process exactly 3 jobs of them. Otherwise, the load of the machine processing more than 3 jobs is at least 8, which is greater than the upper threshold 7, contradicting the fact that p_7 instead of p_6 is the critical job. Assume that p_1 is assigned to M_1. Then p_2 and p_3 are assigned to M_2 by PAR. Moreover, p_4 is assigned to M_1. Otherwise, $p_1 > p_2 + p_3 > 4$, and thus $T > p_1 + \sum_{i=2}^{7} p_j > 16$, a contradiction. Hence

$$p_1 < p_2 + p_3. \tag{2}$$

There are only two situations for the assignment of p_5 and p_6 as follows:

$$
\begin{array}{ll}
M_1\ \{p_1,\ p_4,\ p_6\} & M_1\ \{p_1,\ p_4,\ p_5\} \\
M_2\ \{p_2,\ p_3,\ p_5\} & M_2\ \{p_2,\ p_3,\ p_6\} \\
\quad\quad \text{A1} & \quad\quad \text{A2}
\end{array}
$$

We consider A1 first. Thus we have

$$C^{PT(1/8)} \le p_1 + p_4 + p_6 + p_7, \quad C^{PT(1/8)} \le p_2 + p_3 + p_5 + p_7. \tag{3}$$

As p_7 is the critical job, we have

$$p_1 + p_4 + p_6 < 7, p_2 + p_3 + p_5 < 7, p_1 + p_4 + p_6 + p_7 > 9, p_2 + p_3 + p_5 + p_7 > 9, \tag{4}$$

By similar reasoning to obtain (2), we have

$$p_1 + p_4 > p_2 + p_3, \quad p_1 + p_4 < p_2 + p_3 + p_5. \tag{5}$$

As $PT(1/8)$ terminates by ASR, either $p_1 + p_2 < 7$ or $p_1 + p_2 > 9$ holds. If $p_1 + p_2 > 9$, $T > p_1 + p_2 + \sum_{j=3}^{7} p_j > 19$, a contradiction. Therefore,

$$p_1 + p_2 < 7. \tag{6}$$

Similarly, we have

$$p_2 + p_3 + p_4 < 7, \quad p_1 + p_4 + p_5 < 7, \quad p_2 + p_3 + p_5 + p_6 > 9. \tag{7}$$

The inequalities corresponding to A2 can be established similarly. Next, consider the lower bounds of C^{OPT} for A1. If in an optimal schedule there exists one machine processing at least 6 jobs of the first 7 jobs, $C^{OPT} \ge \sum_{j=2}^{7} p_j \ge C^{PT(1/8)}$ (due to (3)). If the first 7 jobs are distributed on two machines in the optimal schedule as 5:2, two possible assignments are as follows, and it is easily verified that for the remaining assignments we have $C^{PT(1/8)} \le C^{OPT}$.

M_1 $\{p_1, \; p_2\}$ M_1 $\{p_1, \; p_3\}$
M_2 $\{p_3, \; p_4, \; p_5, \; p_6, \; p_7\}$ M_2 $\{p_2, \; p_4, \; p_5, \; p_6, \; p_7\}$
        C1   C2

If the first 7 jobs are distributed on two machines in the optimal schedule as 4:3, we can easily verify that except the following 11 assignments, denoted by C3-C13, all remaining result in $C^{PT(1/8)} \le C^{OPT}$.

M_1 $\{p_1, p_2, p_3\}$ M_1 $\{p_1, p_2, p_4\}$ M_1 $\{p_1, p_2, p_5\}$ M_1 $\{p_1, p_2, p_6\}$
M_2 $\{p_4, p_5, p_6, p_7\}$ M_2 $\{p_3, p_5, p_6, p_7\}$ M_2 $\{p_3, p_4, p_6, p_7\}$ M_2 $\{p_3, p_4, p_5, p_7\}$
        C3   C4   C5   C6

M_1 $\{p_1, p_2, p_7\}$ M_1 $\{p_1, p_3, p_4\}$ M_1 $\{p_1, p_3, p_5\}$ M_1 $\{p_1, p_3, p_6\}$
M_2 $\{p_3, p_4, p_5, p_6\}$ M_2 $\{p_2, p_5, p_6, p_7\}$ M_2 $\{p_2, p_4, p_6, p_7\}$ M_2 $\{p_2, p_4, p_5, p_7\}$
        C7   C8   C9   C10

M_1 $\{p_1, p_3, p_7\}$ M_1 $\{p_1, p_4, p_5\}$ M_1 $\{p_2, p_3, p_4\}$ M_1 $\{p_2, p_3, p_5\}$
M_2 $\{p_2, p_4, p_5, p_6\}$ M_2 $\{p_2, p_3, p_6, p_7\}$ M_2 $\{p_1, p_5, p_6, p_7\}$ M_2 $\{p_1, p_4, p_6, p_7\}$
        C11   C12   C13   C12'

For C1, we have

$$C^{OPT} \ge p_1 + p_2, \quad C^{OPT} \ge p_3 + p_4 + p_5 + p_6 + p_7. \tag{8}$$

For any one of C2-C13, corresponding inequalities can be established similarly.

Next for A2, by similar arguments, the possible assignments of the optimal schedule may be C1-C11, C12' and C13 to avoid $C^{PT(1/8)} \le C^{OPT}$.

To obtain the worst-case ratio of the situation where the schedule yielded by $PT(1/8)$ is in the form of A1 and the optimal schedule is in the form of C1, we construct a linear program as follows: the objective is to maximize

$C^{PT(1/8)} - \frac{9}{8}C^{OPT}$, the constraints are (2)-(8) and variables are $C^{PT(1/8)}, C^{OPT}$, $p_1, p_2, \cdots, p_7$. By any software to solve linear programming, we know that its optimal value is $-9/4 < 0$. For any combination of A1 with every one of C2-C13, and A2 with every one of C1-C11 and C12', C13, we can also construct the corresponding linear programs in the same way, and it can be shown their optimal values are at most 0. Hence we conclude that $C^{PT(1/8)} \leq 9C^{OPT}/8$ for $t = 7$.

Case 2 $t = 6, 5$. The proof is similar to Case 1. For detail see the full paper.
Case 3 $t \leq 4$. It is trivial to get the worst-case ratio by direct deduction.

Finally, the instance $p_1 = p_2 = 3$, $p_3 = \cdots = p_7 = 2$ shows that the tight bound of the algorithm class PT with LDC cannot be smaller than $9/8$. □

During above process, we have solved totally 36 linear programs which cover all possible situations. In fact, as we only need to consider the assignment of $p_1, \cdots, p_t$, the numbers of linear programs, the variables and constraints in a given programs must be finite. Note that each linear program corresponds to one possible situation which may occur when analyzing algorithm. Recall that in our proof, we exclude several possible assignments, for example, the ones yield $C^{PT(1/8)} \leq C^{OPT}$. In fact, we can also construct their corresponding linear programs, and their objective function values are less than 0, too. We prefer to this analysis instead of solving linear programs for the purpose of theoretical conciseness, since an easily manual analysis excluded several situations. Furthermore, when constructing linear programs, we do not need to find out all possible constraints of a linear program. In fact, those which can make the optimal objective values at most 0 are sufficient. In other words, each linear program solved here is only a relaxation of the considered situation.

4   Dual Threshold Algorithm Class DTm

In this section, we present a new dual threshold algorithm class DTm for $Pm||C_{\max}$.

Dual Threshold Algorithm $DTm(\epsilon)$
While there exists at least one unassigned job, and p is the first such job, **we do:**

1. **(NSR)** If there exists i, $i = 1, 2, \cdots, m$, such that by assigning job p to M_i the number of machines with new loads in $[(1 - \frac{\epsilon}{m-1})\frac{T}{m}, \frac{(1+\epsilon)T}{m}]$ becomes $m - 1$, then assign p to M_i, and all remaining jobs to the unique machine with current load smaller than $(1 - \frac{\epsilon}{m-1})\frac{T}{m}$. Stop.

2. **(ASR)** If any machine will have a new load greater than $\frac{(1+\epsilon)T}{m}$ by assigning job p to it, then assign p and all remaining jobs according to LS rule. Stop.

3. **(DAR)** Denote by $\mathcal{M}_0$ the subset of machines M_i whose new load of M_i will not greater than $\frac{(1+\epsilon)T}{m}$ if assigning p to it. Assign p to the machine with the largest current load among $\mathcal{M}_0$.

Note that in the Step 3 of the algorithm, $\mathcal{M}_0$ must be not empty. Otherwise, the algorithm would stop at Step 2 (by ASR). Moreover, before the algorithm enters Step 2, no machine has a current load greater than the upper threshold.

The following result can also be proved by computer-aided proof method, we choose to use traditional analysis for conciseness and exploring the idea.

Theorem 5. *For any $m \geq 3$, the tight bound of DTm without LDC for $Pm||C_{\max}$ is $2 - \frac{2}{m+1}$.*

Proof. Denote $C = C^{DTm(\frac{m-1}{m+1})}$ for short. We first show that $C/C^{OPT} \leq 2 - \frac{2}{m+1}$. By normalization we assume $T = m(m+1)$. Then the upper and lower thresholds are $2m$ and m, respectively. Similarly to Section 3, suppose that $DTm(\frac{m-1}{m+1})$ stops by ASR and thus $p_t > m$.

Case 1 On the arrival of p_t, a machine not processing any job exists.

Note that assigning the critical job p_t to any machine, its new load would be greater than the upper threshold $2m$. Hence, from the fact that there exists an empty machine, we have $p_t > 2m$. It implies $C^{OPT} \geq p_t > 2m$. Suppose that $C/C^{OPT} > 2 - \frac{2}{m+1}$. Then there exists a machine with a load greater than $C > (2 - \frac{2}{m+1})C^{OPT} > \frac{4m^2}{m+1}$. Let p be the last job assigned to this machine. Then p is assigned by LS rule. Hence the remaining $m-1$ machines except one processing p have loads at least $C - p$. Summing all machine loads, we have

$$(m-2)(C-p) + C + p_t \leq T, \qquad (9)$$

i.e., $p \geq \frac{(m-1)C+p_t-T}{m-2}$. Substituting $C > \frac{4m^2}{m+1}$, $p_t > 2m$ and $T = 2m$ into this inequality, we obtain

$$p > \frac{m(3m^2 - 4m + 1)}{(m-2)(m+1)}. \qquad (10)$$

At the same time, (9) implies

$$C - p \leq \frac{T - C - p_t}{m - 2} < \frac{m(m+1) - \frac{4m^2}{m+1} - 2m}{m - 1} = \frac{m(m^2 - 4m - 1)}{(m-2)(m+1)}. \qquad (11)$$

Combining (10) and (11), we obtain a contradiction as follows.

$$\frac{C}{C^{OPT}} \leq \frac{C}{p} = 1 + \frac{C-p}{p} \leq 1 + \frac{m^2 - 4m - 1}{3m^2 - 4m + 1} < 2 - \frac{2}{m+1}.$$

Case 2 On the arrival of p_t, all machines have processed at least one job.

Denote p_{j_i} be the first job assigned to M_i, $1 \leq i \leq m$. W.l.o.g., we assume $1 = j_1 < j_2 < \cdots < j_m$. We prove by induction that after assigning p_{j_l} by $DTm(\frac{m-1}{m+1})$, there are at least $l - 1$ machines among $\{M_1, M_2, \ldots, M_l\}$ with loads in $[m, 2m]$. The claim is obviously true for $l = 1$. Now assume that it is true for $l = k$. If on the arrival of $p_{j_{k+1}}$, the loads of $M_1, M_2, \ldots, M_k$ are all greater than m, we have done. Otherwise, there exists a machine, denoted by M_0, with a current load $L(M_0) < m$. As $p_{j_{k+1}}$ is not assigned to M_0 by DAR, we have $L(M_0) + p_{j_{k+1}} > 2m$, i.e. $p_{j_{k+1}} > m$. Hence M_{k+1} processing $p_{j_{k+1}}$ has a load greater than m, and there are at least $k - 1 + 1 = k$ machines among $\{M_1, \ldots, M_k, M_{k+1}\}$ with loads in $[m, 2m]$, which implies that the claim is true.

By the above claim, we know that after the assignment of p_{j_m}, there are at least $m-1$ machines with loads in $[m, 2m]$. Since $T = m(m+1)$, we deduce that the algorithm terminates by NSR, contradicting the fact that p_t is a critical job.

The following instance shows that the tight bound of DTm cannot be smaller than $2 - \frac{2}{m+1}$: $p_1 = \cdots = p_m = 1$, $p_{m+1} = \cdots = p_{2m} = m$. □

By Theorem 5, the worst-case ratio of $DTm(\frac{m-1}{m+1})$ for semi-online problem $Pm|sum|C_{\max}$ is also $2 - \frac{2}{m+1}$, which is smaller than $2 - \frac{1}{m-1}$, the worst-case ratio of an algorithm in [6].

The following theorem can be shown by a similar analysis as above, but more complicated and lengthy, and omitted here. However if we incorporate with the computer-aided proof method introduced in Section 3, the proof can be easy.

Theorem 6. *The tight bound of $DT3$ with LDC for $P3||C_{\max}$ is 15/13.*

By Theorem 6, we know that the worst-case ratio of $DT3(2/13)$ with LDC is smaller than and equal to those of LPT and $MULTIFIT$, respectively, whereas the complexity is smaller than those of latter two algorithms.

We conjecture that the tight bound of DTm with LDC must be strictly smaller than that of DTm without LDC for any $m > 4$, too.

5   Concluding Remarks

In this paper, we gave a comprehensive discussion of a kind of threshold algorithms. Three threshold algorithm classes with linear time complexity were studied carefully. Future research may include the following issues. First, by extending the basic idea of PT, we can obtain PTm for m machine problems. Then the tight bounds of the new class with or without LDC is open. Since PT with LDC has a smaller tight bound than that of DT with LDC, we conjecture that this property may remain for PTm and DTm. Second, as mentioned above, in [5] and [1], the authors obtained linear time approximation algorithms with very small worst-case ratios through combinations of Dual Threshold Algorithm, we conjecture that through combinations of PTm, DTm (PT, DT) and their variants, we can obtain improved approximation algorithms with lower time complexities and smaller worst-case ratios for offline parallel machine scheduling problems. It is also interesting to modify the threshold algorithms to obtain the best possible algorithm for other semi-online problems, just like in [11]. We highlight that the computer-aided proof method may be powerful for these issues.

References

1. R. E. Burkard, Y. He, H. Kellerer, A linear compound algorithm for uniform machine scheduling, *Computing*, **61**, 1-9(1998).
2. E. G. Coffman, M. R. Garey, D. S. Johnson, An application of bin-packing to multiprocessor scheduling, *SIAM Journal on Computing*, **7**, 1-17(1978).

3. R. L. Graham, Bounds for certain multiprocessor anomalies, *Bell Systems Technical Journal*, **45**, 1563-1581(1966).
4. R. L. Graham, Bounds on multiprocessing timing anomalies, *SIAM Journal on Applied Mathematics*, **17**, 416-429(1969).
5. Y. He, H. Kellerer, V. Kotov, Linear compound algorithms for the partitioning problem, *Naval Research Logistics*, **47**, 593-601(2000).
6. Y. He, Q. F. Yang, Z. Y. Tan, E. Y. Yao, Algorithms for semi on-line multiprocessor scheduling, *Journal of Zhejiang University Science*, **3**, 60-64(2002).
7. Y. He, G. C. Zhang, Semi on-line scheduling on two identical machines. *Computing*, **62**, 179-187(1999).
8. D. S. Hochbaum, E. L. Shmoys, Using dual approximation algorithms for scheduling problems: Theoretical and practical results, *Journal of the ACM*, **34**, 144-162(1987).
9. D. S. Hochbaum, E. L. Shmoys, A polynomial approximation scheme for scheduling on uniform processors: Using the dual approximation approach, *SIAM J. on Computing*, **17**, 539-551(1988).
10. H. Kellerer, V. Kotov, M. G. Speranza, Z. Tuza, Semi on-line algorithms for the partition problem, *Operations Research Letters*, **21**, 235-242(1997)
11. Z. Y. Tan, Y. He, Semi-on-line problems on two identical machines with combined partial information, *Operations Research Letters*, **30**, 408-414(2002).

A New Method for Retrieval Based on Relative Entropy with Smoothing*

Hua Huo[1,2], Junqiang Liu[2], and Boqin Feng[1]

[1] Department of Computer Science,
Xi'an Jiaotong University,
Xi'an, P.R. China
hhuo@mail.xjtu.edu.cn
[2] School of Electronics and Information,
Henan University of Science and Technology,
Luo yang, P.R. China

Abstract. A new method for information retrieval based on relative entropy with different smoothing methods has been presented in this paper. The method builds a query language model and document language models respectively for the query and the documents. We rank the documents according to the relative entropies of the estimated document language models with respect to the estimated query language model. While estimating a document language, the efficiency of the smoothing method is considered, we select three popular and relatively efficient methods to smooth the document language model. The feedback documents are used to estimate a query model by the approach that we assume that the feedback documents are generated by a combined model in which one component is the feedback document language model and the other is the collection language model. Experimental results show that the method is effective and performs better than the basic language modeling approach.

1 Introduction

In recent years, an approach based language modeling has been successfully applied to the problem of retrieval [1, 3, 4, 6]. The basic idea behind the approach is extremely simple - estimate a language model for each document, and rank documents by the likelihood of the query according to the language. The relative simplicity and effectiveness of the language modeling approach make it an attractive framework of text retrieval methodology. Although it has performed well empirically, a significant amount of performance increase is often due to smoothing method and feedback [6]. In the most of existing work, both query

---

* This research is supported by the Natural Science Foundation Program of the Henan Provincial Educational Department in China(200410464004) and the Science Research Foundation Program of Henan University of Science and Technology in China(2004ZY041).

N. Megiddo, Y. Xu, and B. Zhu (Eds.): AAIM 2005, LNCS 3521, pp. 183–191, 2005.

and document are assumed being generated from the same document language model, the difference between query and document is not considered in them.

And only some simple smoothing methods such as linear interpolation method etc. have been used to solve "zero probability" problem. Feedback has so far only been deal with heuristically within the language modeling approach, and it has been incorporated in an unnatural way: by expanding a query with a set of terms [10]. All these have become obstacles to further improve retrieval performance.

We propose a new retrieval method based on relative entropy with different smoothing methods. We build the query language model and document language models for the query and the documents respectively, then rank the documents according to the relative entropies of the estimated document language models with respect to the estimated query language model. While estimating a document language, the efficiency of the smoothing method is considered, so we select three popular and relatively efficient methods to smooth the document language model. When estimating a query model, we develop a natural approach to perform feedback, in which we assume the feedback documents are generated by a combined model in which one component is the feedback document language model and the other is the collection language model.

The rest of the paper is organized as follows. We discuss the retrieval based on language model and the problem of zero probability in section 2. A narrate of the concept of relative entropy and the rank criteria with relative entropy are given in section 3.1 and 3.2. Then we give the computing method of $p(t|\theta_d)$ and $p(t|\theta_q)$ in section 3.3 and 3.4. Experiments and results are given in section 4, and conclusions of the study are presented in Section 5.

2   Retrieval Based on Language Model and the "Zero Probability" Problem

Applied to information retrieval, language modeling refers to the problem of estimating the likelihood that the different query and document can be generated by the same language model, given the language model of the document [1]. The general idea is to build a language model θ_d for each document d , and rank the documents according to how likely the query q can be generated from each of these document models, i.e. $p(q|\theta_d)$. In different models, the probability is calculated in different ways. There are two typical methods for doing it. For example, Ponte and Croft [6] treat the query q as a set of unique terms, and use the product of two probabilities - the probability of producing the query terms and the probability of not producing other terms - to approximate $p(q|\theta_d)$. The formula is

$$p(q|\theta_d) = \prod_{t \in q} p(t|\theta_d) \prod_{tnot \in q} (1.0 - p(t|\theta_d)) \tag{1}$$

Song and Croft [7] treat the query as a sequence of independent terms, taking into account possibly multiple occurrences of the same term. Thus the query probability can be obtained by multiplying the individual term probabilities, and the formula can be written as

$$p(q|\theta_d) = \prod_{i=1}^{n} p(t_i|\theta_d) \tag{2}$$

where t_i is the ith term in the query. The simplest method to estimate $p(t|\theta_d)$ is the maximum likelihood estimator, simply given by relative counts

$$p(t|\theta_d) = P_{ml}(t|\theta_d) \tag{3}$$

$$= \frac{tf(t,d)}{\sum_{t'} tf(t',d)} \tag{4}$$

where $tf(t,d)$ is the number of times the term t occurs in the document d, $\sum_{t'} tf(t',d)$ is the total number of times all terms occur in the document d, it is essentially the length of the document d.

We can find that one obstacle in applying language modeling to information retrieval is the problem of zero probability. Documents are not very large. As a result, many terms are missing in a document. If such a term is used in a query, we would always get a zero probability for the entire query. To address the problem of zero probability, we must use a smoothing method in the retrieval based on language model. The term smoothing refers to the adjustment of the maximum likelihood estimator of a language model so that it will be more accurate [10].

We also use language modeling for retrieval, but it is different from the above approaches in which both the query and the document are assumed to be generated from the document language model. We assume that the query is generated from the query language model while the document is generated from the document language model. We rank the documents according to the relative entropies of the estimated document language models with respect to the estimated query language model. But there is still the problem of zero probability while we computing the term probability by using document language in our approach, so we will use different smoothing method to solve the problem.

3   Retrieval Based on Relative Entropy

3.1   Relative Entropy

The relative entropy is a measure of the distance between two distributions. In statistics, it arises as an expected logarithm of the likelihood ratio[2]. The relative entropy $D(p||q)$ is a measure of the inefficiency of assuming that the distribution is q when the true distribution is p.

Definition 1. *The relative entropy between two probability mass functions $p(x)$ and $q(x)$ is defined as $D(p||q) = \sum_{x} p(x) \log \frac{p(x)}{q(x)}$*

In the above definition, we use the convention (based on continuity arguments) that $0 \log \frac{0}{0} = 0$ and $p \log \frac{p}{0} = \infty$.

3.2   Ranking Criteria with Relative Entropy

The basic idea of the ranking model with relative entropy is to score a document with respect to a query based on the relative entropy between an estimated document language model and an estimated query language model, and, then rank the documents according to their scores. Let's suppose that a query q is generated by a generative model $p(q|\theta_q)$ with θ_q denoting the parameters of the query unigram language model. Similarly, assume that a document d is generated by a generative model $p(d|\theta_d)$ with θ_d denoting the parameters of the document unigram language model. Let $\hat{\theta}_q$ and $\hat{\theta}_d$ be the estimated query language model and document language model respectively, the relevance value of d with respect to q can be measured by the following function:

$$R(\hat{\theta}_q||\hat{\theta}_d) = -D(\theta_q||\theta_d) \tag{5}$$

$$= -\sum_t p(t|\theta_q) \log \frac{p(t|\theta_q)}{p(t|\theta_d)} \tag{6}$$

$$= \sum_t p(t|\theta_q) \log p(t|\theta_d) - \sum_t p(t|\theta_q) \log p(t|\theta_q) \tag{7}$$

The second term of the formula (7) is a query-dependent constant, or more specifically, the entropy of the query model θ_q. It can be ignored for the purpose of ranking documents, so we have a ranking formula such as

$$R(\hat{\theta}_q||\hat{\theta}_d) \propto \sum_t p(t|\theta_q) \log p(t|\theta_d) \tag{8}$$

From the formula (8), we can observe that how to estimate the values of $p(t|\theta_d)$ and $p(t|\theta_q)$ will influences the retrieval accuracy of the ranking model.

3.3   Computing $p(t|\theta_d)$ By Different Smoothing Methods

As the description in section 2, the problem of zero probability will occur if we use the maximum likelihood estimator to estimate $p(t|\theta_d)$ simply. We must use a smoothing method to solve it. Some smoothing methods, such as Good-Turing method, Jelinek-Mercer method, absolute discounting method, and Bayesian method using Dirichlet priors etc., have been proposed, mostly in the context of speech recognition tasks. In general, all smoothing methods are trying to discount the probabilities of the terms seen in the document, and then to assign the extra probability mass to the unseen terms according to some "fallback" model [8]. It makes much sense, and is very common, to exploit the collection language model as the fallback model. Because a retrieval task typically requires efficient computations over a large collection of documents, our study is constrained by the efficiency of the smoothing method. In this paper, we select the Jelinek-Mercer method, absolute discounting method, and Bayesian method using Dirichlet priors which are popular and relatively efficient to implement [9].

1) The $Jelinek-Mercer\ method$ involves a linear interpolation of maximum likelihood model with the collection model, using a coefficient λ to control the influence of each model [8].The method is given by

$$p(t|\theta_d) = (1 - \lambda)p_{ml}(t|\theta_d) + \lambda p(t|C) \qquad (9)$$

$$p(t|C) = \frac{tf(t,C)}{\sum\limits_{t'} tf(t',C)} \qquad (10)$$

where $P_{ml}(t|\theta_d)$ is given in the formula (4),$tf(t,C)$ is the number of times the term t occurs in the document collection C, $\sum\limits_{t'} tf(t',C)$ is the total number of times all terms occur in the document collection C.

2) The idea of the $absolute\ discounting\ method$ is to lower the probability of seen words by subtracting a constant from their counts. It is similar to the Jelinek-Mercer method, but differs in that it discounts the seen word probability by subtracting a constant instead of multiplying it by $1 - \lambda$. The model is given by

$$p(t|\theta_d) = \frac{max(tf(t,d) - \delta, 0)}{\sum\limits_{t'} tf(t',d)} + \sigma p(t|C) \qquad (11)$$

where $\delta \in [0,1]$ is a discount constant and $\sigma = \frac{\delta|d|_u}{|d|}$, so that all probabilities sum to one. Here, $|d|_u$ is the number of unique terms in the document d , and $|d| = \sum\limits_{t'} tf(t',d)$ is the total count of terms in the document.

3) In the $Bayesian\ method\ using\ Dirichlet\ priors$, a language model is a multinomial distribution, for which the conjugate prior for Bayesian analysis is the Dirichlet distribution with parameters $(\mu p(t_1|C), \mu p(t_2|C), ..., \mu p(t_n|C))$. Thus, the model is given by

$$p(t|\theta_d) = \frac{tf(t,d) + \mu p(t|C)}{\sum\limits_{t'} tf(t',d) + \mu} \qquad (12)$$

where μ is the smoothing parameter.

3.4   Computing $p(t|\theta_q)$

We assume the feedback documents are generated by a combined model in which one component is the feedback document language model and the other is the collection language model. Let q_0 be the original query, and $p(t|\theta_{q0})$ be the original query language model,q be the updated query, and $p(t|\theta_q)$ be the updated query language model. We assume that $F = (f_1, f_2, ..., f_n)$ is the set of feedback documents which are judged to be relevant by a user, or which are the top documents from an initial retrieval, and $p(t|\theta_F)$ is the language model of the set F. Then, the updated query model $p(t|\theta_q)$ is

$$p(t|\theta_q) = (1 - \alpha)p_{ml}(t|\theta_{q_0}) + \alpha p(t|\theta_F) \tag{13}$$

where α controls the influence of the feedback documents set model to $p(t|\theta_q)$. For estimating $p(t|\theta_F)$, we assume that the feedback documents are generated by a probabilistic model $p(F|\theta_F)$. Specifically, assume that each term in F is generated independently according to θ_F by a generative model which is a unigram language model. That is,

$$p(F|\theta_F) = \prod_i \prod_t p(t|\theta_F)^{tf(t,f_i)} \tag{14}$$

where $tf(t, f_i)$ is the number of times term t occurs in the document f_i.

Some information may be "background noise". The "background noise" is not considered in the above model, so, it is not very reasonable. A more reasonable model would be a combined model that generates a feedback document by combining the feedback document model $p(t|\theta_F)$ with a collection language model $p(t|\theta_C)$. We have

$$p(F|\theta_F) = \prod_i \prod_t ((1 - \beta)p(t|\theta_F) + \beta p(t|C))^{tf(t,f_i)} \tag{15}$$

where β is a parameter that indicates the amount of "background noise" in the feedback documents, and that needs to be set empirically.

We can use EM (Expectation Maximum) algorithm to compute the maximum likelihood estimate of θ_F. The estimated θ_F is as

$$\hat{\theta_F} = argmax_{\theta_F} \log p(F|\theta_F) \tag{16}$$

We use the result $p_\beta(t|\theta_F)$ of using EM algorithm as a substitute for $p(t|\theta_F)$ of the formula (13). Then we can obtain the value of $p(t|\theta_q)$ by using the formula (13).

4   Experiments and Results

4.1   Data Sets

We experiment over three data sets taken from TREC (Text Retrieval Conference). They are the Wall Street Journal (WSJ) 1987-1989 with queries 51-100, San Jose Mercury News (SJMN) 1991 with queries 51-150, and the Los Angeles Times (LA) (1989, 1990) with queries 301-400. Queries are taken from the title field of TREC topics. Relevance judgments are taken from the judged pool of top retrieved documents by various participating retrieval system from previous[3].

4.2   Experimental Setup

For the convenience of describing the experiments, we call our retrieval approaches with the Jelinek-Mercer method, absolute discounting method, and Bayesian method using Dirichlet priors JMM, ADM, and BDM respectively. And we call the retrieval method in [5] based on traditional language modeling (with feedback) TLM.

Two sets of experiments are performed in this paper. The first set of experiments is to compare the performances of JMM, ADM, and BDM with the performance of TLM. We get different results by using different smoothing parameter settings in JMM, ADM, and BDM, but we use the best results of them while comparing their performances. The second set of experiments investigates whether the performances of JMM, ADM, and BDM are sensitive to the smoothing parameters λ, δ, and μ.

4.3   Experimental Results

Results of the first set of experiments are shown in Table 1. From the average precision in Table 1, we can observe that the performances of JMM, ADM, and BDM are all better than that of TLM on the three data sets. We think that the improvements in average precision of JMM, ADM, and BDM over TLM are mainly attributed to that the relatively efficient smoothing methods and the methods of performing feedback in JMM, ADM, and BDM are more compatible with the essence of the language modeling approach than that in TLM. We also note that ADM performs better in performance than JMM, and BDM performs better than ADM again. We think they are the results of that the Absolute discounting smoothing method performs better than the Jelinek-Mercer smoothing method and the Bayesian smoothing method performs better than the Absolute discounting smoothing method.

Table 1. Average precision of TLM, JMM, ADM, and BDM on WSJ, SJMN, and LA

Data set	TLM	JMN	ADM	BDM	Chg1	Chg2	Chg3
WSJ	0.245	0.271	0.275	0.285	+10.6%	+12.2%	+16.4%
SJMN	0.273	0.296	0.301	0.311	+8.4%	+10.3%	+14%
LA	0.251	0.282	0.287	0.294	+12.4%	+14.4%	+17.1%

Results of the second set of experiments are presented in Fig.1. The plots of part (a) show the average precision of JMM for different settings λ on the three data sets, the plots of part (b) show the average precision of ADM for different settings of δ, and the plots of part (c) show the average precision for different settings of μ. We can observe that the average precision of JMM, ADM, and BDM is quite sensitive to the settings of $\lambda,\delta,$and μ. JMM performs well when the value of λ approximates to 0.65, ADM gives better performance while the value of δ approximates to 0.8. The performance of BDM is much more sensitive to μ especially when the values of μ are small. However, the optimal values of μ seem to vary from data set to data set, though in most they are around 2000.

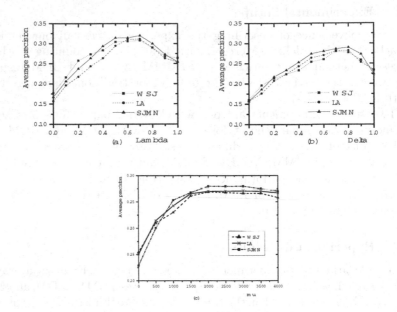

Fig. 1. Sensitivity of average precision to λ, δ, and μ

5 Conclusions

Our approach builds the query model and the document model respectively for the query and the document. We rank the documents according to the relative entropies of the estimated document language models with respect to the estimated query language model. Three popular and relatively efficient methods to smooth the document language model are used while estimating a document, and the feedback documents are used to estimate a query model. Experimental results show that the method is effective and performs better than the basic language modeling approach. Analysis of the results indicates that the performance is sensitive to the smoothing parameters, so we must pay attention to setting appropriate smoothing parameters when using the method.

References

1. Berger,A., Lafferty,J.: Information retrieval as statistical translation. In processing of the 1999 ACM SIGIR Conference on Research and Development in Information Retrieval, (1999)222-119.
2. Cover, T.M., Thomas, J.A.: Elements of Information Theory. Beijing, Tsinghua University Press, (2003)15-18.
3. Croft, W.B., Lafferty,J.: Language modeling for information retrieval. In Kluwer International Series on Information Retrieval, Volume 13, Kluwer Academic Publishers. (2003)125-128.

4. Miller, D.H., Leek,T. and Schwartz, R.: A hidden Markov model information retrieval system. In proceedings of ACM SIGIR'99 , (1999)214-221.
5. Ponte, J.: Language Models for Relevance Feedback. In W.B.Croft (Ed.), Advances in Information Retrieval: Recent Research from the CIIR. Kluwer Academic Publishers, chapter3, (2000)73-95.
6. Ponte,J., Croft,W.B.: A language modeling approach to information retrieval. In proceedings of ACM SIGIR'98, (1998)275-281.
7. Song, F., Croft,W.B.: A general language model for information retrieval. In proceedings of the 22nd annual international ACM-SIGIR'99,(1999)279-280.
8. Zaragoza,H., Hiemstra,D. and Tipping,M.: Bayesian extension to the language model for ad hoc information retrieval. In proceedings of ACM SIGIR'03.(2003)325-327.
9. Zhai,C., Lafferty,J.: A study of smoothing methods for language models applied to ad hoc information retrieval. In Proceeding of SIGIR'01, (2001)334-342.
10. Zhai,C., Lafferty,J.: Model-based feedback in the language modeling approach to information retrieval. In proceding of SIGIR'01, (2001)403-410.

Airplane Boarding, Disk Scheduling and Space-Time Geometry

Eitan Bachmat[1], Daniel Berend[2],
Luba Sapir[3], and Steven Skiena[4]

[1] Department of Computer Science, Ben-Gurion University,
[2] Departments of Mathematics and Computer Science, Ben-Gurion University,
{ebachmat, berend}@cs.bgu.ac.il
[3] Department of Industrial Engineering and Management, Ben-Gurion University,
lsapir@bgumail.bgu.ac.il
[4] Department of Computer Science, SUNY at Stony Brook,
skiena@cs.sunysb.edu

Abstract. We show how the process of passengers boarding an airplane and the process of optimal I/O scheduling to a disk drive with a linear seek function can be asymptotically modeled by 2-dimensional space-time geometry. We relate the space-time geometry of the models to important quantities such as total boarding time and total service time. We show that the efficiency of a boarding policy depends crucially on a parameter k which depends on the interior design of the airplane. Policies which are good for small values of k are bad for large values of k and vice versa.

1   Introduction

The process of airplane boarding is experienced daily by millions of passengers worldwide. Airlines have adopted a variety of boarding strategies in the hope of reducing the gate turnaround time for airplanes. Significant reductions in gate delays would improve on the quality of life for long-suffering air travelers, and yield significant economic benefits from more efficient use of aircraft and airport infrastructure [11], [9], [13].

The most pervasive strategy employed today links boarding time to seat assignment, implemented by announcements of the form "Passengers from rows 30 and above are now welcome to board the plane". It is not clear *a priori* how to analyze such strategies or to determine which policies are most effective at minimizing the expected boarding time under realistic probability distributions.

Previously, airplane boarding has been studied mainly through discrete event simulations, [11], [9], [13], [6]. In addition, the problem of minimizing airplane boarding has been considered via a related non-linear integer programming problem to which various heuristics have been applied [13].

A seemingly unrelated problem is disk scheduling. In disk scheduling one is presented with n I/O requests to a disk drive. The goal is to service the

N. Megiddo, Y. Xu, and B. Zhu (Eds.): AAIM 2005, LNCS 3521, pp. 192–202, 2005.

requests in an order which will minimize total service time. Following some experimental and heuristic studies, [8], [12], [7], Andrews et al [1] provided the first comprehensive analysis of this problem. In particular, they presented an optimal scheduling algorithm (process) in the special case in which the seek function of the disk head is assumed to be linear. we will call their algorithm, the *ABZ process*. Average case analysis of the number of disk rotations required by the ABZ process for servicing all requests was provided in [3], by relating disk scheduling to increasing subsequences in permutations.

The longest increasing subsequence in a permutation can be found by a simple procedure known as *patience sorting*. Patience sorting was apparently first invented as a card game procedure for manually sorting cards, [2]. It was reinvented several times and was shown to provide an optimal algorithm for concurrently finding the longest increasing subsequence in a permutation and a minimal partition of the permutation into decreasing subsequences. A comprehensive treatment of patience sort is provided in [2].

Finally, space-time geometry, a.k.a Lorentzian geometry, was introduced by H. Minkowski to mathematically model special relativity theory. It was subsequently used by A. Einstein to formulate general relativity theory. While Lorentzian geometry has been used to model various physical systems [14], it is still mainly associated with relativity theory. In fact, the motivation for considering these other models was to gain a better understanding of cosmology using lab experiments.

The purpose of this paper is to reveal the connections between airplane boarding, disk scheduling, patience sorting and 2-dimensional space-time geometry. In particular, we will use 2-dimensional space-time models to compute and analyze the effects on boarding time of various boarding strategies and airplane designs. To the best of our knowledge, this provides the first application of Lorentzian geometry outside of physics.

This research started from an attempt to understand disk scheduling. Then it was noticed that the same techniques can be applied to airplane boarding. A closer study of airplane boarding provided better insights into the disk scheduling problem. Thus, this study exemplifies the mutual benefits which can be gained by considering related problems from industrial engineering and computer science.

We now state the main results of the paper:

- We introduce a new discrete random process which we call *airplane boarding*. As the name suggests, the process is designed to model the process of passengers boarding an airplane.
- We explain the relations between airplane boarding, the ABZ process, patience sorting and space-time geometry. We show that passengers, disk I/O requests and elements in a permutation can be naturally considered as points in various 2-dimensional Lorentzian spaces. The ABZ process and patience sorting can then be viewed as natural "peeling" processes on the points with respect to the induced causality partial order, while airplane boarding which generalizes patience sorting provides an approximation to the peeling process. The volume form on the Lorentzian space is related to the probability

distribution on row number/queueing time of passengers and location of I/O requests, respectively.

- We show that the associated Lorentzian spaces can be used to study the asymptotics of the various processes. For instance, the length of the longest geodesic (as measured using Lorentzian length) in the Lorentzian space can be interpreted as total boarding time, number of rotations needed to service all I/O requests and size of the longest increasing subsequence in a permutation respectively.
- We demonstrate that the effectiveness of a boarding strategy depends crucially on a local congestion parameter k which is related to interior design parameters of the airplane, namely, the distance between rows and the number of passengers per row. A phase transition in the effectiveness of conventional airplane boarding strategies occurs when $k = 1$. In particular, given realistic values of this parameter (about 3.5) back-to-front boarding policies as currently practiced by airlines are unfortunately ineffective, being on the wrong side of the phase transition.

Related work: As previously mentioned, the airline boarding problem has been the subject of few prior studies [9, 11, 6, 13], the first three being purely simulation based. In particular, van den Briel et al [13] considered the problem of optimization of airplane boarding time via a related non-linear integer programming problem. The conclusions of these previous studies are compatible with our analytical results. However, previous studies yield no conceptual understanding of the problem, in particular how the effects of different parameters interact with each other.

When we assume the width of passengers to be 0, the boarding process reduces to patience sorting. In this case our modeling result is a combination and reinterpretation of results of D. Aldous and P. Diaconis [2] and of J.D. Deuschel and O. Zeitouni [4].

Organization: We define the basic parameters of our airplane boarding process and introduce the process itself in Section 2. We apply this geometric model to airplane boarding and disk scheduling in Section 3. Some consequences for airplane boarding are presented in Section 4. The appendix briefly covers some basic material on Lorentzian geometry.

2   Modeling the Airplane Boarding Process

In this section we define a discrete process which models the airplane boarding process. We assume that passengers are assigned seats in the airplane in advance of the boarding process. We assume boarding from the front of the plane, and the front row is row 1.

The general parameters of our model include:

- n denotes the number of passengers. For simplicity, we assume the airplane is full throughout this paper.
- d denotes the number of passengers per row. Thus there are $R = n/d$ rows in the airplane.
- w denotes the distance between consecutive rows in the airplane.
- u is the average width of the passengers.
- D denotes the probability distribution for *sit-down time*; how long does a passenger block the aisle while they are settling into a seat. This distribution is affected by seat type (e.g. window, aisle), amount of carry on-baggage, passenger arrival time (e.g. increased probability of requiring another passenger to move), etc. For simplicity of exposition, unless otherwise stated, we will assume that X is a constant distribution.
- We represent passengers as points (q, r) in the plane, where r is the row assigned to the passenger and q represents the time at which the passenger joins the boarding queue. For convenience, we shall normalize both variables; r is replaced by r/R, and the total amount of time allotted to joining the queue is taken as the unit of time. Via these normalizations, the point (q, r) will lie in the unit square.
- P denotes the airline's boarding policy, which is given by a function $F_P(r)$ representing the time at which passengers from row r can begin joining the queue.
- $p(q, r)$ denotes the passenger density distribution. It is the result of interaction between the airline boarding policy P and the passenger reaction model which governs how passengers react to the opportunity to join the queue. In the leisurely passenger reaction model, passengers board uniformly within the range of allowed times. In the attentive reaction model with attentiveness parameter t, passengers board uniformly within a time window of size t within the initial allowable boarding time, $F_P(r)$. Thus, $1/t$ is a measure of the passengers attentiveness. Although our methods apply equally to the leisurely reaction model, we will assume the attentive reaction model.

Our assumption of a constant distribution for X leads us to a synchronous boarding process, which we describe in terms of rounds. At first all passengers line up in a queue in front of the airplane gate. In the first round, all passengers who can walk unobstructed all the way to their assigned row do so. Those who cannot reach their seat, due to another passenger with a smaller row number obstructing their way, proceed as far as possible and wait. At the end of the round, all passengers who have reached their assigned row sit down simultaneously. Once the first round of passengers is seated, the remaining passengers can advance forward again, beginning a second round of movement. The process repeats until everyone is seated. The number of rounds needed times D will be taken as our measure of total boarding time.

We define a natural partial order on passengers. A passenger A *blocks* another passenger B, and we denote $A < B$, if the latter may sit only after the former has done so. Formally, let $A = (q_A, r_A)$, $B = (q_B, r_B)$. Let Z be the set of passengers with coordinates (q, r) satisfying $q_A < q < q_B$ and $r > r_A$. Then $A < B$ if $q_B > q_A$ and $S(r_A - r_B) \leq k|Z|$. The relation is then extended by transitivity.

In terms of the partial order, the boarding process is a well known "peeling" process, which can be traced to the work of Cantor on ordinal arithmetic. The process peels the partially ordered set by successively eliminating (in rounds) the minimal elements in the partial order. This process provides simultaneously a minimal decomposition of the poset into independent sets and the longest chain in the poset. In the boarding process each passenger (not in the first round) can be assigned a pointer which points to the last passenger who blocked his/her way to the assigned row. Following the trail of pointers starting from a passenger who sat in the last round identifies a longest chain in the partial order. In particular, the number of rounds needed is the size of the longest chain in the partial order.

Example 1. When $k = 0$ the partial order condition becomes $A < B$ if and only if $r_B > r_A$ and $q_B > q_A$, and thus chains are increasing subsequences. The boarding process is then identified with patience sorting.

3   Modeling with Lorentzian Geometry

In this section we show how we can use Lorentzian geometry to model airplane boarding and the ABZ process from disk scheduling.

For an introduction and general mathematical facts on Lorentzian geometry we refer the reader to [10]. Chapter 42 of [5] can be used as a starting point for exploring the relations of Lorentzian geometry to relativity theory.

We begin with airplane boarding. The key parameter k is defined as $k = ud/w$. As will be seen later, k plays an important role in the analysis of boarding policies. Given a density $p(q, r)$ and k, let $\alpha(q, r) = \int_r^1 p(q, u) du$, and let $g(q, r)$ be the metric associated to the quadratic form $ds^2 = 4D^2 p(q, r)(dqdr + k\alpha(q, r)dq^2)$. Explicitly, the metric is given by $g_{q,q} = 4D^2 p(q, r)k\alpha$, $g_{r,q} = g_{q,r} = 2X^2 p(q, r)$, $g_{r,r} = 0$. Let M be the *support* of the density p, that is, the closure of the set of points for which $p(q, r) > 0$. By abuse of notation we also let $M = M_{k,p}$ denote the Lorentzian space (space-time) consisting of the set M, equipped with the Lorentzian metric $g_{k,p}$. It is easy to verify that the vector $(1, 1)$ is time-like throughout M. We choose the time-like cone containing the vector $(1, 1)$ as the cone of future pointing time-like vectors throughout M. As we shall see below, the metric is constructed so that the partial orders, which are induced by the boarding process on the one hand and causality in M on the other, are locally asymptotically the same. The metric is then scaled conformally so that it asymptotically computes for two nearby points A and B the number of points in a maximal chain between A and B divided by $\sqrt{n}$.

Given a point $x \in M$, let the *height* $h(x)$ of x be the supremum of the lengths of curves in M ending at x. It can be shown that the supremum is in fact attained. The *diameter* $d(M)$ of M is the maximum of all $h(x)$, $x \in M$. We define the *level curve* L_h to consist of all points of height h.

Example: Let $k = 0$ and p be the uniform distribution. The corresponding metric gives rise to $ds^2 = 4drdq$. This is the metric of Minkowski space in a coordinate system rotated by $\pi/4$. the form of the metric is not standard since the coordinates dq and dr are light-like. We can define a time coordinate $t = (q+r)$ and space coordinate $x_1 = q - r$ which lead to the usual representation of the metric $ds^2 = dt^2 - dx^2$.

It is well known that geodesics with respect to this metric are straight lines. The longest geodesic ending at a point starts at $(0,0)$. Thus, the height of a point (q,r) is $2\sqrt{qr}$. The level curves L_h of the height function lie on the hyperbola $rq = h^2/4$. The hyperbola is a space-like curve (negative derivative), and hence we can measure its length with respect to the metric $-g$. The length of the level curve of points of height h is $2h\log(\frac{2}{h})$. The points with $r = 0$ or $q = 0$ form the points of height 0. Given a passenger A, let $t(A)$ denote the time at which passenger A sits. By our conventions, $t(A) = R(A)X$, where $R(A)$ is the round in which the passenger sits. We can now state the main modeling result for airplane boarding, where we use the notation w.h.p for an event whose probability tends to 1 as the number of passengers tends to infinity.

Theorem 3.1

A) Let $\varepsilon > 0$. Then, w.h.p, for any passenger $A = (q,r)$ we have

$$(h(A) - \varepsilon)\sqrt{n} \leq t(A) \leq (h(A) + \varepsilon)\sqrt{n}$$

In addition, the chain of blocking passengers, obtained by tracing back pointers in the boarding process starting from passenger A is contained in an ε-neighborhood of a geodesic curve of length $h(A)$ ending at A. In particular, the total boarding time is asymptotically equal to $d(M)\sqrt{n}$.

B) Let $\varepsilon > 0$. For $\theta > 0$, let $N(\theta)$ be the number of passengers boarding the plane within the first $\theta\sqrt{n}$ rounds. Then, w.h.p $|\frac{N(\theta)}{(1/2)\int_0^\theta l(L_h)dh} - 1| < \varepsilon$,

where L_h is the level curve of points of height h and $l(L_h)$ denotes its length computed with respect to $-g$.

Remark: When $k = 0$, part A is a restatement in terms of Lorentzian geometry and airplane boarding of [4–Th. 2]. When we assume in addition that p is the uniform distribution, part B is a restatement of [2–Th. 12].

Sketch of proof: We prove part A by reduction to theorem 2 of [4]. Given a curve C, we wish to estimate w.h.p the size of the longest chain, with respect to the partial order $<$ on the boarding process, in a small neighborhood of C. We notice that a sequence of blocking passengers is sorted in the q-coordinate, and thus we may assume that C is parameterized by q and given by $(q, \phi(q))$. To obtain a lower bound, we subdivide the curve at points $V_i = (idq, r(idq)) \in C$, where dq is small. Let $A = (q, r)$ be a point. Consider a nearby point B of the form $(q + \delta q, r + \delta r)$. We would like to have a high probability criterion for the relation $A < B$. We consider the passengers queuing behind A. The number of such passengers is by construction, asymptotically, $(\int_u^1 p(q, u)du\delta q)n = \alpha\delta qn$. The difference in row numbers between passengers A and B is by construction δrn. By definition of the relation $A < B$, it will hold if $\delta r \geq 0$ or when, asymptotically, $k\alpha\delta qn > -\delta rn$, which is equivalent to $k\alpha > -\delta r/\delta q$. This is precisely the condition for (dq, dr) to be a time-like vector. We conclude that w.h.p the boarding process relation $>$ can be replaced by the causality relation $>_g$. It can be shown that the approximation of $>$ is by $>_g$ is good enough to preserve asymptotically the length of the longest chain. Next we note that the density distribution $p(q, r)dqdr$ is proportional to volume form of the Lorentzian metric. The computation is thus reduced to computing the expected length of the longest chain w.r.t the causal relation $>_g$ for n points, in a bounded Lorentzian domain, which are chosen with respect to the normalized volume form of the Lorentzian metric. It is well known that two dimensional Lorentzian metrics are conformally flat, hence, there is a coordinate transformation $x = f(q, r)$, $y = g(q, r)$ after which the metric takes the form $ds^2 = u(x, y)dxdy$ for some scaling function u. Finally we observe that for such metrics the relation $>_g$ coincides with the increasing subsequences relation of example 2.1 and part A, becomes a translation to the language of Lorentzian geometry of theorem 2 of [4].

To prove the second part of the theorem, note that, at any point (q, r), the tangent to the fixed height curve L_h (which exists generically) must be orthogonal to the (generically unique) geodesic of length h ending at (q, r). This means that, given dh, the Lorentzian area bounded by the curves L_h and L_{h+dh} is equal to $l(L_h)$ (measured with respect to $-g$ which is the length of the base times the Lorentzian height dh). Let μ denote the standard Lebesgue measure. It is known that the Lorentzian area element is given by $|detg|^{1/2}d\mu$. In all our models $|g_{q,r}|^{1/2}$ equals $2p(q, r)$. The number of passengers in a given area is given locally by $p(q, r)\mu$, thus the number of passengers in the region bounded by L_h and L_{h+dh} is approximately $(1/2)l(L_h)dh$, which yields the second part of the theorem. q.e.d

We can also model using a Lorentzian manifold the problem of disk drive scheduling with linear seek functions. We refer to [1] and [3] for terminology and a description of the ABZ process. We assume that the disk drive is presented with n I/O requests chosen with respect to a location density distribution $p(r, \theta)$, and that the seek function of the disk is given by $f(\theta) = c\theta$. Here r represents the radial location of a piece of data and θ is the angular location. Note that $(r, 0)$ and $(r, 1)$ are identified. The associated model M is that of a cylinder with coordinates (r, θ) and metric such that $ds^2 = p(r, \theta)(dr^2 - c^2d\theta^2)$. We choose the left time-like cone containing the vector $(1, 0)$ as future pointing throughout the manifold. We note that the cylinder equipped with the metric $-g$ is an example of a Lorentzian space with no causal structure. If A denotes an I/O request, let $R(A)$ denote the rotation in which request A will be serviced according to the ABZ scheduling process.

The relation between the ABZ scheduling process and the peeling process applied to the I/O requests with respect to the causal structure are related by the following result.

Theorem 3.2 *Let $A = (r, \theta)$ be an I/O request which is part of a batch of n I/O requests presented to the disk. Let $R_{cau}A$ denote the round of A in the peeling process with respect to the causal structure. Let $R(A)$ be the rotation in which the ABZ scheduling process services request A. Then*

$$R_{cau}(A) - \frac{1}{c} \leq R(A) \leq R_{cau} + \frac{1}{c}.$$

Using this result we can state the analogue of Theorem 3.1 for disk scheduling.

Theorem 3.3

A) *For all $\varepsilon > 0$, w.h.p, for any request A we have*

$$(h(A) - \varepsilon)\sqrt{n} < R(A) < (h(A) + \varepsilon)\sqrt{n}.$$

B) *Let $\varepsilon > 0$. For $\theta > 0$ let $N(\theta)$ denote the number of I/O requests which are serviced in the first $\theta\sqrt{n}$ disk rotations, then w.h.p*

$$\left|\frac{N(\theta)}{(1/2) \int_0^\theta l(C_h)dh} - 1\right| < \varepsilon$$

where $l(C_h)$ denotes the length of C_h with respect to the metric $-g$.

Part A of Theorem 3.3 is essentially a restatement of a theorem in [3], while part B is new. The proof follows the lines of the proof of theorem 3.1 but is simpler since the partial order to which the ABZ peeling process is applied coincides with the causal structure rather than being asymptotically the same as in the airplane boarding case.

Data layouts on disk usually dictate that $p(r, \theta)$ is usually independent of θ. Nonetheless space-time arising from metrics of the form $ds^2 = p(r)(dr^2 - c^2 d\theta^2)$ are still very interesting and include the analogues for space-time of hyperbolic and spherical geometry.

4   Analysis and Optimization of the Boarding Process

The space-time models which were introduced in section 3 allow us to compute analytically the asymptotic boarding time (for a large number of passengers) of essentially all the airline policies presented examined via simulations in [6, 9, 11, 13]. After some practice the boarding time for a specific policy, given a congestion parameter k can be calculated by hand with the aid of a calculator in about 1 minute. With a modern day processor the same calculation is expected to take a few nano seconds, far faster then any simulation. In general there is very good agreement between all simulation studies and our analytical computations regarding the ranking of various policies. In this section we will emphasize a crucial behavioral feature which helps to explain the results. More precisely we aim to clarify the relation between the efficiency of standard boarding policies and the value of the parameter k which is given in terms of interior design parameters of the airplane.

We consider *back-to-front* policies. Such policies enforce row i passengers to board no earlier than row $i + 1$ passengers. In terms of the policy function, this means that $F = F_P(r)$ is decreasing. We will also assume that $F(r)$ is continuously differentiable. Policies of this type may be implemented by placing a wall display at the terminal which shows which row (and rows above it) is allowed to board at any given moment. the numbers on the display will gradually decrease according to the function F, showing row r at time $F(r)$. This scenario is a bit different from the more common situation in which rows which are permitted to board are announced by a gate agent, but the qualitative behavior of both types of policies turns out to be similar, so either one can be considered. Back to front policies attempt to minimize boarding time by reducing the time lost by aisle blocking.

In our analysis we shall keep k fixed. We consider the attentive passenger reaction model, in which passengers board uniformly within t time units from the time of being allowed, i.e., of time $F(r)$. As a result of this assumption, the corresponding density p will take the value $1/t$ in the range $F(r) \leq q < F(r) + t$ and 0 otherwise.

we claim that a phase transition occurs in the boarding problem when the value of k passes through 1 (or other integer values).

Theorem 4.1 *Let P be any back-to-front boarding policy, and let $M_{P,k,t}$ denote the Lorentzian space corresponding to P, with congestion parameter k and an attentive passenger reaction model with attentiveness parameter t. If $k < 1$, then*
$$\lim_{t \to 0} diam(M_{P,k,t}) = 0, \text{ whereas if } k > 1 \text{ then}$$
$$\lim_{t \to 0} diam(M_{P,k,t}) = \infty.$$
Thus, when $k < 1$ back to front boarding policies can be very effective, while for $k > 1$ they can be very detrimental.

Sketch of proof: When $k > 1$ we consider the curve which forms the lower boundary of the domain of the Lorentzian model, namely, $(F(r), r)$. We claim that as t approaches 0, this curve becomes time-like. Looking at the definition of the metric we note that $\alpha(q, r) = \int_r^1 p(q, u) du$. We have $p(q, u) = 1/t$ as long as $F(u) + t \geq F(r)$ and zero otherwise. After some simple manipulations this is seen to be equivalent as t becomes small to the range $r \leq u \leq -(t/F'(r)) + r$, which yields that $\alpha(q, r)$ is asymptotically equal to $-1/F'(r)$. Since we are considering the curve $(F(r), r)$ we have $dq = d(F(r)) = F'(r) dr$. Plugging into the metric we have $ds^2 = \frac{4}{t}(dq^2((k-1)/F'(r)))$. Recaling that $F'(r) < 0$ and integrating over dq (using the inverse function F^{-1}) we obtain that the length of the curve tends to infinity at a rate of $O(\sqrt{1/t})$ as t goes to zero. When $k < 1$, we observe that the argument above actually shows that $\alpha(q, r) \leq -1/F'(r)$. Using this fact it is easy to see that any time-like curve will have Euclidean length of order of magnitude $O(t)$. Since the contribution of the density to ds has order of magnitude $O(1/\sqrt{t})$ we see that the length of any time like will be $O(\sqrt{t})$ as required. $Q.E.D$

This result explains why current airline policies which are typically very similar to the back to front policies which we consider are ineffective or even detrimental. This has also been observed in the simulation studies [9, 6, 11, 13]. The same studies suggest the use of *multiple class* policies in which passengers are divided into classes according to seat type (window, aisle or middle) or row type (even, odd) and only allowing passengers from a single class to board at any given time. Such policies effectively reduce k to k/m, where m is the number of classes, thus allowing them in some cases to be more effective than random passenger queueing. In particular the computations suggest that a policy which orders passengers according to the following 6 groups will be rather effective. The order is: Window passengers from the back of the plane board first, followed by window passengers from the front, then middle seat passengers from the back, middle seat passengers from the front, aisle seats in the back and finally aisle seats in the front. This policy has also been suggested in [13].

Acknowledgments

The authors wish to thank Percy Deift for several helpful conversations and much encouragement.

References

1. M. Andrews, M. Bender, and L. Zhang, New algorithms for the disk scheduling problem, Algorithmica, 32(2), 277–301, 2002. Conference version, Proceedings of FOCS, 580–589, October 1996.
2. D. Aldous and P. Diaconis, Longest increasing subsequences: From patience sorting to the Baik-Deift-Johansson theorem, Bull. AMS, 36(4), 413–432, 1999.
3. E. Bachmat, Average case analysis for batched disk scheduling and increasing subsequences, Proceedings of STOC 2002, 277–286, Montreal, Canada.
4. J.D. Deuschel and O. Zeitouni, Limiting curves for iid records, Annals of Probability, 23, 852–878, 1995.
5. R.P Feynman, The Feynman lectures on physics, Vol. II, Addison Wesley, 1970.
6. P. Ferrari and K. Nagel Robustness of efficient passenger boarding in airplanes, preprint, 2004.
7. G. Gallo, F. Malucelli and M. Marre, Hamiltonian paths algorithms for disk scheduling, University of Pisa technical report, 20/94, 1994.
8. D. Jacobson and J. Wilkes, Disk scheduling algorithms based on rotational position, HP labs technical report, HPL-CSP-91-7rev1, 1991.
9. S. Marelli, G. Mattocks and R. Merry, The role of computer simulation in reducing airplane turn time, Boeing Aero Magazine, Issue 1, 2000.
10. R. Penrose, Techniques of differential yopology in relativity, 1972.

202   E. Bachmat et al.

11. H. van Landeghem and A. Beuselinck, Reducing passenger boarding time in air-planes: A simulation approach, European J. of Operations Research, 142, 294–308, 2002.
12. M. Seltzer, P. Chen, and J. Ousterhout, Disk scheduling revisited, Proceedings of the Usenix technical conference, 313–324, winter 1990.
13. M. van den Briel, J. Villalobos, and G. Hogg, The Aircraft Boarding Problem, Proc. of the 12'th Industrial Eng. Res. Conf., IERC, CD ROM only, paper nr. 2153, 2003.
14. M. Visser, C. Barcelo and S. Liberati, Analogue models of and for gravity, General Relativity and Gravitation, 34, 1719–1734, 2002.

Portfolio Selection: Possibilistic Mean-Variance Model and Possibilistic Efficient Frontier

Wei-Guo Zhang[1,2] and Ying-Luo Wang[1]

[1] School of Management, Xi'an Jiaotong University,
Xi'an,710049, P.R. China
wgzhang@scut.edu.cn, zhwg610263.net
[2] College of Business, South China University of Technology,
Guangzhou, 510641, P.R. China

Abstract. There are many non-probabilistic factors that affect the financial markets. In this paper, the possibilistic mean-variance model of portfolio selection is presented under the assumption that the returns of assets are fuzzy numbers, which can better integrate the experts' knowledge and the managers' subjective opinions to compare with conventional probabilistic mean-variance methodology. The possibilistic efficient frontier is derived explicitly when short sales are not allowed on all risky assets and a risk-free asset.

1   Introduction

The mean-variance methodology for the portfolio selection problem, posed originally by Markowitz[1], has played an important role in the development of modern portfolio selection theory. It combines probability and optimization techniques to model the behavior investment under uncertainty. The return is measured by mean, and the risky is measured by variance, of a portfolio of assets. In Markowitz's mean-variance model for portfolio selection, it is necessary to estimate the probability distribution, strictly speaking, a mean vector and a covariance matrix. It means that all mean returns, variances, covariances of risky assets can be accurately estimated by an investor. The basic assumption for using Markowitz's mean-variance model is that the situation of asset markets in future can be correctly reflected by asset data in the past, that is, the mean and covariance of assets in future is similar to the past one. It is hard to ensure this kind of assumption for real ever-changing asset markets.

Indeed, it is well-known that the returns of risky assets are in a fuzzy uncertain economic environment and vary from time to time, so the future states of returns and risks of risky assets cannot be predicted accurately. Recently, a few of authors such as Watada [6], Tanaka and Guo [7,8], Wang and Zhu [14] etc., studied the fuzzy portfolio selection problem. Watada [6] presented portfolio selection models using fuzzy decision theory. Tanaka and Guo [7,8] proposed the portfolio selection models based on fuzzy probabilities and possibilistic distributions. Zhang [11] introduced the admissible efficient portfolio model under the

N. Megiddo, Y. Xu, and B. Zhu (Eds.): AAIM 2005, LNCS 3521, pp. 203–213, 2005.

assumption that the expected return and risk of asset have admissible errors. This paper discusses portfolio selection problem based on possibilistic mean and possibilistic variance of fuzzy numbers. We present the possibilistic mean-variance model and introduce the notion of the possibilistic efficient portfolio and efficient frontier similar to probabilistic efficient portfolio and efficient frontier. The probabilistic mean and variance in Markowitz's mean-variance model are replaced by the possibilistic mean and variance, respectively. The possibilistic efficient frontier is derived explicitly when short sales are not allowed on all risky assets and there exists a risk-free investment.

2   Possibilistic Mean and Possibilistic Variance

A fuzzy number A is a fuzzy set of the real line $\mathcal{R}$ with a normal, fuzzy convex and continuous membership function of bounded support. The family of fuzzy numbers will be denoted by $\mathcal{F}$.

Let A be a fuzzy number with $\gamma-$ level set $[A]^\gamma = [a(\gamma), b(\gamma)](\gamma > 0)$. Moreover, a function $f : [0, 1] \rightarrow \mathcal{R}$ is said to be a weighting function if f is non-negative, monotone increasing and satisfies the normalization condition $\int_0^1 f(\gamma)d\gamma = 1$.

Fullér and Majlender [10] defined the weighted possibilistic mean value of A as

$$\overline{M}_f(A) = \int_0^1 f(\gamma)\frac{a(\gamma) + b(\gamma)}{2}d\gamma = \frac{M_f^L(A) + M_f^U(A)}{2},$$

where

$$M_f^L(A) = \int_0^1 f(\gamma)a(\gamma)d\gamma = \frac{\int_0^1 a(\gamma)f(Pos[A \leq a(\gamma)])d\gamma}{\int_0^1 f(Pos[A \leq a(\gamma)])d\gamma},$$

$$M_f^U(A) = \int_0^1 f(\gamma)b(\gamma)d\gamma = \frac{\int_0^1 b(\gamma)f(Pos[A \geq b(\gamma)])d\gamma}{\int_0^1 f(Pos[A \geq b(\gamma)])d\gamma},$$

$$Pos[A \leq a(\gamma)] = \Pi((-\infty, a(\gamma)]) = \sup_{u \leq a(\gamma)} A(u) = \gamma,$$

$$Pos[A \geq b(\gamma)] = \Pi([b(\gamma), +\infty]) = \sup_{u \geq b(\gamma)} A(u) = \gamma.$$

We easily get the following conclusion.

Theorem 2.1. Let $A_1, \ldots, A_n$ be n fuzzy numbers, and let $\lambda_1, \ldots, \lambda_n$ be n nonnegative real numbers. Then

$$\overline{M}_f(\sum_{i=1}^n \lambda_i A_i) = \sum_{i=1}^n \lambda_i \overline{M}_f(A_i),$$

where the addition of fuzzy numbers and the multiplication by a scalar of fuzzy number are defined by the sup-min extension principle [9].

Proof.   Suppose $[A_i]^\gamma = [a_i(\gamma), b_i(\gamma)](0 < \gamma < 1)$.
According to $\lambda_i \geq 0, i = 1, \ldots, n$, it holds that

$$[\lambda_i A_i]^\gamma = \lambda_i [A_i]^\gamma = \lambda_i [a_i(\gamma), b_i(\gamma)] = [\lambda_i a_i(\gamma), \lambda_i b_i(\gamma)],$$

$$[\sum_{i=1}^n \lambda_i A_i]^\gamma = \sum_{i=1}^n [\lambda_i A_i]^\gamma = [\sum_{i=1}^n \lambda_i a_i(\gamma), \sum_{i=1}^n \lambda_i b_i(\gamma)].$$

From the definition of the weighted possibilistic mean value and the equations above, we have

$$\overline{M}_f(\sum_{i=1}^n \lambda_i A_i) = \int_0^1 f(\gamma)(\sum_{i=1}^n \lambda_i \frac{a_i(\gamma) + b_i(\gamma)}{2})d\gamma = \sum_{i=1}^n \lambda_i \overline{M}_f(A_i).$$

We introduce the notations of the weighted possibilistic variance and covariance of fuzzy numbers.

Definition 2.1.   *The weighted possibilistic variance of A with $[A]^\gamma = [a(\gamma), b(\gamma)]$ is defined as*

$$\overline{Var}_f(A) = \int_0^1 f(\gamma)([M_f^L(A) - a(\gamma)]^2 + [M_f^U(A) - b(\gamma)]^2)d\gamma.$$

Definition 2.2.   *The weighted possibilistic covariance of $A, B \in \mathcal{F}$ is defined as*

$$\overline{Cov}_f(A, B) = \int_0^1 f(\gamma)[(M_f^L(A) - a_1(\gamma))(M_f^L(B) - a_2(\gamma)) + (M_f^U(A) - b_1(\gamma))(M_f^U(B) - b_2(\gamma))]d\gamma,$$

where $[A]^\gamma = [a_1(\gamma), b_1(\gamma)]$ and $[A]^\gamma = [a_2(\gamma), b_2(\gamma)]$.

The following theorems show properties of the weighted possibilistic variance.

Theorem 2.2.   *Let $A \in \mathcal{F}$ and let θ be a real number. Then*

$$\overline{M}_f(A + \theta) = \overline{M}_f(A) + \theta, \overline{Var}_f(A + \theta) = \overline{Var}_f(A).$$

Proof.   Let $[A]^\gamma = [a(\gamma), b(\gamma)], \gamma \in [0, 1]$.
From the relationship

$$[A + \theta]^\gamma = [a(\gamma) + \theta, b(\gamma) + \theta],$$

we get

$$M_f^L(A + \theta) = \int_0^1 f(\gamma)(a(\gamma) + \theta)d\gamma = M_f^L(A) + \theta,$$

$$M_f^U(A + \theta) = \int_0^1 f(\gamma)(b(\gamma) + \theta)d\gamma = M_f^U(A) + \theta.$$

Form the definitions, it follows that

$$\overline{M}_f(A + \theta) = \overline{M}_f(A) + \theta, \overline{Var}_f(A + \theta) = \overline{Var}_f(A).$$

The proof of the theorem is ended.

Theorem 2.3. Let $A_1, \ldots, A_n$ be n fuzzy numbers, and let $\lambda_1, \ldots, \lambda_n$ be n nonnegative real numbers. Then

$$\overline{Var}_f(\sum_{i=1}^{n} \lambda_i A_i) = \sum_{i=1}^{n} \lambda_i^2 \overline{Var}_f(A_i) + 2 \sum_{i>j=1}^{n} \lambda_i \lambda_j \overline{Cov}_f(A_i, A_j),$$

where the addition and multiplication by a scalar of fuzzy numbers are defined by the sup-min extension principle [9].

Proof. Suppose $[A_i]^\gamma = [a_i(\gamma), b_i(\gamma)] (0 < \gamma < 1)$.
Then

$$[\sum_{i=1}^{n} \lambda_i A_i]^\gamma = [\sum_{i=1}^{n} \lambda_i a_i(\gamma), \sum_{i=1}^{n} \lambda_i b_i(\gamma)].$$

By Theorem 2.1, we have

$$\overline{Var}_f(\sum_{i=1}^{n} \lambda_i A_i) = \int_0^1 f(\gamma)[M_f^L(\sum_{i=1}^{n} \lambda_i A_i) - \sum_{i=1}^{n} \lambda_i a_i(\gamma)]^2 d\gamma +$$

$$\int_0^1 f(\gamma)[M_f^U(\sum_{i=1}^{n} \lambda_i A_i) - \sum_{i=1}^{n} \lambda_i b_i(\gamma)]^2 d\gamma$$

$$= \int_0^1 f(\gamma)[\sum_{i=1}^{n} \lambda_i(M_f^L(A_i) - a_i(\gamma))]^2 d\gamma +$$

$$\int_0^1 f(\gamma)[\sum_{i=1}^{n} \lambda_i(M_f^U(A_i) - b_i(\gamma))]^2 d\gamma$$

$$= \sum_{i=1}^{n} \lambda_i^2 \overline{Var}_f(A_i) + 2 \sum_{i>j=1}^{n} \lambda_i \lambda_j \overline{Cov}_f(A_i, A_j).$$

Let $A_1, \ldots, A_n$ be n fuzzy numbers and let $c_{ij} = \overline{Cov}_f(A_i, A_j), i, j = 1, \ldots, n$. Then the matrix

$$\mathbf{Cov_f} = (c_{ij})_{n \times n}$$

is called as the possibilistic covariance matrix of fuzzy vector $(A_1, A_2, \ldots, A_n)$. We can prove that the possibilistic covariance matrix have the same properties as the covariance matrix in probability theory.

Theorem 2.4. $\overline{\mathbf{Cov_f}}$ is a nonnegative definite matrix.
Proof. From the definitions of possibilistic covariance, it follows that

$$\overline{Cov}_f(A_i, A_j) = \overline{Cov}_f(A_j, A_i), i, j = 1, \ldots, n.$$

Therefore, $\overline{\mathbf{Cov_f}}$ is a real symmetric matrix.
Especially,

$$c_{ii} = \overline{Cov}_f(A_i, A_i) = \overline{Var}_f(A_i), i = 1, 2, \ldots, n.$$

Let $[A_i]^\gamma = [a_i(\gamma), b_i(\gamma)], i = 1, \ldots, n.$

For any $t_i \in \mathcal{R}(i = 1, \ldots, n),$

$$\sum_{i=1}^{n}\sum_{j=1}^{n} c_{ij}t_it_j$$

$$=\sum_{i=1}^{n}\sum_{j=1}^{n}\int_0^1 f(\gamma)([M_f^L(A_i)-a_i][M_f^L(A_j)-a_j]+[M_f^U(A_i)-b_i][M_f^U(A_j)-b_j])t_it_jd\gamma$$

$$=\int_0^1 f(\gamma)[\sum_{i=1}^{n} t_i(M_f^L(A_i)-a_i(\gamma))]^2d\gamma+\int_0^1 f(\gamma)[\sum_{i=1}^{n} t_i(M_f^U(A_i)-b_i(\gamma))]^2d\gamma \geq 0.$$

Hence, $\overline{\mathbf{Cov}_f}$ is the nonnegative definite matrix.
This concludes the proof of the theorem.

3   Possibilistic Mean-Variance Model of Portfolio Selection

We consider a portfolio selection problem with n risky assets and a risk-free asset in this paper. Here, the return rate r_j for risky asset j is considered a fuzzy number, $j = 1, 2, \ldots, n$. Let r_0 be the return rate of risk-free asset and x_j be the interest rate of the risky asset j. Then the return associated with a portfolio $(x_1, x_2, \ldots, x_n)$ is

$$r = \sum_{i=1}^{n} x_ir_i + r_0(1 - \sum_{i=1}^{n} x_i).$$

From Theorems 2.1 and 2.2, the possibilistic mean value of r is given by

$$\overline{M}_f(r) = \sum_{i=1}^{n}\overline{M}_f(x_ir_i) + r_0(1 - \sum_{i=1}^{n} x_i) = \sum_{i=1}^{n} x_i\overline{M}_f(r_i) + r_0(1 - \sum_{i=1}^{n} x_i).$$

From Theorems 2.2 and 2.3, the possibilistic variance of r is given by

$$\overline{Var}_f(r) = \sum_{i=1}^{n} x_i^2\overline{Var}_f(r_i) + 2\sum_{i>j=1}^{n} x_ix_j\overline{Cov}_f(r_i, r_j).$$

In order to describe conveniently, we introduce the following notations:

$$\mathbf{x} = (x_1, x_2, \ldots, x_n)',$$
$$\mathbf{r} = (r_1, r_2, \ldots, r_n)',$$
$$\mathbf{F} = (1, 1, \ldots, 1)',$$
$$\mathbf{M} = (\overline{M}_f(r_1), \overline{M}_f(r_2), \ldots, \overline{M}_f(r_n))',$$
$$\mathbf{C} = (\overline{Cov}_f(r_i, r_j))_{n\times n}.$$

$\mathbf{M}$ is the possibilistic mean vector, $\mathbf{C}$ is the possibilistic covariance matrix. Thus, the possibilistic mean value of r is rewritten as

$$\overline{M}_f(r) = \mathbf{M}'\mathbf{x} + r_0(1 - \mathbf{F}'\mathbf{x}). \tag{1}$$

The possibilistic variance of r is rewritten as

$$\overline{Var}_f(r) = \mathbf{x}'\mathbf{Cx}. \tag{2}$$

Analogous to Markowitz' mean-variance methodology for the portfolio selection problem, the possibilistic mean value correspond to the return, while the possibilistic variance correspond to the risk. From this point of view, the possibilistic mean-variance model of portfolio selection can be formulated as

$$\min \mathbf{x}'\mathbf{Cx}$$
$$s.t. \quad \mathbf{M}'\mathbf{x} + r_0(1 - \mathbf{F}'\mathbf{x}) \geq \mu, \tag{3}$$
$$\mathbf{F}'\mathbf{x} \leq 1, \quad \mathbf{x} \geq \mathbf{0}.$$

From the definition of efficient portfolio, we introduce the concepts of the possibilistic efficient portfolio similar to probabilistic efficient portfolio as follows.

Definition 3.1   *The optimal solution of (3), $\mathbf{x}^*$, is called as the possibilistic efficient portfolio.*

The possibilistic efficient portfolios for all possible μ construct the possibilistic efficient frontier. Solving (3) for all possible μ, the possibilistic efficient frontier is derived explicitly.

4   Analytic Derivation of the Possibilistic Efficient Frontier

In (3), it should be noted that the mean vector and covariance matrix in Markowitz's mean-variance model are replaced by the possibilistic mean vector and covariance matrix, respectively. There exist a number of studies for finding efficient portfolio from solving mean-variance model (see, e.g. [2-5]), which would be useful to compute the possibilistic efficient frontier. On the other hand, it would be very difficult to obtain the efficient frontier in closed form when short sales aren't allowed. In this section we give an analytic derivation of possibilistic efficient frontier based on some assumptions.

Assumption 4.1   *(i) $\mathbf{M} = (m_1, \ldots, m_n) \neq k\mathbf{F}$, for any $k \in \Re$, (ii) $\mathbf{C} = (c_{ij})_{n \times n}$ is a positive definite matrix.*
Set

$$e = \mathbf{M}'\mathbf{C}^{-1}\mathbf{M}, \quad f = \mathbf{F}'\mathbf{C}^{-1}\mathbf{F}, \quad d = \mathbf{M}'\mathbf{C}^{-1}\mathbf{F},$$
$$\delta = ef - d^2, \quad \alpha_0 = (e - dr_0)/(d - fr_0), \quad \beta_0 = 1/(fr_0^2 - 2dr_0 + e). \tag{4}$$

Using Lagrangian multiplier method, we obtain the following conclusion: If Assumption 4.1(ii) be satisfied, then $\mathbf{x} = \mathbf{C}^{-1}\mathbf{F}/f$ is the unique optimal solution of the portfolio problem $\min\{\mathbf{x}'\mathbf{Cx}|\mathbf{F}'\mathbf{x} = 1\}$ and satisfies $\mathbf{M}'\mathbf{x} + r_0(1 - \mathbf{F}'\mathbf{x}) = d/f$.
The following assumption is natural.

Assumption 4.2 $r_0 < d/f$.

Some properties of $e, f, d, \beta_0, \alpha^0$, and δ are given in the following proposition.

Proposition 4.1   *Let Assumptions 4.1 and 4.2 be satisfied. Then the following results hold:*

(i) $e > 0, f > 0$, *(ii)* $\delta > 0, \beta_0 > 0$, *(iii)* $d/f < \alpha_0$.

The following Lemma 4.1 is obvious.

Lemma 4.1 $\max\{\mathbf{M}'\mathbf{x} + r_0(1 - \mathbf{F}'\mathbf{x})|\mathbf{F}'\mathbf{x} \leq 1, \mathbf{x} \geq 0\} = \max\{m_i, 1 \leq i \leq n\}$.

Lemma 4.1 means that the maximum possibilistic return of the portfolio that an investor can get is $\max\{m_i, 1 \leq i \leq n\}$ under constraints $\mathbf{F}'\mathbf{x} \leq 1$ and $\mathbf{x} \geq 0$. This result can be obtained when the investment rate to the asset with the maximum possibilistic mean is 1 and others are 0.

Before proceeding with the solving (3), it is helpful to introduce the following problem

$$\min \mathbf{x}'\mathbf{C}\mathbf{x}$$
$$s.t. \quad \mathbf{M}'\mathbf{x} + r_0(1 - \mathbf{F}'\mathbf{x}) \geq \mu, \tag{5}$$
$$\mathbf{F}'\mathbf{x} \leq 1.$$

The optimal solution for (5) is formulated in the following theorem.

Theorem 4.1   *Let Assumptions 4.1 and 4.2 be satisfied. Then the optimal solution to (5) is :*

(i) $\mathbf{x} = \beta_0(\mu - r_0)\mathbf{C}^{-1}(\mathbf{M} - r_0\mathbf{F})$ *if* $r_0 < \mu \leq \alpha_0$,
(ii) $\mathbf{x} = \mathbf{C}^{-1}[e\mathbf{F} - d\mathbf{M} + (f\mathbf{M} - d\mathbf{F})\mu]/\delta$ *if* $\mu > \alpha_0$.

Proof.   Since $\mathbf{C}$ is a positive definite matrix, the K-T conditions are both necessary and sufficient for optimality(see Yuan Y.X. et al ([12])). Solving the problem (5) is equal to doing $\mathbf{x} \in \Re^n$, $\lambda_i \in \Re(i = 1, 2)$ such that

$$\begin{cases} \mathbf{C}\mathbf{x} = \lambda_1(\mathbf{M} - r_0\mathbf{F}) - \lambda_2\mathbf{F}, \\ \lambda_1[(\mathbf{M} - r_0\mathbf{F})'\mathbf{x} + r_0 - \mu] = 0, \\ \lambda_2(-\mathbf{F}'\mathbf{x} + 1) = 0, \lambda_1 \geq 0, \lambda_2 \geq 0. \end{cases} \tag{6}$$

If $r_0 \leq \mu \leq \alpha^0$, then there exist multipliers $\lambda_1 = \beta_0(\mu - r_0) \geq 0, \lambda_2 = 0$ and

$$\mathbf{x} = \beta_0(\mu - r_0)\mathbf{C}^{-1}(\mathbf{M} - r_0\mathbf{F})$$

such that (6) holds.

If $\mu > \alpha_0$, then $f\mu - d > f\alpha_0 - d = \delta/(d - fr_0) > 0$ and $\mu(d - fr_0) - (e - dr_0) \geq 0$. Correspondingly, there exist multipliers $\lambda_1 = (f\mu - d)/\delta > 0, \lambda_2 = [\mu(d - fr_0) - (e - dr_0)]/\delta > 0$ and $\mathbf{x} = \mathbf{C}^{-1}[e\mathbf{F} - d\mathbf{M} + (f\mathbf{M} - d\mathbf{F})\mu]/\delta$ such that (6) holds. Thus, parts (i) and (ii) hold.

The proof of the theorem is completed.

The relation to the optimal solutions between (3) and (5) is described by the following Theorem 4.2

Theorem 4.2   Let Assumption 4.1(ii) be satisfied and let μ be constant. If $\mathbf{x}^* = (x_1^*, \ldots, x_n^*)'$ is the optimal solution of (5) with $x_{i_1}^* < 0, \ldots, x_{i_k}^* < 0$ and others $x_j^* \geq 0$, and if $\mathbf{x}^{1*} = (x_1^{1*}, \ldots, x_n^{1*})'$ is the optimal solution of (3), then

$$\mathbf{x}^{1*} \in \bigcup_{s=1}^{k} \{\mathbf{x} = (x_1, \ldots, x_n)' | x_{i_s} = 0, x_i \geq 0, i \neq i_s\}.$$

Proof.   With the help of reduction to absurdity. Let $x_{i_s}^{1*} \neq 0$ for all $s = 1, \ldots, k$. Then $x_{i_s}^{1*} > 0$ for all $s = 1, \ldots, k$ and others $x_j^{1*} \geq 0$.

Assumption 4.1(ii) is satisfied, so $V(\mathbf{x}) = \mathbf{x}'\mathbf{C}\mathbf{x}$ is a strictly convex function. We set $F(t) = V(t\mathbf{x}^{1*} + (1-t)\mathbf{x}^*)$ and $\mathbf{x}^{**}(t) = t\mathbf{x}^{1*} + (1-t)\mathbf{x}^*$ for $t \in [0, 1]$.

According to Theorem 3.2.4 of [13], $F(t)$ is a strictly convex function of t for $t \in [0, 1]$.

It can be shown that $\mathbf{x}^{**}(t) = (x_1^{**}(t), \ldots, x_n^{**}(t))'$ for all $t \in [0, 1]$ is a feasible solution of (5), so $F(0) = V(\mathbf{x}^*) < V(t\mathbf{x}^{1*} + (1-t)\mathbf{x}^*) = F(t)$ for all $t \in (0, 1]$.

Thus, $F(t)$ is a strictly increasing function of t for $t \in (0, 1]$. Setting $t_{i_s} = -x_{i_s}^*/(x_{i_s}^{1*} - x_{i_s}^*)$ for $s = 1, \ldots, k$ and using $x_{i_s}^* < 0$, we obtain $t_{i_s} \in (0, 1)$ and $x_{i_s}^{**}(t_{i_s}) = 0$ for $s = 1, \ldots, k$. Define $t_0 = max\{t_{i_s} | s = 1, \ldots, k\}$ and take $t_0 < t_1 < 1$. Since $x_{i_1}^{**}(t), \ldots, x_{i_k}^{**}(t)$ are strictly increasing functions, we have $\mathbf{x}^{**}(t_1) \geq \mathbf{0}$. This implies that $\mathbf{x}^{**}(t_1)$ is also a feasible solution of (3). $V(\mathbf{x}^{**}(t_1)) = F(t_1) < F(1) = V(\mathbf{x}^{1*})$, which is in contradiction with the assumption of $\mathbf{x}^{1*}$. Thus, the proof of the theorem is concluded.

In next section we give the procedure for solving the optimal solution $\mathbf{x}$ of the model (3). In order to describe conveniently, we introduce the following notations:

$$\mathbf{C}^{-1}\mathbf{F} = (g_1, g_2, \ldots, g_n)', \mathbf{C}^{-1}\mathbf{M} = (a_1, a_2, \ldots, a_n)', I = \{1, 2, \ldots, n\},$$

where I denotes the set of n risky assets.

We discuss the optimal solution of (3) from two cases: $\mathbf{C}^{-1}\mathbf{M} \geq r_0\mathbf{C}^{-1}\mathbf{F}$; $\mathbf{C}^{-1}\mathbf{M} \not\geq r_0\mathbf{C}^{-1}\mathbf{F}$.

Case 1: $\mathbf{C}^{-1}\mathbf{M} \geq r_0\mathbf{C}^{-1}\mathbf{F}$

If $\mu \geq r_0$, then $\beta_0(\mu - r_0)\mathbf{C}^{-1}(\mathbf{M} - r_0\mathbf{F}) \geq \mathbf{0}$.

For $r_0 < \mu \leq \alpha_0$, the optimal solution of (5) is also a feasible one of (3).

Hence, the optimal solutions of (3) for $r_0 < \mu \leq \alpha_0$ is given by Theorem 4.1(i).

For $\mu \geq \alpha_0$, Theorem 4.1(ii) can be described by

$$x_k = \frac{1}{\delta}[eg_k - da_k + (fa_k - dg_k)\mu], k = 1, \ldots, n. \tag{7}$$

Since $\delta > 0$ and (7) satisfies $\mathbf{F}'\mathbf{x} = \sum_{k=1}^{n} x_k = 1$, there exists $h \in I$ such that

$$fa_h - dg_h < 0.$$

Define

$$\alpha = min\{\frac{da_k - eg_k}{fa_k - dg_k} | fa_k - dg_k < 0, k = 1, \ldots, n.\}$$

Without loss of generality we assume that there exists the unique $h \in I$ such that

$$\alpha = \frac{da_h - eg_h}{fa_h - dg_h}.$$

Especially, at $\mu = \alpha_0$,

$$\mathbf{C}^{-1}[e\mathbf{F} - d\mathbf{M} + (f\mathbf{M} - d\mathbf{F})\mu]/\delta = \frac{1}{d - fr_0}\mathbf{C}^{-1}(\mathbf{M} - r_0\mathbf{F}) \geq 0.$$

For $\alpha_0 \leq \mu \leq \alpha$, according to the definition of α and considering the continuity of variables it is sufficient to see that (7) satisfies $x_k \geq 0, k = 1, \ldots, n$, i.e., the unique optimal solution of (3) is given by (7).

For $\alpha < \mu \leq \max_{1 \leq i \leq n} \{m_i\}$, the optimal solution of (5) satisfies

$$x_h = \frac{1}{\delta}[eg_h - da_h + (fa_h - dg_h)\mu] < 0.$$

Using to Theorem 4.2, the optimal solution of (3) satisfies $x_h = 0$ for $\alpha < \mu \leq \max_{1 \leq i \leq n} \{m_i\}$.

We erase x_h in (3) and consider the new model with $n - 1$ assets:

$$
\begin{aligned}
&\min \mathbf{x}_1' \mathbf{C}_1 \mathbf{x}_1 \\
&s.t. \quad \mathbf{M}_1' \mathbf{x}_1 + r_0(1 - \mathbf{F}_1' \mathbf{x}_1) \geq \mu, \\
&\qquad \mathbf{F}_1' \mathbf{x}_1 \leq 1, \quad \mathbf{x}_1 \geq 0,
\end{aligned}
\tag{8}
$$

where

$$
\begin{aligned}
\mathbf{x}_1 &= (x_1, \ldots, x_{h-1}, x_{h+1}, \ldots, x_n)', \\
\mathbf{M}_1 &= (m_1, \ldots, m_{h-1}, m_{h+1}, \ldots, m_n)', \\
\mathbf{F}_1 &= (1, 1, \ldots, 1)', \\
\mathbf{C}_1 &= (c_{ij})_{(n-1) \times (n-1)}, i \neq h, j \neq h.
\end{aligned}
$$

We continue to solve (8) by the same way as shown above.

Similarly, there exists

$$\alpha_1 = \min \left\{ \frac{d_1 a_{1,k} - e_1 g_{1,k}}{f_1 a_{1,k} - d_1 g_{1,k}} \;\middle|\; f_1 a_{1,k} - d_1 g_{1,k} < 0, k \in I_1 \right\}$$

such that the unique optimal solution of (3) for $\alpha < \mu \leq \alpha_1$ is

$$x_k = [e_1 g_{1,k} - d_1 a_{1,k} + (f_1 a_{1,k} - d_1 g_{1,k})\mu]/\delta_1 \text{ for all } k \in I_1, x_h = 0,$$

where

$$
\begin{aligned}
(g_{1,1}, \ldots, g_{1,h-1}, g_{1,h+1}, \ldots, g_{1,n})' &= \mathbf{C}_1^{-1}\mathbf{F}_1, \\
(a_{1,1}, \ldots, a_{1,h-1}, a_{1,h+1}, \ldots, a_{1,n})' &= \mathbf{C}_1^{-1}\mathbf{M}_1, \\
e_1 = \mathbf{M}_1' \mathbf{C}_1^{-1}\mathbf{M}_1, \quad f_1 &= \mathbf{F}_1' \mathbf{C}_1^{-1}\mathbf{F}_1, \\
d_1 = \mathbf{M}_1' \mathbf{C}_1^{-1}\mathbf{F}_1, \quad \delta_1 = e_1 f_1 - d_1^2, \quad I_1 &= I \backslash \{h\}.
\end{aligned}
$$

If $\alpha_1 < \mu \leq \max_{1 \leq i \leq n} \{m_i\}$, then we still erase the variable being zero at α_1 and determine the point α_2 for the new subset I_2 such that $x_k > 0$ for $k \in I_2, x_k = 0$ for $k \in I \backslash I_2, \alpha_1 < \mu \leq \alpha_2$.

We continue the same process as shown above, until $\alpha_t = \max\limits_{1 \le i \le n} \{m_i\}$.

The unique optimal solution of (3) at $\mu = \alpha_t$ is $x_q = 1$, $x_k = 0$ for all $k \ne q$, where q satisfies $m_q = \max\limits_{1 \le i \le n} \{m_i\}$.

Thus, we obtain all optimal solutions of (3) for $\mu \le \max\limits_{1 \le i \le n} \{m_i\}$.

Case 2: $C^{-1}M \not\ge r_0 C^{-1}F$

There exists at least $k \in I$ such that $a_k < r_0 g_k$.

According to Assumption 4.2, $a_k < r_0 g_k < dg_k/f$. Then $fa_k - dg_k < 0$.

For $r_0 < \mu \le \alpha_0$, the optimal solution of (5) satisfies

$$x_k = \beta_0(\mu - r_0)(a_k - r_0 g_k) < 0.$$

For $\alpha_0 < \mu \le \max\limits_{1 \le i \le n} \{m_i\}$, the optimal solution of (5) satisfies

$$\begin{aligned}
x_k &= [eg_k - da_k + (fa_k - dg_k)\mu]/\delta \\
&\le [eg_k - da_k + (fa_k - dg_k)\alpha_0]/\delta \\
&= (a_k - r_0 g_k)/(d - r_0 f) < 0.
\end{aligned}$$

Theorem 4.2 implies that the optimal solution of (3), $\mathbf{x}$, is not a positive vector. In other words, the optimal solution of (3) cannot contain all risky assets if $C^{-1}M \not\ge r_0 C^{-1}F$. Hence, we need to erase a risky asset each time, and so on, until the condition that $C^{-1}M \ge r_0 C^{-1}F$. Thus, we can obtain all optimal solutions of (3) by the same procedure as Case 1.

References

1. H. Markowitz.: Portfolio selection: efficient diversification of Investments. Wiley, New York, 1959
2. A.F. Perold.: Large-scale portfolio optimization. Management Science, 30(1984) 1143-1160
3. J.S. Pang.: A new efficient algorithm for a class of portfolio selection problems. Operational Research, 28(1980) 754-767
4. J. VÖRÖS.: Portfolio analysis-An analytic derivation of the efficient portfolio frontier. European journal of operational research, 203(1986) 294-300
5. M.J. Best, J. Hlouskova.: The efficient frontier for bounded assets. Math.Meth.Oper.Res. 52(2000) 195-212
6. J. Watada.: Fuzzy portfolio selection and its applications to decision making. Tatra Mountains Mathematical Publication, 13(1997) 219-248
7. H. Tanaka, Guo. P.: Portfolio selection based on upper and lower exponential possibility distributions. European Journal of Operational Research, 114(1999) 115-126
8. H. Tanaka, Guo. P, I.Burhan Türksen.: Portfolio selection based on fuzzy probabilities and possibility distributions. Fuzzy sets and systems, 111(2000) 387-397
9. L.A. Zadeh.: Fuzzy Sets. Inform. and Control, 8 (1965) 338-353
10. R. Fullér, P. Majlender.: On weighted possibilistic mean and variance of fuzzy numbers. Fuzzy Sets and Systems, 136 (2003) 363–374
11. W.G. Zhang, Z.K. Nie.: On admissible efficient portfolio selection problem. Applied Mathematics and Compution, 159(2004) 357-371
12. Y.X. Yuan, W.Y. Sun.: Optimal theory and methodology. Science and Technology Press, China,1997

13. Q.L. Wei, R.S. Wang, B. Xu.: Mathematical programming theory. Beijing University of Aeronautics and Astronautics Press, China, 1991
14. S.Y. Wang, S.S. Zhu. On fuzzy portfolio selection problems. Fuzzy Optimization and Decision Making, 1(2002) 361-377

Design DiffServ Multicast with Selfish Agents

WeiZhao Wang[1], Xiang-Yang Li[1,*], and Zheng Sun[2,**]

[1] Illinois Institute of Technology,
Chicago, IL, USA
{lixian, wangwei4}@iit.edu
[2] Hong Kong Baptist University,
Hong Kong, China
sunz@comp.hkbu.edu.hk

Abstract. Differentiated service (DiffServ) is a mechanism to provide the Quality of Service (QoS) with a certain performance guarantee. In this paper, we study how to design DiffServ multicast when the participants (*i.e.*, relay links) are selfish. We assume that each link e_i is associated with a cost coefficient a_i such that the cost of e_i to provide a multicast service with bandwidth demand x is $a_i \cdot x$. We first show that a previous approximation algorithm does not directly induce a truthful mechanism. We then give a new polynomial time 8-approximation algorithm to construct a DiffServ multicast tree. Based on this tree, we design a truthful mechanism for DiffServ multicast, *i.e.*, we give a polynomial-time computable payment scheme to compensate all chosen relay links such that each link e_i maximizes its profit when it reports its privately cost coefficient a_i truthfully.

Keywords: DiffServ, multicast, selfish agents, algorithm mechanism design, approximation algorithms.

1   Introduction

The Differentiated Services framework (DiffServ) [1, 2] has been proposed to provide multiple Quality of Service (QoS) classes over IP networks. DiffServ is built upon a simple model of traffic conditioning and policing at the links of the network in addition to classifying flows into different service classes. The traffic is forwarded using simple differentiated treatments, called per-hop behaviors (PHBs), in the core of the network. This differential treatment results in differential pricing [3], which is one of the motivating factors for adopting DiffServ by major network providers and ISPs.

Multicast has been a popular mechanism for supporting group-based applications. In a multicast, different receivers could have different bandwidth demands. Each link of the network may have different cost of providing multicast with different bandwidth dedication [4]. Due to the heterogeneity in receivers' demand requirements, different

---

* The research of the author was supported in part by NSF under Grant CCR-0311174.
** The research of the author was supported in part by Grant FRG/03-04/II-21 and Grant RGC HKBU2107/04E.

N. Megiddo, Y. Xu, and B. Zhu (Eds.): AAIM 2005, LNCS 3521, pp. 214–223, 2005.
© Springer-Verlag Berlin Heidelberg 2005

links in a multicast tree will carry different traffic loads such that the demand requirements of the downstream receivers are satisfied. The cost of a link in a multicast tree is then the cost needed to dedicate a certain bandwidth for downstream receivers; it is typically determined by the maximum bandwidth required by downstream receivers, as well as the cost coefficient of the link (which we will define later). Thus, the DiffServ multicast problem is to find a *tree* and the bandwidth reservation at each link such that the receivers' bandwidth demands are met. Note that the traditional Steiner tree problem for link weighted graph [5, 6], an NP-hard problem, is a special case of the problem of computing a DiffServ multicast tree with the minimum cost.

What introduces an additional degree of complexity to DiffServ multicast is that the relay links may be *non-cooperative*, instead of *cooperative* as assumed by previous protocols. This means that every relay link will aim to maximize its own benefit instead of the whole network's performance. Usually, each link is first asked to report its relay cost, then a payment to this link is computed (typically by the source) based on a certain payment scheme. It is not often in the best interests of these relay links to report their costs truthfully when they are paid whatever they ask for. Thus, instead of paying the links their *reported* costs, we should design a payment scheme that can ensure that all links reveal their true costs out of their own interests, which is known as *truthfulness*. The truthful mechanism for traditional multicast has been previously addressed in [7, 8]. However, unlike the traditional multicast in which every link has a *fixed* cost in the multicast transmission, each link may incur different costs for different bandwidth demands in DiffServ multicast. In summary, in this paper, we study two different aspects of the DiffServ multicast: the construction of the multicast tree that has low cost, and a truthful payment scheme.

Our main contributions are as follows. First, we show that a previous approximation algorithm does not directly induce a truthful mechanism, and we give an alternative 8-approximation algorithm to construct a DiffServ multicast tree. We then characterize the necessary and sufficient condition for the existence of a truthful payment scheme based on a given multicast tree construction method. Finally, we design a truthful mechanism for DiffServ multicast based on our 8-approximation construction method.

2 Preliminaries and Previous Works

2.1 Algorithmic Mechanism Design

In a standard model of algorithm mechanism design, there are n agents $\{1, 2, \cdots, n\}$. Each agent $i \in \{1, \cdots, n\}$ has some *private* information t_i, called its *type*, (*e.g.*, the cost to forward a packet for a node/link in a network environment). The types of all agents define a *profile* $t = (t_1, t_2, \cdots, t_n)$. Each agent i reports a valid type τ_i', which may be different from its actual type t_i, and the strategies of all agents define a *reported type vector* $\tau = (\tau_1, \cdots, \tau_n)$. A mechanism $M = (\mathcal{O}, \mathcal{P})$ is composed of two parts: an allocation method $\mathcal{O}$ that maps a reported type vector τ to an output o and a *payment* method $\mathcal{P}$ that decides the monetary payment $p_i = \mathcal{P}_i(\tau)$ for every agent i. Each agent i has a valuation function $w_i(t_i, o)$ that expressed its preference over different outcomes. Agent i's *utility* or called *profit* is $u_i(t_i, o) = w_i(t_i, o) + p_i$, given output o and payment p_i. An agent i is said to be *rational* if it always chooses its strategy τ_i that maximizes its utility u_i. Let $\tau_{-i} = (\tau_1, \cdots, \tau_{i-1}, \tau_{i+1}, \cdots, \tau_n)$, *i.e.*, the strategies of all other agents except i, and let $\tau|^i t_i = (\tau_1, \tau_2, \cdots, \tau_{i-1}, t_i, \tau_{i+1}, \cdots, \tau_n)$, *i.e.*, agent i reports t_i. Here, we are only interested in mechanisms $M = (\mathcal{O}, \mathcal{P})$ that satisfy the following three conditions: (1) **Incentive Compatibility (IC)**: $\forall i, \forall \tau$, $w_i(t_i, \mathcal{O}(\tau|^i t_i)) + p_i(\tau|^i t_i) \geq w_i(t_i, \mathcal{O}(\tau)) + p_i(\tau)$, *i.e.*, revealing its true type t_i will maximize its utility *regardless* of what other agents do; (2) **Individual Rationality (IR)** (a.k.a., Voluntary Participation): Each agent must have a non-negative utility, *i.e.*, $w_i(t_i, \mathcal{O}(\tau|^i t_i)) + p_i(\tau|^i t_i) \geq 0$; and (3) **Polynomial Time Computability (PC)**: $\mathcal{O}$ and $\mathcal{P}$ are computed in polynomial time. A mechanism is *truthful* (or called *strategyproof*) it satisfies both IC and IR properties.

2.2   Problem Statement

DiffServ multicast Tree Construction: We assume that there is a connected network $G = (V, E)$ with vertex set V, link set E, where $|V| = n$ and $|E| = m$. Every link e_i incurs a cost $c_i = a_i x$ if x is the bandwidth e_i dedicated to a multicast transmission. Hereafter a_i is called the *cost coefficient* of the link e_i. All links' coefficients define a vector $a = (a_1, a_2, \cdots, a_m)$. There is a source node s and a set of receivers $R \subset V$ that request the multicast service. Every receiver $r_i \in R$ has a bandwidth demand d_i that specifies the minimum bandwidth it needs.

The DiffServ multicast problem consists of two parts: 1) a network topology rooted at the sender s that spans all receivers in the receiver set; 2) a bandwidth allocation for each link of the tree. The tree topology and bandwidth reservation should satisfy that for any receiver r_i, each link on the tree path between r_i and s has a bandwidth reservation not smaller than d_i. Thus, for a link e_i, the reserved bandwidth should not be smaller than the maximum bandwidth demand of its downstream receivers. The weight of a multicast topology T with link bandwidth reservation vector $b = \{b_1, b_2, \cdots, b_m\}$ is $\omega(T, b) = \sum_{e_i \in T} c_i = \sum_{e_i \in T} a_i \cdot b_i$. Given the cost coefficients vector a of all links and the bandwidth demand d of all receivers, the DiffServ multicast problem is to construct a tree T and a bandwidth reservation b with the minimum cost $\omega(T, b)$. It is called Quality of Service Steiner Tree (QoSST) problem in [9].

Payment Computation: Throughout this paper, we assume all the links are selfish and rational. After designing a method $\mathcal{O}$ to construct a multicast tree, we need to design a payment scheme $\mathcal{P}$ for the links such that the mechanism $M = (\mathcal{O}, \mathcal{P})$ is truthful.

2.3   Literature Review of Steiner Tree Construction

Given a homogeneous bandwidth demand $d = \{d_1, d_1, \cdots, d_1\}$, the weight of a tree T is $\omega(T, d) = \sum_{e_i \in T} a_i \cdot d_1 = d_1 \cdot \sum_{e_i \in T} a_i = d_1 \cdot \omega(T, \langle 1 \rangle)$. Then we can normalize the demand of every receiver to 1, *i.e.*, the problem becomes the standard link weighted Steiner tree problem, which enjoys several constant approximation methods [5, 6]. We briefly review a 2-approximation method given in [5]. It works as follows. It iteratively selects the shortest path to the nearest receiver, sets the costs of the links on the selected path to 0, and removes the nearest receiver from the receiver set until no receiver left. We call the constructed tree the *Link Weighted Steiner Tree* (LST), denoted as $LST(R, c)$.

For multicast with a homogeneous bandwidth demand, Li and Wang [8] proved that the VCG mechanism [16, 17, 18] is not truthful if LST is used. In light of the failure of the VCG mechanism, they proposed a truthful payment scheme for all round-based methods for constructing a multicast tree, including LST structure.

The DiffServ multicast problem was studied before in several contexts. Maxemchuk [4] proposed a heuristic algorithm for its solution. Some results for the case of few rates were obtained in [10, 11]. For example, for the case of two non-zero rates, a $\frac{4}{3}\alpha$-approximation algorithm was proposed [11], where $\alpha \simeq 1.549$ is the currently best approximation ratio [6] of an algorithm for the Steiner tree problem. Recently, Charikar *et al.* [12] gave the first constant-factor approximation algorithm for an unbounded number of rates. They achieved an approximation ratio of 4α using rounding and $e\alpha \simeq 4.211$ using randomized rounding. Recently, Karpinski *et al.* [9] gave algorithms with improved approximation factors. They achieved an approximation ratio of 1.960 when there are two non-zero rates and an approximation ratio of 3.802 when there is an unbounded number of rates. Calinescu *et al.* [13] gave a Primal-Dual algorithm with approximation ratio 4.311. Xue *et al.* [14] and Kim *et al.* [15] studied the Grade of Service Steiner Tree Problem (GOSST) in Euclidean planes.

For DiffServ multicast, the method by Charikar *et al.* [12] works as follows. Given an instance of the DiffServ multicast, it first constructs the rounded-up instance by rounding up all receivers' demands to the nearest power of 2. It then solves the standard Steiner tree problem for the receivers of each different demand separately by applying any of the well-known heuristics. Finally, it does a "clean-up" process to transform the union of these Steiner trees into a tree. They proved that this simple approach yields a $4\alpha_{ST}$ approximation of the optimal cost, where α_{ST} is the approximation factor of the Steiner tree heuristic used. Our algorithm presented later is similar to this approach at the first glance, but it has some key differences, which will be described later.

3   A New Approximation Algorithm

We first provide motivations for this section by showing that, for DiffServ multicast, the following straightforward approach of combining the algorithm of [12] with a truthful payment scheme for homogeneous multicast does not give a truthful mechanism. Let $T_1, T_2, \cdots, T_k$ be the k different homogeneous multicast trees constructed for each distinct bandwidth demand rate in the rounded-up instance. Let $p_{e,i}$ be the payment to link e based on tree T_i, and let $b_{e,i}$ be the bandwidth reservation (the maximum demand of its downstream receivers) on link e based on T_i. The final multicast structure is defined to be the union $\bigcup_{i=1}^{k} T_i$ of all these trees, while the payment p_e and bandwidth reservation b_e of each link e are defined to be $\max_{i=1}^{k} p_{e,i}$ and $\max_{i=1}^{k} d_{e,i}$, respectively.

Although this approach (*i.e.*, taking the union of partial outcomes and using the maximum payment of each agent over all partial outcomes as its final payment) works for binary selection problems (see [19, 8] for more details), we can show by example that for DiffServ multicast the resulting payment scheme is no longer truthful. The reason why this approach does not work for DiffServ multicast (and why mechanism design for DiffServ multicast is more difficult than many of problems previously studied in the literature) is that, even if a link is selected, its cost is no longer a fixed number: it depends on the outcome of the game. Thus, a link could influence the outcome of the game by carefully choosing its reported cost to reduce its final cost. As a consequence, it still increases its utility by falsely reporting its cost. Furthermore, the union $\bigcup_{i=1}^{k} T_i$ is not necessarily a tree, but instead a *mesh*, which is not desirable for practical reasons. Thus, even if we can define a truthful mechanism for $\bigcup_{i=1}^{k} T_i$, it is still not clear how to extend it to a truthful mechanism for the output computed by "clean-up". Actually, we will show that *no* truthful mechanism exists for $\bigcup_{i=1}^{k} T_i$ for the bandwidth allocation determined according to b_e.

In this section, we present an alternative DiffServ multicast tree construction algorithm. In the next section, we show how to design a truthful payment scheme based on this algorithm. Given a network G, a receiver set R, a cost coefficient vector a and a bandwidth demand vector d, the following algorithm shows how to find a DiffServ multicast tree $\overline{DMT}(a, d)$ and its corresponding bandwidth allocation $\overline{B}$ with low weight. We also call this algorithm $\overline{DMT}$ if no confusion is caused.

Algorithm 1. Construct DiffServ Multicast Tree

1: Sort all receivers according to their bandwidth demands in an descending order, say $R = \{r_1, r_2, \cdots, r_k\}$.
2: Initialize the tree T to empty and index $t = 1$.
3: For each link e_i, label it as WHITE and set $\overline{B}_i = 0$.
4: **repeat**
5:     Let r_j be the first receiver in the receiver set R.
6:     Find the maximal index k such that $d_k \geq \frac{d_j}{2}$.
7:     Set the cost of each WHITE link e_i as $c_i = a_i \cdot d_j$ and each BLACK link as 0.
8:     Let $R_t = \{r_j, \cdots, r_k\}$ and find the spanning tree $T_t = LST(R_t, c)$.
9:     Remove R_t from R and mark all links in tree T_t as BLACK.
10:    Set $T = T \bigcup T_t$.
11:    For each link $e_i \in T_t$, if $\overline{B}_i = 0$ then set $\overline{B}_i = d_j$.
12:    Set $t = t + 1$.
13: **until** the receiver set R is empty.
14: Output T as $\overline{DMT}$ and bandwidth vector $\overline{B}$.

The major difference between this algorithm and the algorithm in [12] is that, instead of computing several trees *independently* and then combining them to make the final DiffServ multicast tree, we construct a single tree directly. The receiver set is divided into subsets, each containing receivers with demands in a particular range. These subsets are handled in multiple rounds, in a descending order according to their bandwidth demand ranges. In each round, all receivers in a subset are connected to the Diff-Serv multicast tree being built. The links picked in earlier rounds will be used in later rounds, without additional costs involved, to connect receivers with lower demands.

Notice that, as indicated by Line 11 of Algorithm 1., for each link e_i added into T in round t the bandwidth allocation of e_i is set to be the maximum bandwidth demand among all receivers in R_t. This may be more than necessary; after all, e_i will not relay packets for all of them. Indeed, one can design the following Algorithm 2., which constructs the same tree as Algorithm 1. does, and yet allocates less bandwidth on each link e_i by setting the bandwidth allocation to be maximum bandwidth demand of e_i's downstream receivers. In order to distinguish these two algorithms, we use DMT to denote the tree constructed by Algorithm 2.. As minor (and harmless) as this modification seems to be, Algorithm 2. does not induce a truthful payment scheme. In the next section, we will use this algorithm as an example to show how to use a general criterion to determine the truthfulness of a payment scheme induced by a given algorithm.

Algorithm 2. Construct DiffServ Tree $DMT(a, d)$ with Less Bandwidth Allocation

1: Compute a multicast tree T using Algorithm 1..
2: **for** each link e_i in tree T **do**
3:    Find the maximal bandwidth demand of e_i's downstream receivers, say r_j.
4:    Allocate link e_i a bandwidth $B_i = d_j$.
5: Output T as DMT and bandwidth vector B.

Theorem 1. *Either of Algorithm 1. and Algorithm 2. constructs a tree whose weight is at most 8 times the weight of the minimal cost DiffServ multicast tree.*

Although there are only subtle differences between these two algorithms presented here and the one in [12], the proof of Theorem 1 is not as obvious as that one. We omit the proof due to space limit.

4   Payment for Selfish Links

Instead of simply presenting a truthful payment scheme for a specific DiffServ multicast tree construction algorithm (such as Algorithm 1.), we study a general framework to design a truthful payment scheme for any given tree construction algorithm. We fist give a necessary and sufficient condition for the existence of a truthful payment scheme for a given tree construction method. In the meanwhile, we also present a truthful payment scheme if it exists. We then apply this general framework to the DiffServ multicast tree constructed by Algorithm 1. and design a truthful payment scheme.

4.1   General Framework

From the definition of the truthfulness, we can fix the graph G, the receiver set R and bandwidth demand d. Thus, for our notational convenience, we use $b(\mathcal{A}, a) = \{b_1(\mathcal{A}, a), \cdots, b_m(\mathcal{A}, a)\}$ to denote the bandwidth reservation vector computed by an algorithm $\mathcal{A}$, where $b_i(\mathcal{A}, a)$ is the bandwidth reserved at link e_i. We assume that $b_i(\mathcal{A}, a)$ is *piecewise continuous* with respect to any variable a_j, i.e., a finite number of piece-wise linear functions. The only possible types of discontinuities for a piece-wise continuous function are removable and step discontinuities. In the following we give a definition that is critical to the presentation of our general framework.

Definition 1 (Monotone Non-increase Property (MNP)). *An algorithm $\mathcal{A}$ is said to satisfy the* monotone non-increasing property *if for every link e_i and any two of its possible coefficients $a_{i_1} < a_{i_2}$, $b_i(\mathcal{A}, a|^i a_{i_1}) \geq b_i(\mathcal{A}, a|^i a_{i_2})$.*

Theorem 2. *For a given algorithm $\mathcal{A}$, there exists a payment scheme $\mathcal{P}$ such that the mechanism $M = (\mathcal{A}, \mathcal{P})$ is truthful if and only if $\mathcal{A}$ satisfies MNP.*

This theorem is similar to the forklore for the binary demand games and its proof is omitted here due to space limit. Here we construct a truthful payment scheme $\mathcal{P}$ for an allocation method $\mathcal{A}$. For a link e_i, fix a_{-i} and use x to denote cost vector $a|^i x$ if no confusion is caused. Since $\mathcal{A}$ satisfies MNP, function $b_i(\mathcal{A}, x)$ is non-increasing. Recall that $b_i(\mathcal{A}, x)$ is a piecewise continuous function. We let $x_1 < x_2 \cdots < x_m$ be the points at which $b_i(\mathcal{A}, x)$ is not continuous, and introduce a dummy point $x_{m+1} = \infty$. We define a function $\kappa_i(x)$ as: for $x_p < x \leq x_{p+1}$, $\kappa_i(x) = x \cdot b_i(\mathcal{A}, x) + \int_x^{x_{p+1}} b_i(\mathcal{A}, y) dy + \sum_{j=p+1}^m \int_{x_j}^{x_{j+1}} b_i(\mathcal{A}, y) dy$. Given an algorithm $\mathcal{A}$ and a coefficient vector a, we compute the payment based on algorithm $\mathcal{A}$ as follows: for each link i, the payment to a selected link e_i is

$$\mathcal{P}_i(\mathcal{A}, a) = \kappa_i(a_i). \tag{1}$$

We then summarize the general framework to design a truthful payment scheme $\mathcal{P}$, such that $M = (\mathcal{A}, \mathcal{P})$ is truthful, for a given output algorithm $\mathcal{A}$ that constructs a DiffServ multicast tree and outputs the bandwidth allocation for DiffServ multicast.

1. Check whether the bandwidth allocation of algorithm $\mathcal{A}$ satisfies MNP. If not then there is no payment scheme $\mathcal{P}$ such that $M = (\mathcal{A}, \mathcal{P})$ is truthful, else continue.
2. Find the bandwidth reservation $b(\mathcal{A}, a)$.
3. Design the payment scheme according to Equation (1).

4.2   Design Truthful Mechanism

Lemma 1. *Algorithm 2. does not satisfy MNP.*

Proof. We prove it by presenting an example here. A network G has three receivers r_1, r_2, r_3 with bandwidth demand $d_1 = d_2 = 1$ and $d_3 = 2$. The coefficient of the link is described in Figure 1 (a). When we apply Algorithm 2. to network G, we obtain a tree shown in Figure 1 (b). Let agent 2 be link $v_2 v_3$. The bandwidth allocation of link $e_2 = v_2 v_3$ is 2. Consider the scenario when the coefficient of link e_2 changes from 1.1 to 0.9 while other coefficients remain the same. The new spanning tree is shown in Figure 1 (c). The bandwidth allocation of e_2 becomes 1, which decreases by half compared with the bandwidth reservation with coefficient 1.1. ▣

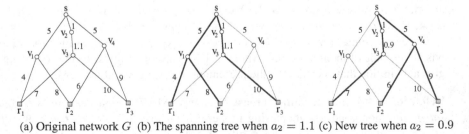

(a) Original network G (b) The spanning tree when $a_2 = 1.1$ (c) New tree when $a_2 = 0.9$

Fig. 1. Algorithm 2. does not satisfy MNP

Theorem 3. *There is no truthful mechanism M based on Algorithm 2..*

The above example also shows that there is no truthful mechanism for the DiffServ multicast tree construction method presented in [12]. Meanwhile, we can show that there exists a truthful payment scheme for Algorithm 1. with the following theorem.

Theorem 4. *Algorithm 1. satisfies MNP.*

The proof of this theorem is straightforward and thus is omitted here due to space limit. To find the truthful payment for Algorithm 1., we should find the bandwidth allocation $b_k(\overline{DMT}, a|^k x)$ for every link e_k first. Recall that for every link e_i, the allocated bandwidth could only be a real value that $R_j^{\max}$ for some index j. Here $R_j^{\max}$ is the maximum demand of the jth group of receivers. Let $x_1^k < x_2^k < \cdots < x_q^k$ be the points at which $b_k(\overline{DMT}, a|^k x)$ is not continuous, then the bandwidth allocation function $b_k(\overline{DMT}, a|^k x)$ should be a constant, say y_j^k in (x_j^k, x_{j+1}^k). In order to find the values of these discontinuous points, we first need to compute the truthful payment for standard Steiner tree problem. We use $\tau(c_{-i}, R)$ to denote the payment computed for a link e_i based on a LST tree heuristic [8]. Algorithm 3. shows how to find the bandwidth-allocation function $b_k(\overline{DMT}, a|^k x)$. Algorithm 4. illustrates our truthful payment scheme by following the general framework. The proof of the correctness of these algorithms are either straightforward or omitted here due to space limit.

Algorithm 3. Bandwidth Allocation Function for Algorithm 1.

1: Apply Algorithm 1.. Let ℓ be the number of iterations in Algorithm 1..
2: **for** every link e_k in $DMT(a, d)$ **do**
3:     Set $c_k = \infty$ and apply Algorithm 1. again.
4:     At the beginning of each iteration i, compute the value $\tau_k^i(a_{-k}, R_i)$.
5:     Initialize the list $X^k = \emptyset, Y^k = \emptyset, up = 0$, and $q = 0$.
6:     **for** $i = 1$ to ℓ **do**
7:       **if** $\tau_k^i(a_{-k}, R_i) > up$ **then**
8:         $q = q + 1$; Set $x_q^k = \tau_k^i(a_{-k}, R_i)$ and $y_q^k = R_i^{\max}$; Add x_q^k to set X^k and y_q^k to Y^k.
9:     Set $x_0^k = 0$ and $x^{q+1} = \infty$.
10:     **for** $i = 1$ to $q + 1$ **do**
11:       Set $b_k(\mathcal{A}, a|^k x) = y_i^k$ for $x_{i-1}^k \leq x < x_i^k$.

Algorithm 4. Payment Scheme for Algorithm 1.

1: Compute the multicast tree $\overline{DMT}$ by applying Algorithm 1..
2: Compute the bandwidth allocation function for tree $\overline{DMT}$ by applying Algorithm 3..
3: **for** each link e_k **do**
4:     **if** e_k is in tree $\overline{DMT}$ **then**
5:       Find i such that $x_i^k < a_k \leq x_{i+1}^k$. Then the payment is $\mathcal{P}_k(a) = \sum_{j=i+1}^{|X^k|-1} y_j^k \cdot (x_{j+1}^k - x_j^k) + (x_{i+1}^k - a_k) \cdot y_i^k$.
6:     **else**
7:       $\mathcal{P}_k(a) = 0$.

4.3   Performance Improvement and Special Case

In essence, Algorithm 1. converts the original instance of the DiffServ multicast problem to a "rounded-up" one, with bandwidth demand vector forming a geometric sequence of ratio 2. According to the result of Charikar *et al.* [12], the approximation ratio of 8 of Algorithm 1. can be improved (while still using LST structure for computing approximately optimal Steiner trees) if the "randomized bucketing" technique is used. Specifically, a number y is picked randomly with a uniform distribution in the range $[0, 1]$, and the (non-zero) bandwidth demands of all receivers are rounded up to the nearest e^{y+i}. (Note that the ratio of the geometric sequence is e instead of 2.) The *expected* approximation ratio is $e \cdot 2 \simeq 5.437$.

Here we argue that we can also convert the mechanism described above for DiffServ multicast to a randomized one with an expected approximation ratio of 5.437, while maintaining the truthfulness of the mechanism. First of all, it is easy to see that using a "start point" of e^y for some fixed y and replacing the ratio of 2 by e for the geometric sequence (of rounded up bandwidth demands) should not affect truthfulness. Furthermore, the randomized process also does not encourage untruthfulness of the links: if for any fixed start point e^y, the links find no incentive to report false cost, nor will they find incentives to report false cost when such start point is randomly selected.

Charikar *et al.* [12] also proposed a de-randomized process to replace the above random selection of start point e^y. For each distinct bandwidth demand d_i, the same algorithm is invoked with $y_i = \ln d_i - \lfloor \ln d_i \rfloor$. It is claimed that there is at least one y_i such that the solution for $y = y_i$ has a cost no more than the expected cost of the solution for a randomly picked y. Therefore, we can simply pick the best solution (with the minimum cost) among all solutions computed using different y. A similar technique is used for the case with only two non-zero rates for bandwidth demands [11], improving the approximation bound to $\frac{4}{3} \cdot 2 \simeq 2.667$. The common characteristic of the two algorithms is to compute multiple DiffServ multicast trees using different methods (or same method but with different parameters), and pick the one with the smallest cost. Although this approach (*i.e.*, taking the best output of several outcomes and using a certain combination of the payments for these separated games as its final payment) works for binary selection problems under certain conditions [20, 19], a problem arises when it comes to determining the payments to the links for DiffServ multicast. We can show by example that the tree construction algorithm in [11] violates the MNP property, which implies that no truthful payment exists.

5   Conclusion

We studied the DiffServ multicast problem in a game theoretic context, where the network links are selfish agents who would demand payments to at least cover their costs when relaying data packets, and may falsely report their actual costs in order to maximize their gains. We show that a naive conversion of the previously known 8-approximation algorithm does not work. We then propose an alternative approximation algorithm for DiffServ multicast with the same approximation bound. We also introduced a general method to convert any DiffServ multicast algorithm satisfying the

Monotone Non-increasing Property to a truthful mechanism, and applied it to the algorithm we proposed. The truthful payment scheme is not the end story for designing protocols for DiffServ multicast. A natural question to ask is how these payments can be split among the receivers, which is known as the *multicast payment sharing* problem. Several criteria for the *fairness* of sharing have been proposed in previous work, and we would like to design payment sharing schemes that are considered to be fair with these criteria.

References

1. K. Nichols, S. Blake, D.B.: Definition of the differentiated services field (ds field) in the IPv4 and IPv6 headers. In: IETF RFC 2474. (1998)
2. K. Nichols, S.B.: Differentiated services operational model and definitions. In: Networking. (1993)
3. Wang, X., Schulzrinne, H.: Pricing network resources for adaptive applications in a differentiated services network. In: INFOCOM. (2001) 943–952
4. Maxemchuk, N.F.: Video distribution on multicast networks. IEEE JSAC (1997)
5. Takahashi, H., Matsuyama, A.: An approximate solution for the steiner problem in graphs. Mathematical Japonica **24** (1980) 573–577
6. Robins, G., Zelikovsky, A.: Improved steiner tree approximation in graphs. In: Proceedings of ACM/SIAM SODA. (2000) 770–779
7. Wang, W., Li, X.Y., Sun, Z., Wang, Y.: Design multicast protocols for non-cooperative networks. In: IEEE INFOCOM. (2005).
8. Wang, W., Li, X.Y., Wang, Y.: Truthful mutlicast in selfish wireless networks. In: ACM MobiCom. (2004)
9. Karpinski, M., Mandoiu, I.I., Olshevsky, A., Zelikovsky, A.: Improved approximation algorithms for the quality of service steiner tree problem. In: WADS. (2003)
10. Balakrishnan, A., Magnanti, T., Mirchandani, P.: Modeling and heuristic worst-case performance analysis of the two-level network design problem. Management Science (1994) 846–867
11. Balakrishnan, A., Magnanti, T., Mirchandani, P.: Heuristics, lps, and trees on trees: Network design analyses. Operations Research (1996) 478–496
12. Charikar, M., Naor, J.S., Schieber, B.: Resource optimization in qos multicast routing of real-time multimedia. In: IEEE INFOCOM. (2000)
13. Calinescu, G., Fernandes, C., Mandoiu, I., Olshevsky, A., Yang, K., Zelikovsky, A.: Primal-dual algorithms for qos multimedia multicast. In: GlobeCom. (2003)
14. Xue, G., Lin, G.H., Du, D.Z.: Grade of service steiner minimum trees in the euclidean plane. Algorithmica (2001) 479–500
15. Kim, J., Cardei, M., Cardei, I., Jia, X.: A polynomial time approximation scheme for the grade of service steiner minimum tree problem,. Jour. of Global Optim. **24** (2002) 439–450
16. Vickrey, W.: Counterspeculation, auctions and competitive sealed tenders. Journal of Finance (1961) 8–37
17. Clarke, E.H.: Multipart pricing of public goods. Public Choice (1971) 17–33
18. Groves, T.: Incentives in teams. Econometrica (1973) 617–631
19. Kao, M.Y., Li, X.Y., Wang, W.: Towards Truthful Mechanisms for Binary Demand Games: A General Framework. In proceedings ACM EC (2005).
20. Mu'alem, A., Nisan, N.: Truthful approximation mechanisms for restricted combinatorial auctions: extended abstract. In: Proceedings of the 18th National Conference on Artificial Intelligence(2002) 379–384

Competitive Analysis of On-line Securities Investment*

Shuhua Hu[1], Qin Guo[1], and Hongyi Li[2]

[1] School of Management, Xi'an Jiaotong University, Xi'an, China
shuhuahu@hotmail.com, yourqinr@163.com
[2] Dept of Decision Sciences and Managerial Economics,
The Chinese University of Hongkong, Hongkong, China
hongyi@baf.msmail.cuhk.edu.hk

Abstract. Based on the unidirectional conversion model, we investigate a practical buy-and-hold trading problem. This problem is useful for long-term investors, we use competitive analysis and game theory to design some trading rules in the securities markets. We present an online algorithm, Mixed Strategy, for the problem and prove its competitive ratio $1 + \frac{(n-1)t}{2}$, where n is the trading horizon and t is the daily fluctuations of securities prices. The Dynamic-Mixed Strategy is also presented to further reduce the competitive ratio. An investing example is simulated with the Mixed Strategy and Dollar Average Strategy based on the actual market data.

1   Introduction

In many situations, we are forced to choose between different alternatives without knowledge of each alternative's future worth. The finance management problem has attracted substantial attention lately due to the coexistence of high returns and high risks. Like many other ongoing financial activities, securities investing must be carried out in an on-line fashion, with no secure knowledge of future events. Faced with this lack of knowledge, the on-line investors of this financial activity often use models based on assumptions about the future distribution of relevant quantities such as securities prices, and aim for acceptable results on the average. The on-line algorithm and competitive analysis[1] are important tools for making decision. The approach we follow here is to use competitive analysis, which was first applied to on-line algorithms by Sleator and Tarjan in[2].

Competitive analysis has been extensively discussed in the domain of financial management since Cover[3] first investigated the portfolio problem with it in 1991. Cover presented a simple on-line strategy that dynamically changed the distribution of its current wealth among the stocks based on the market history. El-Yaniv[4][5][6] explored the on-line algorithm for the unidirectional conversion problem under the assumption that the whole exchange rates, instead of the daily

---

* Supported by NSF Grant No.70471035 and No.10371094.

N. Megiddo, Y. Xu, and B. Zhu (Eds.): AAIM 2005, LNCS 3521, pp. 224–232, 2005.

ones, are between a pair of upper and lower bounds, no matter how erratically or unfortunately the rates vary from day to day. Xu et al.[7] further studied the same setting in a framework with commission and interest rate. Chou[8] investigated the bidirectional currency trading problem against a weak statistical adversary to make a money-making strategy.

In most studies, it is inevitable for the researchers to make some assumptions about future events to design a competitive strategy to maximize the future gain. Although much research has been devoted to the investing problem, little research has been done based on the real trading rules. Instead of knowing about the distribution of future market prices, an online strategy might be based only on the knowledge of the daily bounds on possible market prices. In this paper we use the framework of the daily price constraints model, in which the next day's price e' depends on the current day's price e with $e \cdot (1 - t) \le e' \le e \cdot (1 + t)$ for some fixed $t < 1$, to investigate the buy-and-hold trading problem using game theory and competitive analysis based on the *unidirectional conversion model* in [4]. We obtain an online algorithm Mixed Strategy for the problem and prove its exact competitive ratio $1 + \frac{(n-1)t}{2}$. We then present the Dynamic-Mixed Strategy to further reduce the competitive ratio. The Mixed Strategy and Dollar Average Strategy are also compared and an investing example is simulated based on the actual market data of CBM in 2003. Our results could be beneficial to the long-term trading players in the securities market to make a decision.

This paper is organized as follows. In Section 2, we present the on-line securities investment problem, and derive the Mixed Strategy and the Dynamic-Mixed Strategy using the Game Theory and Competitive Analysis. Section 3 refers to compare the Mixed Strategy and the popular Dollar Average Strategy, and an investing example is stimulated using the two strategies based on the actual market data of CBM in 2003. Conclusions and work-in-progress are reported in section 4.

2   On-line Securities Investment, OSI

2.1   Problem Statement

Suppose that an investor has initial capital of $w_0 = 1$, and invests it for a certain security in n days, which is referred to as the trading horizon (assume $n \ge 2$ to avoid triviality). We also assume that, on each trading day, the security has only one price, which refers to the number of shares of the securities one unit capital could buy. After each price is realized, the investor executes one transaction for that day and invests all or part of his capital. All the capital must be invested on the n-th trading day, and the holding securities are not permitted converting back to the capital. The performance of the investing is the accumulation of shares of the securities that the investor holds at the end of the investment.

In many securities markets, it is regulated that the next day's closing price e_{i+1} depends on the current day's closing price e_i with $e_{i+1} \in [e_i(1-t), e_i(1+t)]$, for some fixed $t < 1$, and it means the daily price fluctuation. Let E denote the set of all the feasible price sequences, O denote the optimal off-line trading algorithm, and A be the investor's on-line trading algorithm. On the i-th trading day, the investor must decide the investing amount with only the information of e_i. In this model, the optimal off-line return is $O(\bar{e}) = \max\limits_{1 \leq i \leq n} 1/e_i$, and the on-line investor's return is $A(\bar{e}) = \sum_{i=1}^{n} a_i/e_i$, where a_i is the amount of capital that the investor invests on the i-th trading day.

Let S_i be the trade-once algorithm for $i \in Z_n$, i.e., the investor trades all his initial capital on the i-th trading day, and his return is $S_i(\bar{e}) = 1/e_i$. Note that S_i is static because converting the holding securities back to cash is prohibited once the trade realized. Let S be a randomized static algorithm, and s_i be the expected amount of capital invested by S on the i-th day. For all i, $s_i \geq 0$ and $\sum_{i=1}^{n} s_i = 1$. Let S' denote the deterministic static algorithm that invests s_i on the i-th day. Since the value of s_i defines a probability density function in $\Phi(Z_n)$, let S'' denote the randomized static algorithm that applies S_i with probability s_i.

Lemma 1. [9] S, S', and S'' are equivalent in the sense that for all $\bar{e} \in E$, $S(\bar{e}) = S'(\bar{e}) = S''(\bar{e})$.

Let r_s^* be the smallest possible competitive ratio of the on-line static algorithm, then according to Lemma 1,

$$r_s^* = \inf_{f \in \Phi(Z_n)} \sup_{\bar{e} \in E} \frac{O(\bar{e})}{\sum_{i=1}^{n} f(i) S_i(\bar{e})}. \tag{1}$$

2.2 Mixed Strategy and Competitive Analysis, MIX

The OSI problem could be considered as a two-person game $G(m, n)$ [4]. For any integer $k > 0$, we set $Z_k = \{1, 2, \cdots, k\}$. The maximizing player is the on-line investor, whose pure strategies are the deterministic on-line algorithm A_i of G indexed with $i \in Z_m$. The minimizing player is the adversary in G, whose pure strategies are the input price sequences $\bar{\sigma}_j$ of G with $i \in Z_n$. If the vectors of the payoff matrix are denoted by the reciprocal of the competitive ratio, then the payoff matrix is

$$H(i, j) = \frac{A_i(\bar{\sigma}_j)}{O_i(\bar{\sigma}_j)} > 0, i \in Z_m, j \in Z_n. \tag{2}$$

Let $\Phi(Z_k)$ be the set of all probability density functions defined on Z_k. For $k = m$, or $k = n$, each $h \in \Phi(Z_k)$ could be considered as a point in the k-dimensional Euclidean space and represents a Mixed Strategy that applies the l-th pure strategy indexed by Z_k with probability $h(l)$. Let r^* denote the smallest possible competitive ratio of any randomized on-line algorithm for $G(m, n)$, by equation (1)

$$r^* = \min_{f \in \Phi(Z_m)} \max_{j \in Z_n} \frac{O(\bar{\sigma}_j)}{f(i) A_i(\bar{\sigma}_j)}. \tag{3}$$

A randomized on-line algorithm is *optimal* if its competitive ratio is r^*.

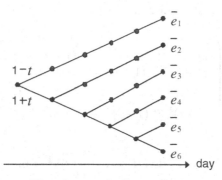

Fig. 1. The adversary's dominating sequences

The buy-and-hold trading of *OSI* is an infinite game because the adversary's pure strategies are infinite. However, the on-line investor has n pure strategies S_i. Using the strictly inferior method of the game theory and the worst-case analysis, we can delete the non-worst-case price sequences of the adversary and make the game be finite. In other words, we neglect all the sequences that dose not rise to or lower to the limit. For $j = 1, \cdots, n$, let $\bar{e}_j$ be the adversary's dominating sequences, then

$$\bar{e}_j = [\underbrace{(1-t), (1-t)^2, \cdots, (1-t)^j},_{j}$$

$$\underbrace{(1-t)^j(1+t), (1-t)^j(1+t)^2 \cdots, (1-t)^j(1+t)^{n-j}}_{n-j}]. \qquad (4)$$

See figure 1 for its illustration.

Lemma 2. [9] Given a static algorithm S, each $\bar{e} \in E$ is dominated by $\bar{e}_j$, i.e., $\frac{O(\bar{e})}{S(\bar{e})} \leq \frac{O(\bar{e}_j)}{S(\bar{e}_j)}$, where $e_j = \min_{1 \leq i \leq n} e_i$.

Theorem 1. If the investor know the trading horizon n and the daily price fluctuation t at the beginning, there is **Nash Balance** of the Mixed Strategy

$$x_i^* = \begin{cases} \frac{1+t}{(n-1)t+2} & i = 1 \\ \frac{t}{(n-1)t+2} & i \in [2, n-1] \\ \frac{1}{(n-1)t+2} & i = n \end{cases}$$

and $\sum_{i=1}^{n} x_i^* = 1$, which is the optimal on-line investing strategy. Its competitive ratio is $1 + \frac{t}{2}(n-1)$.

Proof. Let S_i be the investor's i-th pure strategy and $\bar{e}_j$ be the j-the pure strategy of the adversary. According to equations (2) and (4), the payoff matrix of the on-line game is defined by $H(i,j) = \frac{S_i(\bar{e}_j)}{O(\bar{e}_j)}$, where $O(\bar{e}_j) = (1-t)^{-j}$ and

$S_i(\bar{e}_j) = (1-t)^{-i}$ if $i \le j$ or $S_i(\bar{e}_j) = (1-t)^{-j}(1+t)^{j-i}$ otherwise, i.e.,

$$H = \begin{pmatrix} 1 & (1-t) & (1-t)^2 & \cdots & (1-t)^{n-1} \\ (1+t)^{-1} & 1 & (1-t) & \cdots & (1-t)^{n-2} \\ (1+t)^{-2} & (1+t)^{-1} & 1 & \cdots & (1-t)^{n-3} \\ \vdots & \vdots & \vdots & \ddots & \vdots \\ (1+t)^{1-n} & (1+t)^{2-n} & (1+t)^{3-n} & \cdots & 1 \end{pmatrix} \tag{5}$$

for $n \ge 2$,

$$det(H) = (\frac{2t}{1+t})^{n-1} > 0. \tag{6}$$

Let $X = (x_1, x_2, \ldots, x_n)^T$ denote the Mixed Strategy for the on-line investor, and π be the payoff of the algorithm. The investor could get the same payoff, i.e., the same competitive ratio, with the Mixed Strategy whatever strategies the adversary choose

$$\pi_1 = \pi_2 = \cdots = \pi_n. \tag{7}$$

The payoff of the algorithm is

$$\begin{aligned} \Pi = HX &= (\pi_1, \pi_2, \cdots, \pi_n)^T \\ &= \begin{pmatrix} 1 \cdot x_1 & + & (1-t) \cdot x_2 & + \cdots + (1-t)^{n-1} \cdot x_n \\ (1+t)^{-1} \cdot x_1 + & & 1 \cdot x_2 & + \cdots + (1-t)^{n-2} \cdot x_n \\ (1+t)^{-2} \cdot x_1 + & (1+t)^{-1} \cdot x_2 & + \cdots + (1-t)^{n-3} \cdot x_n \\ \cdots\cdots\cdots\cdots\cdots\cdots\cdots\cdots\cdots\cdots\cdots\cdots\cdots\cdots\cdots\cdots \\ (1+t)^{1-n} \cdot x_1 + (1+t)^{2-n} \cdot x_2 + \cdots + & 1 \cdot x_n \end{pmatrix} \end{aligned} \tag{8}$$

The undermentioned solution follows from equation (7).

$$x_i = \begin{cases} \frac{1+t}{(n-1)t+2} & i = 1 \\ \frac{t}{(n-1)t+2} & i \in [2, n-1] \\ \frac{1}{(n-1)t+2} & i = n \end{cases} \tag{9}$$

and $\sum_{i=1}^{n} x_i = 1$. Equation (9) means that the on-line investor should invest $\frac{1+t}{(n-1)t+2}$ unit on the first trading day, $\frac{1}{(n-1)t+2}$ unit on the last trading day and $\frac{t}{(n-1)t+2}$ on each of the other trading days starting with one unit capital initially. The payoff of the algorithm is

$$\pi_1 = \pi_2 = \cdots = \pi_n = \frac{2}{(n-1)t+2}. \tag{10}$$

Combining $r = \frac{1}{\pi}$, it suffices to show that

$$r_{mix} = \frac{(n-1)t+2}{2} = 1 + \frac{t}{2}(n-1). \tag{11}$$

The unique optimality of the Mixed Strategy among the static algorithms for the *price constraints* model follows from Lemma 2. □

According to Theory 1, the on-line investor's investing decision depends on the trading horizon n. Though the competitive ratio and the daily investing amount take $O(1)$ time to evaluate, $\lim_{n \to \infty} r_{mix} = \lim_{n \to \infty} [1 + \frac{t}{2}(n-1)] = \infty$. The algorithm is not competitive when the trading horizon goes to infinite. However, the algorithm is a practical buy-and-hold strategy for a fund manager who decides to change the position of some portfolio, and those who invest in the stock-index fund.

2.3   Dynamic-Mixed Strategy, D-MIX

The Dynamic-Mixed Strategy is an improved strategy against a clumsy adversary. The simple investment strategy above is overly pessimistic since it fixes the competitive ratio based on the assumption of worst-case input sequences of prices, and does not change it thereafter. However, the on-line investor could delete the not-occurred worst-case sequences mentioned in the equation 6 according to the history data whenever some sequences are realized. Thus, we could strictly improve the off-line to on-line ratio by recalculating the achievable competitive ratio, and this is a dynamic process. At the start of each trading day, the on-line player has some number w' of capital. The investor knows the number n' of days remaining, and is given a price e'. The investor acts as if the current day were the first trading day of an n'-day trading horizon in which the adversary starts with one unit capital and the player with w' unit capital. By arguments similar to those used in section 2.2, an expression for the competitive ratio is determined and maximized over the remaining market prices. For the k-th realizing sequence the amount x'_j to invest is given by

$$
x'_j = \begin{cases} \frac{1+t}{(n'-1)t+2} \cdot w'_k & j = 1 \\ \frac{t}{(n'-1)t+2} \cdot w'_k & j \in [2, n'-1] \\ \frac{1}{(n'-1)t+2} \cdot w'_k & j = n' \end{cases} \tag{12}
$$

The corresponding competitive ratio is

$$
r_{dm} = 1 + \frac{(n'-1)t}{2}. \tag{13}
$$

2.4   Dollar Average Strategy, DA

The Dollar Average Strategy is a static algorithm which invests an equal amount of capital, i.e., $\frac{1}{n}$ unit capital on each trading day. The following theory gives the competitive ratio of the static algorithm.

Theorem 2. $r_{da} = \frac{nt}{1-(1-t)^n}$.

Proof. By Theorem 1, DA is the uniform Mixed Strategy for the on-line investor in the game $G(m, n)$. That is to say, the on-line investor's investing strategy is $\tilde{X} = (\tilde{x}_1, \tilde{x}_2, \cdots, \tilde{x}_n) = (\frac{1}{n}, \frac{1}{n}, \cdots, \frac{1}{n})^T$ with one unit capital initially. The associated payoff is $\tilde{\Pi} = (\tilde{\pi}_1, \tilde{\pi}_2, \cdots, \tilde{\pi}_n)^T$. The payoff matrix of DA is

$$\tilde{\Pi} = H\tilde{X} = \frac{1}{n} \begin{pmatrix} 1 & (1-t) & (1-t)^2 & \cdots & (1-t)^{n-1} \\ (1+t)^{-1} & 1 & (1-t) & \cdots & (1-t)^{n-2} \\ (1+t)^{-2} & (1+t)^{-1} & 1 & \cdots & (1-t)^{n-3} \\ \vdots & \vdots & \vdots & \ddots & \vdots \\ (1+t)^{1-n} & (1+t)^{2-n} & (1+t)^{3-n} & \cdots & 1 \end{pmatrix} \qquad (14)$$

The competitive ratio of the Dollar Average Strategy is

$$r_{da} = max(\frac{1}{\tilde{\pi}_1}, \frac{1}{\tilde{\pi}_2}, \cdots, \frac{1}{\tilde{\pi}_n}) = \frac{nt}{1-(1-t)^n}. \qquad (15)$$

□

3   Simulation

The competitive ratios of DA and MIX were plotted against trading horizon in Figure 2 for $t = 0.1$. It is obvious that the Mixed Strategy is better than the Dollar Average Strategy with increase of the trading horizon.

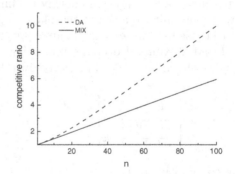

Fig. 2. Competitive ratios of MIX and DA (t=0.1)

This seems inconsistent with El-Yani's[4] result, who demonstrated that the Dollar Average Strategy is the best plan of the threat-based algorithm. One reason for this could be that the two models were based on different constraints of the input price sequences. El-Yani's model fixes a static upper bound and a static lower bound on the daily exchange rates for the entire investing horizon, while our model sets new dynamic price constraints for every trading day. According to the constraints of daily price fluctuation in many securities markets, the latter one should be more practical.

Based on the market data of China Merchants Bank Co., Limited (CMB) in 2003, we simulated an investing example using MIX and DA. Figure 3 shows the daily closing prices of CMB in 2003. All stock prices are quoted in RMB. One investment plan is executed to buy the shares of CBM in one year with an initial capital of one RMB. For easy comparison, the daily accumulation is expressed by the summation of the remaining cash and the currency values of the holding share. The currency values of the holding share are evaluated by the price of the current trading day. Figure 4 shows the daily accumulations of MIX, DA and the optimal off-line strategy based on the data of CBM in 2003. At the end of the plan, the return of the off-line strategy, MIX, and DA are 1.34, 1.054 and 1.044, respectively. The corresponding realized competitive ratio of MIX and DA are 1.27 and 1.28, respectively. The Mixed Strategy outperforms the Dollar Average Strategy as a whole but the difference between the two strategies is not distinct. The main reason is that the real price sequence of CMB in 2003 is not given in the worst-case, but one of the infinite non-worst-cases.

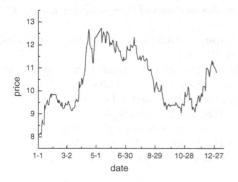

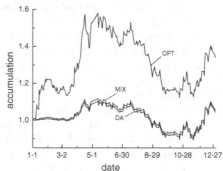

Fig. 3. CMB's daily closing stock prices in 2003

Fig. 4. The accumulations of OPT, MIX and DA (t=0.1)

4   Conclusions

Considering the trading rules of daily price constraints in securities markets, we obtained an on-line algorithm, the Mixed Strategy, and proved its optimality among the static algorithms in the daily price constraints model for the buy-and-hold trading problem in a discrete case. This method can provide a valuable and quantitative analysis tool for securities investment. However, the Mixed Strategy is pessimistic because of the assumption of the worst-case input sequences and the invariability of the competitive ratio. Taking the not-occurred worst-cases into account, the Mixed Strategy is improved further to the Dynamic-Mixed Strategy. Based on the market data of China Merchants Bank Co., Limited (CMB) in 2003, we compared the Mixed Strategy and the Dollar Average Strategy. The results show that the former strategy outperforms the latter one as a whole. But the difference is not so obvious as in the worst-case.

However, it must be pointed out that the Mixed Strategy is not competitive when the trading horizon goes to infinity. Therefore, how to design a competitive algorithm not related to n is still an open problem for this model. In addition, it would be interesting to investigate the bidirectional trading problem and the trading problem with transaction cost using this model.

References

1. Borodin A, R.El-Yaniv. Online computation and competitive analysis. *Cambridge University Press*, 1998.
2. Sleator, D.and Tarjan, R.E.. Amortized efficiency of list update and paging rules. *Commun.ACM*, 28:202-208, 1985.
3. T.M.Cover. Universal Portfolio. Mathematics Finance, 1(1):1-29, 1991.
4. R.El-Yaniv. Competitive Solutions for Online Financial Problems. *ACM Computing Surveys*, 30:28-68, 1998.
5. R.El-Yaniv, A.Fiat, R.M.Karp, and G.Turpin. Competitive Analysis of Financial Games. *Proc. 33rd Annual Symposium on Foundations of Computer Science*, pp.327-333, 1992.
6. R.El-Yaniv, A.Fiat, R.M.Karp, and G.Turpin. Optimal Search and One-way Trading Online Algorithms. *Algorithmica*, 30:101-139, 2001.
7. Zhu Zhijun, Xu Yinfeng and Jiang Jinhu. Competitive Analysis of One-way Trading with Interest and Transaction Cost. *Forecasting*, 22(4):51-56, 2003.
8. A.Chou, J.R. Cooperstock, R. El-Yaniv, M.Kugerman and T.Leighton. The Statistical Adversary Allows Optimal Money-making Trading Strategies. *In the Proceedings of the 6th Annual ACM-SIAM Symposium on Discrete Algorithms* 1995.
9. Gen-Huey Chen, Ming-Yang Kao, Yuh-Dauhlyuu and Hsing-kuo Wong. Optimal Buy-and-Hold Strategies for Financial Markets with Bounded Daily Returns. *SIAM J, COMPUT*, 31:447-459, 2000.

Perfectness and Imperfectness of the kth Power of Lattice Graphs

Yuichiro Miyamoto[1] and Tomomi Matsui[2]

[1] Sophia University, Kioicho 7-1, Chiyoda-ku,
Tokyo 102-8554, Tokyo, Japan
y-miyamo@sophia.ac.jp
[2] University of Tokyo, Hongo 7-3-1, Bunkyo-ku,
Tokyo 111-1111, Japan
tomomi@misojiro.t.u-tokyo.ac.jp

Abstract. Given a pair of non-negative integers m and n, $S(m,n)$ denotes a square lattice graph with a vertex set $\{0,1,2,\ldots,m-1\} \times \{0,1,2,\ldots,n-1\}$, where a pair of two vertices is adjacent if and only if the distance is equal to 1. A triangular lattice graph $T(m,n)$ has a vertex set $\{(x e_1 + y e_2) \mid x \in \{0,1,2,\ldots,m-1\},\ y \in \{0,1,2,\ldots,n-1\}\}$ where $e_1 \overset{\text{def.}}{=} (1,0)$, $e_2 \overset{\text{def.}}{=} (1/2, \sqrt{3}/2)$, and an edge set consists of a pair of vertices with unit distance. Let $S^k(m,n)$ and $T^k(m,n)$ be the kth power of the graph $S(m,n)$ and $T(m,n)$, respectively. Given an undirected graph $G = (V, E)$ and a non-negative vertex weight function $w : V \to Z_+$, a multicoloring of G is an assignment of colors to V such that each vertex $v \in V$ admits $w(v)$ colors and every adjacent pair of two vertices does not share a common color.

In this paper, we show necessary and sufficient conditions that [$\forall m$, the $S^k(m,n)$ is perfect] and/or [$\forall m$, $T^k(m,n)$ is perfect], respectively. These conditions imply polynomial time approximation algorithms for multicoloring $(S^k(m,n), w)$ and $(T^k(m,n), w)$.

1 Introduction

Given a pair of non-negative integers m and n, $S(m,n)$ denotes a square lattice graph with a vertex set $\{0,1,2,\ldots,m-1\} \times \{0,1,2,\ldots,n-1\}$, where a pair of two vertices is adjacent if and only if the distance is equal to 1. A triangular lattice graph $T(m,n)$ has a vertex set $\{(x e_1 + y e_2) \mid x \in \{0,1,2,\ldots,m-1\},\ y \in \{0,1,2,\ldots,n-1\}\}$ where $e_1 \overset{\text{def.}}{=} (1,0)$, $e_2 \overset{\text{def.}}{=} (1/2, \sqrt{3}/2)$, and an edge set consists of pairs of vertices with unit distance. The kth power of a graph G is a graph whose vertex set is equivalent to that of G and a pair of two vertices is adjacent if and only if there is a path of length at most k between the pair. Clearly, $G^1 = G$ holds. We denote the kth power of $S(m,n)$ and $T(m,n)$ by $S^k(m,n)$ and $T^k(m,n)$, respectively.

Given an undirected graph $H = (V, E)$ and a non-negative integer vertex weight function $w : V \to Z_+$, a *multicoloring* of (H, w) is an assignment of colors

N. Megiddo, Y. Xu, and B. Zhu (Eds.): AAIM 2005, LNCS 3521, pp. 233–242, 2005.

to vertices of H such that each vertex v admits $w(v)$ colors and every adjacent pair of two vertices does not share a common color. A *multicoloring problem* on (H, w) finds a multicoloring of (H, w) which minimizes the required number of colors.

In this paper, we study weighted kth power of the square lattice graphs and weighted kth power of the triangular lattice graphs. First, we show a necessary and sufficient condition that $[\forall m,\ S^k(m,n)$ is perfect], and a necessary and sufficient condition that $[\forall m,\ T^k(m,n)$ is perfect]. If a graph is perfect, we can solve the multicoloring problem easily. Next, we propose polynomial time approximation algorithms for multicoloring $S^k(m,n)$ and $T^k(m,n)$. Our algorithm is based on the well-solvable cases that the given graph is perfect. For any $k \geq 3$, our algorithm finds a multicoloring of $S^k(m,n)$ which uses at most

$$\left(1 + \frac{k}{\lfloor \frac{k+3}{2} \rfloor}\right) w + \mathrm{O}(k^3)$$

colors, where w denotes the weighted clique number of $S^k(m,n)$. Table 1 shows the value of the above approximation ratio in case that k is small. For any $k \geq 2$,

Table 1. Approximation ratios in case of the square lattice

k	3	4	5	6	7	8	9	10	11	12	$\cdots$	∞
ratio	2	7/3	9/4	5/2	12/5	13/5	15/6	8/3	18/7	19/7	$\cdots$	3

our algorithm finds a multicoloring of $T^k(m,n)$ which uses at most

$$\left(\frac{2k+1}{k+1}\right) w + \mathrm{O}(k^3)$$

colors, where w denotes the weighted clique number of $T^k(m,n)$. Table 2 shows the value of the above approximation ratio in case that k is small.

Table 2. Approximation ratios in case of the triangular lattice

k	2	3	4	5	6	7	8	9	10	11	$\cdots$	∞
ratio	5/3	7/4	9/5	11/6	13/7	15/8	17/9	19/10	21/11	23/12	$\cdots$	2

The multicoloring problem has been studied in several context. When a given graph is the kth power of the triangular lattice graph, the problem is related to the radio channel (frequency) assignment problem. McDiarmid and Reed [6] showed that the multicoloring problem on triangular lattice graph $T^1(m,n)$ is

NP-hard. Some authors [6, 10] independently gave (4/3)-approximation algorithms for this problem. We discussed perfectness and imperfectness of unit disk graphs on triangular lattice points [9]. In case that a given graph H is the square lattice graph $S^1(m, n)$, the graph H becomes bipartite and so we can obtain an optimal multicoloring of (H, w) in polynomial time (see [6] for example). In our previous paper [8], we showed that the multicoloring problem on $S^2(m, n)$ is NP-hard, and we proposed (4/3)-approximation algorithm for the problem. Halldórsson and Kortsarz [4] studied planar graphs and partial k-trees. For both classes, they gave a polynomial time approximation scheme (PTAS) for variations of multicoloring problem with min-sum objectives. These objectives appear in the context of multiprocessor task scheduling.

2   Perfectness and Imperfectness

In this section, we discuss some well-solvable cases such that the multicoloring number is equivalent to the weighted clique number. An undirected graph G is *perfect* if for each induced subgraph H of G, the coloring number of H, denoted by $\chi(H)$, is equal to its clique number $\omega(H)$. The following theorems are main results of this paper.

Theorem 1. *When $k \geq 3$, we have the following; [$\forall m \in Z_+$, $S^k(m, n)$ is perfect] if and only if $k \geq 2n - 3$.*

The perfectness of $S^2(m, n)$ is discussed in our previous paper [8]. Table 3 shows the perfectness and imperfectness of $S^k(m, n)$ for small k and n.

Table 3. Perfectness and imperfectness of the kth power of the square lattice graph

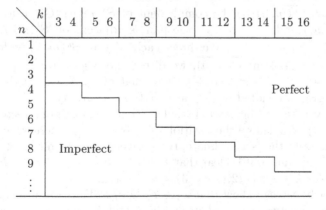

Theorem 2. *The graph $T^1(m, 3)$ is perfect. When $k \geq 2$, we have the following; [$\forall m \in Z_+$, $T^k(m, n)$ is perfect] if and only if $k \geq n - 1$.*

Table 4. Perfectness and imperfectness of the kth power of the triangular lattice graph

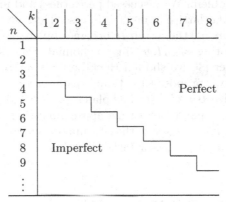

Table 4 shows the perfectness and imperfectness of $S^k(m, n)$ for small k and n.

To show the above theorems, we introduce some definitions. We say that an undirected graph has a transitive orientation property, if each edge can be assigned a one-way direction in such a way that the resulting directed graph (V, F) satisfies that $[(a, b) \subset F$ and $(b, c) \in F$ imply $(a, c) \in F]$. An undirected graph which is transitively orientable is called *comparability graph*. The complement of a comparability graph is called *co-comparability graph*. It is well-known that every co-comparability graph is perfect.

Lemma 1. *For any integer $n \geq 2$, if $k \geq 2n - 3$, then $S^k(m, n)$ is a co-comparability graph.*

Proof. It suffices to show that the complement of $S^k(m, n)$ is a comparability graph. Let $\overline{S^k(m, n)}$ be the complement of $S^k(m, n)$. Given a pair of vertices $u = (x_1, y_1)$ and $v = (x_2, y_2)$, the pair is adjacent on $S^k(m, n)$ if and only if $|x_1 - x_2| + |y_1 - y_2| \leq k$. We direct each edge in $\overline{S^k(m, n)}$ as follows. For any edge $e = \{v_1, v_2\}$ in $\overline{S^k(m, n)}$, we direct the edge e from v_1 to v_2 when the x-coordinate of v_1 is strictly less than that of v_2. We show that the obtained directed graph, denoted by G', satisfies the transitivity.

Clearly, G' is acyclic. Assume that G' contains a pair of directed edges (v_1, v_2) and (v_2, v_3). We denote the position of v_i by (x_i, y_i) where x_i and y_i are the x-coordinate and the y- coordinate, respectively. The definition of G' implies that $x_1 < x_2 < x_3$. Then, it is clear that $x_2 - x_1 \geq k + 1 - |y_2 - y_1| \geq k + 1 - (n - 1) = k + 2 - n \geq k + 2 + (1/2)(-k - 3) = k/2 + 1/2$.

Similarly, we can show that $x_3 - x_2 > k/2$. Thus we have $x_3 - x_1 > k$ and the distance between v_1 and v_2 on the graph $S(m, n)$ is greater than k. From the definition of G', the digraph G' contains the edge (v_1, v_3).

In the similar way, we can show the following.

Lemma 2. *For any integer* $n \geq 2$, *if* $k \geq n - 1$, *then* $T^k(m, n)$ *is a co-comparability graph.*

The following lemma deals with the remained special case that $n = 3$, $k = 1$.

Lemma 3. *For any* $m \in Z_+$, *the graph* $T^1(m, 3) = T(m, 3)$ *is perfect.*

Proof. Let H be an induced subgraph of $T^1(m, 3)$. Clearly, $\omega(H) \leq \omega(T^1(m, 3))$ $= 3$. When $\omega(H) \leq 2$, H has no 3-cycle. Then it is easy to show that H has no odd cycle and thus $\chi(H) = \omega(H)$, since H is bipartite. If $\omega(H) = 3$, then it is clear that $3 = \omega(H) \leq \chi(H) \leq \chi(T^1(m, n)) = 3$, since $T^1(m, n)$ has a trivial 3-coloring.

Note that though the graph $T^1(m, 3)$ is perfect, the graph $T^1(m, 3)$ is not co-comparability graph.

From the above, the perfectness of a graph satisfying the conditions of Theorems 1 and 2 is clear. In the following, we discuss the inverse implication. We say that an undirected graph G has an *odd-hole*, if G contains an induced subgraph isomorphic to an odd cycle whose length is greater than or equal to 5. It is obvious that if a graph has an odd-hole, the graph is not perfect.

Lemma 4. *For any integer* $n \geq 4$, *if* $k \leq 2n - 4$, *then* $\exists \in Z_+$, $S^k(m, n)$ *is imperfect.*

Proof. In the following, we show that $\forall n \geq 4$, if $3 \leq k \leq 2n - 4$, then $\exists m \in Z_+$, $S^k(m, n)$ has at least one odd-hole, by induction on n. When $n = 4$ and $k = 3$, a subgraph induced by

$$\{ (0,0), (0,3), (2,3), (3,3), (3,0) \}$$

is a 5-hole. When $n = 4$ and $k = 4$, a subgraph induced by

$$\{ (0,0), (0,3), (2,3), (4,3), (4,0) \}$$

is a 5-hole.

Now we consider the case that $n = n' \geq 5$ under the assumption that if $3 \leq k \leq 2n' - 6$, then $\exists m' \in Z_+, S^k(m', n' - 1)$ has at least one odd-hole. If $3 \leq k \leq 2n' - 6$, then $S^k(m', n')$ has at least one odd-hole, since $S^k(m', n' - 1)$ is an induced subgraph of $S^k(m', n')$. In the remained case that $k = 2n' - 5$ and $k = 2n' - 4$, subgraphs induced by

$$\{ (0,0), (0, n' - 1), (n' - 2, n' - 1), (2n' - 5, n' - 1), (2n' - 5, 0) \}$$

and

$$\{ (0,0), (0, n' - 1), (n' - 2, n' - 1), (2n' - 4, n' - 1), (2n' - 4, 0) \}$$

are 5-holes, respectively.

Lemma 5. *The graph $T^1(4,4)$ is imperfect. For any integer $n \geq 4$, if $k \leq n-2$, then $\exists \in Z_+$, $T^k(m,n)$ is imperfect.*

Proof. In the following, we show that $\forall n \geq 4, 1 \leq k \leq n-2, \exists m \in Z_+, T^k(m,n)$ has at least one odd-hole, by induction on n. In this proof, we denote a point $(xe_1 + ye_2)$ by $\langle x, y \rangle$. When $n = 4$ and $k = 1$, a subgraph induced by

$$\{ \langle 2,0 \rangle, \langle 1,1 \rangle, \langle 0,2 \rangle, \langle 0,3 \rangle, \langle 1,3 \rangle, \langle 2,3 \rangle, \langle 3,2 \rangle, \langle 3,1 \rangle, \langle 3,0 \rangle \}$$

is a 9-hole. When $n = 4$ and $k = 2$, a subgraph induced by

$$\{ \langle 0,2 \rangle, \langle 1,3 \rangle, \langle 3,2 \rangle, \langle 3,0 \rangle, \langle 2,0 \rangle \}$$

is a 5-hole.

Lastly we consider the case that $n = n' \geq 5$ under the assumption that if $1 \leq k \leq n' - 3$, then $\exists m' \in Z_+, T^k(m', n' - 1)$ has at least one odd-hole. If $1 \leq k \leq n' - 3$, then $T^k(m', n')$ has at least one odd-hole, since $T^k(m', n' - 1)$ is an induced subgraph of $T^k(m', n')$. In the remained case that $k = n' - 2$, a subgraph induced by

$$\{ \langle 0, n'-2 \rangle, \langle n'-3, n'-1 \rangle, \langle 2n'-5, n'-2 \rangle, \langle 2n'-5, 0 \rangle, \langle n'-2, 0 \rangle \}$$

is a 5-hole.

Lemma 4 shows the imperfectness of every graph which violates the condition of Theorem 1. Lemma 5 shows the imperfectness of every graph which violates the condition of Theorem 2. Thus, we completed proofs of Theorems 1 and 2. From the above lemmas, the followings are immediate.

Corollary 1. *Let $k \geq 3$ be an integer. Then, $S^k(m,n)$ is a co-comparability graph, if and only if $n \leq (k+3)/2$.*

Corollary 2. *Let $k \geq 2$ be an integer. Then, $T^k(m,n)$ is a co-comparability graph, if and only if $n \leq k + 1$.*

In the rest of this section, we discuss some algorithmic aspects. Assume that we have a co-comparability graph G and related digraph H which gives a transitive orientation of the complement of G. Then each independent set of G corresponds to a chain (directed path) of H. The multicoloring problem on G is essentially equivalent to the minimum size chain cover problem on H. Every clique of G corresponds to an anti-chain of H. Thus the equality $\omega(G) = \chi(G)$ is obtained from Dilworth's decomposition theorem [1]. It is well-known that the minimum size chain cover problem on an acyclic graph is solvable in polynomial time by using an algorithm for minimum-cost circulation flow problem (see [11] for example). Though the graph $T^1(m,3)$ is not a co-comparability graph, we proposed a simple strongly polynomial time algorithm for multicoloring $(T^1(m,n), w)$ (see Appendix: Algorithm for Multicoloring $(T^1(m,3), w)$).

3 Multicoloring

In this section, we propose approximation algorithms for multicoloring the graph $(S^k(m,n), w)$ and $(T^k(m,n), w)$. The basic idea of our algorithm is similar to the shifting strategy [5].

McDiarmid and Reed [6] proposed an approximation algorithm for $(T^1(m,n), w)$, which finds a multicoloring with at most $(4/3)\omega(T^1(m,n), w) + 1/3$ colors. In our previous paper [8], we proposed an approximation algorithm for $(S^2(m,n), w)$, which finds a multicoloring with at most $(4/3)\omega(S^2(m,n), w) + 4$ colors.

We describe our algorithm in a proof of the following theorem.

Theorem 3. *When $k > 1$, there exists a polynomial time algorithm for multicoloring $(S^k(m,n), w)$ such that the number of required colors is bounded by*

$$\left(1 + \frac{k}{\left\lfloor \frac{k+3}{2} \right\rfloor} \right) \omega(S^k(m,n), w) + \left\lfloor \frac{k+3}{2} \right\rfloor \chi(S^k(m,n)).$$

Proof. We describe an outline of the algorithm. For simplicity, we define $K_1 = \left\lfloor \frac{k+3}{2} \right\rfloor$ and $K_2 = \left\lfloor \frac{k+3}{2} \right\rfloor + k$.

First, we construct K_2 vertex weight functions w_i' for $i \in \{0, 1, \ldots, K_2 - 1\}$ by setting

$$w_i'(x, y) = \begin{cases} 0, & y \in \{i, i+1, \ldots, i+k-1\} \,(\mathrm{mod}\, K_2), \\ \left\lfloor \frac{w(x,y)}{K_1} \right\rfloor, & \text{otherwise.} \end{cases}$$

Next, we exactly solve K_2 multicoloring problems defined by K_2 weighted graphs $(S^k(m,n), w_i')$, $i \in \{0, 1, \ldots, K_2 - 1\}$ and obtain K_2 multicolorings. We can solve each problem exactly in polynomial time, since every connected component of the graph induced by the set of vertices with positive weight is a perfect graph discussed in the previous section. Thus $\chi(S^k(m,n), w_i') = \omega(S^k(m,n), w_i')$ for any $i \in \{0, 1, \ldots, K_2 - 1\}$. Put $w'' = w - \sum_{i=0}^{K_2-1} w_i'$. Then each element of w'' is less than or equal to $K_1 - 1$. Thus we can construct a multicoloring of $(S^k(m,n), w'')$ from the direct sum of $K_1 - 1$ trivial colorings of $S^k(m,n)$. The obtained multicoloring uses at most $(K_1-1)\chi(S^k(m,n))$ colors. Lastly, we output the direct sum of $K_2 + 1$ multicolorings obtained above. The definition of the weight function w_i' implies that $\forall i \in \{0, 1, \ldots, K_2 - 1\}$, $K_1\,\omega(S^k(m,n), w_i') \leq \omega(S^k(m,n), w)$. Thus, the obtained multicoloring uses at most $(K_2/K_1)\omega(S^k(m,n), w) + (K_1 - 1)\chi(S^k(m,n))$ colors.

In a similar way, the following theorem is obtained.

Theorem 4. *When $k > 1$, there exists a polynomial time algorithm for multicoloring $(T^k(m,n), w)$ such that the number of required colors is bounded by*

$$\left(\frac{2k+1}{k+1} \right) \omega(T^k(m,n), w) + k\chi(T^k(m,n)).$$

It is easy to see that when m, n are sufficiently large, both $\chi(S^k(m,n))$ and $\chi(T^k(m,n))$ are bounded by $O(k^2)$.

4   Discussion

Our main theorem implies approximation algorithms for fractional multicoloring problems and maximum weight stable set problems.

4.1   Fractional Multicoloring and Imperfection Ratio

Given an undirected graph $G = (V, E)$ and a non-negative integer vertex weight function $w : V \to Z_+$, a *fractional multicoloring problem* is the linear programming problem:

$$\text{minimize} \sum_{S \in \mathcal{S}} y_S$$

$$\text{subject to} \sum_{S \ni v} y_S \geq w(v), \ \forall v \in V,$$

$$y_S \geq 0, \ \forall S \in \mathcal{S},$$

where $\mathcal{S}$ is the family of all the stable sets in G. The optimal value of the above problem is called the *fractional multicoloring number* of (G, w) and denoted by $\chi_f(G, w)$. From our main theorems, the followings are immediate.

Theorem 5. *When $k > 3$, there exists a polynomial time*
$\left(1 + \frac{k}{\lceil \frac{k+3}{2} \rceil}\right)$ *-approximation algorithm for the fractional multicoloring problem on $(S^k(m, n), w)$.*

We can prove this theorem by modifying our proof of Theorem 3. We only need to put

$$w_i'(x, y) = \begin{cases} 0, & y \in \{i, i+1, \ldots, i+k-1\} \pmod{K_2}, \\ \frac{w(x,y)}{K_1}, & \text{otherwise.} \end{cases}$$

and $w'' = 0$. Then we can obtain a feasible solution of the above linear programming by combining fractional multicolorings of K_2 weighted graphs $(S^k(m, n), w_i'), \ i \in \{0, 1, \ldots, K_2 - 1\}$.

Theorem 6. *When $k > 1$, there exists a polynomial time*
$\left(\frac{2k+1}{k+1}\right)$ *-approximation algorithm for the fractional multicoloring problem on $(T^k(m, n), w)$.*

The *imperfection ratio* $\text{Imp}(G)$ of an undirected graph G is defined by

$$\text{Imp}(G) \overset{\text{def.}}{=} \max_w \left\{ \frac{\chi_f(G, w)}{\omega(G, w)} \right\}$$

where the maximum is over all nonzero integral weight function w [2, 3, 7]. From Theorems 5 and 6, the followings are immediate.

Corollary 3.

$$1 \leq \mathrm{Imp}(S^k(m,n)) \leq 1 + \frac{k}{\lfloor \frac{k+3}{2} \rfloor}$$

Corollary 4.

$$1 \leq \mathrm{Imp}(T^k(m,n)) \leq \frac{2k+1}{k+1}$$

4.2   Maximum Weight Stable Set Problem

Given an undirected graph $G = (V, E)$ and a non-negative integer vertex weight function $w : V \to Z_+$, a *maximum weight stable set problem* on (G, w) finds a stable set $S \subseteq V$ whose weight $\sum_{v \in S} w(v)$ is maximized. The maximum weight stable set problem is clearly NP-hard on kth power of square lattice graphs and on kth power of triangular lattice graphs.

From our main theorems, the followings are immediate.

Theorem 7. *When $k > 3$, there exists a polynomial time* $\left(\frac{\lfloor \frac{k+3}{2} \rfloor}{k + \lfloor \frac{k+3}{2} \rfloor} \right)$ *-approximation algorithm for the maximum weight stable set problem on $(S^k(m,n), w)$.*

Theorem 8. *When $k > 1$, there exists a polynomial time* $\left(\frac{k+1}{2k+1} \right)$ *-approximation algorithm for the maximum weight stable set problem on $(T^k(m,n), w)$.*

References

1. R. P. Dilworth: A decomposition theorem for partially ordered sets, *Annals of Mathematics*, **51** (1950) 161–166
2. S. Gerke and C. McDiarmid: Graph Imperfection. *Journal of Combinatorial Theory*, **B 83** (2001) 58–78.
3. S. Gerke and C. McDiarmid: Graph Imperfection II. *Journal of Combinatorial Theory*, **B 83** (2001) 79–101.
4. M. M. Halldórsson and G. Kortsarz: Tools for Multicoloring with Applications to Planar Graphs and Partial k-Trees. *Journal of Algorithms*, **42** (2002) 334–366.
5. D. S. Hochbaum: Efficient bounds for the stable set, vertex cover and set packing problems, *Discrete Applied Mathematics*, **6** (1983) 243–254.
6. C. McDiarmid and B. Reed: Channel Assignment and Weighted Coloring. *Networks*, **36** (2000) 114–117.
7. C. McDiarmid: Discrete Mathematics and Radio Channel Assignment. appears in *Recent Advances in Algorithms and Combinatorics* (Springer-Verlag, 2003).
8. Y. Miyamoto and T. Matsui: Linear Time Approximation Algorithm for Multicoloring Lattice Graphs with Diagonals. *Journal of the Operations Research Society of Japan*, **47** (2004) 123–128.

9. Y. Miyamoto and T. Matsui: Multicoloring Unit Disk Graphs on Triangular Lattice Points. *Proceedings of the Sixteenth ACM-SIAM Symposium on Discrete Algorithms*, (to appear).
10. L. Narayanan and S. M. Shende: Static Frequency Assignment in Cellular Networks. *Algorithmica*, **29** (2001) 396–409.
11. A. Schrijver: *Combinatorial Optimization* (Springer-Verlag, 2003).

Appendix: Algorithm for Multicoloring $(T^1(m,3),w)$

In the following, we describe a strongly polynomial time algorithm for multicoloring an weighted graph $(T^1(m,3),w)$. We denote the set of colors by $C^* = \{1,2,\ldots,\omega^*\}$ where $\omega^* = \omega(T^1(m,3),w)$. Let $V(G)$ be a set of vertices of a graph G. The following algorithm finds an assignment of colors $c\colon V(T^1(m,3)) \to 2^{C^*}$ such that $\forall v \in V(T^1(m,3))$, $|c(v)| = w(v)$ and for every edge $\{u,v\} \in T_{m,3}(1)$, $c(u) \cap c(v) = \emptyset$.

For any $x \in \{0,1,\ldots,m-1\}$, we denote the points xe_1+2e_2, xe_1+1e_1, xe_1+0e_2 by t_{x+1}, u_{x+1}, v_x, respectively. Thus $\{t_1,t_2,\ldots,t_m\}$, $\{u_1,u_2,\ldots,u_m\}$ and $\{v_0,v_1,\ldots,v_{m-1}\}$ form a partition of $V(T^1(m,3))$. Without loss of generality, we can assume that $w(v_0) = w(t_m) = w(u_m) = 0$. Our algorithm assigns colors to vertices in the following manner. Assume that we have a multicoloring $c\colon P' \to 2^C$ where $P' = \{t_1,t_2,\ldots,t_j\} \cup \{u_1,u_2,\ldots,u_j\} \cup \{v_0,v_1,\ldots,v_j\}$ satisfying that $\forall i \in \{1,2,\ldots,j\}$, $c(t_i) \subseteq c(v_i)$ or $c(t_i) \supseteq c(v_i)$. Next, we assign colors to u_{j+1}. Since $w(v_0) = w(t_m) = w(u_m) = 0$, we can assume that $w(t_j) \geq w(v_j)$ without loss of generality. Since $\{u_j,t_j,u_{j+1}\}$ is a 3-clique, $|c(u_j)|+|c(t_j)|+w(u_{j+1}) \leq w^*$. Thus there exists a subset of colors C_1 with $|C_1| = w(u_{j+1})$ and C_1 is disjoint with $c(u_j) \cup c(t_j)$. Then we set $c(u_{j+1})$ to C_1. Next, we assign colors to t_{j+1}. There exists a set of colors $C_2 \subseteq C^* \setminus (c(t_j) \cup c(u_{j+1}))$ with $|C_2| = w(t_{j+1})$, since $\{t_j,u_{j+1},t_{j+1}\}$ is a 3-clique. Then set $c(t_{j+1})$ to C_2. Lastly we assign colors to v_{j+1}.

(Case 1) If $w(t_{j+1}) \geq w(v_{j+1})$, then put $c(v_{j+1})$ be a subset of $c(t_{j+1})$ whose cardinality is $w(v_{j+1})$.

(Case 2) Consider the case that $w(t_{j+1}) < w(v_{j+1})$ and $w(t_j) + w(t_{j+1}) \geq w(v_j) + w(v_{j+1})$. Then there exists a subset of colors $C_3 \subseteq c(t_j) \setminus c(v_j)$ whose cardinality is $w(v_{j+1}) - w(t_{j+1})$. We set $c(v_{j+1}) = c(t_{j+1}) \cup C_3$.

(Case 3) Consider the case that $w(t_{j+1}) < w(v_{j+1})$ and $w(t_j) + w(t_{j+1}) < w(v_j)+w(v_{j+1})$. Then we set $c(v_{j+1}) = c(t_{j+1}) \cup (c(t_j) \setminus c(v_j)) \cup C_4$ where C_4 and $v(t_j) \cup c(u_{j+1}) \cup c(t_{j+1})$ are disjoint and $w(v_{j+1}) = w(t_{j+1})+(w(t_j)-w(v_j))+|C_4|$. Since $\{v_j,u_{j+1},v_{j+1}\}$ is a 3-clique, it is easy to see that there exists such a subset of colors.

A naive implementation of the above procedure gives a pseudo polynomial time algorithm, since the algorithm maintains the set of colors C^* explicitly. If we represent the assigned set of colors by the union of some intervals and implement the above procedure carefully, we can obtain a polynomial time algorithm with respect to m.

An Approximation Algorithm for Weak Vertex Cover Problem in Network Management

Zhiping Cai, Jianping Yin, Xianghui Liu, and Shaohe Lv

School of Computer, National University of Defense Technology,
Changsha, 410073, China
xiaocai@163.net, jpyin@nudt.edu.cn,
liuxh@tom.com, chi.shaohe@gmail.com

Abstract. Link-bandwidth utilization and flow information are obviously critical for numerous network management tasks. The problem of efficiently monitoring the network flowing based on flow-conservation could be reduced to Weak Vertex Cover problem, which is NP-hard. In this paper, using the primal-dual method, we give an approximation algorithm with approximation ratio 2 to solve the problem. It is a near-optimal algorithm as it is very difficult to get an approximation algorithm with approximation ratio lower than 2 for Weak Vertex Cover problem. The effectiveness of our monitoring algorithm is validated by simulations evaluation over a wide range of network topologies. The practices indicate that our work is valuable to solve Weak Vertex Cover problem and its application in network management.

1   Introduction

Knowledge of the up-to-date bandwidth utilizations is critical for numerous important network management tasks, including identifying and relieving congestion points, proactive and reactive resource management and traffic engineering, as well as providing and verifying QoS guarantees for end-user applications. Some novel tools and infrastructures for measuring network bandwidth have been developed and proposed by researchers and industries, like as SNMP and RMON measurement probes [1], Cisco's NetFlow tools [2], the IDMaps [3], [4] and packet-pair algorithms for measuring link bandwidth [5], [6].

These measurement tools periodically query and collect detailed traffic data on packet flows for monitoring and measuring network flows and bandwidth usage. Unfortunately, processing queries can adversely impact routers performance and monitoring data transfers can result in significant volumes of additional network traffic [7]. In particular, as the network monitoring process requires more data to be collected and at much higher frequencies, the overhead that a polled monitoring agent imposes on the underlying router can be significant and can adversely impact the router's throughput.

The number of placed monitors of a monitoring system should be kept as small as possible in order to reduce the deployment cost and the actual monitoring operating cost [8]. Several measurements over backbone routers show each

N. Megiddo, Y. Xu, and B. Zhu (Eds.): AAIM 2005, LNCS 3521, pp. 243–251, 2005.

IP router satisfies a flow-conservation law that, the sum of the traffic flowing into router is approximately the same as those of the traffic flowing out [7]. The flow-conservation law could be applied to reduce the number of activated monitor agents used to monitor link bandwidth usage. Thereby the application results in a substantial reduction in the monitoring method impacting on the underlying router's throughput and performance.

The problem of efficiently monitoring the network flowing based on flow-conservation could be reduced to the Weak Vertex Cover (WVC) problem, which is NP-hard. In this paper, using the primal-dual method, we give an approximation algorithm with approximation ratio 2 to solve the problem. It is a near-optimal algorithm as getting an approximation algorithm with approximation ratio being lower than 2 for the weak vertex cover problem is very difficult [12].

The paper is structured as follows. The weak vertex cover problem is brought forward, and some approximation results for the weak vertex cover problem are listed in the section 2. In next section, we give an approximation algorithm to solve the weak vertex cover problem with approximation ratio 2. The effectiveness of our monitoring algorithm is validated by simulations evaluation over a wide range of network topologies in section 4. And we depict our further research in the last section.

2   Weak Vertex Cover Problem

2.1   Problem Formulation and Related Work

The problem of efficiently monitoring the network flow based on the flow conservation law is to find the minimum Flow Monitoring Setof a graph [7], [9].

Definition 1 (Flow Monitoring Set). *Given an undirected graph $G = (V, E)$, where V denotes the set of nodes, E represents the edges between two nodes, we say $S \subseteq V$ is a flow monitoring set of G, if monitoring the flow of those edges that are incident on nodes in S is sufficient to infer the flow of every edge in E. And the following two constraints must be satisfied:*

(1)$\forall v \in V, d(v) \geq 2$, where $d(v)$denotes the degree of node v;
(2)$\forall v \in V, \sum_{u \in V} f(u, v) = 0$, where$f(u, v)$denotes the flow from node u to node v.

The minimum flow monitoring set is a flow monitoring set that contains minimum number of nodes of the graph. The problem of finding the minimum flow monitoring set from an underlying network could be abstracted to the problem of finding the minimum Weak Vertex Cover set (WVC) of a graph [7], [9]. Hence we could solve the problem of finding the minimum flow monitoring set by solving the problem of finding the minimum weak vertex cover set for a given graph.

Definition 2 (Weak Vertex Cover). *Given an undirected graph $G = (V, E)$, where $\forall v \in V$, $d(v) \geq 2$ holds, we say $S \subseteq V$ is a Weak Vertex Cover Set of G, if and only if every edge in G can be marked by performing the following three steps:*

(1) *Mark all edges that are incident on vertices in S;*
(2) *Mark the edge if it is the only unmarked edge among all of the edges that are incident on the same vertex;*
(3) *Repeat step (1) until no new edge can be marked.*

For solving the problem of finding the minimum weak vertex cover set, Xianghui Liu et al. [9] brought forward a greedy approximation algorithm which gives an approximation ratio $2(1+\ln d)$, where $d = \max_{v \in V}\{d(v)\}$. And Xianghui Liu et al. [10] proved that the weak vertex cover problem is NP-complete. Yong Zhang et al. [11] gave an approximation algorithm with approximation ratio $1 + \ln d$. Zhiping Cai et al. [12] gave an approximation preserving reduction from the vertex cover problem to the weak vertex cover problem. Due to this reduction, it implied that it is difficult to get an approximation algorithm with approximation ratio smaller than 2.

2.2   Property of the Weak Vertex Cover

The Weak Vertex Cover problem is a generalization of the well-known Vertex Cover problem, which is also NP-hard [13]. It is not hard to notice that every Vertex Cover of a graph G is also a Weak Vertex Cover of G, but not necessarily a minimal one.

We have some properties of the Weak Vertex Cover as follows.

Proposition 1. [9] *Given an undirected graph $G = (V, E)$, where $\forall v \in V$, $d(v) \geq 2$, the set $S \subseteq V$, is a weak vertex cover, if and only if $G' = (V', E')$ is a forest, where $V' = V - S$ and $E' = \{(u, v)|(u, v) \in E \wedge u \in V' \wedge v \in V'\}$.*

Corollary 1. [12] *Given an undirected graph $G = (V, E)$, where $\forall v \in V$, $d(v) \geq 2$, the set $S, S \subseteq V$, is a weak vertex cover, if and only if $G' = (V', E')$ is acyclic, where $V' = V - S$ and $E' = \{(u, v)|(u, v) \in E \wedge u \in V' \wedge v \in V'\}$.*

3   Approximation Algorithm for Weak Vertex Cover

We give a 2-approximation algorithm for the weak vertex cover by using the primal-dual method for approximation algorithms, which has been used to derive approximation algorithms for network design problems [13-15].

3.1   Integer Programming Formulation

For a given graph $G = (V, E)$, we let $\tau(V)$ denote the cardinality of the smallest WVC for G. Let $d(v)$ denote the degree of vertex v in G. Given a subset S of vertices, let $E[S]$ denote the subset of edges that have both endpoints in S. Let $G[S]$ denote the subgraph $(S, E[S])$ induced by G, and let $d_s(v)$ denote the degree of v in $G[S]$. We let $b(S) = |E[S]| - |S| + 1$ and $b(V) = |E| - |V| + 1$. We say that a WVC F is *minimal* if for any $v \in F$, $F - v$ is not a WVC.

Then we we give some inequalities that will be needed in giving the integer programming formulation.

Theorem 1. *Let F denote any WVC of a graph $G = (V, E)$, where $\forall v \in V, d(v) \geq 2$ holds. Then*

$$\sum_{v \in F} [d(v) - 1] \geq b(V) \tag{1}$$

$$\sum_{v \in F} d(v) \geq b(V) + \tau(V). \tag{2}$$

Proof. We prove the inequality (1) at first. And we consider two cases. If $F = V$, we have that $\sum_{v \in V} [d(v) - 1] = 2|E| - |V|$. Then $|E| \neq 0$ ensures that inequality holds. And due to Corollary 2, if $F \neq V$, it follows that the removal of F from G gives an acyclic subgraph. The number of edges in this subgraph is thus less than its number of vertices, i.e. this subgraph contains at most $|V| - |F| - 1$ edges. Moreover, by removing the vertices in F, we have removed at most $\sum_{v \in F} d(v)$ edges. The total number of edges being $|E|$, therefore we derive that $|V| - |F| + 1 + \sum_{v \in F} d(v) \geq |E|$. Rearranging the terms gives the desired inequality (1).

As $\tau(V)$ denote the cardinality of the smallest WVC of the graph G, we could get (2) by rearranging the terms of (1).   □

Observe that if F is a weak vertex cover for G, then $F \cap S$ is clearly a weak vertex cover for $G[S]$. Hence we have the following corollary of inequality (2).

Corollary 2. *Let F be any weak vertex set. Then for any $S \subseteq V$, $E[S] \neq 0$,*

$$\sum_{v \in F \cap S} d_s(v) \geq b(S) + \tau(S). \tag{3}$$

By Corollary 2, the integer programming formulation of the Weak Vertex Cover problem is the following:

$$\text{Min} \sum_{v \in V} w_v x_v$$

Subject to:

$$(IP) \qquad \sum_{v \in S} d_s(v) x_v \geq b(S) + \tau(S) \qquad S \subseteq V, E[S] \neq 0$$

$$x_v \in \{0, 1\} \qquad\qquad v \in V.$$

3.2   A Primal-Dual Algorithm

We construct a feasible solution to the dual of the linear programming relaxation of (IP). The linear programming relaxation is

$$\text{Min} \sum w_v x_v$$

Subject to:

(LP) $$\sum_{v \in S} d_s(v)x_v \geq b(S) + \tau(S) \qquad S \subseteq V, E[S] \neq 0$$

$$x_v \geq 0 \qquad v \in V.$$

And its dual is

$$\text{Max} \sum_S (b(S) + \tau(S))y_s$$

Subject to:

(D) $$\sum_{S:v \in S} d_s(v)y_s \leq w_v \qquad v \in V$$

$$y_s \geq 0 \qquad S \subseteq V, E[S] \neq 0.$$

Then we give a primal-dual 2-approximation algorithm as follows. The primal-dual structure of this algorithm is the same as that used for solving rather different problems, such as the feedback vertex set problem [13-15].

Algorithm $(G = (V, E))$:

1. $y = 0, F = 0, l = 0$
2. $V' = V; E' = E$
3. While F is not a WVC for G
   (a) $l = l + 1$
   (b) Recursively remove degree one vertices and incident edges from V' and E'
   (c) Increase $y_{V'}$ until $\exists v_l \in V', s.t. \sum_{T:v_l \in T} d_T(v_l)y_T = w_{v_l}$
   (d) $F = F \cup \{v_l\}$
   (e) Remove v_l from V' and attached edges from E'.
4. For $(j = l; j > 0; j - -)$
   (a) if $F - \{v_j\}$ is a WVC then $F = F - \{v_j\}$
5. $F' = F$

It is not hard to see that this algorithm is effectively equivalent to the following: start with $F = 0$ and the graph G. Recursively remove any degree one vertices and associated edges from the graph. Pick the vertex v that achieves the minimum $\varepsilon = \min_{v \in V'} w_v/d(v)$. Add v to F, and set $w_u = w_u - \varepsilon d(u)$ for all $u \in V$. Remove v from the graph, and repeat until F is a WVC. When F is a WVC, the algorithm goes through the vertices of F in the reverse of the order in which they were added, and removes any extraneous vertices. A straightforward implementation of this algorithm takes $O(mn)$ time, where m is the number of edges in the graph and n is the number of vertices.

For proving the algorithm is a 2-approximation algorithm, we give a theorem as follows at first:

Theorem 2. *Let F_m is any minimal WVC, then*

$$\sum_{v \in F_m} d(v) \leq 2(b(V) + \tau(V)) - 2. \qquad (4)$$

Proof. Let k be the number of connected components of $G[V - F_m]$. Due to *Proposition*1, every connected components of $G[V - F_m]$ must be a tree, so that the edges in $G[V - F_m]$ contribute exactly $2(|V| - |F| - k)$ to $\sum_{v \notin F_m} d(v)$. Let $\delta(S)$ is the set of edges with exactly one endpoint in S. As $\sum_{v \notin F_m} d(v) = |\delta(F_m)| + 2(|V| - |F| - k)$ and $\sum_{v \notin F_m} d(v) + \sum_{v \in F_m} d(v) = 2|E|$, inequality (4) can be rewritten by rearranged terms as

$$|\delta(F_m)| \geq 2|F_m| + 2k - 2\tau(V). \tag{5}$$

To prove the inequality (5), we construct a weighted bipartite graph H: shrink every connected component of $G[V - F_m]$ to a vertex of H, and remove all the edges of $G[F_m]$ in G. The weight of an edge in H is the number of edges from the respective node in F_m to the nodes in the respective connected component of $G[V - F_m]$. For the inequality, we need to show that the total weight of this bipartite graph H is at least $2|F_m| + 2k - 2\tau(V)$.

We first observe that for each vertex $v \in F_m$, there must be some witness cycles C_v of G such that $C_v \cap F_m = \{v\}$; otherwise F_m would not be minimal. Thus in H there must be at least one edge of weight at least 2 incident to every vertex of F_m; we designate one such edge for each vertex in F_m and call it a primary edge.

Let T be a maximum collection of vertex-disjoint witness cycles in G, and let $\tilde{F}$ denote the vertices of F_m whose witness cycles are not in T. By the properties of T, it must be the case that for every connected component of $G[V - F_m]$ adjacent to a vertex $v \in \tilde{F}$ via a primary edge, it must also be adjacent to a vertex of $F_m - \tilde{F}$ via a primary edge, since otherwise we would be able to add the witness cycle of v to T. Thus if we remove the primary edges adjacent to the vertices of $\tilde{F}$ from H, there still must be adjacent to each component of $G[V - F_m]$. Hence the weight of edges in H is at least $2|\tilde{F}| + 2k$.

As $|F_m| = |\tilde{F}| + |T|$ and $|T| \leq \tau(V)|$, the total weight of edges in H is at least $2|F_m| + 2k - 2\tau(V)$. $\square$

Then we prove the algorithm is a 2-approximation algorithm. Notice that for any feasible solution y for the dual program (D), $\sum_S (b(S) + \tau(S))y_S$ is a lower bound on the value of the optimal integer solution.

Theorem 3. *The primal-dual algorithm constructs a WVC F' and a solution y feasible for (D) such that*

$$\sum_{v \in F'} w_v \leq 2 \sum_S (b(S) + \tau(S))y_S - 2 \sum_S y_S. \tag{6}$$

Hence the algorithm is a 2-approximation algorithm.

Proof. We reduce the proof of the theorem to inequality (4). By construction of the algorithm,

$$\sum_{v \in F'} w_v = \sum_{v \in F'} \sum_{S: v \in S} d_s(v)y_s = \sum_S \sum_{v \in S \cap F'} d_s(v)y_s. \tag{7}$$

Thus if we can show that for any $y_s > 0$ then

$$\sum_{v \in (S \cap F')} d_s(v) \leq 2(b(S) + \tau(S)) - 2, \qquad (8)$$

then the theorem statement will follow. In order to apply (4), it is sufficient to argue that $S \cap F'$ is a minimal WVC for the graph $G[S]$. By construction of the algorithm F' is a minimal WVC. And at the point in time when the algorithm chooses the vertex v_l, none of the vertices in $F' \cap S$ is currently in F; they are added at some later point in the algorithm. Therefore, because the final step of the algorithm deletes redundant vertices in the reverse of the order in which they were added, $F' - F$, for the current value of F, must be a minimal WVC for the current graph (V', E'). Thus, we have that $F' \cap S$ is a minimal WVC for $G[S]$. $\qquad \square$

4   Simulations

In this section, we present simulation results of comparing the performance of the various algorithms that solve the weak vertex cover problem. The main objective of the simulations is to demonstrate that our proposed algorithmic solutions are not only theoretically sound but also they could give significant benefits over naive solutions in practice for a wide variety of realistic network topologies. The simulations are based on network topologies generated using the Waxman Model [16], which is a popular topology model for networking research. Different network topologies are generated by varying three parameters: (1)n, the number of nodes in the network graph; (2)α, a parameter that controls the density of short edges in the networks; and (3)β, a parameter that controls the average node degree.

We compare the performance of three algorithms: the 2-approximation algorithms for Vertex Cover [13], the greedy algorithm with approximation ratio $2(1 + \ln d)$ for Weak Vertex Cover [9], and our primal-dual algorithm. The comparison is in terms of the number of nodes that need to run SNMP [1] in order to measure the bandwidth of each link in the generated network graphs. We denote the number of SNMP activations for these algorithms by N_2^{VC}, N_{greedy}^{WVC}, and N_2^{WVC} respectively.

Table 1 presents one set of simulation results; we have obtained similar results for other parameter settings. The third and fourth columns in the table represent the maximum and average degree of the nodes in the generated network graph respectively. Our results indicate that using our approximation algorithm can reduce the number of SNMP activations as much as 71% over the naive approach which activate an SNMP agent on every network node [7]. And the result of our algorithm is better than the other two algorithms.

Table 1. Comparisons of Monitoring Algorithms

n	α	β	Maximum Degree	Average Degree	N_2^{VC}	N_{greedy}^{WVC}	N_2^{WVC}	$\frac{N_2^{WVC}}{n}$
400	0.1	0.08	9	2.800	254	174	113	0.283
400	0.4	0.02	12	3.115	297	211	132	0.330
400	0.4	0.08	25	5.685	336	247	157	0.393

5 Conclusion

In this paper, we have addressed the problem of efficiently monitoring bandwidth and flow in network management. This problem could be abstracted to Weak Vertex Cover problem, which is NP-hard. We have proposed a primal-dual algorithm with approximation ratio 2 to solve Weak Vertex Cover problem. As it is very difficult to get an approximation algorithm with approximation ratio lower than 2 for Weak Vertex Cover problem, the proposed algorithm is near optimal. Finally, we have verified the effectiveness of our approximation algorithms through simulations evaluation. This work is helpful to solve Weak Vertex Cover problem and the application in network management.

Further research would be conducted to develop novel algorithms and apply it to network management.

References

1. W. Stallings: SNMP, SNMPv2, SNMPv3, and RMON 1 and 2. Addison-Wesley Longman, Inc., 1999.
2. Cisco Systems: NetFlow Services and Applications, White Paper, 1999
3. P. Francis, S. Jamin, V. Paxson, L. Zhang, D.F.Gryniewicz, and Y. Jin: An Architecture for a Global Internet Host Distance Estimation Service. In Proc. IEEE INFOCOM 1999.
4. S. Jamin, C. Jin, Y. Jin, Y. Raz, Y. Shavitt, and L. Zhang: On the Placement of Internet Instrumentation. In Proc. IEEE INFOCOM 2000.
5. J.C.Bolot: End-to-End Packet Delay and Loss Behavior in the Internet. In. Proc. ACM SIGCOMM 1993.
6. K. Lai and M. Baker: Measuring Bandwidth. In Proc. IEEE INFOCOM 1999.
7. Breitbart Y., Chan CY., Garofalakis M., Rastogi R., Siberschatz A.: Efficiently Monitoring Bandwidth and Latency in IP Networks. In Proc. IEEE INFOCOM 2001.
8. Kyoungwon Suh, Yang Guo, Jim Kurose, and Don Towsley. Locating Network Monitors: Complexity, Heuristics, and Coverage. In Proc. IEEE INFOCOM 2005.
9. Xianghui Liu, Jianping Yin, Lele Tang: Analysis of Efficient Monitoring Method for the Network Flow. *Journal of Software*, 2003,14(2): 300-304(in Chinese with English abstract).
10. Xianghui Liu, Jianping Yin, Xicheng Lu: A Monitoring Model for Link Bandwidth Usage of Network Based on Weak Vertex Cover. *Journal of Software*, 2004,15(4): 545-549(in Chinese with English abstract).

11. Yong Zhang and Hong Zhu: Approximation Algorithm for Weighted Weak Vertex Cover. *Journal of Computer Science and Technology*, 2004,19(6): 782-786.
12. Zhiping Cai, Jianping Yin, Xianghui Liu: Approximation Preserving Reduction between Weak Vertex Cover Problem and Other NP-hard Problems, Manuscript, 2005
13. Dorit S. Hochbaum: Approximation Algorithm for *NP*-Hard Problems. PWS Publishing Company,1997.
14. A. Becker and D. Geiger: Approximation Algorithms for the Loop Cutest Problem. In Proc. 10th Conference on Uncertainty in Artificial Intelligence.
15. F. A. Chudak, M. X. Goemans, D. S. Hochbaumn, and D. P. Williamson: A Primal-Dual Interpretation of Two 2-Approximation Algorithms for the Feedback Vertex Set Problem in Undirected Graphs. *Operations Research Letters*, 1998, 22 :111-118.
16. B.M.Waxman: Routing of Multipoint Connections. IEEE Journal on Selected Areas in Communications, 1988, 6(9):1617-1622.
17. Zhiping Cai, Wentao Zhao, Jianping Yin and Xianghui Liu: Using Passive Measuring to Calibrate Active Measuring Latency. In Proc. ICOIN2005, *Lecture Notes in Computer Science 3391*, C.Kim(eds.), Springer-Verlag, 2005.
18. Jianping Yin, Zhiping Cai, Wentao Zhao and Xianghui Liu: Passive Calibration of Active Measuring Latency. In Proc. ICN2005, *Lecture Notes in Computer Science 3421*, P.Lorenz and P.Dini(eds.), Springer-Verlag, 2005.

Constructing Correlations in Attack Connection Chains Using Active Perturbation*

Qiang Li, Yan Lin, Kun Liu, and Jiubin Ju

JiLin University, ChangChun JiLin 130012, China
sckextjg@mail.jlu.edu.cn

Abstract. Usually network attackers conceal their real attacking paths by establishing interactive connections along a series of intermediate hosts (stepping stones) before they attack the final target. We propose two methods for detecting stepping stones by actively perturbing inter-packet delay of connections. Within the attacker's perturbation range, the average value of the packets in the detecting window is set to increase periodically. The methods can construct correlations in attacking connection chains by analyzing the change of the average value of the inter-packet delay between the two connection chains. The methods can reduce the complexity of correlation computations and improve the efficiency of detecting stepping stones.

Keywords: Traceback; Connection Chain; Active Delay.

1   Introduction

Usually network attackers conceal their real attacking paths by establishing interactive connections along a series of intermediate hosts (stepping stones) before they attack the final target[1]. To identify the real source of attack, tracer can execute a complex tracebacking process from the last host of connection chain using each host's logs. But this approach is not available because attackers usually destroy the tail. Tracer can also install a passive connection traffic monitor in the networks and construct correlations through analyzing input or output traffic of each host. The key problem of connection chain tracebacking is connection correlation in the intermediate hosts (stepping stones)[1, 2]. However, the correlation process is more difficult because traffic's encrypt or compression changes the connection content and delay changes the connection time. And the correlation process on the stepping stones must be quick because the network intrusion often happens in high speed networks. Timing-based is one of the efficient correlation approach of encrypting connection, but existing timing-based correlation approaches are excessively depend on the packets' timing characteristics. In particular, the attacker can perturb the timing characteristics of a

---

* Supported by NSFC(90204014).

N. Megiddo, Y. Xu, and B. Zhu (Eds.): AAIM 2005, LNCS 3521, pp. 252–260, 2005.

connection through the stepping stones. These passively timing-based correlation approaches need collect all the network traffic to construct correlation, so the computation work is too big.

In this paper, we propose two methods for detecting stepping stones by actively perturbing inter-packet delay of connections. Within the attacker's perturbation range, the methods analyzes the activity degree of the correlation windows and monitors periodical characteristic of inter-packets delay. The stepping stone connection in each time window can have a unique periodical characteristic through changing a part of the packets' arrival delay at the network's ingress. The methods can construct correlations in attacking connection chains through detecting these characteristics at the network exgress.

The remainder of the paper is organized as follows. In section 2, we propose an attack connection correlation approach based on increasing inter-packet delay. In section 3, we propose an attack connection correlation approach based on active delay. In section 4, we give a summary of related works. In section 5, we conclude with summary of our findings.

2   Satisfying the Increasing Characteristic

2.1   Method Description

We assume an ingress/egress node of each network exists while all input or output traffic must transmit through this node. In order to construct correlation, we can monitor and perturb the traffic in this node and detect input and output connection, which belongs to the same connection. We assume that the packets in the attacking connection keep their original sequence after through the stepping stones and there are no dropped and reordered packets. We only consider the situation that the attackers do not change the number of packets.

We use t_i and t'_i to represent the arrival and departure times , respectively, of the i th packet. We define the arrival inter-packet delay of the i th packet as $d_i = t_{i+1} - t_i$ and the departure inter-packet delay as $d'_i = t'_{i+1} - t'_i$. We further define the perturbation by the attacker as c_i. Then we have $t'_{i+1} = t'_i + c_i + u$. In this paper, u represents the delay of system(such as processing time, waiting time, etc). Assume the delay range that the attacker can add is $[-D, D]$.

Let $d_{i,k}$ and $d_{j,k}$ be the random variables that denote the random delays added by the attacker to packets $P_{i,k}$ and $P_{j,k}$ respectively for k=1,...,m. Let $x = d_{j,k} - d_{i,k}$ be the random variable that denotes the impact of these random delays on k th inter-packet delay and X be the random variable that denotes the overall impact of random delay on the average of inter-packet delay. Then we have $X = \frac{1}{m} \sum_{k=1}^{m} (d_{j,k} - d_{i,k}) = \frac{1}{m} \sum_{k=1}^{m} X_k$. Similarly we define the probability that the impact of the timing perturbation by the attacker is out of the tolerable perturbation range $(-s/2, s/2]$[3] as $Pr(|X| < s/2)$. They show the probability can be reduced to be arbitrarily close to 0 by increasing m and s.

The method for satisfying the increasing characteristic by adjusting the inter-packet delay is responsible for both incremental rule injection and detection. To achieve this, actively perturbation is exerted on the average inter-packet delay sequence of the being-guarded connection chain at ingress, by which certain of incremental characteristic is injected, while still maintain a certain robustness when the attacker perturbs the timing characteristics of the attacking connection traffic.

Supposed that, in the incoming connection chain, the packet's arrival time sequence is denoted as $\{t_1, t_2, t_3, ...\}$ and the outgoing $\{t'_1, t'_2, t'_3, ...\}$. When monitor the ingress, for each m+1 received packets, average IPD is computed, and an average IPD array is obtained, denoted as $\{\overline{d_1}, \overline{d_2}, ..., \overline{d_{n-1}}, \overline{d_n}\}$. In this array active perturbation is performed, to each $\overline{d_i}$, we make it satisfy the inequation of $\overline{d_{i+1}} - \overline{d_i} \geq s, (i > 1)$, by which an incremental rule is injected actively. Also it is needed to limit the increase, at where factor P is defined, according to which the active perturbation is reset by every P times to sustain the synchronization between the characteristic-injected traffic and the original non-injected traffic.

2.2   Adjusting the Inter-packet Delay

While timing adjustment is performed, all the packets are pushed into a waiting stack, by which a small delay is exerted to the sequence. Where P is defined as a cycle factor, referring to which a reset for adjustment of $m+1$ packets is pursued at the beginning of each cycle. If there is any packet in the waiting stack, send it out as soon as possible, otherwise, keep the transmit characteristic as what it is before. Also where H is defined as a referential delay factor (with an init of the average IPD of the preceding $m + 1$ packets), and g the comparative factor (init as s), f amendment factor(init as 0). In order that every sequence's average IPDs is s bigger than the preceding one, each of the IPD in this sequence must be s bigger than the corresponding one in the preceding sequence. So at here, every departure time is adjusted to make that the delay is bigger by a quantity of g. But in other occasions, the IPD turns to be large enough, and no delay is needed. To decrease the influence on the connection exerted by us, the excess delay is cumulated to the next periods, where f is used to control the amendment factor. The algorithm is described as the following:

1. Set increase count factor p.
2. Let g=s, f=0, i=1. For the first m+1 packets, if there are packets remained in the delay queue, forwarding them as soon as possible; if there is none packet remained in the queue, forwarding the packets according to its original rule. At the same time record the IPDs of the fist $m + 1$ packets, denote as $d_{1,1}, d_{1,2}, d_{1,3}, ..., d_{1,m}$.
3. i++; Adjust the IPD of the m packets in the next cycle.
    3.1 For the first packet, none adjustment is pursued. When it is not in the delay queue, then simply forwarding the packet according to its original characteristic; if it is in the delay queue, then forward the packet directly.
    3.2 initialize factor j with 1, which is utilized to denote the index of the IPDs.
        3.2.1 When a packet is received, compute the IPD between this packet and the preceding one, which is denoted as $d_{i,j}$.

3.2.2 compare $d_{i,j}$ with $d_{i-1,j} + g$. a) if $d_{i,j} \geq d_{i-1,j} + g$, then none perturbation is committed, and let $f = (d_{2,1} - (d_{1,1} + g))/q$, where q denote the count of packets that need to be adjusted but not yet(eg. If $m = 20$,and the preceding 5 packets have been dealt, then $q = 20 - 5 = 15$),$g = g - f$; b) else if $d_{i,j} \neq d_{i-1,j} + g$, then $d_{i,j}$ shall be delayed, and the delay time is $d_{i-1,j} + g - d_{i,j}$
  3.2.3 j++; if $j \neq m + 1$, then go to 3.2.1
3.3 if $i = p$,then go to 2; else go to 3.

2.3   Detecting the Incremental Delay

1. When receiving the packets, compute the preceding packet's IPD, denoting as $d_1, d_2, ...$
2. Compute the IPD in turn.
  2.1 From every m+1 packets, m IPDs can be computed. Let $T_{1,1}$ denotes the average IPD of the packets selection of $\{P_1, P_2, ..., P_m, P_{m+1}\}$, and $T_{1,2}$ of $\{P_2, P_3, ..., P_{m+1}, P_{m+2}\}$, ..., and so on. So $T_{i,j}$ denotes
$$\{P_{m(i-1)+j}, P_{m(i-1)+j+1}, ..., P_{mi+j-1}, P_{mi+j}\}$$
  2.2 From the above definition, we get the arithmetic as
$$T_{i,j} = \sum_{j=(m+1)(i-1)+j}^{(m+1)i-2+j} d_j.$$
3. Detect incremental characteristic in $T_{i,j}$ array.
  3.1 If incremental characteristic is detected, then the tentative synchronization point is the real synchronization point.
  3.2 Perform the correlation detection, if the following IPDs still satisfy the incremental rule, then the connection chain is correlated chain that is being sought for; else go to 3.3
  3.3 Forward the sensitive synchronization point to the next position, go to 3.If the tentative synchronization point has been moved for m times, then it turns to be decided that this connection chain is not a correlation connection chain.

2.4   Evaluation

We derive test data from over 49 million packet headers of the Bell Labs-1 Traces of NLANR[4]. It contains 121 SSH flows that have at least 600 packets and are 120 seconds long at least. We use these 121 SSH flows for 30 times to evaluate active delay approach.

In Figure 1, the results are computed when s is 200ms and m is 20, 15, 10, 5 respectively. It is shown that there are high true positives when m is 20, 15, 10 and it can detect almost all of the correlation connections. But the effect is a little poor when m=5. The experiments made when s=100ms and m is 20, 15, 10 respectively are shown in Figure 2. ¿From this Figure 2, we can find out the effect is best when m=20 where the true positive reaches 100%,while there are two occasions in which the deviation turns to be a bit larger.

¿From the two experiments above, it is shown that the effect is distinctive as the value of m and s is different. It obvious that the effect of big s is better

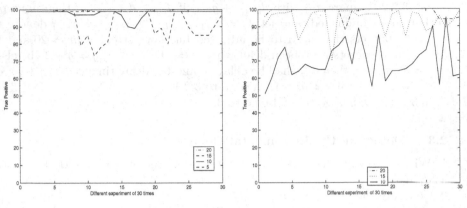

Fig. 1. TP for s=200ms   **Fig. 2.** TP for s=100ms

than that of smaller s and the effect becomes good when m is increased. But if s is increased, we will change the delay too much. While the value of m is large, the real time performance is bad and also the change made on delay increases. We can observe the effect is good when m=200ms, s=10.

In Figure 3, the experiments are made as the value of p is 4 and 8 respectively. Although the effect of p=8 is better than that of p=4, the change made on the delay is too small. So we can achieve the intention of detection well when choosing p=4. In Figure 4, the perturbation is different. It obvious that the true positive

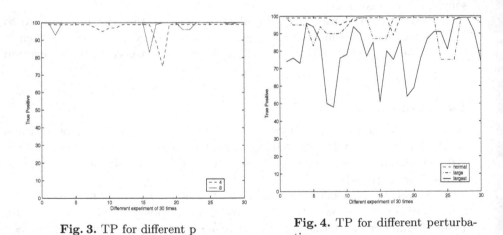

Fig. 3. TP for different p

Fig. 4. TP for different perturbation

decrease while the perturbation is bigger. In fact, the perturbation introduced by the attacker isn't too big; otherwise it will influence the normal transmission. So the method can detect all of the correlation when the perturbation introduced by the attacker isn't too big.

3   Satisfying the Periodical Characteristic

3.1   Method Description

For a connection A, the timing sequence of packet arriving at the network egress is defined as $T = \{t_1, t_2, ..., t_{n-1}, t_n\}$, the delay of inter-packet arriving is defined as $D = \{d_1, d_2, ..., d_{n-1}, d_n\}$, where $d_i = t_{i+1} - t_i$. We define w packets around one connection as a window with length w. The sub sequence read according the sequence of arriving is taken as a window sequence. Notes as $W_k = \{d_{k+1}, d_{k+2}, ..., d_{k+w-1}, d_{k+w}\}$. We define M as the appearance location with maximum inter-packet delay. If $d_i = MAX\{d_{k+1}, d_{k+2}, ..., d_{k+w-1}, d_{k+w}\}$, then $M = i$, where $M \in (1..w]$. If the probability $P(M)$ is average distribution, we say that this timing connection has not periodical characteristic. If we insert a delay into the window $W = \{W_1, W_2, .., W_k, ..\}$ at the location j actively and periodically and let $d_{k+j} = MAX\{d_{k+1}, d_{k+2}, ..., d_{k+w-1}, d_{k+w}\}$ when any window sequence W_k is given, it shows that the probability $P(M)$ is approximately normal distribution, and its expectation $\theta(M) = j$. We say the connection has periodical characteristic, and the periodical signal is M.

3.2   Active Delay Insertion

For a connection, we read its window sequence in turn W. To any window sequence W_k, we define $max\_d = MAX\{d_{k+1}, d_{k+2}, .., d_{k+w-1}, d_{k+w}\}$, if $max\_d + \tau < T_{out}$, where τ is a constant, T_{out} is maximum delay time, then a delay is inserted at place w in order to let $d_{k+w} = max\_d$; if $max\_d + \tau \geq T_{out}$, then $d_{k+w} = max\_d$.

We can make the insertion of delay more reasonable. For a connection, we read its window sequence in turn W. To any window sequence W_k, we insert a delay actively to let $d_{k+w} = \dfrac{\Gamma}{w-1} \sum\limits_{i=1}^{w-1} d_{k+i}$ (where $\Gamma \in [1..3]$ is a constant factor).

3.3   Active Delay Detection

After insertion of active delay at the ingress, the periodical characteristic can be detected at the egress and the correlation can be constructed either. For a connection, we read its window sequence in turn W. To any window sequence W_k and the statistic array of the maximum delay inserted by detector $S = \{s_1, s_2, ..., s_{w-1}, s_w\}$ and the initial value $s_i = 0, i \in (1, .., w)$, we detect the place where the periodical signal M appears.

Iff $d_{k+i} \geq MAX\{d_{k+i-(w-1)}, d_{k+i-(w-2)}, .., d_{k+i-2}, d_{k+i-1}\}$, we let $s_i = \alpha S_i + \lambda$; otherwise $s_i = \alpha S_i$, where α is a constant factor of reduction, λ is a constant factor of increase, and $\alpha \in (0..1)$.

In the given T_{period}, we calculate the statistic array $S = \{s_1, s_2, ..., s_{w-1}, s_w\}$ in turn. If $K = \dfrac{MAX\{s_1, s_2, ..., s_{w-1}, s_w\}}{\sum\limits_{i=1}^{w} s_i} \geq \kappa$, $\kappa \in (0..1)$ is a constant factor of periodical detection, we say that this connection in T_{period} is a periodical connection or there is a periodical signal in T_{period}.

For another inserted delay, iff $\mid d_{k+i} - \frac{\Gamma}{w-1} \sum_{j=1}^{w-1} d_{k+i+j} \mid \leq \sigma'$, we let $s_i = \alpha S_i + \lambda$, σ' is a constant as small as possible; otherwise $s_i = \alpha S_i$.

3.4   Analysis

From those methods mentioned above, it is the key work for us to choose a suitable different activity degree constant factor γ to construct correlation and exclude the different perturbed connection. Choose a right constant factor of reduction α and a constant factor of increase λ and a constant factor of periodical detection κ from experiment to detect periodical signal. Choose a right constant factor Γ and τ and to ensure the performance of process.

Assume the input window sequence W_k already has periodical signals. After attacker changes inter-packet delay randomly, the output window sequence $W'_k = \{d'_{k+1}, d'_{k+2}, \cdots, d'_{k+w-1}, d'_{k+w}\}$. The change process is described as $d'_i = \mu d_i$. Assuming μ is average distribution in area (0.5,1.5), and $\alpha = \frac{n}{n+1}$, $\lambda = \frac{1}{n+1}$, n is total number of the connection packets received presently, initial value is 0. When d_{k+i} fits for the condition of periodical signal, $s'_i = \frac{n}{n+1}s_i + \frac{1}{n+1}$; otherwise $s'_i = \frac{n}{n+1}s_i$ and $\sum_{i=1}^{w} s_i = 1$.

3.5   Evaluation

We observe the effect of different K on true positive rate and false positive rate, and find out the optimal K value. In Figure 5, we calculate the correlation of the 121 SSH flows with different K, they are 0.3, 0.4, 0.5 respectively. We find out the true positive is the biggest and the false positive is the smallest when $\kappa_1 = 0.3$, and true positive is the smallest and the false positive is the biggest when $\kappa_3 = 0.5$. But when the number of received packets in window is more than 50, the true positive and false positive are both approaching to 0.

4   Related works

Existing tracing approaches for a connection chain can be divided into two categories[5] based on tracing object. 1. Host-based: The host-based approaches[1, 6] are restricted the ability of hosts processing because they utilize hosts as information collect point. Too many authentications and communications between the hosts result in more processing time. 2. Network-based: One fundamental problem with passive network-based approaches[2, 5, 7, 8, 9] is its computational complexity. Because it passively monitors and compares network traffic, it needs to record all the concurrent incoming and outgoing connections even when there is no intrusion to trace. The irrelevant traffic wastes much computation time and needs long time to collect. Active approaches[10, 11] differ from passive approaches that they can perturb connection actively and analyze correlations to reduce tracing time and overhead.

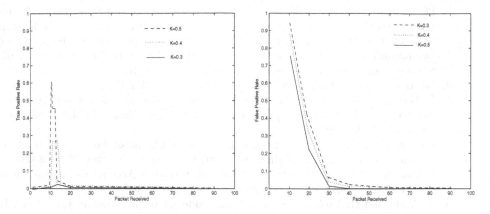

Fig. 5. True positive rate and false positive rate

Our active approach need not modify packet overlay and cooperates between stepping stones, it can be applied on encrypting connection. Compared with FootFall project[11],in our method, the process of active perturbation in the part of input connections is simple; the size of the monitor window can be changed according to the real time request. The little modification of inter-packet delay results the computing complexity of this method is small. The processing of packets is less because it only needs to modify the packets' delay with plus and subtract operation.

5 Conclusions

In this paper, we propose a method for detecting stepping stones by actively perturbing inter-packet delay of connections. The method can construct correlations in attacking connection chains through detecting these increases at the network egress. The method uses actively perturbed correlation algorithm based on passively monitoring the network egress, it can reduce the complexity of correlation computations and improve the efficiency of detecting stepping stones when the attackers use the encrypting connection and timing perturbation.

References

1. S. C. Lee and C. Shields, "Tracing the Source of Network Attack: A Technical, Legal and Societal Problem", Proceedings of the 2001 IEEE Workshop on Information Assurance and Security, June 2001.
2. Y. Zhang and V. Paxson, "Detecting Stepping Stones", Proceedings of 9th USENIX Security Symposium, August 2000.

3. D. Donoho, A.G. Flesia, U. Shanka, V. Paxson, J. Coit and S. Staniford. "Multiscale Stepping Stone Detection: Detecting Pairs of Jittered Interactive Streams by Exploiting Maximum Tolerable Delay". In Proceedings of the 5th International Symposium on Recent Advances in Intrusion Detection RAID 2002, October, 2002. Springer Verlag Lecture Notes in Computer Science, 2516.
4. NLANR Trace Archive. http://pma.nlanr.net/Traces/long/.
5. X. Wang, D. Reeves, S. F. Wu, and J. Yuill, "Sleepy Watermark Tracing: An Active Network-Based Intrusion Response Framework", Proceedings of IFIP Conference. on Security, Mar. 2001.
6. Kwong, H. Yung. " Detecting Long Connection Chains of Interactive Terminal Sessions". In Proceedings of the RAID 2002 Conference. October 16-18, 2002.
7. S. Staniford-Chen, L. T. Heberlein. "Holding Intruders Accountable on the Internet", In Proceedings of IEEE Symposium on Security and Privacy, 1995.
8. K. Yoda and H. Etoh, "Finding a Connection Chain for Tracing Intruders", In F. Cuppens, Y. Deswarte, D. Gollamann, and M. Waidner, editors, 6th European Symposium on Research in Computer Security - ESORICS 2000 LNCS -1985, Toulouse, France, Oct 2000.
9. X.Wang, D.Reeves, and S.F Wu, "Inter-Packet Delay Based Correlation for Tracing Encrypted Connections Through Stepping Stones", Proc. of European Symposium on Research in Computer Security ESORICS 2002.
10. Active Network Intrusion Detection and Response project, http://www.pgp.com/research/nailabs/adaptive-network/active-networks.asp, 2001.
11. X. Wang and D. S. Reeves. "Robust Correlation of Encrypted Attack Traffic Through Stepping Stones by Manipulation of Interpacket Delays". Proc. of ACM Conference on Computer and Communications Security CCS 2003, October 2003.

Sequence Jobs and Assign Due Dates with Uncertain Processing Times and Quadratic Penalty Functions

Yu Xia[1], Bintong Chen[2], and Jinfeng Yue[3]

[1] Department of Management and Marketing,
Fort Hays State University, Hays, KS 67601, USA
[2] Department of Management and Decision Sciences,
Washington State University, Pullman, WA 99164, USA
[3] Department of Management and Marketing,
Middle Tennessee State University, Murfreesboro, TN 37132, USA

Abstract. This paper considers due date assignment and sequencing for multiple jobs in a single machine shop. The processing time of each job is assumed to be uncertain and is characterized by a mean and a variance with no knowledge of the entire distribution. The objective is to minimize the combination of three penalties: penalty on job earliness, penalty on job tardiness, and penalty associated with long due date assignment. The earliness and tardiness penalties and the penalty associated with long due date assignment are all expressed quadratic functions. Heuristic procedures are developed for the objective function. The due dates and sequences obtained by these procedures depend not only on means but also variances of the job processing times. Our numerical examples indicate that the variance information of job processing times can be useful for sequencing and due date assignment decisions. In addition, the performance of the procedures proposed in this paper are robust and stable with respect to job processing time distributions.

Keywords: scheduling, sequencing, due date assignment, heuristics.

1 Introduction

Job sequencing in a single machine shop has been a popular topic in the operations research literature. Early research focuses on job sequencing with predetermined due dates. However, due date assignment is also an important decision in practice. In fact, it is more cost effective to consider due date assignment and job sequencing together. This idea has been actively pursued by Kanet (1981), Hall (1986), Bagchi *et al.* (1986, 1987), Cheng (1987), Ow *et al.* (1989), Szwarc (1989), and Baker *et al.* (1989). See also Raghavachari (1988), Cheng *et al.* (1989), and Baker and Scudder (1990) for excellent surveys on the related literature published before 1990. The majority of the papers mentioned above consider a common due date assignment, where all jobs in a single machine shop

N. Megiddo, Y. Xu, and B. Zhu (Eds.): AAIM 2005, LNCS 3521, pp. 261–269, 2005.
© Springer-Verlag Berlin Heidelberg 2005

share the same due date. The key result for the common due date sequencing is the V-shaped optimal schedule, i.e., jobs finished before the due date are sequenced in descending order of their processing times, and those finished after the due date are sequenced in ascending order of their processing times.

Due to the growing adoption of the just-in-time concept for various operation systems, jobs often have different due dates in practice. Seidmann *et al.* (1981) consider a distinct due date assignment problem where all jobs have deterministic processing times and an extra penalty on setting a late due date. They prove that it is optimal to sequence jobs in nondecreasing order of their processing times and then determine the due date accordingly. Hall (1986) defines the class of scheduling problems, for which due dates are specified in term of the positions in which jobs appear in an ordered sequence rather than by the identities of jobs. Gordon and Strusevich (1998) consider adding a common positive slack to maintain individual due dates. Their objective is to explore the trade-off between the size of the slack and the arising holding costs for the early orders. Qi *et al.* (2002) introduce a flexible method in assign a set of given due dates while sequencing the jobs.

Although the majority of due date assignment papers assume deterministic job processing times, Some recent papers consider stochastic job processing times. Cheng (1986) initiates the study. In Cheng (1991), he considers a distinct due date assignment and job sequencing problem where the processing time of each job is a random variable with known mean and variance. The objective is to minimize the penalty associated with the deviation of the completion time of each job from its due date and the penalty of assigning late due dates. The model implicitly assumes an equal penalty for earliness and tardiness. More recent papers, such as Al-Turki *et al.* (1996), Cai *et al.* (1996), and Qi *et al.* (2000a, 2000b), take into consideration the possibility of machine breakdowns. For the common due date assignment problem, they either prove that the V-shaped sequence is optimal or provide conditions for the conclusion to be true, for various of objective functions. The paper by Al-Turki *et al.* (1996) also considers distinct due date assignments. The results of the above four papers, however, depend on, explicitly or implicitly, the job processing time distributions.

This paper extends the work of Seidmann *et al.* (1981), Cheng (1991), and Al-Turki *et al.* (1996). It considers the job sequencing and due date assignment problem for a single machine shop with the following features:

- The processing time of each job is a random variable with known mean and variance but not the entire distribution.
- Different due dates are assigned to different jobs.
- The objective function consists of quadratic penalties associated with early jobs, late jobs, and quadratic/linear long due date assignments.

The rest of the paper is organized as follows. Section 2 describes the job sequencing and due date assignment model as well as the notation to be used. The objective functions are introduced. Section 3 proposes heuristic approach on the objective function. The numerical tests on the proposed procedures are reported in Section 4 and conclusions are drawn in Section 5.

2   Model Description

Consider a single machine shop with n jobs, indexed by $i = 1, \ldots, n$, waiting to be processed. The processing time of job i, denoted by p_i, is assumed to be a mutually independent random variable that follows certain probability distribution, with finite mean μ_i and variance σ_i^2. The complete information about the probability distribution, however, is unknown. Let S be the set of all possible sequences for the n jobs. Denote $[j]$ as the job in position j of a sequence $s \in S$. Then the completion time for job $[j]$, denoted by $c_{[j]}$, can be expressed as $c_{[j]} = \sum_{i=1}^{j} p_{[i]}$. Since the processing time of each job is independent from those of the other jobs, the mean and variance of job $[j]$'s ($j = 1, 2, \ldots, n$) complete time are

$$C_{[j]} = \text{Exp}(c_{[j]}) = \text{Exp}(\sum_{i=1}^{j} p_{[i]}) = \sum_{i=1}^{j} \mu_{[i]} \quad \text{and} \quad V_{[j]} = \sum_{i=1}^{j} \sigma_{[i]}^2,$$

respectively. Denote $D_{[j]}$ as the unique due date assigned to job $[j]$ and D_s as the vector of the due date assignments for all jobs for a given sequence $s \in S$. The expected earliness of job $[j]$ is then given by

$$E_{[j]} = \text{Exp}(D_{[j]} - c_{[j]})_+,$$

where x_+ represents the positive part of x. Similarly, the expected tardiness of job $[j]$ is

$$T_{[j]} = \text{Exp}(c_{[j]} - D_{[j]})_+.$$

For the objective function, we assume that the job earliness and tardiness are penalized at different levels, although the same level of penalty applies to all jobs to be sequenced. Let α and β be the unit penalties associated with job earliness and tardiness for all jobs, respectively. In addition, we also consider a penalty associated with a long due date assignment, since most customers prefer an earlier due date quote. Let γ be the unit cost for the length of the quoted due date. Our paper considers the following three objective functions, which represent various combinations of linear and quadratic functions for the above mentioned penalties:

For the objective function, we assume that the job earliness and lateness are penalized at different parameters, although the same parameters apply to all jobs to be sequenced. Let α and β be the unit penalties associated with job earliness and tardiness, respectively. In addition, we also consider a penalty associated with a long due date assignment, since most customers prefer to be quoted an earlier due date. Let γ be the unit cost for the length of the quoted due date. Our paper considers the following objective function:

$$F(s, D_s) = \sum_{j=1}^{n} (\alpha E_{[j]}^2 + \beta T_{[j]}^2 + \gamma D_{[j]}^2), \tag{1}$$

The sequencing and due date assignment problems with the above objective function do not have simple analytical optimal solutions in general when the job processing time is uncertain. In fact, we are not aware of any optimal algorithms even when the probability distributions for all job processing times are available. Therefore, we are interested in developing heuristic procedures for this problem and hopefully obtaining good robust solutions. The next section explains the heuristic procedure we propose.

To obtain the upper and lower bounds for the objective functions, we need the following results from distribution-free optimization:

Theorem 1. *Let x be a random variable satisfying a certain probability distribution with mean μ and variance σ^2. Denote*

$$F(q) - t_1\mathrm{Exp}(q - x)_+^2 + t_2\mathrm{Exp}(x - q)_+^2,$$

where $t_1, t_2 > 0$ are fixed constants. The following inequality holds for all q:

$$\min\{t_1,t_2\}\sigma^2+t_1(q-\mu)_+^2+t_2(\mu-q)_+^2 \leq F(q) \leq \max\{t_1,t_2\}\sigma^2+t_1(q-\mu)_+^2+t_2(\mu-q)_+^2.$$

In addition, any of the above upper and lower bounds is tight (becoming equality) for a two-point probability distribution, see Gallego et al (1993), Yue (2000) for detail.

In the about theorem, the upper bound for $k = 1$ was obtained by Scarf (1958) and a simplified proof was found in Gallego et al. (1993). The remaining bounds were discovered by Yue (2000). It has been shown in Yue (2000) theoretically and numerically that the averages of the respective upper and lower bounds serve as good approximations of functions $F_k(q)$, $k = 1, 2$, for various probability distributions and values of q.

3   Heuristic Procedures for Objective Functions $F(s, D_s)$

We approximate $F(s, D_s)$ by the average of its upper and lower bounds. Let $s \in S$ be a given sequence and D_s be a set of due date assignments for all jobs. Applying Theorem 1, we approximate $F(s, D_s)$ by $\tilde{F}(s, D_s)$ as follows:

$$\tilde{F}(s, D_s) = \sum_{j=1}^n \left[\frac{1}{2}(\alpha + \beta)V_{[j]} + \beta(D_{[j]} - C_{[j]})^2 + \gamma D_{[j]}^2\right]. \tag{2}$$

Let D_s be the optimal due date assignments of all jobs for the given sequence s based on the approximation. It can be obtained by setting the partial derivative of (2) equal to zero:

$$\frac{\partial \tilde{F}(s, D_s)}{\partial D_{[j]}} = 2(\beta + \gamma)D_{[j]} - 2\beta C_{[j]} = 0, \quad j = 1,\ldots,n.$$

It follows that

$$D_{[j]} = \frac{\beta}{\beta+\gamma} C_{[j]} = \frac{\beta}{\beta+\gamma} \sum_{i=1}^{j} \mu_{[i]}, \quad j = 1, \dots, n. \tag{3}$$

Substituting D_s back into $\tilde{F}(s, D_s)$, we have

$$\tilde{F}(s) = \sum_{j=1}^{n} \left[\frac{1}{2}(\alpha + \beta)V_{[j]} + \frac{\beta\gamma}{\beta+\gamma} C_{[j]}^2 \right]$$

$$= \sum_{j=1}^{n} \left[\lambda \sum_{i=1}^{j} \sigma_{[i]}^2 + \eta (\sum_{i=1}^{j} \mu_{[i]})^2 \right], \tag{4}$$

where

$$\lambda = \frac{\alpha+\beta}{2} \text{ and } \eta = \frac{\beta\gamma}{\beta+\gamma}.$$

To sequence the jobs, we have the following result.

Theorem 2. *Let m and l be any two jobs among the n jobs to be sequenced. In any optimal sequence that minimizes $\tilde{F}(s)$, job m should be placed in front of job l if the following conditions is satisfied:*

1. *$\mu_m < \mu_l$, and $\eta\mu_m^2 + \lambda\sigma_m^2 < \eta\mu_l^2 + \lambda\sigma_l^2$;*
2. *$\sigma_m^2 < \sigma_l^2$, and $\eta(\mu_m + \sum_{i=1}^{n} \mu_i - \min\{\mu_i\})^2 + \lambda\sigma_m^2 < \eta(\mu_l + \sum_{i=1}^{n} \mu_i - \min\{\mu_i\})^2 + \lambda\sigma_l^2$.*

To derive the heuristic procedure, we use the following lemma.

Lemma 1. *Let $x > y \geq 0$ and $K_1 > K_2 \geq 0$. Then*

$$(K_1 + x)^2 - (K_1 + y)^2 > (K_2 + x)^2 - (K_2 + y)^2$$

We now describe a heuristic procedure that sequences all jobs according to our objective function. The sequence generated is not always optimal, but it does satisfy the properties of the optimal sequence stated in Theorem 2.

Heuristic Sequencing Procedure

Step 0. Let T be the set of jobs remaining to be sequenced and j be the job position currently under consideration. Set $T = \{1, \dots, n\}$ and $j = 1$.

Step 1. Find a job $k \in T$ that minimizes $\eta(\sum_{i \notin T} \mu_i + \mu_k)^2 + \lambda\sigma_k^2$. Place job k in position j in the sequence. Delete job k from set T.

Step 2. If $T = \emptyset$, stop; otherwise, set $j = j + 1$ and repeat Step 1.

Proof. We prove that condition 1 stated in Theorem 2. The proofs for the second one is similar. Since we set due dates $D_{[j]} = \frac{\beta}{\beta+\gamma} C_{[j]} < C_{[j]}$, by Lemma 1, we have

$$\lambda(\sigma_m^2 - \sigma_l^2) + \eta[(K + \mu_m)^r - (K + \mu_l)^r] \leq \lambda(\sigma_m^2 - \sigma_l^2) + \eta(\mu_m^r - \mu_l^r) < 0$$

for all $K \geq 0$. It follows that

$$\eta(\sum_{i \notin T} \mu_i + \mu_m)^r + \lambda\sigma_m^2 < \eta(\sum_{i \notin T} \mu_i + \mu_l)^r + \lambda\sigma_l^2 \quad \text{for any set } T.$$

In view of Step 1 of the heuristic procedure, job l will not be placed ahead of job m.

4   Numerical Tests

Our numerical tests aim at answering the following questions:

- How robust is the heuristic procedure with respect to the processing time probability distributions?
- How much does the information of processing time variances contribute to the sequencing and due date assignment decisions?

4.1   Design of Numerical Test

The data of our test examples are generated as follows:

- The job shop has 50 jobs to be processed.
- The mean and the variance of each job processing time are generated randomly. The mean follows a uniform distribution between 100 and 150. The variance follows another uniform distribution between 400 and 1600.
- The actual processing time of each job is generated based on four probability distributions: Uniform, Normal, Gamma, and Two-Parameter Exponential, respectively, where all distributions share the same mean and variance generated in the previous step.
- The following penalty parameter combinations are considered: both α and β range from 0.1 to 1 with a step size of 0.1 and γ ranges from 0.01 to 0.09 with a step size of 0.01. All together, there are 900 penalty parameter combinations for the objective function and processing time distribution.
- For each objective function, penalty parameter combination, and job processing time distribution, we repeat the simulation 500 times.
- The numerical experiment is implemented by using SPLUS programs.

To show the effectiveness of our heuristic procedures, ideally, we should compare them with the optimal sequencing and due date assignment decisions derived from the probability distributions of job processing times. However, we are not aware of any procedure in the scheduling literature that is capable of obtaining these optimal decisions efficiently. As a compromise, we compare our heuristic procedures with the corresponding sequencing and due date assignment decisions derived from the mean job processing times only, i.e., those procedures suggested by Seidmann et al. (1981). We measure the performance of our heuristics by calculating the percentage of improvement of the corresponding objective functions.

4.2   Numerical Results and Explanations

Since the output of our numerical tests is quite extensive, we summarize most of our findings and display only a sample of representative results for illustration purpose.

Robustness to Processing Time Distributions. For the same job processing time means and variances and for each objective function, we compare the objective functions under Uniform, Gamma, and Exponential job processing time distributions, respectively, to that under Normal job processing time. We calculate the ratios and average them over 900 penalty parameter combinations. The results are summarized in Table 1.

Table 1. Average Ratios of Objective Function Values Under Various Distributions

	Normal	Uniform	Gamma	Exponential
F	1	1.0922	1.0023	1.0738

Our heuristic procedures are fairly robust to job processing time probability distributions. The differences are less than 10% with Normal distribution performing the best and the other three distribution performs close. It is because Normal distribution is centralized and symmetric, first moments (mean and variance) can fully represent the distribution.

Contribution of Variance Information. Our heuristic procedures, which utilize both mean and variance information of job processing times, outperform the sequencing and due date assignment decisions that based on mean processing times only in all cases. The average percentage improvements (over 900 penalty parameter combinations) as well as the ranges of percentage improvements are summarized in Table 2.

Table 2. Percentage Improvements and Their Ranges of Heuristic Procedures

	Normal		Uniform		Gamma		Exponential	
	Average	Range	Average	Range	Average	Range	Average	Range
F	0.15	(0.00,0.90)	0.14	(0.00,0.90)	0.15	(0.00,0.91)	0.15	(0.00,0.92)

Based on Table 2, by utilizing the information of processing time variances, our heuristic procedures outperform their deterministic counterparts in all cases. The improvements can be as significant as more than 90% in some cases and the average improvement is about 15%.

5   Conclusion

In this paper, we study a sequencing and due date assignment problem for a single machine shop with uncertain processing times. The heuristic procedure that utilizes both means and variances of the process times is proposed for the quadratic objective functions.

Our numerical experiments indicate that the heuristic procedures are quite robust to job processing time probability distributions. Furthermore, the additional information on processing time variances does improve the performance of the sequencing and due date assignment decisions.

In conclusion, our procedure successfully transfers the additional variance information into the stable penalty function decrease which is robust to specific distributions of the processing times. Our procedure is hence valuable and practical since the first two moments (mean and variance) are much more easy to estimate in reality than the exactly distribution of processing times, not to mention that a specific distribution is very difficult and sometimes impossible to find in real life cases.

References

1. U.M. Al-Turki, J. Mittenthal, M. Raghavachari, 1996. The single-machine absolute-deviation early-tardy problem with random completion times. Naval Research logistics 43, 573-587.
2. U. Bagchi, Y. Chang, R. Sullivan, 1986. Minimizing mean absolute deviation of completion times about a common due date. Naval Research Logistics 33, 227-240.
3. U. Bagchi, Y. Chang, R. Sullivan, 1987. Minimize absolute and squared deviation of completion times with different earliness and tardiness penalties and a common due date. Naval Research Logistics 34, 739-751.
4. K.R. Baker, G.D. Scudder, 1989. On the assignment of optimal due dates. Journal of Operational Research Society 40, 93-95.
5. K.R. Baker, G.D. Scudder, 1990. Sequencing with earliness and tardiness penalties: a review. Operations Research 38, 22-36.
6. X. Cai, F.S. Tu, 1996. Scheduling jobs with random processing times on a single machine subject to stochastic breakdowns to minimize early-tardy penalties. Naval Research Logistics 43, 1127-1146.
7. T.C.E. Cheng, 1986. Optimal due-date assignment for a single machine sequencing problem with random processing times. International Journal of System Science 17, 1139-1144.
8. T.C.E. Cheng, 1991. Optimal Assignment of slack due-dates and sequencing of jobs with random processing times on a single machine. European Journal of Operational Research 51, 348-353.
9. T.C.E. Cheng, 1987. An algorithm for the con due date determination and sequencing problem. Computers and Operations Research 14, 537-542.
10. T.C.E. Cheng, M.C. Gupta, 1989. Survey of scheduling research involving due date determination decisions. European Journal of Operational Research 38, 156-166.
11. G. Gallego, I. Moon, 1993. The distribution free newsboy problem: review and extensions. Journal of Operational Research Society 44, 825-834.
12. V. Gordon, V.A. Strusevich, 1999. Earliness penalties on a single maching subject to precedence constraints:SLK due daassignment, Computers & Operations Research, V.26, 157-177.

13. N. Hall, 1986. Single and multi-processor models for minimizing completion time variance. Naval Research Logistics 33, 49-54.
14. N. Hall, 1986. Scheduling problems with generalized due dates, IIE TRansactions, V.18,220-222
15. J. Kanet, 1981. Minimizing variation of flow time in single machine systems. Management Science 27, 1453-1459.
16. P. Ow, T. Morton, 1989. The single machine early-tardy problem. Management Science 35, 177-191.
17. X.D. Qi, G. Yin, J.R. Birge, 2000a. Scheduling problems with random processing times under expected earliness/tardiness costs. Stochastic Analysis and Applications 18, 453-473.
18. X.D. Qi, G. Yin, J.R. Birge, 2000b. Single-machine scheduling with random machine breakdowns and randomly compressible processing times. Stochastic Analysis and Applications 18, 635-653.
19. X.D. Qi, G. Yu, J.F. Bard, 2002. Single machine schdeulign with assignable due dates, Discrete Applied Mathematics, V.122,211-233.
20. M. Raghavachari, 1988. Scheduling problems with non-regular penalty functions-a review. Operations Research 25, 144-164.
21. H. Scarf, 1958. A min-max solution of an inventory problem. In: Arrow K, Karlin S, Scarf H (eds). Studies in the Mathematical Theory of Inventory and Production. Stanford University Press, California, 201-209.
22. A. Seidmann, S.S. Panwalker, M.L. Smith, 1981. Optimal assignment of due-dates for a single processor scheduling problem. International Journal of Production Research 19, 393-399.
23. W. Szwarc, 1989. Single machine scheduling to minimize absolute deviation of completion times from a common due date. Naval Research logistics 36, 663-673.
24. J. Yue, 2000. Distribution free optimization procedures with business application. PhD Dissertation, Washington State University.

Computation of Arbitrage in a Financial Market with Various Types of Frictions

Mao-cheng Cai[1], Xiaotie Deng[2], and Zhongfei Li[3,*]

[1] Institute of Systems Science,
Chinese Academy of Sciences, Beijing 100080, China
caimc@iss.ac.cn
[2] Department of Computer Science,
City University of Hong Kong, Hong Kong
csdeng@cityu.edu.hk
[3] Lingnan (University) College, Sun Yat-Sen University,
Guangzhou 510275, China
lnslzf@zsu.edu.cn

Abstract. In this paper we study the computational problem of arbitrage in a frictional market with a finite number of bonds and finite and discrete times to maturity. Types of frictions under consideration include fixed and proportional transaction costs, bid-ask spreads, taxes, and upper bounds on the number of units for transaction. We obtain some negative result on computational difficulty in general for arbitrage under those frictions: It is *NP*-complete to identify whether there exists a cash-and-carry arbitrage transaction and it is *NP*-hard to find an optimal cash-and-carry arbitrage transaction.

1 Introduction

No-arbitrage is a generally accepted condition in finance. In general, if there is any arbitrage opportunity, the market force would act as an invisible hand to drive the prices change and bring the market back to equilibria. An underlying assumption behind the general principle is the existence of active profit seeking agents in the financial market. Their restless effort in locating arbitrage possibilities is essential for the no-arbitrage condition to hold. For the above argument to work, it is essential that locating arbitrage possibilities is not a formidable task, computationally.

For frictionless financial markets, the no-arbitrage condition is very well understood. See, for example, Ross (1978), Harrison and Kreps (1979), Green and Srivastava (1985), and Spremann (1986).

In reality, however, financial markets are never short of friction. Investors are required to pay transaction costs, commissions and taxes. Selling and buying prices are differentiated with ask-bid spread. A security is available at a price only for up to a maximum amount. One may buy or sell a stock at an integer

---

* Correspondence author.

N. Megiddo, Y. Xu, and B. Zhu (Eds.): AAIM 2005, LNCS 3521, pp. 270–280, 2005.

number of shares (or an integer number of hundreds of shares). Friction is a de facto matter in financial markets.

Study of arbitrage in frictional markets has attracted more and more attention in recent years. Garman and Ohlson (1981) extended the work of Ross (1978) to markets with proportional transaction costs. Later, Prisman (1986) studied the valuation of risky assets in arbitrage-free economies with taxation. Dermody and Prisman (1993) extended the results of Garman and Ohlson (1981) to markets with increasing marginal transaction costs. Jouini and Kallal (1995) investigated, by means of martingale method, the no-arbitrage problem under transaction costs. Ardalan (1999) showed that, in financial markets with transaction costs and heterogeneous information, the no-arbitrage imposes a constraint on the bid-ask spread. Deng, Li and Wang (2000, 2002) presented necessary and sufficient conditions for no-arbitrage in a finite-asset and finite-state financial market with proportional transaction costs. These results allows ones to use polynomial time algorithms to look for arbitrage opportunities by applying linear programming techniques. These works weer generalized to the case of multiperiod by Zhang, Xu and Deng (2002).

Kabanov, Rásonyi and Stricker (2001) pointed out that, although the literature on models with friction is rapidly growing, arbitrage theory for markets with frictions still contains a number of questions with much less satisfactory answers than in the theory of frictionless markets and there are only a few papers dealing with necessary and sufficient conditions for the absence of arbitrage for markets with frictions. In addition, to the best of our knowledge, works on algorithmic study of arbitrage under friction are rare, although it is a central problem for discrete finite time models in finance. To capture the current price structure, to find out whether there is an arbitrage opportunity, and to price arbitrary cash stream, the study of algorithmic issues of arbitrage with realistic frictions is important, interesting and challenging.

In the present paper we study computational issues of arbitrage with fixed and proportional transaction costs, bid-ask spreads, taxes, and upper bounds on transaction. The fixed transaction costs capture the situation in which an individual investor requests a broker to invest money on the securities exchange, paying a fixed sum for the service. The payment includes for example brokerage fees, fixed investment taxes to access to a market, operational and trade processing costs, information obtaining costs, or opportunity costs of looking at a market or of doing a specific trade, which are independent of the amount invested in each security. The proportional transaction costs are, as most usual, the fees that are proportional to the transaction size of each security. The bid-ask spreads are the difference between bid and ask prices of an individual security. The (income) taxes at every time to maturities are also set to be proportional to the transaction size of each security.

2   Notation and Definitions

Consider a market of n fixed income securities (or bonds) $i = 1, 2, \ldots, n$. Let $0 = t_0 < t_1 < t_2 < \ldots < t_m$ be all the payment dates (or the times to maturities) that can occur, which need not be equidistant. A cash stream is a vector $w = (w_1, w_2, \ldots, w_m)^T$, where ^{T} denotes the transposition of vector or matrix, and w_j is the income received at time t_j and may be positive, zero or negative. Assume that bond i pays the before-tax cash stream $A_i = (a_{1i}, a_{2i}, \ldots, a_{mi})^T$. So we have the $m \times n$ payoff matrix $A = (A_1, A_2, \ldots, A_n)$.

Bond i can be purchased at a current price p_i^a, the so-called ask price. There is also a bid price p_i^b at which bond i can be sold. The difference between these two prices, the so-called bid-ask spread, reflects a type of friction. This friction exists in most economic markets. We form the ask price vector $p^a = (p_1^a, p_2^a, \ldots, p_n^a)^T$ and the bid price vector $p^b = (p_1^b, p_2^b, \ldots, p_n^b)^T$.

The second type of friction considered in this paper is transaction costs including fixed and proportional. We assume that the fixed transaction cost is c_i if bond i is traded and that no fixed transaction cost occurs if no trading of bond i. The c_i is a positive constant regardless of the amount of bond i traded. Denote $c = (c_1, c_2, \ldots, c_n)^T$ the fixed transaction cost vector. Besides the fixed transaction cost, there is additional transaction cost that is proportional to the amount of the bond traded. Let λ_i^a and λ_i^b be such fees if one dollar of bond i is bought and sold respectively. Here $0 \leq \lambda_i^a, \lambda_i^b < 1, i = 1, 2, \ldots, n$. Denote $\lambda^a = (\lambda_1^a, \lambda_2^a, \ldots, \lambda_n^a)^T$ and $\lambda^b = (\lambda_1^b, \lambda_2^b, \ldots, \lambda_n^b)^T$.

The third type of friction incorporated into our model is taxes. Here we concentrate only on a single investor as a member of just one tax class among many. For all investors in this class, the tax amount at time t_j for holding one unit of bond i in long position is assumed to be t_{ji}^a, and the after-tax income at that time is then $a_{ji} - t_{ji}^a$; whereas the tax amount for holding one unit of bond i in short position is t_{ji}^b as a credit against the obligation to pay a_{ji} at time t_j, and the net after-tax payment to be made is then $a_{ji} - t_{ji}^b$. Let T^a be the $m \times n$ matrix whose entries are t_{ji}^a, and T^b the $m \times n$ matrix whose entries are t_{ji}^b.

Every investor in the fixed tax class under consideration will modify his or her position. Let the modification be $x = (x_1, x_2, \ldots, x_n)^T \in \mathbb{R}^n$, called also a portfolio, where x_i is the number of units of bond i modified by the investor. If $x_i > 0$, additional bond i is bought for the amount of x_i; and if $x_i < 0$, additional bond i is sold for the amount of $-x_i$.

Finally, the fourth type of friction considered in our model is bounds. An upper bound $b_i^+ > 0$ (maximum amount of units that can be bought in bond i) and an upper bound $b_i^- > 0$ (maximum amount of units that can be sold in bond i) are set on the modified amount x_i for each bond i. Denote $b^+ = (b_1^+, b_2^+, \ldots, b_n^+)^T$ and $b^- = (b_1^-, b_2^-, \ldots, b_n^-)^T$. If $-b_i^- \leq x_i \leq b_i^+$ for $i = 1, 2, \ldots, n$, we call x a admissible portfolio.

Now, the bond market considered in this paper can be described by the 10-tuple $\mathcal{M} = \{p^a, p^b, \lambda^a, \lambda^b, b^+, b^-, c, A, T^a, T^b\}$.

For convenience, we use the vector notation $x \geq y$ to indicate that $x_i \geq y_i$ for all i. Denote, for $i = 1, 2, \ldots, n$ and $j = 1, 2, \ldots, m$,

$$\tau_i(x) = \begin{cases} (1 + \lambda_i^a)p_i^a x & \text{if } x > 0, \\ (1 - \lambda_i^b)p_i^b x & \text{if } x \leq 0, \end{cases} \quad g_{ji}(x) = \begin{cases} (a_{ji} - t_{ji}^a)x & \text{if } x > 0, \\ (a_{ji} - t_{ji}^b)x & \text{if } x \leq 0, \end{cases}$$

and $\delta(x) = 1$ if $x \neq 0$ or 0 if $x = 0$. If trading a portfolio $x = (x_1, x_2, \ldots, x_n)^T$, the investor pays the cost $f(x) := \sum_{i=1}^{n} \tau_i(x_i) + \sum_{i=1}^{n} c_i \delta(x_i)$ in the present and receive the after-tax gain $g_j(x) := \sum_{i=1}^{n} g_{ji}(x_i)$ at future time t_j for $j = 1, 2, \ldots, m$. The after-tax cash stream of gains generated by the portfolio x is then the vector $G(x) := (g_1(x), g_2(x), \ldots, g_m(x))^T$.

Definition 1. *An after-tax cash stream $w = (w_1, w_2, \ldots, w_m)^T$ is called no future obligations if $\sum_{j=1}^{k} w_j \geq 0, k = 1, 2, \ldots, m$, or, in matrix notation, if $Bw \geq 0$, where B is the lower-triangular $m \times m$-matrix whose diagonal and lower-triangular elements all are ones.*

Definition 2. *A portfolio x is said to be a cash-and-carry arbitrage transaction if it is admissible (i.e., $-b^- \leq x \leq b^+$) and if it has a negative payment (i.e., $f(x) < 0$) and generates an after-tax cash stream that implies no future obligations (i.e., $BG(x) \geq 0$).*

Definition 3. *The market $\mathcal{M}$ is said to exhibit weak no-arbitrage if there exists no cash-and-carry arbitrage transaction.*

3 Characterizations of No-Arbitrage

Theorem 1. *The market $\mathcal{M}$ exhibits weak no-arbitrage if and only if the optimal value of the following nonlinear programming problem is zero:*

$$(P1) \quad \begin{cases} minimize & f(x) \\ subject\ to & BG(x) \geq 0, \ -b^- \leq x \leq b^+. \end{cases}$$

Proof. Sufficiency. Assume that the optimal value of $(P1)$ is zero. Then, for any admissible portfolio x with $BG(x) \geq 0$, x is feasible to $(P1)$ and hence $f(x) \geq 0$. Thus, there exists no admissible portfolio x such that $f(x) < 0$ and $BG(x) \geq 0$. Therefore, the market $\mathcal{M}$ exhibits weak no-arbitrage.

Necessity. Assume that the market $\mathcal{M}$ exhibits weak no-arbitrage. Then, for any x with $BG(x) \geq 0$ and $-b^- \leq x \leq b^+$, it must holds that $f(x) \geq 0$ otherwise a cash-and-carry arbitrage transaction occurs. This means that the objective function of $(P1)$ is nonnegative at any feasible solution. On the other hand, it is clear that $x = 0$ is feasible to $(P1)$ and the objective function vanishes at $x = 0$. Hence, the optimal value of $(P1)$ is zero. □

Problem $(P1)$ states the lowest total cost or gain induced by trading a portfolio that generates a cash stream with no future obligation. Theorem 1 means that this lowest amount is zero if there exists a consistent term structure.

Now we reformulate the model set up in the previous section. For any portfolio $x = (x_0, x_1, \ldots, x_n)^T$, let $x_i^+ = \max\{x_i, 0\}$ be the number of units of bond i bought and $x_i^- = -\min\{x_i, 0\}$ the number of units of bond i sold. Denote $x^+ =$

$(x_1^+, x_2^+, \ldots, x_n^+)^T$, $x^- = (x_1^-, x_2^-, \ldots, x_n^-)^T$, $p^+ = ((1 + \lambda_1^a)p_1^a, \ldots, (1 + \lambda_n^a)p_n^a)$, and $p^- = ((1 - \lambda_1^b)p_1^b, \ldots, (1 - \lambda_n^b)p_n^b)$. Then,

$$x_i = x_i^+ - x_i^-, \quad x_i^+ x_i^- = 0, \quad 0 \le x_i^\pm \le b_i^\pm, \quad i = 1, 2, \ldots, n,$$

$$f(x) = \sum_{i=1}^n (1 + \lambda_i^a)p_i^a x_i^+ - \sum_{i=1}^n (1 - \lambda_i^b)p_i^b x_i^- + \sum_{i=1}^n c_i \delta(x_i^+ - x_i^-),$$

$$g_j(x) = \sum_{i=1}^n (a_{ji} - t_{ji}^a)x_i^+ - \sum_{i=1}^n (a_{ji} - t_{ji}^b)x_i^-, \ j = 1, 2, \ldots, m.$$

Further we have

$$f(x) = p^+ x^+ - p^- x^- + \sum_{i=1}^n c_i \delta(x_i^+ - x_i^-) \quad \text{and} \quad G(x) = (A - T^a)x^+ - (A - T^b)x^-.$$

Hence, problem $(P1)$ can be equivalently formulated as the problem

$$(P2) \quad \begin{cases} \text{minimize} & p^+ x^+ - p^- x^- + \displaystyle\sum_{i=1}^n c_i \delta(x_i^+ - x_i^-) \\ \text{subject to} & B[(A - T^a)x^+ - (A - T^b)x^-] \ge 0 \\ & x_i^+ x_i^- = 0, \ 0 \le x_i^\pm \le b_i^\pm, \ i = 1, 2, \ldots, n. \end{cases}$$

Theorem 2. *The market $\mathcal{M}$ exhibits weak no-arbitrage if and only if the optimal value of problem $(P2)$ is zero.*

Thus, to identify whether the market exhibits weak no-arbitrage we need only to solve problem $(P2)$.

Clearly, a cash-and-carry arbitrage transaction is a solution (x^+, x^-) of the system

$$(S) \quad \begin{cases} p^+ x^+ - p^- x^- + \displaystyle\sum_{i=1}^n c_i \delta(x_i^+ - x_i^-) < 0 \\ B[(A - T^a)x^+ - (A - T^b)x^-] \ge 0 \\ x_i^+ x_i^- = 0, \ i = 0, 1, \ldots, n \\ 0 \le x^\pm \le b^\pm. \end{cases}$$

The negative of optimal value of $(P2)$ can be interpreted as the maximal arbitrage profit. The optimal solutions of $(P2)$ with nonzero objective value are called optimal cash-and-carry arbitrage transactions.

4   Computational Complexity of Arbitrage

In this section, we will discuss the computational complexity of finding an optimal cash-and-carry arbitrage transaction and of identifying whether there exists a cash-and-carry arbitrage transaction. The technique which we use to reach this purpose is a polynomial time transformation of the EXACT COVER BY 3-SETS into an instance of the problem $(P2)$. The EXACT COVER BY 3-SETS (Garey and Johnson (1979)) is as follows:

Given an arbitrary instance $\mathcal{I}$ of EXACT COVER BY 3-SETS with a ground set $S = \{s_1, \cdots, s_{3h}\}$ and a collection $C = \{C_1, \cdots, C_k\}$ of 3-element subsets of S, does C contain an exact cover for S, that is, a subcollection $C' \subseteq C$ such that every element of S occurs in exactly one member of C'?

First we construct a digraph $G = (V, E)$ from the instance $\mathcal{I}$ as follows:

$$V = \{w\} \cup \{u_1, \ldots, u_{3h}\} \cup \{v_1, \ldots, v_k\},$$
$$E = \{(w, u_i), \ldots, (w, u_{3h})\} \bigcup_{j=1}^{k} \bigcup_{i=1}^{3h} (\{(u_i, v_j) | s_i \in C_j\} \cup \{(v_1, w), \ldots, (v_k, w)\}.$$

In this digraph, element s_i corresponds to vertex u_i, and subset C_j corresponds to vertex v_j. Further, there is an arc (u_i, v_j) if and only if $s_i \in C_j$. Clearly, the indegrees $d^-(u_i) = 1$, $d^-(v_j) = 3$ and $d^-(w) = k$; the outdegrees $d^+(u_i) = |\{s_i \in C_j \in \mathcal{C}\}|$, $d^+(v_j) = 1$ and $d^+(w) = 3h$. The numbers of vertices and arcs of G are $|V| = 3h + k + 1$ and $|E| = 3h + 4k$.

Let D denote the incidence matrix of G, that is, the matrix with rows and columns indexed by V and E, respectively, where the entry in position (v, e) is -1, $+1$, or 0, if v is the head of e, the tail of e, or neither, respectively. Further, we assume that the first $3h$ columns of D is indexed by arcs $(w, u_1), \ldots, (w, u_{3h})$.

To simplify expressions, we write

$$B(A - T^a) = R^+ = (r_{ji}^+), \quad B(A - T^b) = R^- = (r_{ji}^-).$$

Theorem 3. *It is NP-hard to find an optimal cash-and-carry arbitrage transaction even if $R^\pm$ are $(0, \pm1)$-matrices, $c_1 = \cdots = c_n = 1$, and there is no constraint $x_i^+ x_i^- = 0$, $i = 1, 2, \ldots, n$.*

Proof. Let us construct a reduction from instance $\mathcal{I}$ of EXACT COVER BY 3-SETS to the problem $(P2)$. For this purpose, set $m = 18h + 10k + 4$ and $n = 3h + 4k + 1$. We compose $m \times n$-matrices R^+ and R^- as follows:

$$R^+ = \begin{pmatrix} 0_4 \\ I_2 \\ -I_2 \end{pmatrix}, \quad R^- = \begin{pmatrix} I_1 & 0_1 & -1 \\ -I_1 & 0_1 & 1 \\ D & 0_2 \\ -D & 0_2 \\ 0_3 & \\ 0_3 & \end{pmatrix}$$

where D is the incidence matrix of G; I_1 and I_2 are the identity matrices of orders $3h$ and n; 0_1, 0_3 and 0_4 are all-zero $3h \times 4k$-, $n \times n$- and $(12h + 2k + 2) \times n$-matrices, respectively; $\mathbf{1}$ and 0_2 are the all-one and all-zero column vectors of dimensions $3h$ and $3h + k + 1$, respectively.

Further put $c = p^+ = (1, \cdots, 1)$, $p^- = (0, \cdots, 0, 7h+2)$, and

$$b_e^\pm = \begin{cases} 3 & \text{if } e = (v_j, w), \ j = 1, \ldots, k, \\ 1 & \text{otherwise.} \end{cases}$$

It is easy to see that the construction above can be accomplished in polynomial time. Then for the specified $R^\pm$, $p^\pm$ and c, it is straightforward to check that problem $(P2)$ becomes $(\hat{P}2)$:

$$\text{minimize} \quad \sum_{i=1}^{n} \delta(x_i^-) - (7h+2)x_n^-$$

$$\text{subject to} \qquad\qquad\qquad x^+ = 0 \qquad\qquad\qquad\qquad (1)$$

$$x_e^- - x_n^- = 0 \qquad \forall e \in \delta^+(w) \qquad (2)$$

$$\sum_{e \in \delta^+(v)} x_e^- - \sum_{e \in \delta^-(v)} x_e^- = 0 \qquad \forall v \in V \qquad (3)$$

$$b^- \geq x^- \geq 0 \qquad\qquad\qquad\qquad (4)$$

where $\delta^+(v) = \{(v,u) \in E\}$ and $\delta^-(v) = \{(u,v) \in E\}$.

Clearly, (1) yields $(x^+)^T x^- = 0$ and $\{x_e^- : e \in E\}$ is a *circulation* in G by (3) and (4). Further we have

Claim. If $x^- \neq 0$ satisfies (2)–(4), then

$$x_n^- > 0, \qquad\qquad\qquad\qquad (5)$$

$$\sum_{i=1}^{n} \delta(x_i^-) \geq 7h+1. \qquad\qquad\qquad (6)$$

Indeed, assume (5) to be false, then $x_e^- = 0$ for all $e \in \delta^+(w)$ by (2). It follows from (3) that $x_e^- = 0$ for all $e \in \delta^+(u_i)$, $i = 1, \ldots, 3h$, implying $x_e^- = 0$ for all $e \in \delta^+(v_j)$, $j = 1, \ldots, k$. Hence $x^- = 0$, a contradiction.

To show (6), $x_e^- > 0$ for all $e \in \delta^+(w)$ by (2) and (5), that is, $x_{(w, u_i)}^- > 0$ for $i = 1, \ldots, 3h$. It follows from (3) that for each u_i there is at least one arc $e \in \delta^+(u_i)$ with $x_e^- > 0$. As $d^-(v_j) = 3$, $j = 1, \ldots, k$, it derives from (3) that there are at least h vertices v_j with $x_{(v_j, w)}^- > 0$. Therefore (6) holds.

Claim. There is $x^- \neq 0$ satisfying (2)–(4) and

$$\sum_{i=1}^{n} \delta(x_i^-) = 7h+1 \qquad\qquad\qquad (7)$$

if and only if the instance $\mathcal{I}$ of EXACT COVER BY 3-SETS has an exact cover C' of S.

First suppose $x^- \neq 0$ satisfies (2)–(4) and (7). Then it follows easily from the proof of (6) that

- there is exactly one arc $e \in \delta^+(u_i)$ with $x_e^- > 0$ for each $i = 1, \ldots, 3h$ and
- there are exactly h vertices v_{j_ℓ}, $\ell = 1, \ldots, h$, with $x_{(v_{j_\ell}, w)}^- > 0$.

For otherwise (7) cannot hold. Set $C' = \{C_{j_\ell} \in C : x_{(v_{j_\ell}, w)}^- > 0\}$. Then C' is an exact cover of S. Indeed, each s_i is in some $C_{j_\ell} \in C'$ as $x_e^- > 0$ for some $e \in \delta^+(u_i)$ and $C_{j_p} \cap C_{j_q} = \emptyset$ for all $1 \leq p < q \leq h$ since $|S| = 3h$, $|C'| = h$ and $\cup\{C_{j_\ell} \in C'\} = S$.

Conversely, suppose that there exists an exact cover $C' = \{C_{j_1}, \ldots, C_{j_h}\} \subseteq C$ of S. We need to find an n-vector x^- satisfying (2)–(4) and (7). Now set

$$x_n^- = 1,$$

$$x_e^- = \begin{cases} 1 & \text{if } e = (w, u_i), \ i = 1, \ldots, 3h, \\ 1 & \text{if } e = (u_i, v_{j_\ell}) \text{ and } u_i \in C_{j_\ell} \in C', \\ 3 & \text{if } e = (v_{j_\ell}, w) \text{ and } C_{j_\ell} \in C', \\ 0 & \text{otherwise.} \end{cases}$$

It is straightforward to verify that the defined x^- satisfies (2)–(4) and (7).

Claim. The optimal value of Problem $(\hat{P}2)$ is either -1 or 0. Moreover, the optimal value is -1 if and only if the instance $\mathcal{I}$ of EXACT COVER BY 3-SETS has an exact cover of S.

Indeed, as $x = 0$ is a feasible solution to $(\hat{P}2)$, the optimal value

$$\text{minimize} \left\{ \sum_{i=1}^{n} \delta(x_i^-) - (7h + 2)x_n^- \right\} \leq 0. \tag{8}$$

If $(\hat{P}2)$ has a optimal solution $\hat{x} = (0, \hat{x}^-)$ with $\hat{x}^- \neq 0$, then by Claims 1 and 2, $\sum_{i=1}^{n} \delta(\hat{x}_i^-) = 7h + 1$ if and only if the instance $\mathcal{I}$ has an exact cover of S, and $\sum_{i=1}^{n} \delta(\hat{x}_i^-) \geq 7h + 2$ otherwise as $\sum_{i=1}^{n} \delta(\hat{x}_i^-)$ is integer. Furthermore, $\hat{x}_n^- = 1$ follows easily from the proof of Claim 2 and the optimality of $\hat{x}$. Therefore $\sum_{i=1}^{n} \delta(\hat{x}_i^-) - (7h + 2)\hat{x}_n^- = -1$ if and only if the instance $\mathcal{I}$ has an exact cover of S, and $\sum_{i=1}^{n} \delta(\hat{x}_i^-) - (7h+2)\hat{x}_n^- \geq 0$ otherwise, implying $\sum_{i=1}^{n} \delta(\hat{x}_i^-) - (7h+2)\hat{x}_n^- = 0$ by (8). So the claim is true.

Now we come to the conclusion that the optimal value of Problem $(\hat{P}2)$ is either -1 or 0 according to whether the instance $\mathcal{I}$ of EXACT COVER BY 3-SETS has an exact cover of S or not. To complete the proof, we have to show

Claim. For the composed matrices R^+ and R^-, there exist matrices A, T^a and T^b satisfying $B(A - T^a) = R^+$ and $B(A - T^b) = R^-$.

Clearly, the following $m \times n$ linear systems

$$\begin{cases} a_{ji} - t_{ji}^a = m_{ji}^+ \\ a_{ji} - t_{ji}^b = m_{ji}^- \end{cases}, \qquad i = 1, 2, \ldots, n, \ j = 1, 2, \ldots, m$$

have feasible solutions, where a_{ji}, t_{ji}^a and t_{ji}^b are variables, and $(m_{ji}^{\pm}) = B^{-1}R^{\pm}$. The proof is completed. $\square$

Theorem 4. *It is NP-complete to identify whether there exists a cash-and-carry arbitrage transaction in the market $\mathcal{M}$.*

Proof. Equivalently we need only to show that it is NP-complete to determine feasibility of system (S). Clearly, the problem is in NP. We transform EXACT COVER BY 3-SETS to the identification problem by the same reduction used in the proof Theorem 3. To prove the theorem, it suffices to show

Claim. There exists a cash-and-carry arbitrage transaction, that is, there is x^- satisfying (2)–(4) with $\sum_{i=1}^{n} \delta(x_i^-) - (7h+2)x_n^- < 0$, if and only if the instance $\mathcal{T}$ of EXACT COVER BY 3-SETS has an exact cover C' of S.

Clearly, the claim is a corollary of Claim 3. The theorem is proved.   □

Note that $m > n$ in the proofs of Theorems 3 and 4. Let us show the theorems to be still true for the case $m \leq n'$.

Indeed, let R' and R'' be $m \times (n'-n)$-matrices whose entries are non-negative, $\check{p}^+$ and $\check{p}^-$ be the all-zero column vectors of dimension n', $\check{c}$ be the all-one column vector of dimension n', and $\check{b}^\pm = (b^\pm, \overbrace{0,\ldots,0}^{n'-n})$. Set

$$\check{B}(\check{A} - \check{T}^a) = (R^+, -R'), \quad \check{B}(\check{A} - \check{T}^b) = (R^-, R''),$$

where R^+ and R^- are the matrices defined in the proof of Theorem 3. Consider the following programming:

$$(\check{P}2) \quad \begin{cases} \text{minimize} & \check{p}^+\check{x}^+ - \check{p}^-\check{x}^- + \sum_{i=1}^{n'} \check{c}_i\delta(\check{x}_i^+ - \check{x}_i^-) \\ \text{subject to} & \check{B}(\check{A} - \check{T}^a)\check{x}^+ - \check{B}(\check{A} - \check{T}^b)\check{x}^- \geq 0 \\ & \check{x}_i^+\check{x}_i^- = 0,\ \check{b}_i^\pm \geq \check{x}_i^\pm \geq 0,\ i = 1, 2, \ldots, n'. \end{cases}$$

It is easy to see that the optimal values of $(\check{P}2)$ and $(P2)$ are equal. Furthermore, for any optimal solution $(\check{x}^+, \check{x}^-)$ of $(\check{P}2)$, clearly $\check{x}_j^+ = \check{x}_j^- = 0$ for $j = n + 1, \ldots, n'$, and $(x^+, x^-) = (\check{x}_1^+, \check{x}_2^+, \ldots, \check{x}_n^+, \check{x}_1^-, \check{x}_2^-, \ldots, \check{x}_n^-)$ is an optimal solution of $(P2)$. Conversely, for any optimal solution (x^+, x^-) of $(P2)$, then

$$(x^+, \overbrace{0, \ldots, 0}^{n'-n}, x^-, \overbrace{0, \ldots, 0}^{n'-n})$$

is an optimal solution of $(\check{P}2)$. As $(P2)$ is NP-hard for $m > n$, so is $(\check{P}2)$.

Theorems 3 and 4 tell us that it is unlikely to find efficient optimal solution procedures and that one has to look for heuristic algorithms for problem $(P2)$.

5   Conclusion

In this paper, we have derived two necessary and sufficient conditions for the weak no-arbitrage in markets with fixed and proportional transaction costs, bid-ask spreads, taxes, and bounds for transaction. These characterizations extend some known results in discrete time frictionless security markets. With the help of the EXACT COVER BY 3-SETS, the computational complexity of the arbitrage problem is showed to be in NP. These motivate us to consider computational complexity in a more general setting of friction or/and time (period). Such extensions require more sophisticated tools and are worthy of being investigated further in future.

Acknowledgements

This work is partially supported by a grant from RGC of Hong Kong (CityU 1156/04E), a NSFC Major Research Program (60496327), a Foundation for the Author of National Excellent Doctoral Dissertation of China (No. 200267), and grants of the National Natural Science Foundation of China (Nos. 70471018, 10171115, 19971001, 10171054).

References

Ardalan, K.: The no-arbitrage condition and financial markets with transaction costs and heterogeneous information. Global Finance Journal **10** (1999) 83–91

Deng, X. T., Li, Z. F., Wang S. Y.: On computation of arbitrage for markets with friction. In: Du, D. Z., et al. (eds.): Computing and Combinatorics. Lecture Notes in Computer Science, Vol. **1858**. Springer-Verlag, Berlin Heidelberg New York (2000) 309–319

Deng, X. T., Li, Z. F., Wang, S. Y.: Computational complexity of arbitrage in frictional security market. International Journal of Foundations of Computer Science **3** (2002) 681–684

Dermody, J. C., Prisman, E. Z.: No arbitrage and valuation in market with realistic transaction costs. Journal of Financial and Quantitative Analysis **28** (1993) 65–80

Garey, M. R., Johnson, D. S.: Computers and Intractability: A Guide of the Theory of NP-Completeness. San Francisco, Freeman (1979)

Garman, M. B., Ohlson, J. A.: Valuation of risky assets in arbitrage-free economies with transactions costs. Journal of Financial Economics **9** (1981) 271–280

Green, R. C., Srivastava, S.: Risk aversion and arbitrage. The Journal of Finance **40** (1985) 257–268

Harrison, J. M., Kreps, D. M.: Martingales and arbitrage in multiperiod securities markets. J. Econom. Theory **20** (1979) 381–408

Jouini, E., Kallal, H.: Martingales and arbitrage in securities markets with transaction costs. Journal of Economic Theory **66** (1995) 178–197

Kabanov, Yu., Rásonyi, M., Stricker, Ch.: No-arbitrage criteria for financial markets with efficient friction. Université de Besançon, Preprint (2001)

Prisman, E. Z.: Valuation of risky assets in arbitrage free economies with fictions. The Journal of Finance **41** (1986) 545–560.

Ross, S. A.: A Simple approach to the valuation of risky streams. Journal of Business **51** (1978) 453–485

Spremann, K.: The simple analytics of arbitrage. In: Bamberg, G., Spremann, K. (eds): Capital Market Equilibria. Springer-Verlag, Berlin Heidelberg New York (1986) 189–207

Zhang, S. M., Xu, C. L., Deng, X. T.: Dynamic arbitrage-free asset pricing with proportional transaction costs. Mathematical Finance **12** (2002) 89–97

Solving SAT Problems with TA Algorithms Using Constant and Dynamic Markov Chains Length

Héctor Sanvicente–Sánchez[1], Juan Frausto–Solís[2],
and Froilán Imperial–Valenzuela[2]

[1] IMTA, Paseo Cuauhnáhuac 8532, Col. Progreso,
C.P. 62550, Jiutepec Morelos, México
hsanvice@tlaloc.imta.mx
[2] ITESM, Campus Cuernavaca, Department of Computer Science,
Av. Paseo de la Reforma 182-A, Col. Lomas de Cuernavaca C.P. 62589
Temixco Morelos, México
{juan.frausto, A00379100}@itesm.mx

Abstract. Since the apparition of Simulated Annealing algorithm (SA) it has shown to be an efficient method to solve combinatorial optimization problems. Due to this, new algorithms based on two looped cycles (temperatures and Markov chain) have emerged, one of them have been called Threshold Accepting (TA). Classical algorithms based on TA usually use the same Markov chain length for each temperature cycle, these methods spend a lot of time at high temperatures where the Markov chain length is supposed to be small. In this paper we propose a method based on the neighborhood structure to get the Markov chain length in a dynamic way for each temperature cycle. We implemented two TA algorithms (classical or TACM and proposed or TADM) for SAT. Experimentation shows that the proposed method is more efficient than the classical one since it obtain the same quality of the final solution with less processing time.

Keywords: Simulated Annealing, Threshold Accepting, Cooling Scheme, Dynamic Markov Chains, Combinatorial Optimization, Heuristic Optimization, SAT problem.

1 Introduction

Nowadays we found a big interest for developing new and efficient algorithms to solve difficult problems, mainly those considered in the complexity theory ($\mathcal{NP}$–complete or $\mathcal{NP}$–hard) [1].

In practice there are two main ways to solve $\mathcal{NP}$–hard problems: the first one is when we try to find the optimal solution using different techniques that require a lot of computational resources (memory, cpu–time, etc.) which sometimes is undesirably in practice. The last one is using approximation methods (heuristic methods) to get a sub–optimal solution [2].

N. Megiddo, Y. Xu, and B. Zhu (Eds.): AAIM 2005, LNCS 3521, pp. 281–290, 2005.
© Springer-Verlag Berlin Heidelberg 2005

Since the apparition of Simulated Annealing algorithm (SA) by Kirkpatrick et al. in 1983 [3] and Cerny independently in 1985 [4] it has shown to be an efficient method to solve combinatorial optimization problems despite to its two main characteristics: easy to implement and fast convergence rate to good solutions.

Due to this, new algorithms based on two looped cycles (temperatures and Markov chain) have emerge, one of them have been called Threshold Accepting algorithm (TA) [5] which is considered as a modification of SA.

The objective of these variations is to find better methods to reduce the computational resources and to increment the quality of the final solution. This is done applying different accelerating techniques such as: variations of the cooling scheme [6, 7, 8], variations of the neighborhood scheme [9] and with parallelization techniques [10, 11].

In this paper we propose an analytic adaptive method to establish the length of each Markov chain in a dynamic way for TA, named TADM. We applied it to solve some SAT instances and we compared its results versus those obtained with a classical TA algorithm that uses the same length for all Markov chains. Experimentation shows that our method is more efficient than the classical one (named TACM), and we noted a similar quality.

2   The Satisfiability Problem (SAT)

SAT was the first problem referred to be as $\mathcal{NP}$–complete [12]. SAT is fundamental to the analysis of the computational complexity of many problems and it is used in different reasoning methods [13].

An instance of SAT is a boolean formula which consists on the next components:

- A set of n variables $x_1, x_2, \ldots, x_n$.
- A set of literals; a literal is a variable x_i or its negation $\neg x_i$.
- A set of m clauses: $C_1, C_2, \ldots, C_m$ linked by the logical connective AND ($\wedge$) where each clause consists of literals linked by the logical connective OR ($\vee$).

This is:

$$\Phi = C_1 \wedge C_2 \wedge \ldots \wedge C_m = \bigwedge_{k=1}^{m} C_k \tag{1}$$

where Φ is the SAT instance and $C_1, C_2, \ldots, C_m$ are the set of clauses.

Thus, SAT problem can be stated as follows:

Definition 1. *Given a finite set $\{C_1, C_2, \ldots, C_m\}$ of clauses, determine whether there is an assignment of truth-values to the literals appearing in the clauses which makes all the clauses true.*

2.1   Application Areas

Satisfiability is widely studied in different areas such as: operations research, planning, circuit test, temporal reasoning, complexity theory, scheduling, cryptology, constraint satisfaction problems, machine vision, computer network design, computer architecture design, and many others (i.e. see [14] for a detailed list).

Cook [12] and Creignou [15] proved that any other $\mathcal{NP}$–complete problem can be transformed to a SAT instance, which implies that:

1. Any $\mathcal{NP}$–complete problem can be solved as a SAT problem.
2. If SAT can be solved efficiently then any other $\mathcal{NP}$–complete problem can be solved efficiently.

This is why SAT has received a lot of interest in computing and engineering.

2.2   Algorithms of Solution

SAT is considered a hard problem due there is no method to solve it in an 'efficient' way. In this problem, a solution method or a solution algorithm[1] should be able to determine if the instance is satisfiable or not. Here we can distinguish two classes of methods: *complete*[2] and *incomplete*[3].

Complete methods usually are based on splitting and resolution techniques while incomplete methods are based on integer programming and/or local search. Next we show some algorithms of these two classes:

- Complete:
  - Davis and Putnam algorithm [16] (here resolution techniques were introduced).
  - The brute–force method. (i.e. truth tables).
  - DPLL (Davis, Putnam, Loveland, Logemann) [17]. A depth–first search algorithm that enumerate all solutions.
- Incomplete:
  - GSAT, Simulated Annealing [18].
  - WalkSAT [19] and
  - UnitWalk [20].

Complete methods require a lot of computational resources and have exponencial executing time that depends on the instance size [21, 22]. Incomplete methods have reasonable executing time but they may get stuck on a local optima.

3   TA Algorithm

Threshold Accepting algorithm (TA) was first introduced by Dueck and Scheuer in 1990 [5]. TA simplifies the Simulated Annealing algorithm since the accepting

---

[1] An algorithm is a well defined procedure to do someting.
[2] Complete methods usually give a definite yes or no answer to the problem.
[3] Incomplete methods sometimes give a yes answer, but in most cases they do not give a definite answer.

probability calculus is not needed. To accept a solution a deterministic parameter named threshold is introduced. TA works with the ideas of SA since it begins with a current solution S_i from which a new solution S_j is generated through a perturbation mechanism. A worse solution is always accepted if the cost difference of S_i and S_j ($\Delta Z = Z(S_j) - Z(S_i)$) is smaller than the threshold (which is reduced through the process). In Fig. 1 the pseudo–code of TA is shown. Here we can see that TA consists on two cycles: the outer cycle (lines 4–12) that controls the threshold value (named temperature too) and the inner cycle (lines 5–9) which makes a stochastic walk for each temperature cycle.

```
1. Begin
2. Initialization (Si = initial state, c = initial temperature)
3. k = 1
4. Repeat
5.        Repeat
6.               Sj = Generate (Si)
7.               If Z(Si) - Z(Sj) < c
8.                    Si = Sj
9.        Until the equilibrium distribution
10.       k = k+1
11.       c = alpha*c
12. Until the stop criteria (the system is frozen)
13. End
```

Fig. 1. Pseudo–code of TA algorithm

To determine the outer cycle (temperatures cycle) we need to establish the next parameters: initial temperature, final temperature, and the cooling function. To determine the inner cycle (Markov chain builder cycle) we need to set the length of each Markov chain.

4   Cooling Scheme

The balance between efficiency and efficacy in any TA algorithm is established by the cooling scheme. Next we show a full analysis of the parameters of the cooling scheme and the way that we set them for the purposes of this paper.

4.1   Initial and Final Temperature

Initial (c_1) and final (c_f) temperatures are the explicit bounds of any TA algorithm since they determine the beginning and the end of the process. At the beginning c_1 must be determined in a way that almost all transitions may be accepted. If c_1 is too high TA will expend a lot of time, and if it is too low the probability to get stuck on a local optima is high. On the other hand, if c_f is set too high TA probably does not explore the desired area of the solution space. If c_f is set to a very low value a lot of time will be expend at the final of the process.

We set c_1 and c_f with the method suggested in [6]; this is, we require a well defined neighborhood structure, and the values of the maximum and minimum cost increment of the objective function that can be get from the neighborhood structure. In this sense, the neighborhood structure can be defined as follow:

Definition 2. *Let* $\{\forall\ S_i \in S,\ \exists\ a\ set\ V_{S_i} \subset S | V_{S_i} = V{:}S \rightarrow S\}$ *be the neighborhood of a solution* S_i, *where* V_{S_i} *is the neighborhood set of* S_i, $V{:}S \rightarrow S$ *is a mapping and* S *is the solution space of the problem being solved.*

From the above definition we can see that the neighbors of S_i only depend on the neighborhood structure V from every particular problem. Thus, the maximum and minimum cost increments produced from this neighborhood structure are:

$$\Delta Z_{V\max} = Max\Big\{ Z(S_j) - Z(S_i) \Big\} \qquad \forall\ S_j \in V_{S_i}, \forall\ S_i \in S, \qquad (2)$$

$$\Delta Z_{V\min} = Min\Big\{ Z(S_j) - Z(S_i) \Big\} \qquad \forall\ S_j \in V_{S_i}, \forall\ S_i \in S . \qquad (3)$$

Finally c_1 and c_f are calculated as follow:

$$c_1 = \Delta Z_{V\max} \qquad (4)$$

and for c_f:

$$c_f \leq \Delta Z_{V\min} \qquad (5)$$

From (4) we can see that this way of determining the initial temperature enable TA to accept any possible transition at the beginning, since c_1 is set to the maximum deterioration in cost that may be produced through the neighborhood structure.

The final temperature in any TA algorithm should be determined in a way that only good solutions may be accepted at the final of the process [3]. From (5) it can be noted that c_f enable TA to have the control of climbing probability of local optimums and allows TA to do a greedy local search at c_f [6].

4.2   Markov Chains and Cooling Function

As can be see in Fig. 1, by the time $c_k \rightarrow c_f$ (k represents the sequence index) when the next cooling function is applied:

$$c_{k+1} = f(c_k) \qquad (6)$$

TA makes a stochastic walk on the solution space. In any TA algorithm this stochastic walk can be modeled as a sequence of homogeneous Markov chains which are constructed for descending values of the control parameter $c_k > 0$.

Definition 3. *Let* L_k *be the length of each Markov chain that must satisfy* $L_k > 0$ *for any temperature cycle* c_k

where c_k must satisfy:

$$\lim_{k \rightarrow \infty} c_k = 0 \qquad (7)$$

$$c_k \geq c_{k+1} \qquad\qquad \forall k \geq 1 .$$

From (6) and (7) we can establish a strong relation between c_k and L_k in a way that when $c_k \to \infty$, $L_k \to 0$ and when $c_k \to 0$, $L_k \to \infty$.

In a similar way that the section 4.1 we can determine the Markov chain length through neighborhood structure V. In this sense, the maximum number of different solutions that can be rejected from S_i is the neighborhood size $|V_{S_i}|$. Then the length L_k of any Markov chain in a TA algorithm is a function of the neighborhood size:

$$L_k = g(|V_{S_i}|) \tag{8}$$

here the function $g(|V_{S_i}|)$ gives the maximum number of samples that must be taken from the neighborhood V_{S_i} in order to evaluate an expected fraction of different solutions in a Markov chain.

In (8) it is shown that the value of L_k only depends on the number of elements of V_{S_i} that will be explored at c_k.

When a new solution S_j is generated from S_i through V_{S_i}, usually TA employees a replacement random sampling function $G(c_k)$ to explore V_{S_i} at a given temperature c_k, $G(c_k)$ is defined as:

$$G(c_k) = G = \begin{cases} 1/|V_{S_i}|, & \forall S_j \in V_{S_i}; \\ 0, & \forall S_j \notin V_{S_i}. \end{cases} \tag{9}$$

Thus, the probability to choose S_j taking N samples when a replacement random sampling method is being used is:

$$P(S_j) = 1 - \exp\left(-(N/|V_{S_i}|)\right) \tag{10}$$

where N can be obtained as:

$$N = -|V_{S_i}| \ln\left(1 - P(S_j)\right) \tag{11}$$
$$= C|V_{S_i}|$$

here C establishes the exploration level to be done, if $N = |V_{S_i}|$ (i.e. C = 1) then TA will explore 63% of the neighborhood. In a similar way, the levels of exploration 86%, 95% and 99% are obtained when $C = 2$, 3 and 4.6 respectively [6].

In general, for a classical TA algorithm the length of the Markov chain is set constant for all temperatures ($L_k = L$). The most commonly Markov chain length used is one or two times the neighborhood size [5]. Thus, with the analysis made above, this way of determining the Markov chain length produces 63% and 86% exploration level of V_{S_i} for each Markov chain.

5   Dynamic Markov Chains

From the strong relation between c_k and L_k given in section 4.2, at the beginning of the process ($c_k = c_1$) in a TA algorithm all solutions have the same probability to be accepted as the current solution; in this sense, the Markov chain length could be small ($L_k = L_1 \approx 1$) and it guarantees that the system reaches the stationary distribution (equilibrium) at c_1. By the time k increases, the value of c_k is incremented until it reaches c_f. Thus, for consecutive values of c_k (k>1), TA should be forced to increment the Markov chain length in order to reach the stationary distribution for each temperature cycle. This is, L_k must be incremented for each c_k (k>1) until it reaches the maximum Markov chain length L_{max} at c_f.

From (11) we can establish L_{max} to be used in any TA algorithm as the number of samples that must be taken in order to evaluate an expected fraction of different solutions from V_{S_i} at c_f, this is:

$$L_{max} = C|V_{S_i}| \tag{12}$$

where C varies from $1 \leq C \leq 4.6$ for a good neighborhood exploration level [6].

From the strong relation between c_k and L_k, the length of the Markov chain must be incremented at any temperature cycle in a similar but inverse way that c_k is decremented, this may be done as follows:
Let

$$c_{k+1} = \alpha c_k \tag{13}$$

be the geometric reduction cooling function proposed by Kirkpatrick [3] and Cerny [4] also used by Dueck and Scheuer [5]. This cooling function is applied to reduce the temperature from c_1 to c_f, this is done after n steps:

$$c_f = \alpha^n c_1 \tag{14}$$

In a similar way that (13), an incremental Markov chain function can be proposed:

$$L_{k+1} = \beta L_k \tag{15}$$

where L_k is the length of the Markov chain at c_k, L_{k+1} represents the length of the Markov chain at c_{k+1} and β is the increment coefficient ($\beta > 1$).

As we said earlier, the length of each Markov chain must be incremented from L_1 to L_{max} by the time c_1 reaches c_f. In this sense, we get:

$$L_{max} = \beta^n L_1 \tag{16}$$

Now, from (14) the number of steps (n) that TA performs from c_1 to c_f is:

$$n = \frac{\ln c_f - \ln c_1}{\ln \alpha} \tag{17}$$

Finally, with (16) and (17) we can obtain the parameter β as follows:

$$\beta = \exp\left(\frac{\ln L_{max} - \ln L_1}{n}\right) \tag{18}$$

here we can see that if we know the parameter β, the length of each Markov chain may be calculated with (15). This way of determining the length of each Markov chain differs from other methods since no experimentation is required, which is a big advantage.

6   Results and Discussion

To test our method we developed two TA algorithms to solve some SAT instances. Table 1 shows the instances obtained from [23, 24] that were tested during the execution of both TA algorithms: TACM where the length of every Markov chain was maintained constant for each temperature cycle ($L = L_k = N = 2|V_{S_i}|$, see section 4.2), and TADM with the method developed in section 5. In TADM the value of L_{max} was set to $L_{max} = 2|V_{S_i}|$.

Table 1. SAT instances tested

Category	Name	Id
Random	unif-r4.25-v600-c2550-03-S1158627995.cnf	r3
Handmade	bqwh.60.1080.cnf	h1
Handmade	genurq30Sat.cnf	h3
Planning	huge.cnf	pp1
Planning	bw-large.a.cnf	pp3
Circuit fault analysis	ssa7552-038.cnf	c1

Table 2 shows the values of c_1 and c_f and the number of variables and clauses for each instance. The value of c_1 and c_f were obtained with the method described in section 4.1.

Table 2. Characteristics of the instances and values of c_1 and c_f

Id	Variables	Clauses	c_1	c_f
r3	600	2550	26	4
h1	6283	53810	31	8
h3	3622	17076	32	8
pp1	459	7054	65	7
pp3	459	4675	57	7
c1	1501	3575	272	1

For the cooling coefficient of the cooling function, two values were used: $\alpha = 0.85$ and 0.95.

Table 3 shows that both algorithms obtain the same solution quality. The main improvement produced for our method to TA is the executing time, i.e. for h1 the processing time was reduced from 584.2 sec to 63.8 sec (89.079%) when $\alpha = 0.85$ and from 1723 sec to 186.9 sec (89.152%) when $\alpha = 0.95$. In general, for all the instances tested, the mean reduction of the processing time that our method produced was 89.702% for $\alpha = 0.85$ and 89.041% for $\alpha = 0.95$.

Table 3. Performance of TACM and TADM with $\alpha = 0.85$, $\alpha = 0.95$ and $C = 2$

Id	Quality (%)				Time (secs)			
	TACM		TADM		TACM		TADM	
	$\alpha = 0.85$	$\alpha = 0.95$	$\alpha = 0.85$	$\alpha = 0.95$	$\alpha = 0.85$	$\alpha = 0.95$	$\alpha = 0.85$	$\alpha = 0.95$
r3	99.247	99.501	97.643	98.905	2	6.2	0.3	0.9
h1	99.224	99.228	98.112	99.128	584.2	1723	63.8	186.9
h3	99.830	99.944	98.743	99.594	127.4	397.9	9.5	46.4
pp1	99.776	99.791	99.244	99.683	6	21.5	0.8	2.7
pp3	99.726	99.775	98.620	99.574	3.6	13.3	0.3	1.5
c1	98.769	98.791	98.366	98.537	33.3	106.7	3.2	10.2

7 Conclusions

In this paper we developed a new analytic method to determine in a dynamic way the length of each Markov chain for TA. Experimentation with some SAT instances shows that a TA algorithm using the proposed method is more efficient than the classical constant Markov chain TA algorithm since the first one obtain the same solution quality with less processing time. For the SAT instances tested the mean processing time was reduced in 89.702% for $\alpha = 0.85$ and 89.041% for $\alpha = 0.95$.

References

1. Crescenzi, P., Kann, V.: How to find the best approximation results a follow–up to garey and johnson., ACM SIGACT, News. 9097 (1998)
2. Sanvicente-Sánchez, H.: Metodología de paralelización del ciclo de temperaturas en algoritmos tipo recocido simulado. PhD thesis, ITESM Campus Cuernavaca, MÉXICO (2003)
3. Kirkpatrick, S., Gelatt, C.D., Vecchi, M.P.: Optimization by Simulated Annealing. Science, Number 4598, 13 May 1983 **220**, **4598** (1983) 671680
4. Cerny, V.: Thermodynamical approach to the traveling salesman problem: An efficient simulation algorithm. Journal of Optimization Theory and Applications **45** (1985) 4151
5. Dueck, G., Scheuer, T.: Threshold accepting: a general purpose optimization algorithm appearing superior to simulated annealing. Journal of Computational Physics (1990) 161175
6. Sanvicente-Sánchez, H., Frausto-Solís, J.: Method to Establish the Cooling Scheme in Simulated Anneling Like Algorithms. In Laganá, A., Gavrilova, M.L., Kumar, V., Mun, Y., Tan, C.K., Gervasi, O., eds.: Computational Science and its Applications ICCSA 2004. Volume 3045., Springer Verlag (2004)
7. Munakata, T., Nakamura, Y.: Temperature control for simulated annealing. The American Physical Society, PHYSICAL REVIEW E **64** (2001)

8. Atiqullah, M.M.: An efficient simple cooling scheme for simulated annealing. In Laganá, A., Gavrilova, M.L., Kumar, V., Mun, Y., Tan, C.K., Gervasi, O., eds.: Computational Science and its Applications ICCSA 2004. Volume 3045., Springer Verlag (2004)

9. Miki, M., Hiroyasu, T., Ono, K.: Simulated annealing with advanced adaptive neighborhood. In: Second international workshop on Intelligent systems design and application, Dynamic Publishers, Inc. (2002) 113118

10. Miki, M., Hiroyasu, T., Kasai, M., Ono, K., Jitta, T.: Temperature parallel simulated annealing with adaptive neighborhood for continuous optimization problem. In: Second international workshop on Intelligent systems design and application, Dynamic Publishers, Inc. (2002) 149154

11. Fleischer, M., Jacobson, S.H.: Cybernetic optimization by simulated annealing: an implementation of parallel processing using probabilistic feedback control. In J. Kelly, I.O., ed.: Metaheuristics: The State of the Art 1995. Proceedings of Meta-heuristics International Conference 1995, Kluwer Academic (1995)

12. Cook, S.A.: The complexity of theorem proving procedures. In: Proceedings of 3rd Annual ACM symposium on the Theory of Computing, ACM (1971) 151158

13. Papadimitriou, C.H.: Computational Complexity. Addison Wesley Longman (1995)

14. GU, J.: Multispace search for satisfiability and np-hard problems. DIMACS Series in Discrete Mathematics and Theoretical Computer Science. Satisfiability Problem: Theory and Applications: Proceedings of a DIMACS Workshop 35 (1996) 407517

15. Creignou, N.: The class of problems that are linearly equivalent to satisfiability or a uniform method for proving np-completeness. Lecture Notes in Computer Science 702 (1993) 115133

16. Davis, M.G., Putnam, H.: A computing procedure for quantification theory. Journal of the ACM 7 (1960) 201215

17. 17. Davis, M., Logemann, G., Loveland, D.: A machine program for theorem-proving. Communications of the ACM 5 (1962) 394397

18. Spears, W.M.: Simulated annealing for hard satisfiability problems. Technical report, Naval Research Laboratory, Washington D.C. (1993)

19. Selman, B., Kautz, H.A., Cohen, B.: Noise strategies for improving local search. In: AAAI 94: Proceedings of the twelfth national conference on Artificial intelligence (vol. 1), American Association for Artificial Intelligence (1994) 337343

20. Hirsch, E., Kojevnikov, A.: UnitWalk: A new SAT solver that uses local search guided by unit clause elimination (2001) PDMI preprint 9/2001, Steklov Institute of Mathematics at St.Petersburg, 2001.

21. Mitchell, D.G., Selman, B., Levesque, H.: Hard and easy distributions of sat problems. Proceedings of AAAI92 (1992) 459465

22. Cook, S.A., Mitchell, D.G.: Finding hard instances of the satisfiability problem: A survey. DIMACS Series in Discrete Mathematics and Theoretical Computer Sciences (1997)

23. Hoos, H.H., Stützle, T.: SATLIB: An Online Resource for Research on SAT. (2000)

24. http://www.satlib.org: Consulting date: December 1th 2004 (2004)

Efficiently Pricing European-Asian Options — Ultimate Implementation and Analysis of the AMO Algorithm

Akiyoshi Shioura and Takeshi Tokuyama

Graduate School of Information Sciences,
Tohoku University, Sendai 980-8579, Japan
{shioura, tokuyama}@dais.is.tohoku.ac.jp

Abstract. We propose an efficient and accurate randomized approximation algorithm for pricing a European-Asian option on the binomial tree model. For an option with the strike price X on an n-step binomial tree and any positive integer k, our algorithm runs in $O(kn^2)$ time with the error bound $O(X/k)$ which is independent of n. Our algorithm is a modification of the approximation algorithm developed by Aingworth, Motwani, and Oldham (2000) into a randomized algorithm, which improves the accuracy theoretically as well as practically.

1 Introduction

1.1 Background

Options are popular financial instruments in world financial markets. One of the simplest options is *European call* option, which is a contract giving its holder the right, but not the obligation, to buy a stock or other financial asset at some point in the future (called the *expiration date*) for a specified price X (called the *strike price*). The *payoff* of an option is the amount of money its holder makes on the contract. Suppose that we have a European option on a stock, and the stock price S is more than the strike price X on the expiration date. Then, we can make some money by *exercising* the option to buy the stock and selling the stock immediately at the market price. Hence, the payoff of a European option is given by $(S - X)^+ = \max\{S - X, 0\}$. The price of the option is usually much less than the actual price of the underlying stock. Therefore, options hedge risk more cheaply than stocks only, and provide a chance to get large profit with a small amount of money if one's speculation is good.

The price of an option is given by the discounted expected value of the payoff. Because of the popularity of options, techniques for computing the option price have extensively been discussed in the literature [1, 2, 5, 6, 7, 8, 9, 10, 11, 13]. A standard method of pricing an option is to model the movement of the underlying financial asset as geometric Brownian motion with drift and then to construct an arbitrage portfolio [4, 11]. This yields a stochastic differential equation, and its solution gives the option price. However, it is often difficult to solve

N. Megiddo, Y. Xu, and B. Zhu (Eds.): AAIM 2005, LNCS 3521, pp. 291–300, 2005.
© Springer-Verlag Berlin Heidelberg 2005

this differential equation for many complex options such as European-Asian option dealt with in this paper, and indeed no simple closed-form solution is known. Therefore, it is widely practiced to simulate geometric Brownian motion by using a discrete model, and use this model to approximate the option price. One such discrete model is the *binomial tree model* [8, 11], where the time period is decomposed into n time steps, and geometric Brownian motion is modeled by using a biased random walk on a graph called a *binomial tree* of depth n. The option price obtained from the binomial tree model converges to the price given by the differential equation if n goes to infinity. In the binomial tree model, the process of the movement of a stock price is represented by a path. An option is said to be *path-dependent* [6, 11] if the option's payoff depends on the path representing the process as well as the current stock price. Although path-dependency is often useful in designing a secure option against risk caused by sudden change of the market, it makes the analysis of option prices quite difficult.

1.2   Our Problem and Result

In this paper, we consider the pricing of *European-Asian* option. European-Asian option is a kind of path-dependent options and its payoff is given as $(A - X)^+$, where A is the average stock price during the time from the purchase date to the expiration date of the option and X is the strike price. It is known to be #P-hard in general to compute the exact price of path-dependent options on the binomial tree model [6]. Therefore, it is desired to design an efficient approximation algorithm with provable high accuracy, and various pricing techniques have been developed so far [1, 2, 6, 7, 9, 13].

A naive method for computing the exact price of European-Asian options, called the full-path method, enumerates all paths in the binomial tree. Unfortunately, the full-path method requires exponential time since there are exponential number of paths in the binomial tree. Hence, the Monte Carlo method that samples paths in the binomial tree is popularly used to compute an approximate value of the exact price. The error bound of the Monte Carlo method, however, depends on the volatility of the stock price when a polynomial number of samples are taken by naive sampling [10].

Aingworth–Motwani–Oldham (AMO) [1] proposed the first polynomial-time approximation algorithm with guaranteed worst-case error bound, which enables us to avoid the influence of volatility to the theoretical error bound. The idea is to prune exponential number of high-payoff paths by using mathematical formulae during the run of an aggregation algorithm based on dynamic programming and bucketing. In each of n aggregation steps the algorithm produces the error bounded by X/k, where k denotes the number of buckets used at each node of the binomial tree. Hence, the error bound of the AMO algorithm is nX/k, and the algorithm runs in $O(kn^2)$ time. While algorithms on the "uniform" model has been mainly considered in the literature [1, 2, 6, 7, 9], the AMO algorithm and its analysis work on the "non-uniform" model where the transition probabilities of the stock price may differ at each node [13].

Then, variants of the AMO algorithm were proposed to achieve a better error bound than nX/k. Akcoglu–Kao–Raghavan [2] presented a recursive version of the AMO algorithm and reduce the error bound to $O(n^{\frac{1+\varepsilon}{2}} X/k)$ by spending almost the same time complexity when the volatility of the stock is small.

The error bound is further improved by Dai–Huang–Lyuu (DHL) [9] and by Ohta–Sadakane–Shioura–Tokuyama (OSST) [13]. While the AMO algorithm uses the same number of buckets at each node of the binomial tree, the DHL algorithm [9] uses different number of buckets at each node. By adjusting the number of buckets at each node appropriately while keeping the time complexity $O(kn^2)$, they achieved the error bound $O(\sqrt{n}X/k)$, where k is the average number of buckets used at each node. Their analysis, however, applies only to the uniform model and does not extend to the non-uniform model. On the other hand, the OSST algorithm [13] uses the idea of randomized rounding in the aggregation steps of the algorithm, and achieves the error bound $O(\sqrt{n}X/k)$ for the non-uniform model. Moreover, it is shown in [13] that for the uniform model the error bound of the OSST algorithm can be reduced to $O(n^{1/4}X/k)$.

In this paper, we further reduce the error bound by giving a randomized approximation algorithm with an $O(kn^2)$ time complexity and an $O(X/k)$ error bound. The error bound of our algorithm is independent of the depth n of the binomial tree, although those of the AMO algorithm and its previous variants [2, 9, 13] are dependent on n. Our algorithm uses the ideas in Dai et al. [9] and Ohta et al. [13]. As in [13], we regard the aggregation steps of the algorithm as a Martingale process with $O(n^2)$ random steps by using novel random variables. It can be shown that the expected value of the output by our algorithm equals the exact price, and that the error in each single step is bounded by a function of the number of buckets at a node of the binomial tree. Thus, we can apply Azuma's inequality [3] to the Martingale process to obtain the error bound. If we choose k as the number of buckets at each node, the algorithm coincides with the one in [13]. To reduce the error bound as much as possible, we adjust the number of buckets at each node and obtain the error bound $O(X/k)$, where k is the average number of buckets used at each node. Since the value X/k can be seen as the "average" of the absolute error produced at each node of the binomial tree, the error bound of our algorithm is the best possible within the framework of the AMO algorithm. We also show the practical quality of the approximate value computed by our algorithm by some numerical experiments.

Although we only consider the pricing of call options on the binomial tree model in this paper, our algorithm can be easily modified to put options and to the trinomial tree model as in [1, 2, 9, 13].

2 Preliminaries

2.1 The Binomial Tree Model

A *binomial tree* of depth n is a leveled directed acyclic graph defined as follows (see Fig. 1). A binomial tree of depth n has $n+1$ levels. There are $i+1$ nodes in

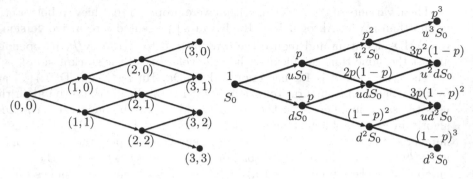

Fig. 1. A binomial tree of depth 3

Fig. 2. The uniform binomial tree model. The probability $\omega(i,j)$ (resp., the stock price $S_i(j)$) at each node is shown above the node (resp., below the node)

the i-th level ($0 \le i \le n$) and each node is labeled as (i,j), where j ($0 \le j \le i$) denotes the numbering of the nodes. The node $(0,0)$ in the 0-th level is called the *root*, and each node (n,j) in the n-th level is called a *leaf*. Each non-leaf node (i,j) has two children $(i+1,j)$ and $(i+1,j+1)$. Therefore, each non-root node (i,j) has two parents $(i-1,j-1)$ and $(i-1,j)$ if $1 \le j \le i-1$, and each of $(i,0)$ and (i,i) has only one parent.

Let us consider a discrete random process simulating the movement of a stock price. We divide the time from the purchase date to the expiration date of an option into n time periods, and the i-th time step means the end of the i-th time period. In particular, 0-th (resp., n-th) time step is the purchase (resp., expiration) date of the option. For $i = 0, 1, \ldots, n$, let S_i be a random variable representing the stock price at the i-th time step, where S_0 is the initial stock price known in advance. The fundamental assumption in the binomial tree model is that in each time step the stock price S either rises to uS or falls to dS, where u and d are predetermined constants satisfying $u > d$ and $u = 1/d$. Thus, we can model the stock price movement by using a binomial tree (see Fig. 2).

Suppose that we are at a non-leaf node (i,j) in the binomial tree model and the current stock price is S. With probability p_{ij}, we move to the node $(i+1,j)$ and the stock price rises to uS; with probability $1 - p_{ij}$, we move to the node $(i+1,j+1)$ and the stock price falls to dS. Thus, the stock price at the node (i,j) is $S_i(j) = u^{i-j}d^j S_0$. The binomial tree model is said to be *uniform* if $p_{ij} = p$ for each node (i,j); otherwise it is *non-uniform*. The uniform model has been widely considered [1, 2, 6, 7, 9] since p is uniquely determined under the non-arbitrage condition of the underlying stock. The non-uniform model, however, is often useful to deal with various stochastic models. For each node (i,j), we denote by $\omega(i,j)$ the probability that the random walk reaches to (i,j). In the uniform model, we have $\omega(i,j) = \binom{i}{j} p^{i-j} (1-p)^j$.

2.2   Options

Let X be the strike price of an option. The *payoff* of an option is the amount of money its holder makes on the contract. We adopt a convention to write F^+ for the value $\max\{F, 0\}$. *European option* is one of basic options, and its payoff is given by $(S_n - X)^+$ which is determined by the stock price S_n on the expiration date. It is quite easy to compute the expected value of the payoff of European options under the binomial tree model. A drawback of European options is that the payoff may be affected drastically by the movement of the stock price just before the expiration date; even if the stock price goes very high during most of time periods, it may happen that the option does not make money at the end. *European-Asian option* is more reliable for the holder than European option, and its payoff is given by $(A_n - X)^+$, where $A_n = (\sum_{i=0}^{n} S_i)/(n+1)$ is the average of the stock prices during n time periods. Let $T_j = \sum_{i=0}^{j} S_i$ be the *running total* of the stock price up to the j-th time step. Once T_j exceeds the threshold $(n+1)X$, the option holder will surely exercise it on the expiration date and obtain the payoff of at least $T_j/(n+1) - X$.

Our aim is to compute the price of European-Asian options. Since the price of an option is given by the discounted expected value of the payoff, it suffices to compute the expected payoff. A simple method is to compute the running total $T_n(\mathcal{P})$ of the stock price for each path $\mathcal{P}$ in the binomial tree together with the probability $\Pr(\mathcal{P})$ that the path occurs, and exactly compute the value

$$E((A_n - X)^+) = \sum\{\Pr(\mathcal{P}) \cdot (\tfrac{T_n(\mathcal{P})}{n+1} - X)^+ \mid \mathcal{P}: \text{ a path from the root to a leaf}\}.$$

We call the expected value of the payoff computed as above the *exact value* of the expected payoff, and denote $U = E((A_n - X)^+)$. This simple method, however, needs exponential time since there are 2^n paths in a binomial tree.

3   A New Algorithm for Pricing European-Asian Options

3.1   A Basic Algorithm

We describe a basic approximation algorithm for the option's expected payoff. This algorithm is a slight generalization of the AMO algorithm, and the previous approximation algorithms in [1, 9, 13] can be seen as specialized versions of this basic algorithm.

As in [1], the basic algorithm uses dynamic programming to compute an approximate value of the option's expected payoff. For a path $\mathcal{P}$ from the root to a node (i, j) in the i-th level, we define the *state* of $\mathcal{P}$ as a pair $(S_i(j), T_i)$ of the stock price $S_i(j) = u^{i-j}d^j S_0$ and the running total T_i. Note that the states of two different paths reaching a node (i, j) can be the same. We define the *weight* of the state $(S_i(j), T_i)$ as the probability that a path $\mathcal{P}$ with the state $(S_i(j), T_i)$ occurs. The basic algorithm is based on a simple observation that if the running total of a current state is above the threshold $(n+1)X$, then the conditional expectation of the payoff at this state can be analytically computed as shown in Lemma 1 below, and such a state can be pruned away.

Lemma 1 ([1, 13]). *Suppose that we are at a node (i, j) in the i-th level and the current state is (S, T), where $T \geq (n+1)X$. Then, the payoff's conditional expectation is given as $\{T + h(i, j)\}/(n+1) - X$, where $h(i, j)$ is defined by the following recursive formula: if $i = n$ then $h(i, j) = 0$, and if $i < n$ then*

$$h(i, j) = p_{ij}\{h(i+1, j) + S_{i+1}(j)\} + (1 - p_{ij})\{h(i+1, j+1) + S_{i+1}(j+1)\}.$$

Hence, we need to consider only the states with running total less than $(n+1)X$, which may be exponentially many. Rather than dealing with each unpruned state individually, we instead aggregate the states by using buckets that divide the interval $[0, (n+1)X)$. At each node (i, j) in the i-th level, the algorithm creates k_{ij} buckets $B_i(j, h)$ $(h = 0, 1, \ldots, k_{ij} - 1)$, each of which corresponds to the interval $[b_h, b_{h+1}) = [\frac{(n+1)X}{k_{ij}}h, \frac{(n+1)X}{k_{ij}}(h+1))$. Each unpruned state of a path terminating at the node (i, j) is stored in one of k_{ij} buckets according to its running total. The algorithm chooses a value $R_i(j, h)$ in the interval $[b_h, b_{h+1})$ appropriately, and approximates all states in the bucket $B_i(j, h)$ by a single state $(S_i(j), R_i(j, h))$, where its weight $w_i(j, h)$ is given by the sum of the weights of all states in $B_i(j, h)$. Then, the algorithm produces two new states $(S_{i+1}(j), R_i(j, h) + S_{i+1}(j))$ and $(S_{i+1}(j+1), R_i(j, h) + S_{i+1}(j+1))$ in the $(i+1)$-st level, and inserts these states in appropriate buckets at the nodes $(i+1, j)$ and $(i+1, j+1)$, respectively, or computes the conditional expectation of the payoff at these states by using Lemma 1.

The error bound and the time and space complexity of the basic algorithm can be analyzed as follows.

Theorem 2. *The basic algorithm computes a value Ψ satisfying $|\Psi - U| \leq X \sum_{i=0}^{n} \sum_{j=0}^{i} \omega(i, j)/k_{ij}$. The time and space complexity of the basic algorithm are $O(\sum_{i=0}^{n} \sum_{j=0}^{i} k_{ij})$.*

Proof. The error obtained by rounding a running total at a node (i, j) is bounded by $(n+1)X/k_{ij}$. Therefore, the contribution in processing one node to the error of the average stock value A_n is at most $X\omega(i, j)/k_{ij}$, and the error in the estimation of A_n is bounded by $X \sum_{i=0}^{n} \sum_{j=0}^{i} \omega(i, j)/k_{ij}$.

Let $b_i(j, h)$ be the number of states inserted in the bucket $B_i(j, h)$. Then, the time and space complexity are written as $O(\sum_{i=0}^{n} \sum_{j=0}^{i} k_{ij} + \sum_{i=1}^{n} \sum_{j=0}^{i} b_i(j, h))$. Since we obtain from each bucket in the i-th level at most two new states in the $(i+1)$-st level, it holds that $\sum_{j=0}^{i+1} b_{i+1}(j, h) \leq 2 \sum_{j=0}^{i} k_{ij}$ for $i = 0, \ldots, n-1$.   □

3.2   Previous Algorithms

We can obtain the algorithms [1, 9, 13] by customizing the number of buckets k_{ij} and the value $R_i(j, h)$.

The AMO algorithm [1] can be obtained by setting $k_{ij} = k$ with a positive integer k for all nodes (i, j) and $R_i(j, h) = \frac{(n+1)X}{k}h$. Note that the AMO algorithm computes a lower bound of the exact value U of the expected payoff; we can also compute an upper bound by setting $R_i(j,h) = \frac{(n+1)X}{k}(h+1)$ instead. We denote by AMO-LB (resp., AMO-UB) the AMO algorithm for a lower (resp., upper) bound of U.

Dai et al. [9] proposed four approximation algorithms nUnifDown, nUnifCvg, nUnifUp, and nUnifSpl, where the first two (resp., the last two) compute lower (resp., upper) bounds of U. All algorithms use k_{ij} values defined as follows:

$$k_{ij} = \left\lceil \frac{k(n+1)(n+2)}{2} \times \frac{\sqrt{\omega(i,j)}}{\sum_{i'=0}^{n}\sum_{j'=0}^{i'}\sqrt{\omega(i',j')}} \right\rceil \quad \text{for all nodes } (i,j),$$

where k is a positive integer corresponding to the average number of buckets at each node. In nUnifDown (resp., nUnifUp) we set $R_i(j,h) = \frac{(n+1)X}{k_{ij}}h$ (resp., $R_i(j,h) = \frac{(n+1)X}{k_{ij}}(h+1)$). The algorithms nUnifCvg and nUnifSpl are modified versions of nUnifDown and nUnifUp by using heuristics; see [9] for details. While the error bounds of nUnifCvg and nUnifSpl are the same as those of nUnifDown and nUnifUp theoretically, they are much better practically.

The OSST algorithm [13] is a randomized algorithm. We set $k_{ij} = k$ with a positive integer k for all nodes (i,j), as in the AMO algorithm. To set the value $R_i(j,h)$, we choose a "representative" state $(S_i(j),T)$ in the bucket $B_i(j,h)$ randomly, where a state with weight w is chosen with probability $w/w_i(j,h)$, and set $R_i(j,h) = T$.

3.3   Our Algorithm and Analysis

Our algorithm is based on the ideas used in Dai et al. [9] and Ohta et al. [13], and can be obtained from the basic algorithm. We set $R_i(j,h)$ in the same way as in the OSST algorithm, i.e., we choose a representative state $(S_i(j),T)$ in the bucket $B_i(j,h)$ randomly, where a state with weight w is chosen with probability $w/w_i(j,h)$, and set $R_i(j,h) = T$. We explain later how to choose k_{ij}.

Let Ψ be the payoff value computed by our algorithm. Since our algorithm is randomized, Ψ is a random variable depending on the random choice of representative states in the buckets. Let $Y_{i,j}$ be a random variable giving the future value of the payoff just after the algorithm processes the node (i,j) in the i-th level, i.e., after the choice of representatives in all buckets has been determined up to the j-th node in the i-th level. By definition, $Y_{0,0} = U$ and $Y_{n,n} = \Psi$. Thus, we have a random process with $\sum_{i=0}^{n}(i+1) = (n+1)(n+2)/2$ steps. The following lemma shows that random variables $Y_{0,0}, Y_{1,0}, \ldots, Y_{n,n}$ constitute a *Martingale sequence*.

Lemma 3. $E(Y_{i,j} \mid Y_{0,0}, Y_{1,0}, Y_{1,1}, \ldots, Y_{i,j-1}) = Y_{i,j-1}$ for $i = 0,1,\ldots,n$, $j = 0,1,\ldots,i$.

Proof. Consider the set $\{a_1, a_2, \ldots, a_q\}$ of states in a bucket at the node (i,j) of the i-th level before selecting a representative. For $l = 1, 2, \ldots, q$, let $Y(a_l)$ be the expected payoff (exactly computed from the model) for a path with the state a_l, and $w(a_l)$ be the weight of a_l. If the state a_l is selected, it contributes

$Y(a_l)W$ to the payoff, where $W = \sum_{l=1}^{q} w(a_l)$. Thus, the expected contribution of the states after the selection is $\sum_{l=1}^{q}(w(a_l)/W)Y(a_l)W = \sum_{l=1}^{q} w(a_l)Y(a_l)$, where the right-hand side is the expected contribution before the selection. □

Lemma 3 also shows that the expected value of the payoff Ψ equals the exact value U of the expected payoff, i.e., $E(Y_{n,n}) = E(\Psi) = U$.

When the algorithm processes a node (i,j), running totals of paths terminating at (i,j) are approximated with the error less than $(n+1)X/k_{ij}$, and the running totals of other paths remain the same. Hence, we have

$$\left.\begin{array}{l} |Y_{i,j+1} - Y_{i,j}| < \frac{X\omega(i,j+1)}{k_{i,j+1}} \ (0 \leq j < i \leq n), \\[2mm] |Y_{i+1,0} - Y_{i,i}| < \frac{X\omega(i+1,0)}{k_{i+1,0}} \ (0 \leq i < n). \end{array}\right\} \tag{1}$$

Thus, Azuma's inequality [3] applies (see also [12–Theorem 4.16]).

Theorem 4 (Azuma's inequality). *Let $Z_0, Z_1, \ldots$ be a Martingale sequence such that $|Z_k - Z_{k-1}| < c_k$ for each k, where c_k is a constant. Then,*

$$\Pr[|Z_t - Z_0| \geq \lambda] \leq 2\exp\left(\frac{-\lambda^2}{2\sum_{k=1}^{t} c_k^2}\right) \qquad (\forall t = 1, 2, \ldots, \ \forall \lambda > 0).$$

Theorem 4 and (1) yield the inequality

$$\Pr[|Y_{n,n} - U| \geq \lambda] \leq 2\exp\left(\frac{-\lambda^2}{2X^2\Gamma}\right), \text{ where } \Gamma = \sum_{i=1}^{n}\sum_{j=0}^{i}\left(\frac{\omega(i,j)}{k_{ij}}\right)^2.$$

Hence, for any positive real number c, our algorithm computes in $O(\sum_{i=0}^{n}\sum_{j=0}^{i} k_{ij})$ time a value Ψ satisfying $|\Psi - U| \leq cX\sqrt{\Gamma}$ with probability at least $1 - 2e^{-c^2/2}$. To minimize the error bound $cX\sqrt{\Gamma}$ while keeping the time complexity $O(kn^2)$, we define the number of buckets at node (i,j) by

$$k_{ij} = \left\lceil \frac{k(n+1)(n+2)}{2} \times \frac{\omega(i,j)}{\sum_{i'=0}^{n}\sum_{j'=0}^{i'} \omega(i',j')} \right\rceil = \left\lceil \frac{k(n+2)}{2}\omega(i,j) \right\rceil.$$

Since $\Gamma \leq 2/k^2$ and $\sum_{i=0}^{n}\sum_{j=0}^{i} k_{ij} \leq (k+1)(n+1)(n+2)/2$, we have the following theorem, showing that the probabilistic error bound is $O(X/k)$.

Theorem 5. *Let k be any positive integer and c be any positive real number. Then, our algorithm computes in $O(kn^2)$ time a value Ψ satisfying $|\Psi - U| \leq \sqrt{2}cX/k$ with probability at least $1 - 2e^{-c^2/2}$.*

3.4   Derandomization

Although the error bound $O(X/k)$ of our algorithm shown in the last section is better than the previous approximation algorithms, our algorithm is randomized

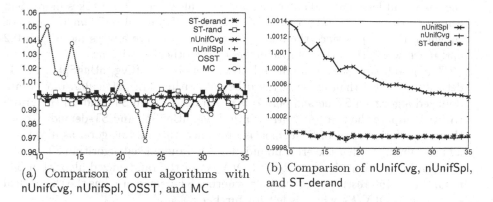

(a) Comparison of our algorithms with nUnifCvg, nUnifSpl, OSST, and MC

(b) Comparison of nUnifCvg, nUnifSpl, and ST-derand

Fig. 3. Relative errors of approximate option prices computed by several algorithms

and therefore the error bound only holds with "high" probability. Hence, it is desired to derandomize our algorithm without losing its accuracy. One idea for derandomization is to take the weighted mean of running totals of the states in each bucket $B_i(j, h)$ as the value $R_i(j, h)$, as in the algorithm nUnifCvg by Dai et al. [9]. Although we have not yet proved the theoretical error bound $O(X/k)$ for this derandomized version, it is experimentally shown that its error bound is better than the original one (see Sect. 3.5).

3.5   Experimental Results

We show some experimental results to illustrate the performance of our randomized approximation algorithm and its derandomized version. In particular, we compare the quality of the option price computed by our algorithms with those by other approximation algorithms. We implemented the full-path method to compute the exact price, and approximation algorithms such as the naive Monte Carlo method (MC), the AMO algorithms (AMO-LB, AMO-UB), the DHL algorithms [9] (nUnifDown, nUnifCvg, nUnifUp, nUnifSpl), and the OSST algorithm [13] (OSST). We denote our randomized and derandomized algorithms by ST-rand and ST-derand, respectively. The experiment is done by a Pentium IV 2.60CGHz PC and all programs are implemented in C++.

In the experiment, we consider a uniform model with $S_0 = X = 100$, $u = 1.1$, $d = 1/u$, $pu + (1 - p)d = (1.06)^{1/n}$. The parameter k is set to 100 in the approximation algorithms except for MC. Recall that the positive integer k denotes the number of buckets used at each node for AMO-LB/UB and OSST while k is the average number of buckets used at each node for the DHL algorithms and ours. The Monte Carlo method MC takes $400n$ sample paths so that it runs in almost the same time as other approximation algorithms. In the experiment, only one trial is made for each algorithm.

Fig. 3 gives the result of the experiment in the range $n \in [10, 35]$, showing the ratio of the approximate prices computed by approximation algorithms to the exact price. The running time of the approximation algorithms are almost the same and less than 0.05 seconds, and the full-path method takes more than 9 hours when $n = 35$. The results of AMO-LB/UB and nUnifDown/Up are not shown in the graphs since the relative errors of these are always more than 0.2 and much worse than the relative errors of the other algorithms.

The graph (a) shows that the relative errors of nUnifCvg, nUnifSpl, and ST-derand are better than those of the other algorithms. In particular, our derandomized algorithm ST-derand performs much better than ST-rand. In the graph (b) we compare the three algorithms nUnifCvg, nUnifSpl, and ST-derand. We see that the relative error of ST-derand is quite accurate and as good as nUnifCvg. This result shows that the error bound of our derandomized algorithm ST-derand is much better than the error bound $O(X/k)$ of the randomized algorithm ST-rand. It is an interesting open question whether ST-derand also has the theoretical error bound $O(X/k)$, which is left for further research.

References

1. D. Aingworth, R. Motwani, and J. D. Oldham, Accurate approximations for Asian options, *Proc. 11th ACM-SIAM Symposium on Discrete Algorithms* (2000), 891–900.
2. K. Akcoglu, M.-Y. Kao, and S. V. Raghavan, Fast pricing of European Asian options with provable accuracy: single-stock and basket options, *Proc. Annual European Symposium on Algorithms 2001, Lecture Notes in Computer Science* **2161** (2001), 404–415.
3. K. Azuma, Weighted sum of certain dependent random variables, *Tohoku Mathematical Journal* **19** (1967) 357–367.
4. F. Black and M. Scholes, The pricing of options and corporate liabilities, *Journal of Political Economy* **81** (1973), 637–654.
5. P. Chalasani, S. Jha, F. Egriboyun, and A. Varikooty, A refined binomial lattice for pricing American Asian options, *Review of Derivatives Research* **3** (1999), 85–105.
6. P. Chalasani, S. Jha, and I. Saias, Approximate option pricing, *Algorithmica* **25** (1999), 2–21.
7. P. Chalasani, S. Jha, and A. Varikooty, Accurate approximation for European Asian options, *Journal of Computational Finance* **1** (1998), 11–29.
8. J. C. Cox, S. A. Ross, and M. Rubenstein, Option pricing: a simplified approach, *Journal of Financial Economics* **7** (1979), 229–263.
9. T.-S. Dai, G.-S. Huang, and Y.-D. Lyuu, Extremely accurate and efficient tree algorithms for Asian options with range bounds, 2002 NTU International Conference on Finance, National Taiwan University, Taiwan, May 2002.
10. P. Glasserman, *Monte Carlo Method in Financial Engineering*, Springer, Berlin, 2004.
11. J. C. Hull, *Options, Futures, and Other Derivatives, Fifth Edition*, Prentice Hall, Upper Saddle River, NJ, 2002.
12. R. Motwani and P. Raghavan, *Randomized Algorithms*, Cambridge Univ. Press, London, 1995.
13. K. Ohta, K. Sadakane, A. Shioura, and T. Tokuyama, A fast, accurate and simple method for pricing European-Asian and Saving-Asian options, *Algorithmica*, to appear.

An Incremental Approach to Link Evaluation in Topic-Driven Web Resource Discovery

Huaxiang Zhang[1] and Shangteng Huang[2]

[1] Information and Management School,
Shandong Normal Univ. Jinan 250014, Shandong, China
huaxzhang@hotmail.com
[2] Department of Computer Science and Technology,
Shanghai Jiaotong Univ. Shanghai 200030, China

Abstract. The key issue concerning with Topic-driven Web resource discovery is how to increase the harvest rate, and the crawler should learn from the crawled online information such as the Web pages and the hyperlink structure. We address this problem by endowing a crawler with an incremental learning ability, and propose an online incremental leaning algorithm (IncL). IncL can effectively utilize the multi-feature characteristics of Web pages to enhance their link evaluation accuracy and reliability. We take into account not only a hyperlink's positive source pages but also its negative source pages in its score that is used to rank the Web pages. Many current crawling approaches ignore the negative pages' effect on the page ranking. Experiments show IncL gets high harvest rate.

1 Introduction

As the information is explosive, searching the Web becomes a very difficult task. In order to improve the resource retrieval performance, ranking the indexed pages by their estimated relevance with respect to user queries is crucial because it heavily influences the perceived effectiveness of a search engine. Early search engines rank pages principally based on their lexical similarity to the query, and evaluating the semantic closeness between the Web pages and a topic becomes a key issue.

It is recognized recently that the structure of hypertext links are powerful new sources of evidence for Web semantics, and machine learning methods have been used for link analysis. So a page's linkage to other pages, together with its content, is com-bined to estimate its relevance to a topic [1]. The PageRank employed by Google and Hits algorithm used in IBM's Clever are two well-known examples of such link analy-sis [2,3], and Google's success has also lead to PageRank's adoption for bibliometrics [4]. Kleinberg's HITS algorithm uses a combination of page content and link struc-tures to identify the most useful pages for the topic matching a search engine user's query. This is based upon the assumption that the overall link structure of the Web is not as important as that

N. Megiddo, Y. Xu, and B. Zhu (Eds.): AAIM 2005, LNCS 3521, pp. 301–310, 2005.

in the locality of the topic of concern. Even though the sophisticated ranking algorithms have been built in universal search engines, it's difficult for search engines to keep up with the growth of the Web, and representing the search with keywords and retrieval information with exact matching lack semantics.

Topical crawlers (also known as focused crawlers) [5,21] respond to the particular information needs expressed by topical queries or interest profiles, and are employed to address the scalability limitations of universal search engines. Topical crawlers support decentralizing the crawling process, which is a more scalable approach [7], and they can also be driven by a rich context such as Web page's content, URL exten-sions and hyperlink structure to classify a page as a more likely candidate for belong-ing to a particular topic and evaluate the links to be visited.

A focused crawler consists of a supervised topic classifier controlling the priority of the unvisited frontier of a crawler. The classifier is pre-trained by topics, seed URLs or other labeled positive samples. The crawler exploits the Web's hyperlink structure to retrieve new pages by traversing links from previously retrieved ones. As Web pages are fetched, their outward links may be added to the unvisited frontier. The goal of the focused crawler is to start from a seed URL in the Web graph and explore links to selectively collect pages similar to the topic, while avoiding fetching irrelevant pages [5]. The algorithms employed to select the next link for traversal are necessarily tied to the goals of the crawler, and the key challenge during the crawling progress is to identify the next most appropriate link to follow from the frontier.

Most proposed algorithms employ pre-trained crawlers with no learning abilities to crawl the Web pages. As the number of samples used for training a topic crawler is quite limited and the crawler may not be well trained, effective crawling strategies should be learned and evolved as the crawling process goes on. A learning task is incremental as the training samples become available over time [9]. If we add the crawled positive Web pages to a crawler's training data set and train the topic classifier again, then the topic crawling can be considered as an incremental task. The crawler learns and crawls concurrently as more Web pages are fetched.

In this paper, we propose an incremental learning Web crawling algorithm, and take the page context together with the hyperlink structure into consideration to in-crease the harvest rate. We set up a virtual positive context and a virtual negative context, and analyze the hyperlink structure of the crawled pages. As more pages are fetched, the virtual positive context and negative context change gradually, the super-vised topic classifier learns from more labeled positive and negative samples. The hyperlink structure can also be analyzed as more pages are visited, and we can get the in-degree of a candidate Web page from the virtual positive and negative context. Taking the above information together with the anchor text into consideration, the topic crawler evaluates the URLs on the crawl frontier and selects the best one to visit.

Our goal in this paper is to improve the harvest rate of the crawling algorithm. We summarize the proposed crawling algorithms in the literature and compare their effectiveness and efficiency in section 2. Virtual context and linkage analysis are introduced in section 3, and the issue of the learning algorithm is discussed in section 4. In section 5, we show the experimental results and present the conclusions in section 6.

2   Related Work

Topic crawlers exploit the Web's hyperlink structure to retrieve new pages by traversing links from previously retrieved ones. As pages are fetched, their outward links may be added to a list of unvisited pages(the crawl frontier). A key challenge during the progress of a topical crawl is to optimize the priority of the crawl frontier and select the next most appropriate link to follow from it. The algorithm to select the next link for traversal is necessarily tied to the goals of the crawler and the crawling efficiency relies on the crawl strategies

Optimizing the priority of unvisited URLs on the crawl frontier for specific crawling goals is not new. The early breadth-first crawler [10] starts the research on crawlers, and a variety of algorithms has emerged after that. Shark Search [11] is a more aggressive variant of Fish Search [12]. In another early paper, Cho et al. [13] relies heavily on link-based criteria and uses the anchor text as a bag of words to guide link expansion to crawl for pages matching a specific keyword query. [14] uses backlink-based context graphs to estimate the likelihood of a page leading to a relevant page, and [5] exploits lexical and conceptual knowledge and uses a hierarchical topic classifier to select links for crawling. [16,17] emphasizes the contextual knowledge for the topic including that received via relevance feedback, and [18,19] point out that the endpoints of a hyperlink are much more similar to each other than two random pages. The IBM's Clever topic distillation systems [20] uses such locality patterns for better semi-supervised learning of topics.

Two important advances have been made beyond the baseline best-first focused crawler: [14] uses the context graphs, and [22] proposes a reinforcement learning approach. Both techniques trained a learner with features collected from paths leading up to relevant nodes. Aggarwal et al. have proposed an "intelligent crawling" frame-work [23] and the classifier is trained as the crawl progresses. InfoSpiders [17] shows that in well-organized portions of the Web, effective crawling strategies can be learned and evolved by agents using neural networks and evolutionary algorithms.

A recent extensive study [21] develops a framework to evaluate different crawling algorithms proposed in the literature, and finds the best performance is achieved by a novel combination of explorative and exploitative bias, and introduces an evolutionary crawler that surpasses the performance of the best non-adaptive crawler after sufficiently long crawls. Experimental results show evolutionary crawlers achieve high efficiency and scalability. [15] takes several evolving prominent features of the Web into consideration, and improve the quality of URL ranking by modeling the growing presence of "link rot" on the Web as more sites and pages fall out of maintenance. It is proved the new ranking methods are more efficient than PageRank.

As more information such as the page context, the anchor text, the linkage structure and etc. is taken into account, the Web classifier can be well trained and the precision can be improved. Much progress has been made in the area. In all the systems mentioned above, the classifier mainly relies on the initial labeled training samples and the hyperlink model learned. It lacks learning ability as the crawl progresses.

We distinguish our work from prior art in the following important ways: The classifier is online in that it continuously refines its estimate of page importance while the Web/graph is visited. Thus it can be used to focus crawling to the most relevant pages. No manual path collection: we employ an agent to analyze the hyperlink structure. Incremental learning: the labeled training samples increase continually and the hyper-link structure changes as the crawl progresses. The crawler evaluates the Web pages more correctly as more pages are labeled positive or negative.

3   Virtual Context and Linkage Analysis

Co-reference and co-citation are two important concepts used to calculate the similarity between papers. Co-reference means if two papers share the same bibliography, we call them bibliographic coupling and they may be relevant. Similarly, if two papers are cited concurrently by a third one, they may be similar. It's pointed out [8] that hyperlink analysis also make assumptions to simplify the problem. If there is a hyperlink from one page to another page, is means the author of the first one recommends the second one. If two pages are connected by a hyperlink, then they might be on the same topic. The above knowledge and assumptions are quite useful in the hyperlink analysis, and should be taken into account in the hyperlink analysis model.

The mainly proposed models for hyperlink analysis are direct graph and matrix model, and they are resource cost. As the topical crawler is trained online, we analyze the hyperlink among the crawled pages, a simplified model is more challengeable. More pages are crawled and labeled positive or negative. Positive pages are topic relevant and negative pages are topic irrelevant. We consider all the positive pages as a positive training sample pool, and call it the virtual positive context. All the negative pages are considered as a negative training sample pool and called as the virtual negative context, and we use the two virtual contexts to re-train the classifier.

As more pages are fetched, the hyperlink information among pages becomes much clear, and we can calculate the outlink number of the virtual positive context and the outlink number of the virtual negative context for each candidate URL in the unvisited frontier. If the outlink number of virtual positive context for a candidate URL is large, it means the Web page pointed by the URL is much relevant to the topic. If the outlink number of virtual negative context for a candidate URL is large, it means the Web page pointed by the URL is much irrelevant to the topic. We adopt the above idea into our crawling algorithm (IncL).

We denote the virtual positive context as p_+ and the virtual negative context as p_-. If the number of inlink of a candidate page from members of p_+ is n_+ , the inlink weight from p_+ to it is n_+; if the number of inlink of a candidate page from members p_- of is n_- , the inlink weight from p_- to it is n_-.

4   Incremental Learning Algorithm

We employ the commonly used vector space model [6] to represent the crawling topics, the page contexts and the anchor texts. In order to evaluate a candidate URL, we propose the following score formula (1)

$$R(u_i) = S_a(u_i) + \gamma_1 (S^+(u_i))^{n^+} - \gamma_2 (S^-(u_i))^{n^-} \tag{1}$$

Where, $R(u_i)$ is the score of hyperlink u_i . We name the source pages of u_i the parent pages and the page linked by u_i the son page. $S_a(u_i)$ is the cosine similarity between p_+ and anchor text, $S^+(u_i)$ is the cosine similarity between p_+ and the combined context of the crawled positive source pages of u_i , $S^-(u_i)$ is the cosine similarity between p_- and the combined context of the crawled negative source pages of u_i. The cosine similarity between two vectors is given by $sim(a, b) = (a \cdot b)/\sqrt[2]{|a|^2 \times |b|^2}$.

Three terms in formula (1) take into account the anchor text, the positive source pages and the negative source pages. As the cosine similarity lies in interval [0,1], $S^+(u_i)^{n^+}$ is no less than $S^+(u_i)$. Two parameters γ_+ and γ_- are used to adjust the effect of the source pages. Taking the crawled outlinks of the parent pages into consideration, we calculate γ_1 and γ_2 as follows

$$\gamma^+ = \frac{l^+ + 1}{l + 1} \tag{2}$$

$$\gamma^- = \frac{m^+ + 1}{m + 1} \tag{3}$$

Where, l is the number of crawled outlinks of the positive parent pages and l^+ the number of son pages relevant to the topic. m is the number of crawled outlinks of the negative parent pages and m^- the number of son pages irrelevant to the topic.

γ^+ and γ^- are updated as the crawl progresses and the effects of the positive and negative parent pages to the son page changes accordingly. If more relevant son pages are crawled, the positive parent pages harness their effect. Otherwise, they lower the effect. Similarly, the negative parent pages enhance or lower their effect as more or less irrelevant son pages are crawled. At the starting crawling stage, l and m may be 0, so we add 1 to them.

As the crawl progresses, we can easily get p^+, p^- , the link structure, l , l^+ , m ,m^-. They all change incrementally online as the crawl progresses, and the Web page classifier is trained continually.

4.1   Generation of Virtual Positive and Negative Context

Virtual positive and negative samples are generated online. At first, we employ a universal Web search engine such as Google to establish the virtual positive samples manually, and the seed URLs are pushed into the unvisited frontier. As crawl progresses, more Web pages are crawled, and whether a Web page is positive or negative is determined online. We use the following formula (4) and (5) to calculate the similarities.

$$r^+(d_i) = (1 - \gamma)s^+(d_i) + \gamma \sum_{j=1,d_i \to d_j}^{n} s^+(d_j)/n \qquad (4)$$

$$r^-(d_i) = (1 - \gamma)s^-(d_i) + \gamma \sum_{j=1,d_i \to d_j}^{n} s^-(d_j)/n \qquad (5)$$

Where, $s^+(d_i)$ and $s^-(d_i)$ are the similarities between d_i and p_+ ,d_i and p_-, n is the number of d_i's son pages crawled, $r^+(d_i)$ and $r^-(d_i)$ and are the probabilities that d_i belongs to p_+ and p_-. We use $\gamma(0 \leq \gamma \leq 1)$ to adjust the weights of the two right terms in (4) and (5), and d_j is the son page of d_i.

The second term in (4) or (5) takes a page's son pages into consideration, and it changes as the son pages are crawled. So re-calculation of $r^+(d_i)$ or $r^-(d_i)$ is executed as a new d_j is crawled. We set two thresholds ε^+ and ε^- for and $r^-(d_i)$ separately. If $r^+(d_i)$ is no less than ε^+ ,then $p^+ = p^+ \cup d_i$, and if $r^-(d_i)$ is no less than ε^- , then $p^- = p^- \cup d_i$. In all other cases, we ignore d_i .

4.2   Learning Algorithm and Analysis

Based on the incremental learning, a topic crawler crawl the Web pages, the URL ranking, and the virtual positive and negative samples are generated continuously. In the crawl process, we take both the context of pages and the hyperlinks into consideration, and incrementally train the classifier and disclose the hyperlink structure. Learning can be characterized as searching a space of hypotheses for one that fits the training samples. The crawler is initially trained on limited labeled positive samples only, and the function learned may not be a proper one. As more Web pages are crawled and the Web structure becomes clearer, adaptive learning to fit the unlabeled samples is necessary, and contributions of different factors change dynamically. Learning and training are concurrent in the crawling, and the pseudo code is given in table 1

The function revaluate() calculates $r^+(d_i)$ and $r^-(d_i)$ again when a page's son page is crawled, and determines whether to eliminate the page from positive/negative training sample pool. Page elimination changes the training samples, so re-training of the classifier is necessary. Crawling a new page also has effect on parameters γ^+ and γ^- . If the newly crawled page is topic relevant, γ^+ increases, and the scores of the frontier's URLs whose source pages are the newly crawled page's parent pages may increase. Similar explanation can be given to γ^- .

Table 1. Incremental learning algorithm (IncL)

Initializing the topics, positive samples, seed URLs, MAX_PAGES;
ranking (topic, seed URLs); //ranking URLs in unvisited frontier
While (visited ¡ MAX_PAGES){
fetch (); // fetch the URL with the highest score
doc = fetch (link); // return the crawled page
enqueue (frontier, extract (doc)); // extract the outlinks of the returned page
pinlink (frontier); //fetch the inlink num. of a URL from positive Web page
ninlink (frontier); //fetch the inlink num. of a URL from negative Web page
determine (doc); // doc joining in positive/negative pool or being ignored
revaluate (); // calculate doc's parents' similarities by (4) and (5), eliminate the
          //parent pages from virtual positive or negative samples if they
          // are less than the corresponding thresholds
training (); //training the classifier again
ranking (topic, unvisited frontier);}//calculating the score of each URL again and
          // ranking them again

As the crawl progresses, some pages may be eliminated from the training
sample pool, others may be added.

5 Experimental Results

The evaluation methodology commonly used in information retrieval is to calculate the recall and precision. But in topic-driven Web resource discovery, it's
impossible to get the number of topic relevant pages online, so recall becomes
unavailable. Precision represents how much percent of all fetched pages is topic
relevant, and it is also called the harvest rate in Web mining.

We apply the incremental learning algorithm to several topics, and compare
its harvest rate with the breadth-first and best-first algorithms. The topics are
Economy, Chinese economy, and Chinese regional economy, and each topic is
more "narrow" than the previous one. We use Google to get the seed URLs and
collect the positive training pages. The number of Web pages crawled is set to
1200, and the data is given in table 2 and the relationships between the harvest
rate and the number of pages crawled are shown in figure 1.

Table 2. The harvest rates of different topics

Topic/Algorithm	Breadth-first	Best-first	IncL
Economy	0.191	0.530	0.762
Chinese economy	0.182	0.452	0.706
Chinese regional economy	0.152	0.348	0.516

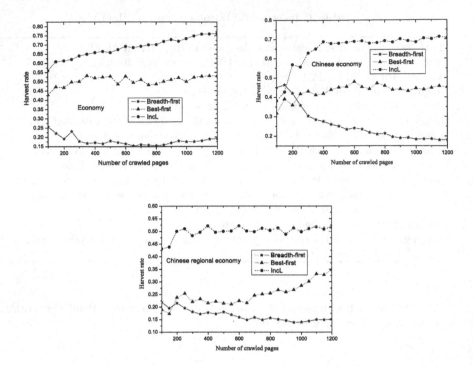

Fig. 1. Trend of the harvest rate on three topics

6   Conclusions and Future Work

Based on the incremental learning, the crawling algorithm can effectively utilize the multi-feature characteristics of Web pages to enhance their link evaluation accuracy and reliability. We take into account not only a hyperlink's positive source pages but also its negative source pages in its score calculation, and the crawled son pages of a hyperlink's source page also contribute to the hyperlink's score. Many current crawling approaches ignore the negative pages' effect on the score.

As more Web pages are labeled positive or negative, re-evaluation of the pages in the training sample pool is executed. Some pages are deleted from the pool, and others are added. Hyperlink structure is disclosed, and the page classifier is re-trained on new samples. The link evaluator also learns incrementally online and adjusts its prediction model. Based on the proposed model, the resulted topical crawler is adaptive to optimize its crawling policies as the crawl progresses.

Comprehensive experiments have been done and the results show that the algorithm proposed in this paper gets better performances than the baseline best-first algorithm.

Exploration and exploitation is a dilemma in crawling policy selection. Some topic relevant Web pages may be linked by an irrelevant hyperlink. In order to crawl this kind of pages, some URL ranking and selection strategies should be used. We will take time in this direction as the future work.

References

1. Kleinberg, J., Lawrence, S.: The structure of the Web. Science 294, 5548 (2001) 1849-1850 item Kleinberg, J.: Authoritative Sources in a Hyperlinked Environment. In: Proc. of the 9th Annual ACM-SIAM Symposium on Discrete Algorithms (1998) 668-677
2. Page, L., Brin S., Motwani, R., Winograd, T.: The PageRank citation ranking: Bringing order to the web. Technical report, Stanford University (1998)
3. Thelwall, M.: Can Google's PageRank be used to find the most important academic Web pages? J. of Documentation 59 (2) (2003c) 205-217
4. Chakrabarti, S., Berg, M. V. V., Dom, B.: Focused crawling: A new approach to topic-specific Web resource discovery. In: Proc. of 8th Int. World Wide Web Conf. (1999)
5. Ricardo B. Y, Berthier R, N.: Modern Information Retrieval. ACM Press Series/Addison Wesley, New York (1999)
6. Pant, G., Menczer, F.: Topical crawling for business intelligence. In: Proc. of the 7th European Conf. on Research and Advanced Technology for Digital Libraries. Lecture Notes in Computer Science, Vol. 2769 (2003)
7. Henzinger M. R.: Hyperlink Analysis for the Web. IEEE Internet Computing, Vol.5, 1 (2001) 45-50
8. Christophe G. G.: A Note on the Utility of Incremental Learning. AI Communications 13, 4 (2000) 215-224
9. Pinkerton, B.: Finding what people want: Experiences with the Web Crawler. In: Proc. of the 2nd Int. World Wide Web Conf., Chicago (1994)
10. Hersovici, M., Jacovi, M., Maarek, Y. S., Pelleg, D., Shtalhaim, M., Sigalit, U.: The shark search algorithm an application: Tailored web site mapping. In: Proc. 7th Int. World Wide Web Conf. (1998)
11. De Bra P., Houben G., Kornatzky Y., Post R., Information retrieval in distributed Hypertexts: making client-based searching feasible. In Proc. 4th RIAO (1994)
12. Cho, J., Garcia-Molina, H., Page, L.: Efficient crawling through URL ordering. In: 7th World Wide Web Conf., Brisbane, Australia (1998)
13. Diligenti, M., Coetzee, F., Lawrence, s., Giles, C. L., Gori, M.: Focused crawling using context graphs. In: Proc. of the 26th Int. Conf. on Very Large Databases. Cairo, Egypt (2000) 527-534
14. Nadav Eiron, Kevin S. McCurley, John A. Tomlin. Ranking the Web Frontier. In: Proc. of the 13th Int. World Wide Web Conf. (2004)
15. Aggarwal, C., Al-Garawi, F., Yu, P.: Intelligent crawling on the World Wide Web with arbitrary predicates. In: Proc. of the 10th Int. World Wide Web Conf. (2001) 96-105
16. Menczer, F., Belew, R.: adaptive retrieval agents: Internalizing local context and scaling up to the Web. Machine Learning 39, 2-3 (2000) 203-242
17. Davison, B. D.: Topical locality in the Web. In: Proc. of the 23rd Annual Int. Conf. on Research and Development in Information Retrieval (SIGIR 2000), Athens, Greece, ACM (2000) 272-279

18. Menczer, F.: Links tell us about lexical and semantic Web content. Technical Report Computer Science Abstract CS.IR/0108004, arXiv.org (2001)
19. Chakrabarti, S., Dom, Gibson, B. E., D. A., Kumar, R., Raghavan, P., Rajagopalan, S., Tomkins, A.: Topic distillation and spectral filtering. Artificial Intelligence Review, 13, 5-6 (1999) 409-435
20. Menczer, F., Pant, G., Srinivasan, P.: Topical Web Crawlers: Evaluating Adaptive Algorithms. ACM Transactions on Internet Technology, Vol. 4, No. 4 (2004) 378-19
21. L. Torgo, J. Gama. Regression by classification. LNAI, Vol. 1159
22. Aggarwal, C., Al-Garawi, F. Yu, P. S.: Intelligent crawling on the World Wide Web with arbitrary predicates. In: World Wide Web Conf., Hong Kong, ACM (2001)

A Continuous Method for Solving Multiuser Detection in CDMA*

Fengmin Xu[1] and Chengxian Xu[2]

[1] School of Science, Xi'an Jiaotong University,
Xi'an,710049, Shaanxi, P.R. China
[2] School of Science, Xi'an Jiaotong University,
Xi'an,710049, Shaanxi, P.R. China
{fengminxu, mxxu}@mail.xjtu.edu.cn

Abstract. This paper mainly focuses on a new continuous method for solving multiuser detection in CDMA using the NCP function. This approach is completely different from the relaxation method in the sense that it is an equivalent conversion. The resulting nonlinear programming problem of multiuser detection is then solved using the augmented Lagrange penalty function method. The convergence property of the proposed algorithm is studied, and numerical experiments show this method is very effective.

Keywords: Max cut problem; Penalty function method; Convex function.

1 Introduction

Multiuser detection plays an important role in suppressing the performance degrading effect of multiuser interference [1]. The constrained maximum likelihood (ML) problem may be described as [2]

$$\bar{d} = \text{argmin}_{d\in\{-1,1\}^K} d^T R d - 2y^T d \tag{1.1}$$

where y is the matched filter output vector, d is the spreading code, R is the correlation matrix, and z is the zero mean Gaussian noise vector with autocorrelation matrix $\delta^2 R$. The problem can be solved by an exhaustive search, however, the exhaustive search is prohibitive for large numbers of users because of its exponentially increasing computational complexity. It is known that the polynomial time algorithms for equation (1.1) exist if the autocorrelation matrix exhibits some special structure. However, in general, it is NP-hard problem [1]. In Tan et al [2] and Ma et al [3], a detection strategy based on a semidefinite relaxation of the CDMA maximum likelihood (ML) problem is investigated. The semidefinite relaxation may solved using interior point methods in polynomial time [4,5]. In

---

* Key project supported by National Nature Science Foundation of China:10231060. This paper is also supported by National Key Laboratory of Mechanical Systems.

N. Megiddo, Y. Xu, and B. Zhu (Eds.): AAIM 2005, LNCS 3521, pp. 311–319, 2005.

this paper we propose the continuous method for multiuser detection in CDMA. This approach is completely different from the relaxation method in the sense that it is an equivalent conversion. As long as the global optimum solution of the resulting nonlinear programming can be found the multiuser detection problem could be resolved.

2 A Continuous Method for Multiuser Detection

Tan et al [2] has proposed a semidefinite programming relaxation for ML detection, we introduce the relaxation as follows.

Let $n = K + 1, x = [\bar{d}^T, \bar{d}_n]^T, (\bar{d}_n = 1)$ and

$$C_1 = \begin{pmatrix} R & -y \\ -y^T & 0 \end{pmatrix},$$

since the cost function is symmetric, $d_n = 1$ need not be maintained explicitly. Problem (1.1) may be formulated as

$$x^* = \mathrm{argmin} x^T C_1 x \quad s.t. \quad x \in \{-1, 1\}^n, \tag{2.1}$$

that is

$$(MD) : \begin{cases} \mu^* = Min \ x^T C_1 x \\ s.t. \ x_i^2 = 1, \quad i = 1, \cdots, n; \end{cases}$$

This optimization problem is well known to be NP-hard and which can be solved by semidefinite programming relaxation. Now we discuss how to get an equivalent problem for the multiuser detection problem by using an NCP function.

2.1 The Continuous NCP Function

The complementary condition is constituted by imposing the constraints $-1 \le x_i \le 1, \ i = 1, \cdots, n$ and observing that $x_i^2 = 1 \Leftrightarrow (x_i + 1)(x_i - 1) = 0$ in the (MD) problem. Therefore

$$\begin{cases} (1 + x_i)(1 - x_i) = 0, \\ (1 + x_i) \ge 0, \qquad\qquad 1 \le i \le n, \\ (1 - x_i) \ge 0, \end{cases} \tag{2.2}$$

We think that the complementary condition (2.2) can be replaced by an equation which uses NCP function. There are two main reasons for this transformation. First, the (MD) problem had absorbed these constraint conditions by not adding the number of the equations, on the other hand, the iterate point is possibly restricted in $[-1, 1]$.

The common NCP functions[6] are:

$$\Phi_F(a, b) = \sqrt{a^2 + b^2} - a - b = 0 \Leftrightarrow ab = 0, a \ge 0, b \ge 0,$$
$$\Phi_M(a, b) = Min\{a, b\} = 0 \Leftrightarrow ab = 0, a \ge 0, b \ge 0.$$

Let $a = 1 - x_i, b = 1 + x_i$, the multiuser detection problem can be described by the following two nonlinear programming based on these two smoothing functions:

$$(CMD1) : \begin{cases} \text{Min} \ \ x^T C_1 x \\ s.t. \ \ \Phi_F(1 - x_i, 1 + x_i) = 0, i = 1, \cdots, n, \end{cases}$$

or

$$(CMD2) : \begin{cases} \text{Min} \ \ x^T C_1 x \\ s.t. \ \ \Phi_M(1 - x_i, 1 + x_i) = 0, i = 1, \cdots, n. \end{cases}$$

Based on the above transformation, we need only solve these two nonlinear programming problems to find a solution for the multiuser detection problem. Next, we make $(CMD1)$ as a example to show how to solve the multiuser detection problem using the multiplier penalty function method.

2.2 The Multiplier Penalty Function Method for Solving the $(CMD1)$ Problem

In this section we present a method for solving the $(CMD1)$ problem using the multiplier penalty function method [7].

The augmented Lagrangian function of the $(CMD1)$ problem has the form

$$P(x, \lambda, \sigma) = f(x) - \sum_{i=1}^{n} \lambda_i \psi_i(x_i) + \frac{\sigma}{2} \sum_{i=1}^{n} \psi_i^2(x_i).$$

where

$$f(x) = x^T C_1 x,$$

$$\psi_i(x_i) = \Phi_F(1 - x_i, 1 + x_i) = \sqrt{2 + 2x_i^2} - 2, \quad i = 1, \cdots, n,$$

$\lambda_i, \ i = 1, 2, \cdots, n$ are Lagrange multipliers, and σ is a penalty factor. If function $\psi_i(x_i) \ \ i = 1, \cdots, n$ is differentiable at the point x_i, the gradient of the function $P(x, \lambda, \sigma)$ with respect to x is given by

$$\nabla_x P(x, \lambda, \sigma) = \nabla f(x) - \sum_{i=1}^{n} (\lambda_i - \sigma \psi_i(x_i)) \nabla \psi_i(x_i). \tag{2.3}$$

Suppose x^* is the optimal solution of the $(CMD1)$ problem and λ^* is the corresponding Lagrange multiplier. According to the KKT condition of problem $(CMD1)$,

$$\nabla f(x^*) - \sum_{i=1}^{n} \lambda_i^* \nabla \psi_i(x_i^*) = 0 \tag{2.4}$$

holds at x^*, λ^*. Observing (2.3) and (2.4) provides the suggestion that the sequence λ^k in the augmented Lagrange penalty method can be generated using the iteration

$$\lambda_i^k = \lambda_i^{k-1} - \sigma \psi_i(x_i^k), \ i = 1, 2, \cdots, n.$$

To ensure the value of penalty factor σ is large enough, the augmented Lagrange penalty function method must be implemented in the following format where $\Psi(x) = (\psi_1(x_1), \cdots, \psi_n(x_n))^T$.

The above algorithm forces the iterates x^k to generate a sequence of $\{\|\Psi(x^k)\|\}$ converging to zero at a rate less than $1/4$. When the condition is not satisfied, the penalty factor σ will be increased.

Algorithm 1. *Multiplier Penalty Function Method*

Step 1. Given $\sigma_1 > 0$, ϵ_0, $\epsilon_1 > 0$, Choose the initial points x^0, λ^0, Set $k = 0$, $t_0 = \|\Psi(x^0)\|_2$;

Step 2. Starting x^k, solve the problem

$$Min_{x \in R^n} P(x, \lambda^k, \sigma_{k+1})$$

to get a solution of x^{k+1}. If termination condition is satisfied, set $x^* = x^{k+1}$ and stop.

Step 3. If $\|\Psi(x^{k+1})\|_2 \leq \frac{1}{4} t_k$, then go to Step4; Else let

$$\sigma_{k+1} = 10\sigma_{k+1}, \quad x^k = x^{k+1}$$

then go to Step2.

Step 4. Choose $t_{k+1} = \|\Psi(x^{k+1})\|_2$,

$$\lambda_i^{k+1} = \lambda_i^k - \sigma_{k+1}\psi_i(x_i^{k+1}),$$

$k = k+1$, $\sigma_{k+1} = \sigma_k$, then go to Step2.

Lemma 1. *Let λ be given, and $x(\sigma_1)$ and $x(\sigma_2)$ be the global solutions of the functions $P(x, \lambda, \sigma_1)$ and $P(x, \lambda, \sigma_2)$ with $\sigma_2 > \sigma_1 > 0$. Then $f(x(\sigma_2)) \geq f(x(\sigma_1))$ holds for sufficient large σ_1.*

Proof It follows from the definition of $x(\sigma)$ that

$$
\begin{aligned}
&f(x(\sigma_1)) - \sum_{i=1}^{n} \lambda_i \psi_i(x_i(\sigma_1)) + \tfrac{\sigma_1}{2}\|\Psi(x(\sigma_1))\|_2^2 \\
&\leq f(x(\sigma_2)) - \sum_{i=1}^{n} \lambda_i \psi_i(x_i(\sigma_2)) + \tfrac{\sigma_1}{2}\|\Psi(x(\sigma_2))\|_2^2;
\end{aligned}
\tag{2.5}
$$

$$
\begin{aligned}
&f(x(\sigma_2)) - \sum_{i=1}^{n} \lambda_i \psi_i(x_i(\sigma_2)) + \tfrac{\sigma_2}{2}\|\Psi(x(\sigma_2))\|_2^2 \\
&\leq f(x(\sigma_1)) - \sum_{i=1}^{n} \lambda_i \psi_i(x_i(\sigma_1)) + \tfrac{\sigma_2}{2}\|\Psi(x(\sigma_1))\|_2^2;
\end{aligned}
\tag{2.6}
$$

Adding equations (2.5) and (2.6) generates

$$\frac{(\sigma_1 - \sigma_2)}{2}(\|\Psi(x(\sigma_1))\|_2^2 - \|\Psi(x(\sigma_2))\|_2^2) \leq 0.$$

Since $\sigma_2 > \sigma_1$, we obtain

$$\|\Psi(x(\sigma_2))\|_2^2 \leq \|\Psi(x(\sigma_1))\|_2^2. \tag{2.7}$$

From equation (2.7), we have

$$f(x(\sigma_1)) \leq f(x(\sigma_2)) + \sum_{i=1}^{n} \lambda_i \psi_i(x_i(\sigma_1)) - \sum_{i=1}^{n} \lambda_i \psi_i(x_i(\sigma_2))$$
$$+ \tfrac{\sigma_1}{2}(\|\Psi(x(\sigma_2))\|_2^2 - \|\Psi(x(\sigma_1))\|_2^2)$$

For sufficient Larger σ_1, the following inequality is true,

$$\sum_{i=1}^{n} \lambda_i \psi_i(x_i(\sigma_1)) - \sum_{i=1}^{n} \lambda_i \psi_i(x_i(\sigma_2))$$
$$+ \frac{\sigma_1}{2}(\|\Psi(x(\sigma_2))\|_2^2 - \|\Psi(x(\sigma_1))\|_2^2) \leq 0.$$

Thus, we have the result $f(x(\sigma_2)) \geq f(x(\sigma_1))$.

It follows from Lemma 2 that the multiplier penalty function method is well-defined, that is, we have the following Theorem 3.

Theorem 2. *For any given λ^k, the condition $\|\Psi(x^{k+1})\|_2 \leq \frac{1}{4}t_k$ in Step 3 of Algorithm 2.1 must be satisfied in a finite number of times increasing the value of σ_k.*

Proof Suppose that conclusion is not true. Then there exists an index $\bar{k}$ such that

$$\|\Psi(x(\hat{\sigma}_j))\| > \frac{1}{4}t_{\bar{k}} = \frac{1}{4}\|\Psi(x^{\bar{k}})\| > 0$$

holds for all $j = 1, 2, \cdots$, where $\hat{\sigma}_{j+1} = 10\hat{\sigma}_j$, and $\hat{\sigma}_1 = 10\sigma_{\bar{k}}$. Since $x(\hat{\sigma}_j)$ is the global minimizer of $P(x, \lambda^{\bar{k}}, \hat{\sigma}_j)$, we have

$$P(x(\hat{\sigma}_j), \lambda^{\bar{k}}, \hat{\sigma}_j) = f(x(\hat{\sigma}_j) - \sum_{i=1}^{n} \lambda_i^{\bar{k}} \psi_i(x_i(\hat{\sigma}_j)) + \frac{1}{2}\hat{\sigma}_j\|\Psi(x(\hat{\sigma}_j)\|^2$$
$$< f(\bar{x}) - \sum_{i=1}^{n} \lambda_i^{\bar{k}} \psi_i(\bar{x}_i) + \frac{1}{2}\hat{\sigma}_j\|\Psi(\bar{x})\|^2 = f(\bar{x}),$$

here $\bar{x}$ is a feasible point of the problem, and we use the fact $\psi_i(\bar{x}_i) = 0$, for $i = 1, 2, \cdots, n$. Let $\hat{\sigma}_j \to \infty$, then it follows from the Lemma 2 and the boundedness of the function $\psi_i(x_i)$ and the fixed $\lambda^{\bar{k}}$, we get

$$\|\Psi(x(\hat{\sigma}_j))\| \to 0.$$

This generates a contradiction and yields the conclusion.

The next theorem indicates that when the criterion

$$\|\Psi(x^k)\| = 0$$

is used to terminate the algorithm, and if an infinite sequence of iterates is generated, then any accumulation point of the sequence is an optimal solution of problem $(CMD1)$.

Theorem 3. *If λ^k is bounded for all k and x^k is the global minimizer of $P(x, \lambda^{k-1}, \sigma_k)$, then any accumulation point x^* of the infinite sequence x^k generated by the algorithm with termination criterion $\|\Psi(x^k)\| = 0$ is the global optimal solution of problem $(CMD1)$.*

Proof   It follows from Theorem 3 that the sequence $\{x^k\}$ generated by the algorithm satisfies $\|\Psi(x^{k+1})\| \le \frac{1}{4}\|\Psi(x^k)\|$. Since $\|\Psi(x^k)\| \ge 0$, the conclusion $\|\Psi(x^k)\| \to 0$ holds. Suppose $\{x^{k_j}\}$ is a convergent subsequence of $\{x^k\}$, and its limit point is x^*. Then from $\|\Psi(x^k)\| \to 0$ and the continuous of the function $\Psi(x)$ we get $\|\Psi(x^{k_j})\| \to \|\Psi(x^*)\| = 0$, which implies x^* is a feasible point. The definition of x^{k_j}, the feasibility of the point x^* gives the inequality

$$f(x^{k_j}) - (\lambda^{k_j-1})^T\Psi(x^{k_j}) + \frac{\sigma_{k_j}}{2}\|\Psi(x^{k_j})\|^2 \le f^*, \qquad (2.8)$$

holds for sufficiently large k_j, where f^* is the global optimal value of problem $(CMD1)$.

On the other hand, using the continuity of the functions $f(x)$ and $\Psi(x)$, and having $k_j \to \infty$ in equation (2.8) generate

$$f(x^*) \le f^*.$$

However, the feasibility of the point x^* gives $f(x^*) \ge f^*$. Thus, we have $f(x^*) = f^*$, that is, x^* is an optimal solution of problem $(CMD1)$.

3   Simulation Results

In this section we run the multiplier penalty function method (MPF) and interior-point method based on the SDP relaxation (SDP) by using SDPpack software from Nayakakuppam et al[8] in the MATLAB 6.1 environment on a 1.6GHz Pentium IV personal computer with 256Mb of Ram.

In our continuous method, R is generated by the random Gold mode and y is produced by randomly b. The simulation results are seen in Figure 1 and Figure 2. BER means the bit error rate. We have implemented our algorithm in a Matlab code. For the minimization of the function $P(x, \lambda^{k-1}, \sigma_k)$ with given values of σ_k and λ^{k-1}, we use the function $fminunc$ in Matlab that employs the BFGS method with a backtracking Armijo line search. All parameters in $fminunc$ keep default values except for the parameter used in termination criterion. Numerical experiments indicate that the accuracy of minimizing $P(x, \lambda^{k-1}, \sigma_k)$ is not crucial, we change the parameter value in termination criterion from the default value 10^{-6} to 10^{-4}. The termination conditions in the Algorithm 1 that we choose are $\|\Psi(x^{k+1})\| \le \epsilon_0$ or $\|x^{k+1} - x^k\| \le \epsilon_1$. In implementing the algorithm , we choose $\epsilon_0 = 0.05$ and $\epsilon_1 = 0.01$, and the initial value for σ is $\sigma_0 = 1$.

From Figure 1, the bit error rate (BER) performance of the detector based on the multiplier penalty function method is lower than the interior-point method based on the SDP relaxation. From Figure 2 we can see that the multiplier penalty function method needs less time. So we can use this algorithm to solve the large scale (MD) problem.

From our investigation we can see that our multiplier penalty function method is an interesting and new technique. We expect that there are other efficient continuous methods similar to the continuous method in this paper. Now we apply ourselves to search for a more convenient and efficient continuous method.

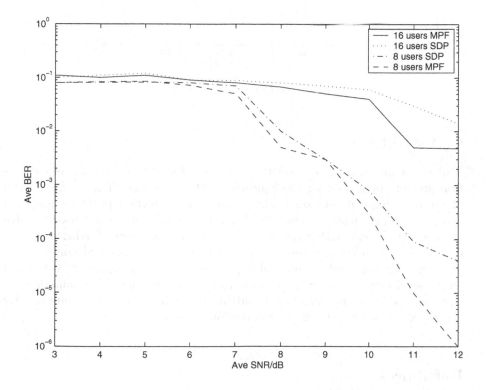

Fig. 1. The Compare of the Ave BER

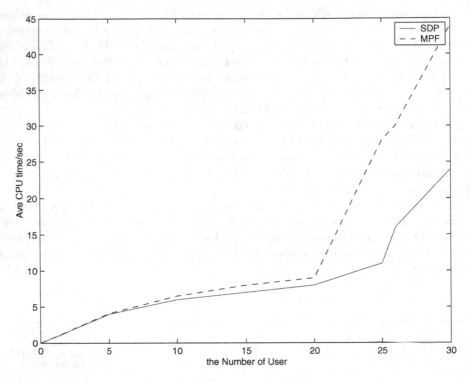

Fig. 2. The Compare of the CPU time

4 Conclusion

This paper presented an equivalent continuation model of the multiuser detection problem by employing two continuous NCP functions. The characteristics of the resulting continuation model of multiuser detection problems are then analyzed, and the famous augmented Lagrange penalty function is used to solve the converted continuation nonlinear optimization problem. Furthermore, the convergence property and termination property of the proposed continuation algorithm are analyzed. Numerical experiments on some randomly generated multiuser detection problems are made to test the efficiency. The numerical results show that the continuation algorithm generates satisfactory solutions for the randomly generated multiuser detection problems.

References

[1] Verdu, S.: Multiuser Detection: Cambridge. Cambridge University Press. (1998).
[2] Tan P., Rasmussen L.: The application of semidefinite programming for detection in CDMA. IEEE. Selected in Communication. **19** (2001) 1442-1449.

[3] Ma W., Davidson T.: Quasi-maximum-likelihood multiuser detection using semidefinite relaxation. IEEE Trans. on Signal Processing. **50** (2002) 912-922.

[4] Alizadeh F., Haeberly J., Overton M.: Primal dual interior point methods for semidefinite programming: convergence rates, stability and numerical results. *SIAM Journal on Optimization.* **8** (1998) 746-768.

[5] Helmberg C., Rendl F., Vanderbei R., Wolkowicz H.: An interior point methods for semidefinite programming. *SIAM Journal on Optimization.* **6** (1996) 342-361.

[6] Fischer A. : A special Newton-type Optimization Method. Optimization. **24** (1992) 269-284.

[7] Yuan Y.: Numerical methods for nonlinear programming. *Shanghai Scientific & Technical Publishers.* (1992).

[8] Nayakakuppam M., Overton M., and Schemita S.: SDPpack user's guide-version 0.9 Beta. Technical Report, Courant Institute of Math. Science. NYU. New York.(1997).

Wavelength Assignment for Satisfying Maximal Number of Requests in All-Optical Networks[*]

Xiaodong Hu and Tianping Shuai

Institute of Applied Mathematics,
Chinese Academy of Sciences,
P. O. Box 2734, Beijing 100080, China
{xdhu, shuaitp}@amss.ac.cn

Abstract. In this paper, we study how to, given a set of pre-routed requests in an all-optical network and a set of wavelengths available on each link, assign a subset of requests with maximal size such that no wavelength constraint on links is violated. While all previous studies on the wavelength assignment problem with the same objective assume the same set of wavelengths available on all links, our work does not make such an assumption. We first prove that this problem is NP-hard even in bus networks, and then propose some approximation algorithms for the problem with guaranteed performance ratios in networks with some special and general topologies.

1   Introduction

All-optical networks employ the technology of Wavelength Division Multiplexing (WDM) that divides the tremendous bandwidth of an optical fiber into many nonoverlapping wavelengths channels. These networks consist of a set of nodes interconnected by bundles of optical fibers, each of them accommodate a few tens of wavelength channels. WDM networks can satisfy the demand of the high-bandwidth applications in the next generation networks and considered as the transport networks of the future.

The nodes of WDM networks have dynamically configurable wavelength switches. They switch the data on a specified input port and wavelength to a specified output port on the same wavelength. In this paper we assume that the switches are incapable of converting the data on one wavelength to another wavelength since eliminating wavelength conversion capability can significantly reduce the cost of the switch. However, empowering the switch to make wavelength conversion can improve the network efficiency because a request can use two more wavelengths when there does not exist one wavelength which is available on each link of its route.

The traffic models usually include static traffic model [2], where a set of call requests is given and routes and wavelengths have to be assigned to requests

---

[*] This work was supported in part by the National Natural Science Foundation of China under Grant No. 70221001 and 60373012.

N. Megiddo, Y. Xu, and B. Zhu (Eds.): AAIM 2005, LNCS 3521, pp. 320–329, 2005.

at one time, and dynamic traffic model [18], where requests arrive and depart from the network over time and decision should be made whether the request is rejected or accepted (in the latter case a route and wavelength have to be assigned to the request one by one without knowledge of future requests). In this paper we assume the static traffic. This model is particularly important in the network design phase, when a candidate network with link capacities is considered and one wants to know how many of the forecasted traffic requirements can be satisfied by the network. Moreover, it is useful in a scenario that supports advance reservation of connections, because then it is possible to collect a number of reservation requests before the admission control is carried out for a whole batch of requests.

The transmission models usually include unicast [2], where the data is transmitted from a source to a destination, and multicast [15], where the data is transmitted from a source to multiple destinations; In this paper we consider unicast in WDM networks, which is supported by lightpaths. A *lightpath* [2] is an all-optical channel which can be used to carry circuit-switched traffic between two communication nodes and it may span multiple fiber links. There are two major issues in the setup of lightpaths: routing the lightpaths in the physical network and assigning a wavelength to each of them.

The objectives for the routing and wavelength assignment problems usually include maximizing network throughput [14], minimizing wavelength costs [11], or call blocking probability [12]. These problems for static traffic are known NP-hard in general, and they have been well studied in the literature over the last ten years. For example, Adamy et al [1] and Erlebach and Jansen [5] studied how to route, for a set of given requests in a ring and tree networks with link capacity constraints on the number of paths that can go through a link, a maximum cardinality subset such that no link constraint is violated; Nomikos et al [10] studied how to, given a set of requests in a ring, assign k wavelengths (available on all links) to a maximum set of requests without causing wavelength conflict; Wilfong and Winkler [16] studied how to route and assign, in a ring network without link capacity constraints, each of given set of requests a wavelength such that the number of wavelengths used is minimal.

In this paper, we assume that the routes for the given requests are prescribed and we focus on the wavelength assignment problem. We study how to, given a set of pre-routed requests in an all-optical network and a set of wavelengths available on a link, assign wavelengths to a subset of requests such that no wavelength constraint on links is violated; the objective is to maximize the size of the subset. In Section 2, we formulate the problem and prove that it is NP-hard even for bus networks. In Section 3, we propose some approximation algorithms for the problem for bus, tree, ring, and general networks. In Section 4, we conclude the paper.

2 Problem Formulation and Its Complexity

We will use a undirected connected graph $G(V, E, w)$ to model a underlying wavelength routed WDM all-optical network, where V denotes the set of routing

322   X. Hu and T. Shuai

nodes in the network, E the set of links between node-pairs in V, and $w(e)$ the set of wavelengths available on link $e \in E$ for any $e \in E$. Assume that a set $W = \{w_1, w_2, \cdots, w_k\}$ of k wavelengths are available for use in the network, but some of them may be occupied in some links of the network, thus $w(e) \neq W$ for some $e \in E$. In particular we call the case of $w(e) = W$ for all $e \in E$ *uniform wavelength distribution*, and other cases *nonuniform wavelength distribution*. A communication request between two nodes s and t is denoted by $r(s,t)$. We assume that the route for request $r(s,t)$ is also given or pre-routed (e.g. the shortest path may be used), which is denoted by $p(s,t)$. We assume further any request must use the same wavelength over all links in its route, that is, no wavelength conversion is allowed.

Given a set of requests $R = \{r(s_i, t_i) \mid i = 1, 2, \cdots, m\}$, a subset $S \subseteq R$ is called a *satisfiable request subset* if there exists a wavelength assignment for S such that two requests in S are assigned two distinct wavelengths if their routes share a common link in the network. In this paper, we shall study how to, given a set R in a network $G(V, E, w)$, find a satisfiable request subset of maximal size. We call it *Maximum Satisfiable Request Subset* (MSRS) problem.

Let us first consider the MSRS problem in bus networks. In this case, $G(V, E)$ is a path, so we can label all vertices in V in a linear order. For the case of uniform wavelength distribution, this problem is reduced to the maximal k-colorable subset problem in interval graphs [17,3], that is, given a set of intervals and k colors, how to find a subset of intervals of maximal size such that they can be colored in such a way that two distinct colors must be assigned to two intervals if they overlap. This problem was proved to be polynomial-time [17], that can be rewritten as the following theorem.

Theorem 1. *For the case of uniform wavelength distribution, the MSRS problem in bus networks is polynomial-time solvable.*

We next consider the MSRS problem for nonuniform wavelength distribution, a much more difficulty case. The following theorem shows that in this case the MSRS problem is NP-hard even for bus networks.

Theorem 2. *For the case of nonuniform wavelength distribution, the MSRS problem in bus networks is NP-hard.*

Proof. Consider the wavelength assignment problem in ring networks, that is how to, given a set of paths in a ring network, assign each of them a wavelength without causing wavelength conflict and the number of wavelengths used is minimal. It is known [17] that this problem is NP-hard. To prove the theorem it suffices to show that any instance of the wavelength assignment problem in ring networks can be transformed to an instance of the MSRS problem in bus networks in polynomial time.

We consider the decision versions of both problems. Let k be the upper bound on the number of wavelengths. Denote a ring network of n nodes by $G(V, E)$ with $V = \{0, 1, 2, \cdots, n-1\}$ and $E = \{(i, i+1) \mid i = 0, 1, 2, \cdots, n-1\}$, where the labels take the module of n. The reduction can be constructed as follows: Delete link

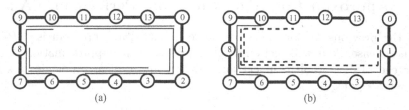

Fig. 1. Reducing the wavelength assignment problem in rings to the MSRS problem in buses

$(n-1,0)$ from ring network G and obtain a bus network $G'(V,E')$ with $E' = E \setminus \{(n-1,0)\}$. Let $L_l(n-1,0) = \{r(s_i, t_i) \mid i = 1, 2, \cdots, g\}$ be the set of requests pass through $(n-1,0)$, $p(s_i, t_i) = (s_i, s_i + 1, \cdots, t_i)$ and $\bar{p}(s_i, t_i) = (t_i, t_i + 1, \cdots, s_i)$. Assume, without loss of generality, that $g \leq k$, and let $W = \{w_1, w_2, \cdots, w_g\}$ and $W' = \{w_{g+1}, w_{g+2}, \cdots, w_k\}$. And then for each link $e \in E'$ define the set of wavelengths available $w(e) = \{w_j \mid e \in \bar{p}(s_j, t_j) \text{ for } 1 \leq j \leq g\} \cup W'$. In the end, define the set of requests $R' = R \setminus L_l(n-1,0)$ and let $k' = |R'|$.

Fig. 1 illustrates the above process. There are five requests in a ring network $G(V,E)$ of 14 nodes as shown in Fig. 1(a) with $k = 3$, where $r_1 = p(9,0)$, $r_2 = p(11,5)$, $r_3 = p(1,7)$, $r_4 = p(3,8)$, and $r_5 = p(6,10)$. Observe that $L_l(13,0) = \{r_1, r_2\}$, $W = \{w_1, w_2, w_3\}$ and $W' = \{w_3\}$. There are three requests in a bus network $G'(V,E')$ as shown in Fig. 1(b), where $R' = \{r_3, r_4, r_5\}$ with $k' = 3$, and $\bar{p}_1 = p(0,9)$ and $\bar{p}_2 = p(5,11)$ (in dashed lines). Notice that $w(i, i+1) = \{w_1, w_3\}$ for $i = 0, 1, \cdots, 4$, $w(i, i+1) = \{w_1, w_2, w_3\}$ for $i = 5, 6, 7, 8$, $w(i, i+1) = \{w_2, w_3\}$ for $i = 9, 10$, and $w(i, i+1) = \{w_3\}$ for $i = 11, 12$. Observe that three wavelengths are sufficient (but two wavelengths are not enough) for these five requests in ring network (wavelength w_1 for requests r_1 and r_3, w_2 for r_2 and r_5, and w_3 for r_4). Meanwhile three wavelengths are sufficient (but two wavelengths are not enough) for these three requests in bus network (wavelength w_1 for request r_3, w_2 for r_5, and w_3 for r_4).

We shall prove that a set R of requests on ring network $G(V,E)$ can be assigned properly using at most k wavelengths if and only if there exists a satisfiable requests subset $S \subseteq R'$ with cardinality at least k'.

"Only-if" part. Suppose that R can be partitioned into k subsets $R_1, R_2, \cdots, R_k$ such that all requests in R_i are assigned wavelength w_i for $i = 1, 2, \cdots, k$. We can assume, without loss of generality, that $L_l(n-1,0) = \{r_i \mid i = 1, 2, \cdots, g\}$ and request r_i is assigned wavelength w_i for $1 \leq i \leq g$. Now let $R'_i = R_i \setminus L_l(n-1,0)$ for $i = 1, 2, \cdots, k$. It is easy to see that each R'_i is a satisfiable request subset of R' by using one single wavelength. In addition, $R' = R'_1 \cup R'_2 \cup \cdots \cup R'_k$ is a satisfiable and $|R'| = k'$.

"If" part. Suppose that R' is a satisfiable request set with $|R'| = k'$. Notice that there are k wavelengths available on links in bus network $G'(V,E')$. Assume now that subset $R'_i \subset R'$ is satisfiable by using wavelength w_i for $i = 1, 2, \cdots, k$. It is easy to see that there exists g wavelengths which can be assigned to g requests in $L_l(n-1,0)$ without causing wavelength conflicts. Thus k wavelengths are enough to satisfy all requests in R.

3   Approximation Algorithms and Performance Analysis

In the previous section, we have proved that MSRS problem is NP-hard in general case. So in this section we shall propose some approximation algorithms for the MSRS problem in networks with some typical topologies (as well as general topology).

3.1   Bus and Tree Networks

Let $E_i = \{e \mid w_i \in w(e), e \in E\}$ for each w_i. Then E_i can be considered as a subgraph of $G(V, E)$ with wavelength w_i available on all edges in E_i. The basic idea of our algorithm is to solve the MSRS problem in E_i with single wavelength k times for $i = 1, 2, \cdots, k$, that is, to find the maximum set of requests in R that can be satisfied by wavelength w_i, after that remove all satisfied requests from R (and remove w_i from W at the same time); and then repeat this process until R is empty or all E_i are considered. In the algorithm we can adopt a greedy rule that each time we choose wavelength w_i which can satisfy the most number of unsatisfied requests. The proposed algorithm is more formally described below.

Algorithm A

Step 1 Initialization

  Sort the requests of $R = \{r(s_i, t_i) \mid i = 1, 2, \cdots, m\}$ with $t_1 \le t_2 \le \cdots \le t_m$.
  Set $W := \{w_j \mid j = 1, 2, \cdots, k\}$ and $S_j := \emptyset$ for $j = 1, 2, \cdots, k$.

Step 2 Finding how many requests that a wavelength can satisfy.

  Set $W' := W$.
  while $W' \ne \emptyset$ **do begin**
    Remove w_j from W'.
    Set $R' := R$.
    while $R' \ne \emptyset$ **do begin**
      Remove the first request $r(s_i, t_i)$ in R'.
      if w_j is available for $r(s_i, t_i)$ and
        $r(s_i, t_i)$ does not share a link with any request in S_j
      then Set $S'_j := S'_j \cup \{r(s_i, t_i)\}$.
    end-while
  end-while
  output $S' := \{S'_j \mid w_j \in W'\}$.

Step 3 Choosing the wavelength that can satisfy the most number of requests.

  Find w_j such that $|S'_j|$ is maximal, Let $S_j = S'_j$
  Set $R := R \setminus S_j$, $W := W \setminus \{w_j\}$ and $S := S \cup S_j$.
  return Step 2.

Fig. 2 illustrates the above algorithm applied to a simple instance. There are eight requests on a bus network of 16 nodes with three wavelengths, where wavelength w_1 is available on every link in the network while w_2 on every link between node 2 and node 7 and w_3 on every link between node 10 and node 15. Eight requests are labelled in the nondecreasing order of their right endpoints $R = \{r_i \mid i = 1, 2, \cdots, 8\}$. The algorithm will first choose w_1 which can satisfy four requests r_1, r_3, r_5 and r_7, and then neither w_2 nor w_3 can satisfy any request left. It then outputs the solution S_1 of size four. However, all eight requests can be satisfied if we assign w_1 to requests $\{r_2, r_4, r_6, r_8\}$, w_2 to requests r_1 and r_3, and w_3 to requests r_5 and r_7. Observe that the ratio of these two solutions is $1/2$. The following theorem shows that this is true for any instance.

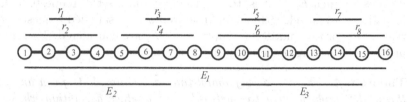

Fig. 2. Illustration of Algorithm A for the MSRS problem in buses

Theorem 3. *Given any instance of the MSRS problem in a bus network, Algorithm A can produce in polynomial time a solution whose size is at least half that of the optimal solution.*

Proof. Assume, without loss of generality, that the selected order of $S \setminus \{i\}$ is $S_1, S_2, \cdots S_k$. Let the maximum satisfiable request subset be $S^* = S_1^* \cup S_2^* \cup \cdots \cup S_k^*$, where all requests in S_i and S_i^* are assigned wavelength w_i. By using the same argument see e.g. in [3] we can show that the proposed algorithm finds the maximum satisfiable request subset S_1 in E_1. Thus we have $|S_1| \geq |S_1^*|$. Similarly, we can obtain $|S_2| \geq |S_2^* \setminus S_1| = |S_2^*| - |S_2^* \cap S_1|$. In general, we have

$$|S_i| \geq |S_i^* \setminus (\cup_{j=1}^{i-1} S_j)| = |S_i^*| - \sum_{j=1}^{i-1} |S_i^* \cap S_j|.$$

Taking the sum of the above inequality over $i = 1, 2, \cdots, k$ yields

$$|S| = \sum_{i=1}^{k} |S_i| \geq \sum_{i=1}^{k} |S_i^*| - \sum_{i=1}^{k-1} \left(\sum_{j=i+1}^{k} |S_j^* \cap S_i| \right)$$

$$\geq |S^*| - \sum_{i=1}^{k-1} |S^* \cap S_i| \geq |S^*| - \sum_{i=1}^{k} |S_i|.$$

Thus we have $|S| \geq |S^*|/2$. The proof is finished.

Observe that the argument in the above proof indeed is not dependent on the order how we choose the wavelengths (as we have specified algorithm A). For example, in Step 3 we can adopt a different greedy rule that each time we choose wavelength w_i such that the load of requests contained in E_i is minimized.

For the MSRS problem in tree networks under the case of uniform wavelength distribution, Wan et al [13] proposed a greedy algorithm that has a guaranteed approximation performance ratio of $1 - 1/e$. For the case of nonuniform wavelength distribution, we can use the same approach as for the case of bus networks. Again, let $E_i = \{e \mid w_i \in w(e), e \in E\}$ for each w_i. Then E_i can be considered as a forest consisting of some trees of $G(V, E)$ with wavelength w_i available on all edges in E_i. It is known that the MSRS problem in each E_i can be solved in polynomial time by using a primal-dual approach [7]. Thus we use this method k times for each E_i, that is, to find the maximum set of requests in R that can be satisfied by wavelength w_i, and then remove all satisfied requests from R and repeat this process until R is empty or all E_i are considered. Therefore, we can prove the following theorem in the same way.

Theorem 4. *There is a polynomial-time algorithm that, given any instance of the MSRS problem in a tree network, can produce a solution whose size is at least half that of the optimal solution.*

3.2   Ring and Tree of Rings Networks

For the MSRS problem in ring networks under the case of uniform wavelength distribution, Nomikos et al [9] proposed a recoloring algorithm that has a guaranteed approximation performance ratio of $2/3$. In this section we will show how to use this algorithm to deal with the case of nonuniform wavelength distribution.

Let us partition the set W of wavelengths into two subsets $W = W_1 \cup W_2$ such that every wavelength in W_2 is available on all links in the network. A simple idea is to first use wavelengths in W_1 to satisfy the most number of requests by using Algorithm A, and then use wavelengths in W_2 to satisfy the most number of requests left by using the recoloring algorithm and algorithm provided by Wan et al [13] for uniform wavelength distribution of ring networks respectively, and then take the best one. The algorithm is more formally stated below.

Algorithm B

Step 1   Apply Algorithm A to the MSRS problem with E_i for $w_i \in W_1$,
            Obtain the satisfiable request set $S_1 \subseteq R$.
Step 2   Apply the recoloring algorithm to the MSRS problem with $W = W_2$
            and $R := R \setminus S_1$, Obtain the satisfiable request set $S_2 \subseteq R \setminus S_1$.
            Apply Wan's Algorithm to the MSRS problem with $W = W_2$
            and $R := R \setminus S_1$, Obtain the satisfiable request set $S_2' \subseteq R \setminus S_1$.
Step 3   **Output** the better solution of $S := S_1 \cup S_2$ and $S' := S_1 \cup S_2'$.

Theorem 5. *Given any instance of the MSRS problem in a ring network, Algorithm B can produce in polynomial time a solution whose size is at least $1/2$ times that of the optimal solution.*

A *tree of rings* is a network that is obtained by interconnecting rings in tree structure such that any two rings share at most one node. Or more formally, it can be defined inductively as follows. (1) A single ring is a trees of rings; (2) If T is a trees of rings, then the graph obtained by adding a node-disjoint ring to

T and then identifying one node of the ring with one node of T is also a trees of rings; (3) No other graphs are trees of rings.

For the MSRS problem in trees of rings with uniform wavelength distribution, applying the technique proposed by Hochbaum and Pathria in [9] and the greedy approach of Erlebach [4] we can design an approximation algorithm with performance ratio of $1 - 1/e^{\frac{1}{4}}$. For the case of nonuniform wavelength distribution, we can just apply the same approach, as have used for bus and tree networks, to design an approximation algorithm with a guaranteed performance ratio.

Theorem 6. *There is a polynomial-time algorithm that, given any instance of the MSRS problem in a trees of rings, can produce a solution whose size is at least one fifth that of the optimal solution.*

3.3   General Networks

In this subsection we will focus on the MSRS problem in general networks. The approach to be used is the same as for special networks.

Let us first consider how to, given a set of paths in a network, find a maximum subset of edge-disjoint paths. A simple method due to Wan and Liu [14], called *Shortest-Path-First* (SPF), is as follows: Each time choose the path p of the shortest length (in terms of the number of links in the path), put into the current solution S and then remove all links in p from the network. Repeat this process until either all paths are considered or no request left whose links are all in the current network. This sequential method is greedy in the sense that if the shortest paths are chosen to be removed from the network, then more links will be left, and as a result more paths could survive. However, this method is not an exact algorithm, by which we mean that for some instances the algorithm can not find the optimal solutions.

Next just as before we use the above method k times for each E_i, that is, to find a set of requests in R that can be satisfied by wavelength w_i using SPF algorithm, and then remove all satisfied requests from R and repeat this process until R is empty or all E_i are considered. The algorithm is given below.

Algorithm C

Set $W := \{w_j \mid j = 1, 2, \cdots, k\}$ and $S_j := \emptyset$ for $j = 1, 2, \cdots, k$.
while $W \neq \emptyset$ **do begin**
    Remove the first wavelength w_j from W.
    while $R \neq \emptyset$ **do begin**
        Choose the shortest path $p(s_i, t_i)$ in R such that $p(s_i, t_i) \in E_j$.
        Set $S_j := S_j \cup \{r(s_i, t_i)\}$,
        Set $E_j := E_j \setminus \{p(s_i, t_i)\}$ and $R := R \setminus \{p(s_i, t_i)\}$
    end-while
end-while
output $S := S_1 \cup S_2 \cup \cdots \cup S_k$.

Theorem 7. *Given an instance of MSRS problem, let $d(E_j)$ be the average length of the paths of requests satisfied by wavelength w_j with respect to the optimal assignment, and $d = \max\{d(E_j) \mid j = 1, 2, \cdots, k\}$. Then the solution produced by Algorithm C has size at least $1/(1 + d)$ times that of the optimal solution.*

Proof. Let the maximum satisfiable request subset be $S^* = S_1^* \cup S_2^* \cup \cdots \cup S_k^*$, where all requests in S_j^* are assigned wavelength w_j. By using the same argument as in [14] we can show that the proposed algorithm finds the satisfiable request subset S_1 in E_1 satisfying $|S_1| \geq |S_1^*|/d(E_1)$. Similarly we can obtain $|S_2| \geq |S_2^* \setminus S_1|/d(E_2) = (|S_2^*| - |S_2^* \cap S_1|)/d(E_2)$. In general, we have

$$|S_j| \geq |S_j^* \setminus (\cup_{i=1}^{j-1} S_i)| = \Big(|S_j^*| - \sum_{i=1}^{j-1} |S_j^* \cap S_i|\Big)/d(E_j).$$

Summing up the above inequalities over $j = 1, 2, \cdots, k$ yields

$$|S| = \sum_{j=1}^{k} |S_j| \geq \frac{1}{d}\Big(\sum_{j=1}^{k} |S_i^*| - \sum_{j=1}^{k-1}\big(\sum_{i=j+1}^{k} |S_i^* \cap S_j|\big)\Big)$$

$$\geq \frac{1}{d}\Big(|S^*| - \sum_{j=1}^{k-1} |S_j|\Big) \geq \frac{1}{d}\Big(|S^*| - |S|\Big).$$

Thus we have $|S| \geq |S^*|/(d + 1)$. The proof is finished.

4   Conclusion

In this paper we have considered a general version of wavelength assignment problem under the wavelength constraint, and proposed some simple approximation algorithms with guaranteed performance ratios (see the following table). Our work is just the first step to study this problem, future work under this topic includes studying the problem for the case when routes of requests are not pre-scribed or wavelength conversion is allowed.

Table 1. Results on the maximum satisfiable request subset problem

	Uniform Distribution	Non-uniform Distribution
Buses	Polynomial-time solvable [17]	$\frac{1}{2}$-approximation
Rings	$\frac{2}{3}$-approximation [10]	$\frac{1}{2}$-approximation
Trees	$(1 - \frac{1}{e})$-approximation [14]	$\frac{1}{2}$-approximation
Trees of Rings	$1 - 1/e^{\frac{1}{4}}$-approximation [4]	$\frac{1}{5}$-approximation
General	$1 - 1/e^{\frac{1}{d}}$-approximation [14]	$\frac{1}{1+d}$-approximation

Acknowledgement

The authors would like to thank the reviewers for their constructive comments which greatly helped us to improve the presentation and quality of the paper.

References

1. U. Adamy, C. Ambuehl, R. S. Anand, and T. Erlebach, Call control in rings, *Lecture Notes in Computer Science*, Vol.2380 (2002), 788-799.
2. D. Banerjee and B. Mukherjee, A practical approach for routing and wavelength assignment in large wavelength-routed Optical Networks, *IEEE Journal of Selected Areas in Communications*, Vol.14 (1996), 903-908.
3. M. C. Carlisle and E. L. Lloyd, On the k-coloring of intervals, *Discrete Applied Mathematics*, Vol.59 (1995), 225-235.
4. T. Erlebach, Approximation Algorithms and Complexity Results for Path Problems in Trees of Rings, MFCS 2001, *Lecture Notes in Computer Science*, Vol.2136 (2001), Springer, 2001, 351-362.
5. T. Erlebach and K. Jansen, The maximum edge-disjoint paths problem in bidirected trees, *SIAM Journal on Discrete Mathematics*, Vol.14(3) (2001), 326-355.
6. T. Erlabach and K. Jansen, Maximizing the number of connections in optical tree networks, *Lecture Notes in Computer Science*, Vol.1533 (1998), 179-188.
7. N. Garg, V. V. Vazirani, and M. Yannakakis, Primal-dual approximation algorithms for integral flow and multicut in trees, *Algorithmica*, Vol.18 (1997), 3-20.
8. U. I. Gupta, D. T. Lee, and J. Y.-T. Leung, An optimal solution for the channel-assignment problem, *IEEE Transactions on Computers*, Vol.C-28(12) (1979), 807-810.
9. D. S. Hochbaum, A. Pathria, Analysis of the greedy approach in problems of maximum k-coverage, *Naval Research quarterly*, Vol.45 (1998), 615-627.
10. C. Nomikos, A. Pagourtzis, and S. Zachos, Satisfying a maximun number of prerouted requests in all-optical rings, *Computer Networks*, Vol.42 (2003), 55-63.
11. R. Ramaswami and K. N. Sivarajan, Routing and wavelength assignment in all-optical networks, *IEEE/ACM Transactions on Networking*, Vol.3(5)(1995), 489-500.
12. A. Sridharran and K. N. Sivarajan, Blocking in all-optical networks, *IEEE/ACM Transactions on Networking*, Vol.12(2) (2004), 384-397.
13. M. Slusarek, A coloring algorithm for interval graphs, *Lecture Notes in Computer Science*, Vol.379 (1989), 471-480.
14. P.-J. Wan and L.-Liu, Maximal throughput in wavelength-routed optical networks, *DIMACS Series in Discrete Mathematics and Theoretical Computer Science*, Vol.48 (1998), 15-27.
15. J. Wang, B. Chen, and R. N. Uma, Dynamic wavelength assignment for multicast in all-optical WDM networks to maximize the network capacity, *IEEE Journal of Select Areas in Communications*, Vol.21(8) (2003), 1274-1284.
16. G. Wilfong and P. Winkler, Ringrouting and wavelength translation, *Proceedings of the 9th Annual ACM-SIAM Symposium on Discrete Algorithms*, 1998, 333-341.
17. M.Yannakakis and F.Gavril, The maximum k-colorable subgraph problem for chordal graphs, *Information Processing Letters*, Vol.24 (1987), 133-137.
18. X. Zhang and C. Qiao, Wavelength assignment for dynamic traffic in multifiber WDM networks, *Proceedings of IEEE International Conference on Computer Communication and Networks*, 1998, 479-485.

An Approximation Algorithm for a Facility Location Problem with Inventories and Stochastic Demands

Adriana F. Bumb and Jan-Kees C.W. van Ommeren

Faculty of Electrical Engineering, Mathematics and Computer Science,
University of Twente, PO Box 217, 7500 AE Enschede, The Netherlands
{a.f.bumb, j.c.w.vanommeren}@ewi.utwente.nl

Abstract. In this article we propose, for any $\epsilon > 0$, a $2(1+\epsilon)$-approximation algorithm for a facility location problem with stochastic demands. At open facilities, inventory is kept such that arriving requests find a zero inventory with (at most) some pre-specified probability. The incurred costs are the expected transportation costs from the demand points to the facilities, the operating costs of the facilities and the investment in inventory.

Keywords: approximation algorithms, stochastic facility location.

AMS Classification: 68W25, 90B06, 60K30.

1 Introduction

Facility location problems have been extensively studied in the OR literature. In a facility location problem, we are given a set of demand points and a set of location where facilities may be opened. The goal is to decide at which location to open facilities and how to assign demand points to facilities such that the total cost of opening facilities and of connecting demand points to facilities is minimized. Variants of this problem can be formulated if one imposes requirements on the set of open facilities or on the assignment of demand points to facilities [1]. Examples of such requirements are a maximum number of facilities that may be opened, a maximum demand that may be served by a facility, or a maximum travel distance from a demand point to an open facility. The facility location problem with its variants has proved to be a very useful tool in modeling many network design or location problems, such as location of plants or warehouses [1, 2] and placement of caches [3].

In this paper we study a variant of the facility location problem where at demand points a stochastic number of requests for items is generated. At open facilities, inventory is kept and, if possible, requests for items are fulfilled immediately. However, since the number of requests is random, it may occur that there is no inventory at the arrival of a request and the request has to be cancelled.

N. Megiddo, Y. Xu, and B. Zhu (Eds.): AAIM 2005, LNCS 3521, pp. 330–339, 2005.
© Springer-Verlag Berlin Heidelberg 2005

An arbitrary request arriving at a facility, should only have a (pre-specified) small probability of being lost. We are interested in the relationship between the problem with stochastic demands and inventory and known facility location problems, in particular from the perspective of approximation algorithms.

We will call a ρ-approximation algorithm a polynomial time algorithm that always finds a feasible solution with **objective function value** within ρ times the optimum. The value ρ is called the *performance (approximation) guarantee* of the algorithm.

The majority of facility location problems for which approximation algorithms are known, are deterministic. The simplest version of a facility location problem, *the metric uncapacitated facility location problem* (UFLP), that is the facility location problem with no restrictions on the facilities or the assignment of demand points and with the transportation costs being a metric, is known to be NP-hard. If the transportation costs are unrestricted, approximating the UFLP is as hard as approximating set cover, and therefore cannot be done better than $O(\log n)$ factor, unless **NP** $\subseteq$ **P̃**. In this article, we assume, for all the facility locations mentioned, that the transportation costs form a metric. The currently known best performance guarantee for the UFLP is 1.52, due to Mahdian, Ye and Zhang [4]. Guha and Khuller [5] and Sviridenko [6] have proved that a better factor than 1.463 for the UFLP is not possible unless **NP** $\subseteq$ **P̃**.

The problem in which each facility has a certain capacity, but more facilities may be opened at a location if the demand exceeds the capacity of one facility, is known as the *soft capacitated* facility location problem. The best approximation algorithm for this problem has an approximation ratio of 2 and was proposed by Mahdian, Ye and Zhang in [7]. In [8] the authors propose a 1.861-approximation algorithm for the variant in which the cost of facilities are concave functions of the number of demand points served. For the *hard capacitated* facility location problem with splittable demands, where each facility has a certain capacity, only one facility may be open at a location and a demand point may be served by several locations, the best approximation algorithm is due to Zhang, Chen and Ye [9], and achieves an approximation ratio between $3 + 2\sqrt{2} - \epsilon$ and $3 + 2\sqrt{2} + \epsilon$, for any given constant $\epsilon > 0$.

Stochastic facility location problems (problems where the demand is stochastic or/and the service offered by facilities is of stochastic nature) were mainly treated in the OR literature [10, 11, 12, 13, 14]. Several heuristics have been proposed to obtain solutions for these problems. To the best of our knowledge, the first approximation algorithm for a stochastic facility location problem was proposed by Ravi and Sinha in [15] and improved by Mahdian in [16]. The latest algorithm is based on the primal-dual technique and has a 3-approximation guarantee. Their approach is scenario-based, i.e. in each scenario all the data are known, including the probability with which each scenario takes place.

The paper is organized as follows. In section 2 we describe the stochastic facility location problem in more detail and formulate it such that it can be reduced to a soft capacitated facility location problem. Based on this reduction, we then propose in Section 3, a $2(1+\epsilon)$-approximation algorithm for our problem. We conclude the section by showing that the same ideas can be applied for designing approximation algorithms for a larger class of problems. Finally, we present some conclusions and remarks on the stochastic facility location problem we have analyzed.

2   The Facility Location Problem with Stochastic Demands

In this section we describe in more detail the stochastic facility location problem in which we are interested. There is a set of demand points $D, |D| = N$ at which requests are generated, and a set of locations, $F, |F| = K$, where facilities may be opened. We assume that the requests at a demand point $j \in D$ are generated according to a Poisson process, independent of the processes at other demand points in D. At each open facility an inventory is kept such that an arriving request finds a zero inventory (and is lost), with probability at most α. We then say that $(1 - \alpha)$ is the *fill rate* of the system . The inventories at the open facilities are restored only at fixed points in time and the period between two such points is called a *reorder period*. The holding cost per unit of inventory at an open facility $i \in F$ is c_i and the cost of keeping a facility open at location $i \in F$ during a reorder period is f_i. The transportation cost per unit of demand from facility $i \in F$ to demand point $j \in D$ is c_{ij}. We assume that the transportation costs are proportional to the distances and form a metric.

The goal is to decide at which locations to open facilities, the level of inventory to be installed at each open facility and how to assign demand points to facilities such that the fill rate is at least $1 - \alpha$ and the average total cost per reorder period is minimized.

Let X_j denote the number of generated requests at demand point j during a reorder period and let $\lambda_j = E(X_j)$. Denote by V_i the inventory order up to level at facility $i \in F$, i.e. the inventory level at the beginning of a reorder period. Let y_i, respectively x_{ij}, be $0 - 1$ variables indicating if a facility at location $i \in F$ is open, respectively if demand point $j \in D$ is assigned to a facility $i \in F$. The facility location problem with stochastic demands given above, is fully described by the following integer program:

$$\min \sum_{i \in F}(f_i + c_i V_i)y_i + \sum_{j \in D}\sum_{i \in F}\lambda_j c_{ij}x_{ij} \qquad (1)$$

$$\text{s.t. } x_{ij} \le y_i, \qquad i \in F, \quad j \in D, \qquad (2)$$

$$\sum_{i \in F} x_{ij} = 1, \qquad j \in D, \qquad (3)$$

$$P\left(\begin{matrix}\text{an arbitrary arriving requests at} \\ \text{facility } i \text{ is lost}\end{matrix}\right) \le \alpha, \quad i \in F, \qquad (4)$$

$$x_{ij}, y_i \in \{0, 1\}, \qquad i \in F, \quad j \in D. \qquad (5)$$

The first term in the objective function includes the costs for keeping facilities open and for the maximum inventory at the facilities during a reorder period, while the second term is the expected transportation cost during such a period. Constraints (2), (3) and (5) guarantee that each demand point is assigned to exactly one open facility and constraints (4) guarantee that the fill rate attained at each open location will be at least $1 - \alpha$.

Next we will give an equivalent formulation of constraints (4). Let $\tilde{X}_i$ be the total demand assigned to location i. Clearly, $\tilde{X}_i = \sum_{j \in D} x_{ij} X_j$. Since the requests generated at demand points during reorder periods are independent Poisson distributed random variables, $\tilde{X}_i$ has a Poisson distribution with mean $E(\tilde{X}_i) = \sum_{j \in D} x_{ij} \lambda_j$. From the theory of regenerative processes (see e.g. [17]), it follows that for location i, the following holds:

$$P\left(\begin{array}{l}\text{an arbitrary arriving requests at}\\\text{facility } i \text{ is lost}\end{array}\right) = \frac{E((\tilde{X}_i - V_i)^+)}{E(\tilde{X}_i)}, \tag{6}$$

where $(a)^+ = \max(0, a)$. Condition (4) can be rewritten as

$$E((\tilde{X}_i - V_i)^+) \le \alpha E(\tilde{X}_i). \tag{7}$$

For a Poisson distributed random variable Y with $E(Y) = \lambda$, define the inventory $V_\alpha(\lambda)$ by

$$V_\alpha(\lambda) = \min\{n | E((Y - n)^+) \le \alpha\lambda\}. \tag{8}$$

Using (7) and (8), our problem can be reformulated as

(P)

$$\min \sum_{i \in F}(f_i + c_i V_\alpha(\sum x_{ij}\lambda_j))y_i + \sum_{j \in D}\sum_{i \in F}\lambda_j c_{ij} x_{ij}$$

$$\text{s.t. } x_{ij} \le y_i, \qquad i \in F, \quad j \in D,$$

$$\sum_{i \in F} x_{ij} = 1, \qquad j \in D,$$

$$x_{ij}, y_i \in \{0, 1\}, \qquad i \in F, \quad j \in D.$$

Note that constraints (4) have moved into the objective function. This will enable us to further reduce the problem to a soft capacitated facility location problem, for which approximation algorithms are known (see e.g. [7]). In the remainder of the paper we will present this reduction in detail.

3 A $2(1+\epsilon)$-Approximation Algorithm for the Stochastic Facility Location Problem

For a facility location problem (P), an instance $\mathcal{I}$ and a feasible solution $\mathcal{S}$ we denote by $cost_{F,\mathcal{I}(P)}(\mathcal{S})$ the cost of opening facilities and by $cost_{T,\mathcal{I}(P)}(\mathcal{S})$ the transportation cost incurred by $\mathcal{S}$. For the sake of simplicity, we will omit to mention the instance.

Definition 1. *We call a polynomial time reduction $\mathcal{R}$ from facility location problem P_1 to P_2 a (σ_F, σ_T)-reduction if $\mathcal{R}$ maps an instance $\mathcal{I}$ of P_1 to an instance $\mathcal{R}(\mathcal{I})$ of P_2 and it has the following properties:*

a) For any feasible solution S_1 for the instance $\mathcal{I}$ of P_1 there is a corresponding solution S_2 for the instance $\mathcal{I}$ of P_2 with

$$cost_{F,P_2}(S_2) \leq \sigma_f cost_{F,P_1}(S_1),$$

and

$$cost_{T,P_2}(S_2) \leq \sigma_c cost_{T,P_1}(S_1).$$

b) For any feasible solution S_2 for the instance $\mathcal{R}(\mathcal{I})$ of P_2, there is a feasible solution S_1 for the instance $\mathcal{I}$ of P_1 with

$$cost_{F,P_1}(S_1) + cost_{T,P_1}(S_1) \leq cost_{F,P_2}(S_2) + cost_{T,P_2}(S_2).$$

Definition 2. *An algorithm is called an (α, β)-approximation algorithm for a facility location problem (P), if for any instance $\mathcal{I}$ of (P), and for any solution S for $\mathcal{I}$ the cost of the solution found by the algorithm is at most $\alpha cost_{F,P}(S) + \beta cost_{T,P}(S)$.*

Remark 1. Note that combining a (σ_F, σ_T)-reduction from P_1 to P_2 and an (α, β)-approximation algorithm for P_2 gives an $(\alpha\sigma_F, \beta\sigma_T)$-approximation algorithm for P_1. Moreover, the approximation guarantee of the algorithm for P_1 is $\max\{\alpha\sigma_F, \beta\sigma_T\}$.

The construction of a $2(1 + \epsilon)$-approximation algorithm for $(\mathbf{P})$, consists of several steps. First we will study the inventory function $V_\alpha(\lambda)$ given by (8). Based on it's properties, we propose a $(2, 1)$-reduction of $(\mathbf{P})$ to a soft capacitated facility location problem, named $(\mathbf{SP_2})$. Finally, we describe a refined soft capacitated problem, $(\mathbf{SP_{1+\epsilon}})$ to which $(\mathbf{P})$ can be $(1 + \epsilon, 1)$-reduced and show that this gives $2(1 + \epsilon)$-approximation algorithm for $(\mathbf{P})$.

Lemma 1. *The function $V_\alpha(\lambda)$ satisfies*

$$V_\alpha(\lambda_1 + \lambda_2) \leq V_\alpha(\lambda_1) + V_\alpha(\lambda_2).$$

Proof. Suppose that two independent Poisson streams with rate λ_1, respectively λ_2, arrive at a location i and that the inventory level at location i is $V_\alpha(\lambda_1) + V_\alpha(\lambda_2)$. Let Y_1 and Y_2 be the number of arrivals in the first, respectively in the second stream. Since

$$(Y_1 + Y_2 - (V_\alpha(\lambda_1) + V_\alpha(\lambda_2)))^+ \leq (Y_1 - V_\alpha(\lambda_1))^+ + (Y_2 - V_\alpha(\lambda_2))^+,$$

it is readily seen that

$$\mathrm{E}(Y_1 + Y_2 - (V_\alpha(\lambda_1) + V_\alpha(\lambda_2)))^+$$
$$\leq \mathrm{E}(Y_1 - V_\alpha(\lambda_1))^+ + \mathrm{E}(Y_2 - V_\alpha(\lambda_2))^+ \leq \alpha(\lambda_1 + \lambda_2).$$

Hence, $V_\alpha(\lambda_1 + \lambda_2) \leq V_\alpha(\lambda_1) + V_\alpha(\lambda_2)$.   □

Remark 2. Note that $V_\alpha(\lambda)$ is a step function, thus not concave. Therefore we cannot directly use the procedure proposed in Mahdian and Pal [18], for solving the facility location problem with concave facility cost functions. Moreover, not even the length of the steps is increasing as function of the height, where the length of a step at level n is defined as $\sup\{\lambda|V_\alpha(\lambda) = n\} - \inf\{\lambda|V_\alpha(\lambda) = n\}$. For example, numerical experiments show that, when $\alpha = 0.1$, the length of the steps is increasing up to level 40 and decreasing above this level.

Next we present a reduction of (**P**) to a soft capacitated facility location problem, which we denote by (**SP₂**). The demand points, their requests and facility locations are the same as in problem (**P**). Let $M = \lceil \log_2(V_\alpha(\sum_{j \in D} \lambda_j)) \rceil$ and let $L = \{1, \cdots, M\}$. We define M types of facilities with capacities $u_\ell = \max\{\lambda|V_\alpha(\lambda) \le 2^\ell\}$, respectively. A facility of type l at location i is denoted by (i, l) and has corresponding cost $f_{il} = f_i + c_i 2^\ell$. At each location $i \in F$, M facilities may be opened.

Let the 0-1 variables y_{il}, x_{ilj}, indicate whether a facility of type l is opened at location i, respectively whether demand point j is assigned to facility (i, l). Then, (**SP₂**) can be formulated as the integer program:

$$\min \sum_{j \in D} \sum_{i \in F} \sum_{\ell \in L} \lambda_j c_{ij} x_{i\ell j} + \sum_{i \in F} \sum_{\ell \in L} f_{i\ell} y_{i\ell}$$

(**SP₂**)

$$\text{s.t.} \sum_{j \in D} \lambda_j x_{i\ell j} \le u_\ell y_{i\ell}, \qquad i \in F, \quad \ell \in L, \qquad (9)$$

$$\sum_{i \in F} \sum_{\ell \in L} x_{i\ell j} = 1, \qquad j \in D, \qquad (10)$$

$$x_{i\ell j}, y_{i\ell} \in \{0, 1\}, \qquad i \in F, \quad j \in D, \quad \ell \in L. \qquad (11)$$

Constraints (9), (10) and (11) insure that each demand point is assigned to one open facility and that no more than demand u_ℓ is assigned to a facility of type ℓ.

Remark 3. Note that although formulated as a hard capacitated facility location problem ($y_{il} \in \{0,1\}$), problem (**SP₂**) is a soft capacitated problem. Suppose that we relax the y variables to be integer. Consider first a $k < M$. The optimal solution of the relaxed version will not choose to open two facilities of type k at a location, since opening a facility of type $k + 1$ is cheaper and has, at least, the same capacity as two facilities of type k. Since one facility of type M can handle all the demand, there will be always at most one facility of type M open in the optimal solution of the relaxed version of (**SP₂**). Thus, (**SP₂**) is a soft capacitated facility location problem.

In the following lemma we describe a $(2,1)$-reduction of (**P**) to (**SP₂**).

Lemma 2.
*(i) For each feasible solution $(\tilde{x}, \tilde{y})$ of (**P**) with facility cost $\text{cost}_{F,\mathbf{P}}(\tilde{x}, \tilde{y})$ and transportation cost $\text{cost}_{T,\mathbf{P}}(\tilde{x}, \tilde{y})$ there exists a feasible solution (x, y) of (**SP₂**) with $\text{cost}_{F,\mathbf{SP_2}}(x, y) \le 2\text{cost}_{F,\mathbf{P}}(\tilde{x}, \tilde{y})$ and $\text{cost}_{T,\mathbf{SP_2}}(x, y) = \text{cost}_{T,\mathbf{P}}(\tilde{x}, \tilde{y}).$*

(ii) For each feasible solution (x, y) of (**SP₂**), *there exists a feasible solution* $(\tilde{x}, \tilde{y})$ *of* (**P**) *of lower cost.*

(iii) There exists a $(2, 1)$-reduction of (**P**) *to* (**SP₂**) .

Proof. (i) Consider a solution $(\tilde{x}, \tilde{y})$ of (**P**). For $i \in F$ with $\tilde{y}_i = 1$ and $\ell \in L$ define $\ell_i = \min\{n| \sum_{j \in D} \tilde{x}_{ij} \lambda_j \leq u_n\}$, set $y_{i\ell} = 1$ for $\ell = \ell_i$, set $y_{i\ell} = 0$ otherwise and set $x_{i\ell j} = \tilde{x}_{ij} y_{i\ell}$ for $j \in D$. For each $i \in F$ with $\tilde{y}_i = 0$, set $x_{i\ell j} = y_{i\ell} = 0$ for $j \in D$ and $\ell \in \{1, \cdots, M\}$ and define $\ell_i = 1$. It can readily be seen that (x, y) is a feasible solution of (**SP₂**) with associated costs

$$\text{cost}_{T,\textbf{SP}_2}(x, y) = \sum_{i \in F} \sum_{j \in D} \sum_{\ell \in L} \lambda_j c_{ij} x_{i\ell j} = \sum_{i \in F} \sum_{j \in D} \lambda_j c_{ij} \tilde{x}_{ij}$$

$$= \text{cost}_{T,\textbf{P}}(\tilde{x}, \tilde{y})$$

and

$$\text{cost}_{F,\textbf{SP}_2}(x, y) = \sum_{i \in F} \sum_{\ell \in L} f_{i\ell} y_{i\ell} = \sum_{i \in F} (f_i + 2^{\ell_i}) y_{i\ell_i}$$

$$\leq 2\text{cost}_{F,\textbf{P}}(\tilde{x}, \tilde{y}),$$

where the inequality follows from the definitions of ℓ_i and u_n.

(ii) For each feasible solution (x, y) of (**SP₂**), define the vector $(\tilde{x}, \tilde{y})$ by $\tilde{x}_{i,j} = \max_{\ell \in \{1, \cdots, M\}} \{x_{i\ell j}\}$ and $\tilde{y}_i = \max_{\ell \in \{1, \cdots, M\}} \{y_{i\ell}\}$. Clearly, $(\tilde{x}, \tilde{y})$ is a feasible solution for (**P**). Moreover, from Lemma 1 follows that $V_\alpha(\sum_{j \in D} \tilde{x}_{ij} \lambda_j) \leq \sum_\ell v_\ell y_{i\ell}$ and so $(\tilde{x}, \tilde{y})$ has a lower cost than the one incurred by (x, y) for (**SP₂**).

(iii) Follows from (i) and (ii) of this lemma.   □

In the following, we prove that one can obtain a $(1 + \epsilon, 1)$-reduction between (**P**) and a slightly modified version of (**SP₂**) by the same reasoning as in Lemma 2. We define this modified version (**SP₁₊ₑ**) as follows.

Define for $\epsilon > 0$ the integer sequence $\tilde{v}_{0,0} = 0$; $v_{m0} = \lfloor (1+\epsilon)(1+v_{m-1,0}) \rfloor$ and $v_{mk} = 2^k v_{m0}$ for $m = 1, 2, \cdots$ and $k = 0, 1, \cdots$. Next, define the integer sequence $v_0 = 0$ and $v_\ell = \min\{\tilde{v}_{mk} > v_{\ell-1}|m = 1, 2, \cdots$ and $k = 0, 1, \cdots\}$ for $\ell = 1, 2, \cdots$ and define $M = \min\{\ell|v_\ell \geq V_\alpha(\sum_{j \in D} \lambda_j)\}$. Since $\tilde{v}_{m0} \geq (1+\epsilon)v_{m-1,0}$, it is easy to find that, for $\epsilon \in (0, 1)$,

$$M \leq \lceil \log_{(1+\epsilon)}(V_\alpha(\sum_{j \in D} \lambda_j)) \rceil \lceil \log_2(V_\alpha(\sum_{j \in D} \lambda_j)) \rceil \leq \frac{4}{\epsilon} \lceil \log_2(V_\alpha(\sum_{j \in D} \lambda_j)) \rceil^2.$$

Furthermore, from the construction of the sequence v_ℓ, we see that $(1 + v_\ell) \leq v_{\ell+1} \leq (1+\epsilon)(1+v_\ell)$. Consider a facility location problem with the same demand points, requests and facility locations as in problem (**P**). At each location $i \in F$, M facilities may be opened, $(i, 1), ...(i, M)$, of costs $f_i + c_i v_\ell$ and capacities $u_\ell = \max\{\lambda|V_\alpha(\lambda) \leq v_\ell\}$.

Let the 0-1 variables y_{il}, x_{ilj}, indicate whether a facility of type l is opened at location i, respectively whether demand point j is assigned to facility (i, l). Then, (**SP₁₊ₑ**) can be formulated as an integer program similar to (**SP₂**).

As in Remark 3, we note that although formulated as a hard capacitated facility location problem, $(\mathbf{SP_{1+\epsilon}})$ is in fact a soft capacitated facility location problem. In order to show this, we prove that, even if we allow more facilities of the same type to be opened at a location, at most one will be opened in the optimal solution. Assume that in the optimal solution, at least one facility of type k at location i is opened. If the cost of facility (i, k) exceeds $f_i + c_i V_\alpha(\lambda)/2$, then opening facility (i, M) (which can handle all demands) is cheaper than opening two facilities (i, k). If the costs of facility (i, k) equals $f_i + c_i v_k$ with $v_k \leq V_\alpha(\lambda)/2$, we see, by the definition of the sequence v_ℓ, that there is also a facility (i, k') with cost $f_i + 2c_i v_k$. By Lemma 1, the capacity of a type k' facility is at least twice the capacity of a type k facility. Hence, in the optimal solution of the relaxed problem of $(\mathbf{SP_{1+\epsilon}})$, at every location at most one facility of type k is opened. Thus, $(\mathbf{SP_{1+\epsilon}})$ is a soft capacitated facility location problem.

Lemma 3. *For any $\epsilon > 0$, the problem $(\mathbf{P})$ can be $(1+\epsilon, 1)$-reduced to $(\mathbf{SP_{1+\epsilon}})$.*

Proof. We follow the proof of Lemma 2. Consider a feasible solution $(\tilde{x}, \tilde{y})$ of $(\mathbf{P})$ and construct a feasible solution (x, y) of $(\mathbf{SP_{1+\epsilon}})$ as follows. Open facility (i, ℓ) at location i only if $\sum_{j \in D} \tilde{x}_{ij} = 1$ and $\ell = \min\{n | \sum_{j \in D} \tilde{x}_{ij}\lambda_j \leq u_n\}$. Since the inventory levels are discrete and $\sum_{j \in D} \tilde{x}_{ij}\lambda_j > u_{\ell-1}$, the inventory at location i satisfies $V_\alpha(\sum_{j \in D} x_{ij}\lambda_j) \geq 1 + v_{\ell-1}$ and therefore the cost of opening facilities in $(\mathbf{SP_{1+\epsilon}})$ is at most $(1 + \epsilon)$ times the facility costs in $(\mathbf{P})$.

Now consider a solution (x, y) of $(\mathbf{SP_{1+\epsilon}})$ and construct a corresponding solution $(\tilde{x}, \tilde{y})$ of $(\mathbf{P})$ by $\tilde{x}_{i,j} = \max_{\ell \in \{1, \cdots, M\}}\{x_{i\ell j}\}$ and $\tilde{y}_i = \max_{\ell \in \{1, \cdots, M\}}\{y_{i\ell}\}$. As in Lemma 2, one can show that $(\tilde{x}, \tilde{y})$ is a feasible solution with the same transportation cost as the one incurred by (x, y) and with less opening facility cost than the one incurred by (x, y). □

Theorem 1. *There is a $2(1+\epsilon)$-approximation algorithm for the facility location problem with stochastic demands $(\mathbf{P})$.*

Proof. Problem $(\mathbf{SP_{1+\epsilon}})$ is a soft capacitated facility location problem with general demands. For the soft capacitated facility location problem with unit demands, a $(2,2)$-approximation algorithm was proposed in [7]. It can easily be shown that their analysis also applies for general demands, thus implying a $(2,2)$-approximation algorithm for $(\mathbf{SP_{1+\epsilon}})$. The existence of a $(2,2)$-approximation algorithm for $(\mathbf{SP_{1+\epsilon}})$, implies, by Lemma 3 and Remark 1, the existence of a $2(1+\epsilon)$-approximation algorithm for the stochastic facility location problem $(\mathbf{P})$. □

Generalization. At the basis of our algorithm lies the property that, for two demand points j and j', with demand λ_j, respectively $\lambda_{j'}$, the inventory which has to be installed at a facility satisfies $V_\alpha(\lambda_j + \lambda_{j'}) \leq V_\alpha(\lambda_j) + V_\alpha(\lambda_{j'})$, i.e., it is more profitable to look at the joint demand than to treat the demands separately. It is easy to see that the same analysis holds for the metric UFLP with the cost of opening facilities depending on the amount served by a facility and satisfying $f_i(\lambda_j + \lambda_{j'}) \leq f_i(\lambda_j) + f_i(\lambda_{j'})$, for each $i \in F$ and $j, j' \in D$. Clearly, concave facility costs have this property.

Remark 4. The same technique can also be used for the following version of the facility location problem with stochastic demands: at facilities an arbitrary number of servers can be placed, which all work at equal speed. At each facility, there is an upperbound on the expected waiting time of a customer. The incurred costs are the transportation costs and the facility costs; the cost of a facility is the sum of the opening cost and the cost for installing servers, which is linear in the number of installed servers.

We model a facility as an $M/M/k$ queue, that is a queue with k servers and exponential interarrival and service times. Without loss of generality, we assume that the expected service time is 1. Let $WT(M_\lambda/M/k)$ denote the expected waiting time at such a queue with arrival rate λ. At an open facility i with arrival rate Λ_i and ki servers, the constraint on the waiting time then is $WT(M_{\Lambda_i}/M/ki) \leq \tau$ for some pre-specified τ. An explicit expression for this expectation can be found in e.g. [19], page 71 Define $N_\tau(\lambda) = \min\{k | WT(M_\lambda/M/k) \leq \tau\}$. It can be shown that $N_\tau(\lambda_1 + \lambda_2) \leq N_\tau(\lambda_1) + N_\tau(\lambda_2)$. Thus, applying a similar reduction as the one described in this section, one obtains a $2(1 + \epsilon)$-approximation algorithm for this problem as well.

4   Conclusions

In this paper we have introduced a facility location problem with inventory and stochastic demands. We proposed a $2(1 + \epsilon)$-approximation algorithm for this model by giving both a $(1 + \epsilon, 1)$-reduction to a soft capacitated facility location problem with general demands and a $(2, 2)$-approximating algorithm for this soft capacitated facility location problem. The same analysis is applied for approximating more general problems.

References

1. G. Cornuejols, G.L. Nemhauser and L.A. Wolsey, The uncapacitated facility location problem, in: P. Mirchandani and R. Francis (Eds.), Discrete Location Theory, John Wiley and Sons, New York, 1990, pp. 119-171.
2. D. Shmoys, E. Tardos, K. Aardal. Approximation algorithms for facility location problems, in: Proceedings of the 29th ACM Symposium on Theory of Computing, 1997, 265-274.
3. S. Guha, A. Meyerson, K. Munagala, A constant factor approximation algorithm for the fault-tolerant facility location problem. J. Algorithms 48(2) (2003) 429-440.
4. M. Mahdian, Y. Ye, J. Zhang, A 1.52 approximation algorithm for the uncapacitated facility location problem, in: Proceedings of the 5th International Workshop on Approximation Algorithms for Combinatorial Optimization, Springer-Verlag LNCS Vol 2462, 2002, 229-242.
5. S. Guha, S. Khuller, Greedy strikes back: Improved facility location algorithms, Journal of Algorithms 31(1) (1999) 228-248.
6. M. Sviridenko. Personal communication. Cited in S. Guha, Approximation algorithms for facility location problems, PhD thesis, Stanford, 2000, (Downloadable from website http://Theory.Stanford.EDU/~sudipto).

7. M. Mahdian, Y. Ye, J. Zhang, A 2-Approximation Algorithm for the Soft-Capacitated Facility Location Problem, RANDOM-APPROX 2003, 129-140
8. M.G. Hajiaghayi, M. Mahdian, V.S Mirrokni, The facility location problem with general cost functions. Networks 42(1) (2003) 42-47.
9. J. Zhang, B. Chen, Y. Ye, A Multi-exchange Local Search Algorithm for the Capacitated Facility Location Problem, IPCO 2004, 219-233
10. R. Batta, A queuing location model with expected service time dependent queuing disciplines, European Journal of Operational Research 39 (1989) 192-205.
11. O. Berman, R. Larson and S. Chiu, Optimal server location on a network operating as a M/G/1 queue, Operations Research 12 (1985) 746-771.
12. O. Berman and K. Sapna, Optimal control of service for facilities holding inventory, Computers & Operations Research 28 (2001) 429-441,
13. V. Marianov, D. Serra, Location-allocation of multiple-server service centers with constrained queues or waiting times, Annals of Operations Research 111 (2002) 35-50.
14. Q. Wang, R. Batta, C. Rump, Algorithms for a facility location problem with stochastic customer demand and immobile servers. Annals of Operations Research 111 (2002) 17-34.
15. R. Ravi, A. Sinha, Hedging Uncertainty: Approximation Algorithms for Stochastic Optimization Problems, IPCO 2004, 101-115.
16. M. Mahdian, Facility Location and the Analysis of Algorithms through Factor-Revealing Programs, Ph.D. Thesis, MIT, June 2004. (available at http://www-math.mit.edu/~mahdian/phdthesis.pdf).
17. W.L. Smith, Regenerative stochastic processes, Proc. Roy Soc. Ser. A 232 (1955) 6-31.
18. M. Mahdian, M. Pal, Universal Facility Location, ESA 2003, 409-421
19. D. Gross and C.M. Harris, Fundamentals of Queueing Theory, 3d ed., John Wiley & Sons, Inc. (1998).

Dynamically Updating the Exploiting Parameter in Improving Performance of Ant-Based Algorithms

Hoang Trung Dinh[1], Abdullah Al Mamun[1], and Hieu T. Dinh[2]

[1] Dept. of Electrical & Computer Engineering,
National University of Singapore, Singapore 117576
{hoang.dinh, eleaam}@nus.edu.sg
[2] Dept. of Computer Science, Faculty of Technology,
Vietnam National University of Hanoi
hieudt@vnuh.edu.vn

Abstract. The utilization of pseudo-random proportional rule to balance between the exploitation and exploration of the search process was shown in Ant Colony System (ACS) algorithm. In ACS, this rule is governed by a parameter so-called exploiting parameter which is always set to a constant value. Besides, all ACO-based algorithm either omit this rule or applying it with a fixed value of the exploiting parameter during the runtime of algorithms. In this paper, this rule is adopted with a simple dynamical updating technique for the value of that parameter. Moreover, experimental analysis of incorporating a technique of dynamical updating for the value of this parameter into some state-of-the-art Ant-based algorithms is carried out. Also computational results on Traveling Salesman Problem benchmark instances are represented which probably show that Ant-based implementations with local search procedures gain a better performance if the dynamical updating technique is used.

Keywords: Ant Colony Optimization, Ant System, Combinatorial Optimization Problem, Traveling Salesman Problem.

1 Introduction

Ant Colony Optimization (ACO) is a metaheuristic inspired by the foraging behavior of real ants. It has applied to combinatorial optimization problems and been able to find fruitfully approximate solutions to them. Examples of combinatorial optimization problems have successfully been tackled by ACO-based algorithms are Traveling Salesman Problem (TSP), Vehicle Routing Problem (VRP), Quadratic Assignment Problem (QAP).

ACO was started out at the time the algorithm *Ant Systems* (AS) was first proposed to solve TSP by Colorni, Dorigo and Maniezzo [3]. Several variants of AS such as Ant Colony System (ACS) [8], Max-Min Ant System (MMAS) [11], Rank-based Ant System (RAS) [2], and Best-Worst Ant System (BWAS)

N. Megiddo, Y. Xu, and B. Zhu (Eds.): AAIM 2005, LNCS 3521, pp. 340–349, 2005.

[4], were then suggested. Claimed by empirical supports, performance of most of those variants is over that of AS. In addition, ACS and MMAS are now counted as two of the most successful candidates among them. Recently, ACO has been extended to a full discrete optimization metaheuristic by Dorigo and Di Caro [6].

In ACS, a state transition rule, which is different from that in AS, namely *pseudo-random proportional rule* playing an important role in the improvement of the solution quality for ACS, is used. This rule can be regarded as an effective technique of the trade-off between the exploitation and exploration of the search process in ACS. In this rule, a parameter of notion q_0 which is henceforth called *exploiting parameter* defines the trade-off exploitation-based exploration. However, in all Ant-based implementations for TSP, this rule has been either omitted or applied with a constant value of q_0. Instances of such implementations are ACS, MMAS, RAS, and BWAS.

More recently, a generalized version for the model GBAS of Gutjahr [10] into which this technique incorporates, proposed by Dinh et al. [5]. In [5], the generalized model called GGBAS is theoretically proven that all convergence properties of GBAS are also held by GGBAS. Based on that convergent results, we carried out a numerical investigation by incorporating this *dynamical updating* trade-off rule into MMAS, ACS and BWAS algorithms on symmetric TSP benchmark instances.

The paper is organized as follows. To let our paper self-contained, the TSP statement and basic operation mode of ACO algorithms will be recalled in section 2. Details of how to dynamically adapt the value of q_0 in ACO algorithms in question will be introduced in section 2 as well. The next section will be devoted to analyze and compare the performance of these modified algorithms with their original version (non-updating dynamically value of q_0). Finally, some concluding remarks and future works will be mentioned in the last section.

2   Ant Colony Optimization

2.1   Traveling Salesman Problem

The TSP is formally defined as: "Let $V = \{a_1, .., a_n\}$ be a set of cities where n is the number of cities, $A = \{(r, s) : r, s \in V\}$ be the set of edges, and $\delta(r, s)$ be the cost measure associated with the edge $(r, s) \in A$. The objective is to find a minimum cost closed tour that goes through each city only once." In the case that all of cities in V are given by their coordinates and $\delta(r, s)$ is the Euclidean distance between any r and s $(r, s \in V)$ then this is so-called an Euclidean TSP problem. If $\delta(r, s) \neq \delta(s, r)$ for at least one edge (r, s) then TSP becomes asymmetric TSP (ATSP).

2.2   ACO Algorithms

A simplified framework of ACO [7] is recalled in Alg. 1:

Following ACO-based algorithms share the same general state transition rule when they are applied to TSP. That is, at a current node r, a certain ant k will make a move to a next node s in terms of the following probability distribution:

Algorithm 1. Ant Colony Optimization (ACO)

1: Initialize
2: **while** termination conditions not met **do**
3:    // at this level, each loop is called an iteration
4:    Each ant is positioned on a starting node
5:    **while** all ants haven't built a complete tour yet **do**
6:      Each ant applies a state transition rule to increasingly build a solution.
7:      Each ant applies a local pheromone updating rule.{optional}
8:    **end while**
9:    Apply the so-called online delayed pheromone trail updating rule.{optional}
10:    Evaporate pheromone.
11:    Perform the deamon actions. {optional: local search, global updating}
12: **end while**

$$p_k(r,s) = \begin{cases} \frac{[\tau_{rs}^{\alpha}] \cdot [\eta_{rs}^{\beta}]}{\sum_{u \in J_k(r)} [\tau_{ru}^{\alpha}] \cdot [\eta_{ru}^{\beta}]}, & \text{if } s \in J_k(r) \\ 0, & \text{otherwise} \end{cases}, \qquad (1)$$

where $J_k(r)$ is the set of nodes which ant k has not visited yet; τ_{rs} and η_{rs} are respectively the pheromone value (or called trail value sometimes) and the heuristic information of the edge (r,s). Brief descriptions of operation of ACS, BWAS, MMAS are shown next.

ACS:

Transition rule: The next node s is chosen as follows:

$$s = \begin{cases} \arg \max_{u \in J_k(r)} \{[\tau_{ru}]^{\alpha} \cdot [\eta_{ru}]^{\beta}\}, & \text{if } q \leq q_0 \\ \mathbf{S}, & \text{otherwise} \end{cases}, \qquad (2)$$

where $\mathbf{S}$ is selected according to Eq. (1), $q_0 \in [0,1]$ is the exploiting parameter mentioned in the previous section, $0 \leq q \leq 1$ is a random variable.

Local updating rule: When an ant visits an edge, it modifies the pheromone of that edge in the following way [1]: $\tau_{rs} \leftarrow (1-\rho) \cdot \tau_{rs} + \rho \cdot \Delta\tau_{rs}$, where $\Delta\tau_{rs}$ is a fixed systematic parameter.

Global updating rule: This rule is done by the deamon procedure which only the best-so-far ant is used to update pheromone values[2].

BWAS:

Transition rule: of BWAS is based on only Eq. (1). But it does not use online pheromone updating rule. The local updating as being used in ACS is discarded in BWAS. Adopting the idea from Population-Based Incremental Learning (PBIL) [1] of considering both current best and worst ants, BWAS

---

[1] Another name is *online step-by-step updating rule*.
[2] It is sometimes called *off-line pheromone updating rule* in other studies.

allows these two ants to perform positive and negative pheromone updating rules respectively according to Eq. (3) and Eq. (5).

$$\tau_{rs} \leftarrow (1 - \rho) \cdot \tau_{rs} + \Delta\tau_{rs} \qquad (3)$$

where

$$\Delta\tau_{rs} = \begin{cases} f(C(S_{\text{global-best}})), & \text{if } (r, s) \in S_{\text{global-best}} \\ 0, & \text{otherwise} \end{cases} \qquad (4)$$

$f(C(S_{\text{global-best}}))$ is the amount of trail to be deposited by the best-so-far ant.

$$\forall(r, s) \in S_{\text{current-worst}} \text{ and } (r, s) \notin S_{\text{global-best}}, \tau_{rs} \leftarrow (1 - \rho) \cdot \tau_{rs} \qquad (5)$$

Restart: A restart of the search progress is done when it gets stuck.

Introducing diversity: BWAS also performs the "mutation" for the pheromone matrix to introduce diversity in the search process. Each component of pheromone matrix is mutated with a probability P_m as follows:

$$\tau'_{rs} = \begin{cases} \tau_{rs} + mut(it, \tau_{\text{threshold}}), & \text{if } a = 0 \\ \tau_{rs} - mut(it, \tau_{\text{threshold}}), & \text{if } a = 1 \end{cases} \qquad (6)$$

$$\tau_{\text{threshold}} = \frac{\sum_{(r,s) \in S_{\text{global-best}}} \cdot \tau_{rs}}{|S_{\text{global-best}}|} \qquad (7)$$

with a being a binary random variable[3], it being the current iteration, and $mut(\cdot)$ being:

$$mut(it, \tau_{\text{threshold}}) = \frac{it - it_r}{Nit - it_r} \cdot \sigma \cdot \tau_{\text{threshold}} \qquad (8)$$

where Nit is the maximum number of iterations and it_r is the last iteration where a restart was done.

MMAS:

Transition rule: of MMAS is the same as BWAS, e.g. it uses only Eq. (1) to choose the next node.

Restart: A restart of the search progress is done when it get stuck.

Introducing bounds of pheromone values: Maximum and minimum values of trail are explicitly introduced. It does not allow trail strengths to get zero value, nor too high value.

---

[3] its value in $\{0, 1\}$.

2.3   Soundness of Incorporation of Trade-Off Technique

Graph-Based Ant System (GBAS) is a proposed Ant-based framework for static combinatorial optimization problems by Gutjahr [10]. In that study, Gutjahr proved that by setting a reasonable value of either the evaporation factor or the number of agents, the probability of which the global-best solution converges to the only optimal solution can be made arbitrarily close to one. However, GBAS framework does not use pseudo-random proportional rule for the state transition to balance between the exploitation and exploration of GBAS's search process. In [5], Dinh et al. proved that adding this rule into the GBAS's state transition rule to form so-called GGBAS framework does not change the convergence properties of GBAS.

The dynamical updating rule to q_0 is governed by the following equation:

$$q_0(t+1) = q_0(t=0) + \frac{(\xi - q_0(t)) \cdot \text{number of current tours}}{\theta \cdot \text{maximum number of generated tours}} \qquad (9)$$

where t is the current iteration, $q_0(t)$ is the value of q_0 at the t-th iteration, parameters ξ and θ are used to control the value range of q_0 to make sure its value always in a given interval. ξ is set to a smaller value than $q_0(0)$ such that

$$\frac{\xi \cdot \text{number of current tours}}{\theta \cdot \text{maximum number of generated tours}} \ll q_0(0). \qquad (10)$$

With (ξ, θ) chosen as in Eq. (10), it is approximately to have $q_0(t) < q_0(0)$ or hence, from Eq. (9)

$$q_0(0) > q_0(t) > q_0(0) \cdot (1 - \frac{1}{\theta}).$$

So, by selecting suitable values for (ξ, θ), we can assure that q_0 receives only values in a certain interval.

The next section will represent an numerical analysis of adding the pseudo-random proportional rule (with q_0 being dynamically adapted according to Eq. (9)) into Ant-based algorithms including MMAS, ACS, and BWAS.

3   Experiments and Analysis of Results

Dynamically updating value of q_0 according to Eq. (9) is carried out either right after all ants finish building their complete tours or at a certain step which they have not finished building those tours yet. To do the later, Eq. (9) must have a little bit modification. For the sake of simplicity, the former is selected.

Because Ant-based algorithms work better when local search are utilized, we will consider the influence of this rule in two cases: using local search or not. For TSP, a well-known local search named *2-opt* is then selected. The other well-known one is the *3-opt* but this local requests a more complex implementation and costs much more runtime than *2-opt* does. Because of these reasons, we select *2-opt* for our purpose of testing. All tests were carried out on a Pentium IV 1.6Ghz with 512MB RAM on Linux Redhat 8.0 platform.[4]

3.1   Without Local Search

MMAS and ACS are the two candidates chosen for this test. The MMAS and ACS algorithms with the new state transition rule (dynamical updating one) are called MMAS-BNL (MMAS-Balance with No Local search) and ACS-BNL correspondingly.

MMAS: In all tests performed by MMAS-BNL, parameters are set as follows: the number of ants $m = n$ with n being the size of instances, the number of iterations = 10,000. The average solutions are computed after 25 independent runs. Computational results of MMAS-BNL and MMAS are shown in Table 1. Here, results of MMAS (without using the trade-off technique) are quoted from [11]. In order to gain a comparison which is as fair as possible, the parameters setting of MMAS-BNL is the same as that of MMAS in [11]. Values in parentheses in this Table are the relative errors between current values (best and average ones) and the optimal solutions. This error is computed as 100%*(current value - optimal value)/optimal value. From Table 1, it shows that performance of MMAS-BNL

Table 1. Computational results of MMAS and MMAS-BNL. There are 25 runs done, and no local search is used in both algorithms. For MMAS-BNL, $\xi = 0.1$, $\theta = 3$, and $q_0(0) = 0.9$. The number attached with a problem name implies the number of cities of that problem. The best results are bolded

Problem	MMAS			MMAS-BNL		
	Best	Avg-best	σ	Best	Avg-best	σ
Eil51	**426 (0.00%)**	**426.7 (0.16 %)**	0.73	**426 (0.00%)**	427.87 (0.44%)	2.0
KroA100	**21282(0.00%)**	**21302.80(0.1%)**	13.69	**21282(0.00%)**	21321.72(0.19%)	45.87
D198	**15963(1.14%)**	**16048.60(1.70%)**	79.72	15994(1.36%)	16085.56(1.93%)	50.37
Att532	**28000(1.13%)**	**28194.80(1.83%)**	144.11	28027 (1.23%)	28234.80 (1.98)	186.30

is worse than that of MMAS. There is no solution quality improvement for any testing instances obtained when the trade-off technique is introduced.

ACS: We carry out experiments for ACS-BNL with parameter settings which are the same as in [9]. The settings are as follows: the number of ants $m = 10$, $\beta = 2.0$, $\rho = \alpha = 0.1$. The number of iterations is computed as $it = 100 * problem\ size$, hence the number of generated tours will be $100*m*problem\ size$, where *problem size* is the number of cities. Except the result of ACS for *pcb442* instance obtained from our implementation, results in Table 2 of ACS on selected testing instances of TSP is recalled from [9]. Values in parentheses in this Table are the relative errors between current values (best and average ones) and the optimal solutions. This error is computed as 100%*(current value - optimal value)/optimal value. Numerical results for ACS and ACS-BNL are shown in Table 2. In comparison with results of ACS which are cited from [9], we see that

---

[4] The software we used is ACOTSP v.1.0 by Thomas Stützle.

Table 2. Computational results of ACS and ACS-BNL. There are 15 runs done, and no local search is used in both algorithms. For MMAS-BNL, $\xi = 0.1$, $\theta = 3$, and $q_0(0) = 0.9$. The number attached with a problem name implies the number of cities of that problem. The best results are bolded

Problem	ACS			ACS-BNL		
	Best	Avg-best	σ	Best	Avg-best	σ
Eil51	**426 (0.00%)**	**428.06 (0.48 %)**	**2.48**	**426 (0.00%)**	428.60 (0.61 %)	3.45
KroA100	**21282(0.00%)**	**21420(0.65%)**	**141.72**	**21282(0.00%)**	21437(0.73%)	234.19
Pcb442*	**50778(0.00%)**	**50778(0.00%)**	**0.0**	**50778(0.00%)**	50804.80(0.05%)	55.48
Rat783	**9015(2.37%)**	**9066.80(2.97%)**	**28.25**	9178(4.22%)	9289.20(5.49%)	70.16

ACS-BNL found the best solutions for small scale instances like *eil51, KroA100, Pcb442* and so did ACS. But the average solutions and values of the standard deviation found by ACS for those instances are better than that by ACS-BNL. Moreover, ACS is over ACS-BNL for *rat783* a large instance in terms of measures of best solution, average solution, and standard deviation. Without using local search, ACS outperforms ACS-BNL in all test instances.

3.2   With Local Search

MMAS and BWAS are the two Ant-based algorithms chosen for this investigation purpose. The MMAS and BWAS algorithms with the new state transition rule are called MMAS-BL (MMAS-Balance with Local search) and BWAS-BL respectively. Results of the original MMAS were taken from [11] while that of the original BWAS were from [4]. Values in parentheses in this Table 3 are the relative errors between current values (best and average ones) and the optimal solutions. This error is computed as 100%*(current value - optimal value)/optimal value.

Table 3. MMAS variants with 2-opt for symmetric TSP. The runs of MMAS-BL were stopped after $n \cdot 100$ iterations. The average solutions were computed for 10 trials. In MMAS-BL, $m = 10$, $q_0(0) = 0.9$, $\rho = 0.99$, $\xi = 0.1$, and $\theta = 3$. The best results are bolded. The number attached with a problem name implies the number of cities of that problem. The best results are bolded

Problem	MMAS-BL	MMAS: $n \cdot 100$ iterations		MMAS: $n \cdot 2500$ iterations	
		10+all-ls	MMAS-ls	10+all-ls	MMAS-ls
KroA100	**21282.00(0.00%)**	21502(1.03%)	21481(0.94%)	**21282(0.00%)**	21282(0.00%)
D198	**15796.20(0.10%)**	16197(2.64%)	16056(1.75%)	15821(0.26%)	15786(0.04%)
Lin318	**42067.30(0.09%)**	43677(3.92%)	42934(2.15%)	42070(0.09%)	42195(0.39%)
Pcb442	**50928.90(0.29%)**	53993(6.33%)	52357(3.11%)	51131(0.69%)	51212(0.85%)
Att532	**27730.50(0.16%)**	29235(5.59%)	28571(3.20%)	27871(0.67%)	27911(0.81%)
Rat783	**8886.80 (0.92%)**	9576 (8.74%)	9171 (4.14%)	9047 (2.74%)	8976 (1.93%)

MMAS: In [11], Stützle studied the importance of adding local search into MMAS with the consideration that either all ants perform a local search or only the best one does so. In addition, in his study, the number of ants is also considered. Thus, there are three versions of MMAS with local search added including: 10 ants used and all ants do local search (named *10+all-ls*), 10 ants used and only the best ant does local search (*10+best-ls*, and the last version which the number of ants used is equal to the number of cities of TSP instance and only the best ant performs local search (named *MMAS+ls*). We mentioned here *10+all-ls* and *MMAS+ls* versions since it was claimed that in long run these two are better than the rest (*10+best-ls*). To make the comparison fairly, all systematic parameters of MMAS-BL were set equally to that of *10+all-ls*. Settings are: number of ants $m = 10$, number of nearest neighbor $= 35$, evaporation factor $\rho = 0.99$, $\alpha = 1.0$, $\beta = 2.0$, all ants are allowed to perform local search. It is noteworthy that the maximum number of iterations of MMAS-BL for an instance of size n is $n \cdot 100$ which implies that the number of generated tours of MMAS-BL is $m \cdot n \cdot 100$. Comparing performance of MMAS-BL with performance of both $MMAS - ls$ and $10 + all - ls$ can be shown in Table 3. For the problem *rat*783, even though only 5000 iterations performed by MMAS-BL, it still outperformed the other two algorithms (much more number of iterations given to those two algorithms). In all tests, both small and large scale instances, performance of MMAS-BL is always over *MMAS-ls* and *10+all-ls* even though the number of generated tours of MMAS-BL is much less than or equal that of the other two.

BWAS: Parameters setting for experiments for BWAS with the trade-off technique (BWAS-BL) is the same that for BWAS in [4]. Let us recall the table of parameters values of BWAS in [4] described in Table 4. Results of BWAS and BWAS-BL are represented in Table 5. Except for *Berlin51*, which performance of BWAS and that of BWAS-BL are the same, from Table 5, it has been seen that despite obtaining the optimal solution, the average solution of BWAS-BL is lightly worse than that of BWAS on small scale instances like *Eil51, KroA100*. Otherwise, on large scale instances, like *att532, rat783, fl1577*, BWAS-BL is over significantly BWAS in terms of measures of best-found solution, average solution, and standard deviation. Except the instance *fl1577* where standard deviation of BWAS-BL is worse than that of BWAS, for other instances the inversion is held.

Table 4. Parameter values and configuration of the local search procedure in BWAS

Parameter	Value
No. of ants	$m = 25$
Maximum no. of iterations	$Nit = 300$
No. of runs	15
Pheromone updating rules parameter	$\rho = 0.2$
Transition rule parameters	$\alpha = 1, \beta = 2$
Candidate list size	$cl = 20$
Pheromone matrix mutation prob.	$P_m = 0.3$
Mutation operator parameter	$\sigma = 4$
% of different edges in the restart condition	5%
No. of neighbors generated per iteration	40
Neighbor choice rule	*1st improvement*
Don't look bit structure	used

Table 5. Compare performance between the BWAS algorithm with its variant utilizing the trade-off technique. In BWAS-BL, $\xi = 0.1$, $\theta = 3$, and $q_0(0) = 0.9$. The optimal value of the corresponding instance is given in the parenthesis. The best results are bolded

Model	Eil51 (426)				Model	Att532 (27686)			
	Best	Average	Dev.	Error		Best	Average	Dev.	Error
BWAS	426	426	0	0	BWAS	27842	27988.87	100.82	1.09
BWAS-BL	426	426.47	0.52	0.11	BWAS-BL	**27731**	**27863.20**	**84.30**	**0.64**
Model	Berlin52 (7542)				Model	Rat783 (8806)			
	Best	Average	Dev.	Error		Best	Average	Dev.	Error
BWAS	7542	7542	0	0	BWAS	8972	9026.27	35.26	2.50
BWAS-BL	7542	7542	0	0	BWAS-BL	**8887**	**8922.33**	**16.83**	**1.32**
Model	KroA100 (21282)				Model	Fl1577 (22249)			
	Best	Average	Dev.	Error		Best	Average	Dev.	Error
BWAS	**21282**	**21285.07**	8.09	0.01	BWAS	22957	23334.53	187.33	4.88
BWAS-BL	**21282**	21286.60	9.52	0.02	BWAS-BL	**22680**	**23051**	351.87	**3.60**

3.3   Discussion

As shown in the above computational results, the trade-off technique or pseudo-random proportional rule with a dynamical updating technique embedded is an efficient and effective tool in improving solution quality of MMAS and BWAS when there is the presence of local search in these algorithms. Indeed, results from Table 3 showed that MMAS-BL presents a better performance than MMAS. It outperformed the other for all six test instances within smaller number of iterations. Also, from Table 5, BWAS-BL proved the effectiveness and usefulness of this modified trade-off technique by outperforming BWAS in large instances.

However, without using local search, Ant-based algorithms incorporating this technique, seem to perform worse than that which are not using this technique. This claim is supported by obtained numerical results. But, it is worth mentioning here that it is said Ant-based algorithms perform very well if local search procedures are utilized. Thus, the solution quality improvement of this trade-off technique with presence of local search is more impressive and worth attentive; and also its failure to improving solution quality when local search procedure is absent can be tolerable.

4   Conclusions

In this paper, we investigated the influence of pseudo-random proportional rule with value of the exploiting parameter being dynamically updating on state-of-the-art Ant-based algorithms like ACS, MMAS, BWAS. Without using local search, performance of these modified algorithms becomes slightly worse than the original ones. However, their solution quality improved significantly when a local search added. In addition, in some test cases, the best solutions were found within a shorter runtime.

Study the dynamic behavior of the exploiting parameter in combination with that of other systematic parameters such as the evaporation parameter is probably an interesting problem.

Acknowledgements

We would like to thank Thomas Stützle for sending his codes of ACOTSP version 1.0 which reduces our time on programming effort, and giving us helpful comments on how to compare fairly our results with that of MMAS.

References

1. S. Baluja and R. Caruana. Removing the genetics from the standard genetic algorithm. In A. Prieditis and S. Rusell, editors, *Machine Learning: Proceedings of the twelfth International Conference.*, pages 38–46. Morgan Kaufmann Publishers, 1995.
2. B. Bullnheimer, R.F. Hartl, and Ch. Strauss. A new rank based version of the ant system: a computational study. *Central European Journal of Operations Research*, 7(1):25–38, 1999.
3. A. Colorni, M. Dorigo, and V. Maniezzo. Distributed optimization by ant colonies. In F.Varela and P.Bourgine, editors, *Proceedings of the First European Conference on Artificial Life.*, pages 134–142. Elsevier Publishing, Amsterdam, 1991.
4. O. Cordón, I. Fernández de Viana, F. Herrera, and Ll. Moreno. A new ACO model integrating evolutionary computation concepts: The best-worst ant system. In M. Dorigo, M. Middendorf, , and T. Stützle, editors, *Abstract Proceedings of ANTS2000 - From Ant Colonies to Artificial Ants: A Series of International Workshops on Ant Algorithms.*, pages 22–29. Universit Libre de Bruxelles, Belgium, 2000.
5. Hoang T. Dinh, A. A. Mamun, and H. T. Huynh. A generalized version of graph-based ant system and its applicability and convergence. In *Proceeding of 4th IEEE International Workshop on Soft Computing as Transdisplinary Science and Technology (WSTST'05)*. Springer-Verlag, 2005.
6. M. Dorigo and G. Di Caro. The ant colony optimization metaheuristic. In D. Corne, M. Dorigo, and F. Glover, editors, *New Ideas In Optimization*. McGraw-Hill, 1999.
7. M. Dorigo, G. Di Caro, and L.M. Gambardella. Ant algorithms for discrete optimization. *Artificial Life*, 5:137–172, 1999.
8. M. Dorigo and L.M. Gambardella. Ant colony system: A cooperative learning approach to the travelling salesman problem. *IEEE Transactions on Evolutionary Computation*, 1:53–66, 1997.
9. L. Gambardella and M. Dorigo. Solving symmetric and asymmetric TSPs by ant colonies. In *IEEE Conference on Evolutionary Computation (ICE'96)*. IEEE Press, 1996.
10. W.J. Gutjahr. A graph-based ant system and its convergence. *Future Gerneration Computer Systems*, 16(9):873 – 888, 2000.
11. T. Stützle and H.H. Hoos. The MAX-MIN ant system and local search for the traveling salesman problem. In T. Bäck, Z. Michalewicz, and X. Yao, editors, *Proceedings of the 4th International Conference on Evolutionary Computation (ICEC'97)*, pages 308–313. IEEE Press, 1997.

Optimal Manpower Planning with Temporal Labor and Contract Period Constraints

Yongjian Li[1], Jian Chen[1], Xiaoqiang Cai[2,3], and Fengsheng Tu[3]

[1] School of Economics and Management,
Tsinghua University, Beijing 100084
[2] Department of System Engineering & Engineering Management,
The Chinese University of HongKong, Shatin N. T., Hong Kong
[3] College of Information Technical Science,
Nankai University, Tianjin 300071
{liyongjian, chenj}@em.tsinghua.edu.cn

Abstract. In this paper, we investigate a manpower planning problem with single job type over a long planning horizon. Dynamic demands of jobs must be fulfilled by allocating enough number of regular and temporal workers and each regular worker has a minimal employment contract period. A cost objective is concerned where costs for workforce include salaries of regular and temporal workers, and recruitment and dismissal costs of regular workers. We first formulate the problem as a multi-period decision model. Then we derive several properties of the optimal solution and develop an improved dynamic programming algorithm with polynomial computational complexity. Finally, numerical results are presented to illustrate several managerial insights.

Keywords: Manpower planning, dynamic program, minimal contract period, temporal labor.

1   Introduction

With the rapid development of economy, manpower planning has become an important problem in today's business world, especially in labor-intensive corporations, where the workforce plays a prominent role in determining the effectiveness and cost of the organization. As such, studies of optimal manpower planning have received extensive attention in the last two decades; see, e.g., [1], [2], and [3]. Considering the dynamic fluctuations of manpower demands, it is natural for an organization to determine the optimal size of its workforce by making proper and dynamic decisions on recruitment and dismissal over different periods of time. Such models, however, have not received much attention in the literature, due to properly the inherent complexity in deriving the optimal dynamic solutions. It is often that such a problem becomes a dynamic optimization model, which requires very sophisticated computational algorithms to search for the optimal or near-optimal solutions. Cai, *et al.* [4] have studied a manpower planning problem with multiple types of jobs, which involves employee recruitment, dismissal

N. Megiddo, Y. Xu, and B. Zhu (Eds.): AAIM 2005, LNCS 3521, pp. 350–359, 2005.

and substitution. However, only variable costs of recruitment and dismissal were considered. Li, *et al.* [5], [6] have studied a manpower planning problem with single employee type when considering setup costs for recruiting and dismissal activities. But, they didn't consider temporal employees.

The manpower planning optimization models can be applied in many areas. Verbeek [7] has suggested a framework for a pilot planning decision support system and described some of the complexities of such a system. Yu *et al.* [8] have provided an advanced optimization model and solution techniques to solve complex, large-scale pilot staffing and training problems. Other applications include military manpower planning (e.g., Grinold *et al.*[9]), and manufacturing manpower planning (e.g., Faalan *et al.*[10]), *etc.*

Our problem addresses the need of decision making in labor-intensive organizations facing dynamic fluctuations in their manpower demands. We investigated a local electronic manufacturer in Hong Kong of China that is competitive for its speed and flexibility. It receives orders from overseas clients. Typically the order comes with materials that the client prepares for his orders, the delivery time is tight - most of time, only 1-2 months. Since the workforce in Hong Kong is not very flexible and many of the workers are women, the company can not use too much overtime. Currently, the company uses only one shift. So some temporal workers need to be employed with a higher salary when peak demand reaches. The decision is to be made regarding whether recruitment/dismissal activity for regular workers occurs, and if an activity takes place, how many regular workers will be recruited/dismissed in each period. The objective is to minimize the overall manpower-related cost for the company.

In this article we will first model the manpower planning problem with dynamic demands as a multi-period decision process with constraints, including 1) demands of jobs in each period must be fulfilled, 2) every regular worker is constrained by a minimal contract period, i.e., a regular worker can be dismissed only after his/her employment time is not less than a given limit value, and 3) temporal workers will be employed to meet the peak demands of jobs if possible. We will then propose an improved dynamic programming algorithm to derive the optimal solution. Our approach will be devised based on analysis on the properties of the optimal decisions.

2   Problem Description

We consider the following manpower planning problem for an organization. Suppose that based on forecast of its business, the number of workers required in each time period has been specified in advance. The organization can decide on the number of the regular workers to be recruited or dismissed at the end of every period, subject to the constraint that the workforce available in the coming periods would meet the demand for manpower. The objective is to find a series of optimal decisions on recruitment/dismissal of regular workers in every time period over the entire planning horizon, so that the total manpower-related cost is minimized.

Notations:

T	The number of time periods being considered
L	The length of the minimal contact period per regular worker
α/γ	The salary per regular/temporal worker in each time period ($\alpha < \gamma$)
β^+/β^-	The recruitment/dismissal costs when a regular worker is recruited / dismissed
D_t	The manpower demand in period t (Assume $D_0 = 0$).
$X[t]/Z[t]$	The number of regular/temporal workers available in period t, which are called the states in period t.
$u[t]$	The number of regular workers being recruited ($u[t] > 0$)/dismissed ($u[t] < 0$) at the end of period t.

The problem under consideration can be formulated as follows.

$$\textbf{MP:} \quad J = \min_{u[t]} \left\{ \alpha X[T] + \gamma Z[T] + \sum_{t=0}^{T-1} \left[\alpha X[t] + \gamma Z[t] + \beta^+ u^+[t] + \beta^- u^-[t] \right] \right\} \quad (1)$$

$$s.t. \qquad X[t+1] = X[t] + u[t], t = 0, 1, \cdots, T-1 \qquad (2)$$

$$X[t] + Z[t] \geq D_t, t = 1, 2, \cdots, T \qquad (3)$$

$$X[t] \geq \sum_{s=t-L}^{t-1} u^+[s], t = 2, 3, \cdots, T \qquad (4)$$

$$X[0] = 0 \qquad (5)$$

where $u^+[t] = \max\{u[t], 0\}$, $u^-[t] = \max\{-u[t], 0\}$.

Constraint (2) shows the dynamics of the workforce available in the organization. Constraint (3) indicates that the demands must be fulfilled by the available workers in every period. Constraint (4) shows that every regular worker must be employed for at least L periods. Note that constraint (4) is a necessary condition, not a sufficient condition for the minimal contact period constraint.

In the following, we call the sequence of $X[t_1], X[t_1+1], \ldots, X[t_2]$ a state trajectory X from the period t_1 to t_2; Similarly, we call the sequence of $D_{t_1}, D_{t_1+1}, \cdots, D_{t_2}$ a demand trajectory D from the period t_1 to t_2.

3   A Standard Dynamic Programming Approach

First, we can easily show that the optimal states satisfy $X^*[t] \leq D_{max}$ and $Z^*[t] = (D_t - X^*[t])^+$, where $D_{max} = \max\{D_0, D_1, \cdots, D_T\}$. Let $f(t, x_t, z_t)$ be the minimum cost from period 0 to t subject to the condition that there are x_t regular workers and z_t temporal workers in period t. Let Φ_t and $U[t]$ be the set of feasible states and feasible decisions in period t respectively, we get

$$\Phi_t = \left\{ (x_t, z_t) | x_t \geq \sum_{s=t-L}^{t-1} u^+[s], z_t = (D_t - x_t)^+, D_t \leq x_t + z_t \leq D_{max} \right\},$$

$$U[t] = \left\{ u[t] | x_t - \max \left\{ D_{t+1}, \sum_{s=t-L+1}^{t} u^+[s] \right\} \leq u[t] \leq D_{max} - D_t \right\}.$$

Then for the problem MP, we can present a forward dynamic programming recursion as follows,

DP:
$$f_0(0) = 0$$

For $\forall (x_{t+1}, z_{t+1}) \in \Phi_{t+1}, 0 \le t < T$

$$f_{t+1}(x_{t+1}, z_{t+1}) = \alpha x_{t+1} + \gamma z_{t+1} + \min_{u[t]} \left\{ \beta^+ u^+[t] + \beta^- u^-[t] \right.$$

$$\left. + f_t(x_{t+1} - u[t], (D_t - x_{t+1} + u[t])^+) | u[t] \in U[t] \right\}.$$

The optimal solution is obtained when getting $f_T^* = \min \left\{ f_T(x_T, z_T) | x_T + z_T \ge D_T, z_T = (D_T - x_T)^+, \sum_{s=T-L}^{T-1} u^+[s] \right\} \le x_T \right\}$. We can easily show that the computational complexity of the approach DP is $O(D_{max}^2 T)$ in the worst case. This time complexity is too high, in particular when we have a large D_{max}. In the following we present our improved algorithm, which requires substantially less computational time (Proofs for the results will be omitted due to the limit of space, which are available upon request).

4   Improved Dynamic Programming Approach

In the algorithm DP, for any given state $(x, z) \in \Phi_t$ we need to compute the value $f(t, x, z)$ over all possible decisions in $U[t]$. This is actually not necessary if we can utilize some properties of the problem MP. Some analysis is given below based on the relationship between the demands in neighboring periods.

The difficulties are how to reduce the number of elements in the set Φ_t without affecting the optimal solution, and how to simplify the computation of the optimal decision $u[t] (\in U[t])$ for each possible state in the set Φ_t. We define $\underline{x}_t$ as the minimum feasible state in period t that is obtained when the problem is considered through period 0 to t, then $\underline{x}_t \ge \sum_{s=t-L}^{t-1} u^+[s]$.

Theorem 1. *When $D_t > D_{t+1}$ during a period $t(t = 0, 1, \dots, T-1)$, we get $\underline{x}_{t+1} = \max\{\underline{x}_t - u^+[t-L], D_{t+1}\}$. We let t_1 and t_2 be the last periods before period t that satisfies $D_{t_1} < \underline{x}_{t+1} \le D_{t_1+1}$ and $D_{t_2+1} < \underline{x}_{t+1} \le D_{t_2}$ respectively, and we compute $L_p = \lceil \frac{\beta^+ + \beta^-}{\gamma - \alpha} \rceil$, where $\lceil x \rceil$ is the smallest integer that is not less than value x. Then we have*
 1) if $2t_2 - t_1 - t > L_p$, then $\Phi_{t+1} = \{(\underline{x}_{t+1}, 0)\}$ and

$$f_{t+1}(\underline{x}_{t+1}, 0) = \alpha \underline{x}_{t+1} + \beta^-(\underline{x}_t - \underline{x}_{t+1}) + f_t(\underline{x}_t); \tag{6}$$

2) if $2t_2 - t_1 - t \le L_p$, then $\Phi_{t+1} = \{(x_{t+1}, 0) | x_{t+1} = \underline{x}_s, \underline{x}_{t+1} \le \underline{x}_s < \underline{x}_t, t_1 \le s \le t+1\}$, $\underline{x}_s = \max\{\underline{x}_{t_1}, x_{t+1}\}$ and $u[s] = 0$ for $t_1 + 1 \le s \le t$, and

$$f_{t+1}(x_{t+1}, 0) = \alpha x_{t+1} + \alpha(t - t_1) \max\{\underline{x}_{t_1}, x_{t+1}\}$$
$$+ \beta^+ (\max\{\underline{x}_{t_1}, x_{t+1}\} - \underline{x}_{t_1}) + \beta^- (\max\{\underline{x}_{t_1}, x_{t+1}\} - x_{t+1})$$
$$+ \gamma \sum_{s=t_1+1}^{t} (D_s - \max\{\underline{x}_{t_1}, x_{t+1}\}) + f_{t_1}(\underline{x}_{t_1}, 0). \tag{7}$$

The theorem shows that when $D_t > D_{t+1}$, some regular workers may be dismissed with the quantity of $\underline{x}_t - \underline{x}_{t+1}$ at the end of period t, or temporal workers may be employed through a period $t_1 + 1 (\leq t)$ to t. Which decision to be chosen depends on the value of L_p and the demand information in the past several periods. The value of L_p indicates a trade-off between the salary cost of a temporal worker and the sum of salary, recruitment and dismissal costs of a regular worker.

When $D_t \leq D_{t+1}$, there is an increasing demand. The increased demand in period $t + 1$ can be fulfilled either by recruiting regular workers at the end of period t, or by reserving some unassigned regular workers that should be dismissed at the end of a period before t. In the second case, although more salary cost needs to be paid, recruiting and dismissal costs will be reduced, so there is a trade-off between the salary cost and the sum of recruiting and dismissal costs of a regular worker and we define $L_v = \lceil \frac{\beta^+ + \beta^-}{\alpha} \rceil$ to indicate the trade-off. Moreover, we let $\underline{d}_t = \max\{\underline{x}_{t-l}, 1 \leq l \leq \min\{L_v, t - 1\}\}$ and $s_v(t) = \arg \max\{\underline{x}_{t-l}, 1 \leq l \leq \min\{L_v, t - 1\}\}$.

Theorem 2. *When $D_t \leq D_{t+1}$ during a period $t(t = 0, 1, \ldots, T - 1)$, we have*
1) if $\underline{x}_t \geq D_{t+1}$, then $\underline{x}_{t+1} = \underline{x}_t$, $\Phi_{t+1} = \{(\underline{x}_{t+1}, 0)\}$ and

$$f_{t+1}(\underline{x}_{t+1}, 0) = \alpha \underline{x}_{t+1} + f_t(\underline{x}_t, 0); \tag{8}$$

2) if $\underline{x}_t \leq D_{t+1} < \underline{d}_{t+1}$, then $\underline{x}_{t+1} = D_{t+1}$, $\Phi_{t+1} = \left\{(\underline{x}_{t+1}, 0)\right\}$ and

$$f_{t+1}(\underline{x}_{t+1}, 0) = \alpha \underline{x}_{t+1}(t + 1 - t_1) + \beta^- (\underline{x}_{t_1} - \underline{x}_{t+1}) + f_{t_1}(\underline{x}_{t_1}, 0), \tag{9}$$

where t_1 is the last period before t that satisfies $\underline{x}_{t_1} > D_{t+1} \geq \underline{x}_{t_1+1}$;
3) if $\underline{x}_t \leq D_{t+1}$ and $D_{t+1} \geq \underline{d}_{t+1}$, then $\underline{x}_{t+1} = D_{t+1}$, $\Phi_{t+1} = \left\{(D_{t+1}, 0)\right\}$.
For $(D_{t+1}, 0) \in \Phi_{t+1}$, $f_{t+1}(D_{t+1}, 0) =$

$$\alpha D_{t+1} + \beta^+ (D_{t+1} - \underline{d}_{t+1}) + \alpha \underline{d}_{t+1}(t - t_2) + \beta^- (\underline{x}_{t_2} - \underline{d}_{t+1}) + f_{t_2}(\underline{x}_{t_2}, 0), \tag{10}$$

where t_2 is the last period before t that satisfies $\underline{x}_{t_2} > \underline{d}_{t+1} \geq \underline{x}_{t_2+1}$.

Since no regular workers are dismissed at the end of period T, we need to compute whether some unassigned regular workers who should be dismissed before period T are reserved till period T. So there is a trade-off between the salary and dismissal cost of a regular worker and then we define $L_{end} = \lceil \frac{\beta^-}{\alpha} \rceil$ to indicate the trade-off. On the other hand, if recruiting regular workers is necessary during the last several periods before period T, it may decrease the total cost when temporal workers are employed instead of recruiting regular ones. So there is also a trade-off between the sum of the salary and recruitment cost of a regular worker and the salary of a temporal one, then we define $R_{end} = \lceil \frac{\beta^+}{\gamma - \alpha} \rceil$.

Theorem 3. When $t = T$, we have $\Phi_T = \left\{ (\underline{x}_s, (D_T - \underline{x}_s)^+) \mid \min\{R_{end}, L_{end}\} \leq s \leq T \right\}$, and for $\forall s : \min\{R_{end}, L_{end}\} \leq s \leq T$,

$$f_T(\underline{x}_s, (D_T - \underline{x}_s)^+) = \alpha(T - s)\underline{x}_s + \gamma \sum_{t=t''}^{T} (D_t - \underline{x}_s)^+ + f_s(\underline{x}_s, 0). \quad (11)$$

When we get the value $f^* = \min\{f_T(x_T, z_T) \mid (x_T, z_T) \in \Phi_T\}$, we get the optimal solution to the problem MP .

From Theorem 1 to 3, we can easily derive the corresponding optimal states through period 0 to $t + 1$. Since the computations in a period t need the values $u[s](t - L \leq s \leq t - 1)$ during the past L periods, we define a set $U_t(x) = \{u_{it}, i = 1, 2, \ldots, L\}$ as a collection of decisions $u[t]$ during the past L periods, where $u[t]$ are optimal decisions in the optimal state trajectory through period 0 to t, subject to the terminal state $X[t] = x$. Let states $X^{(t,x)}[s]$ be the optimal states obtained by our approach through period 0 to t with the terminal state $X^{(t,x)}[t] = x$, then we can derive the following algorithm to compute the optimal value and optimal states over the entire planning horizon.

Algorithm IDP:

Step 1. Let $f_0(0) = 0$, $U_0(0) = \{u_{i0}(= 0), i = 1, 2, \ldots, L\}$.

Step 2. Do loop from Step 3 to Step 8 for $t = 0, 1, \ldots, T - 1$.

Step 3. When $\underline{x}_t > D_{t+1}$, we have $U_{t+1}(\underline{x}_{t+1}) = \{u_{i(t+1)}, i = 1, 2, \ldots, L\}$ and $\underline{x}_{t+1} = \max\{\underline{x}_t - u_{1t}, D_{t+1}\}$. Next, we let t_1 and t_2 be the last periods before period t that satisfies $D_{t_1} < \underline{x}_{t+1} \leq D_{t_1+1}$ and $D_{t_2+1} < \underline{x}_{t+1} \leq D_{t_2}$ respectively, and compute L_p. If $D_t \leq D_{t+1}$ or $(D_t > D_{t+1}$ and $2t_2 - t_1 - t > L_p)$, then go to Step 4; otherwise, if $D_t > D_{t+1}$ and $2t_2 - t_1 - t \leq L_p$, go to Step 5.

Step 4. $u_{L(t+1)} = 0$, $u_{i(t+1)} = u_{(i+1)t}$, $u_{(i+1)t} \in U_t(\underline{x}_t)$, $i = 1, 2, \ldots, L - 1$. $\Phi_{t+1} = \{(\underline{x}_{t+1}, 0)\}$, $f_{t+1}(\underline{x}_{t+1}, 0)$ and $X^{(t+1, \underline{x}_{t+1})}[s]$, $0 \leq s \leq t + 1$ can be computed through equations (6) and

$$X^{(t+1, \underline{x}_{t+1})}[s] = \begin{cases} \underline{x}_{t+1} & , \text{ if } s = t + 1 \\ X^{(t, \underline{x}_t)}[s] & , \text{ if } 0 \leq s \leq t \end{cases}.$$

Step 5. $u_{i(t+1)} = 0$ for $i = L - (t - t_1), \ldots, L$, $u_{i(t+1)} = u_{(i+1)t}$, $u_{(L-i)t_1} \in U_{t_1}(\underline{x}_{t_1})$ for $i = 1, 2, \ldots, L - (t - t_1) - 1$. $\Phi_{t+1} = \{(x_{t+1}, 0) \mid x_{t+1} = \underline{x}_s, \underline{x}_{t+1} \leq \underline{x}_s < \underline{x}_t, t_1 \leq s \leq t + 1\}$; $f_{t+1}(x_{t+1}, 0)$ and $X^{(t+1, x_{t+1})}[s]$, $0 \leq s \leq t + 1$ can be computed through equations (7) and

$$X^{(t+1, x_{t+1})}[s] = \begin{cases} x_{t+1} & , \text{ if } s = t + 1 \\ \max\{\underline{x}_{t_1}, D_{t+1}\} & , \text{ if } t_1 < s \leq t \\ X^{(t_1, \underline{x}_{t_1})}[s] & , \text{ if } 0 \leq s \leq t_1 \end{cases}$$

Step 6. If $\underline{x}_t < D_{t+1}$, then $\Phi_{t+1} = \{(\underline{x}_{t+1}, 0)\}$, $\underline{x}_{t+1} = D_{t+1}$, and $U_{t+1}(\underline{x}_{t+1}) = \{u_{i(t+1)}, i = 1, 2, \ldots, L\}$. We compute L_v and $\underline{d}_{t+1}$. If $\underline{x}_{t+1} < \underline{d}_{t+1}$, then go to Step 7; otherwise, go to Step 8.

Step 7. Let $u_{i(t+1)} = 0$, $i = L, L-1, \ldots, L-(t-t_3)$; $u_{i(t+1)} = u_{(L-i)t_3}$, $i = 1, 2, \ldots, L-(t-t_3)-1$, $u_{it_3} \in U_{t_3}(\underline{x}_{t_3})$, where t_3 is the last period before period t in which $\underline{x}_{t_3+1} \leq \underline{x}_{t+1} \leq \underline{x}_{t_3}$. $f_{t+1}(\underline{x}_{t+1}, 0)$ and $X^{(t+1,\underline{x}_{t+1})}[s]$, $0 \leq s \leq t+1$ can be computed through equations (9) and

$$X^{(t+1,\underline{x}_{t+1})}[s] = \begin{cases} \underline{x}_{t+1} & , \quad \text{if } t_3 < s \leq t+1 \\ X^{(t_3,\underline{x}_{t_3})}[s] & , \quad \text{if } 0 \leq s \leq t_3 \end{cases}.$$

Step 8. Let $u_{L(t+1)} = \underline{x}_{t+1} - \underline{d}_{t+1}$; $u_{i(t+1)} = 0$, $i = L-1, L-2, \ldots, L-(t-t_4)+1$; $u_{i(t+1)} = u_{(L-i)t_4}$, $i = 1, 2, \ldots, L-(t-t_4)$, $u_{it} \in U_t(\underline{x}_t)$, where t_4 is the last period before t that satisfies $\underline{x}_{t_4} > \underline{d}_{t+1} \geq \underline{x}_{t_4+1}$. $f_{t+1}(\underline{x}_{t+1}, 0)$ and $X^{(t+1,\underline{x}_{t+1})}[s]$, $0 \leq s \leq t+1$ can be computed through equations (10) and

$$X^{(t+1,\underline{x}_{t+1})}[s] = \begin{cases} \underline{x}_{t+1} & , \quad \text{if } s = t+1 \\ \underline{d}_{t+1} & , \quad \text{if } t_4 + 1 \leq s \leq t \\ X^{(t_4,\underline{x}_{t_4})}[s] & , \quad \text{if } 0 \leq s \leq t_4 \end{cases}.$$

Step 9. When $t = T$, we compute R_{end} and L_{end}. Then we get $\Phi_T = \left\{ (\underline{x}_{t_5}, (D_T - \underline{x}_{t_5})^+) \mid \min\{R_{end}, L_{end}\} \leq t_5 \leq T \right\}$ and $\forall (x_T, (D_T - x_T)^+) \in \Phi_T$, $U_T(x_T) = \{u_{iT}, i = 1, 2, \ldots, L\}$, where $u_{iT} = 0$ for $i = L - (T - t_5), \ldots, L$ and $u_{iT} = u_{(L-i)t_5}(\in U_{t_5}(\underline{x}_{t_5}))$ for $i = 1, 2, \ldots, L - (T - t_5 - 1)$. $f_T(x_T, 0)$ and $X^{(T,x_T)}[t]$, $1 \leq t \leq T$ can be computed through equations (11) and

$$X^{(T,x_T)}[t] = \begin{cases} x_T & , \quad \text{if } t_5 \leq t \leq T \\ X^{(t_5,x_T)}[t] & , \quad \text{if } 1 \leq t < t_5 \end{cases}$$

where t_5 is the last period in which $\underline{x}_{t_5} = x_T$.

Step 10. Compute $f^* = \min\{f_T(x_T, z_T), (x_T, z_T) \in \Phi_T\}$ and $x_T^* = \mathrm{argmin}\{x_T \mid f_T(x_T, z_T), (x_T, z_T) \in \Phi_T\}$, we get the optimal value f^* and the optimal state $X^*[t] = x^{(T,x_T^*)}[t]$, $Z^*[t] = (D_t - X^*[t])^+$ for $0 \leq t \leq T$.

In the algorithm IDP, the computation is only based on part of the future demand information (*i.e.*, the demands of future $\max\{L_p, L_v\}$ periods), not based on the demand information in all future periods. Moreover, we don't need to know the exact future demand, we only need to know the maximum demand during the future L_v periods and the minimal demand during the future L_p periods.

From Theorem 1, there are at most L_p elements in the set Φ_t that occurs when $D_t > D_{t+1}$. For a given element $(x_t, z_t) \in \Phi_t$, only one computation is necessary to compute the corresponding decision $u[t]$. Let q_t be the quantity of possible states in period t. Obviously, $q_t \leq L_p, \forall t$. From Theorems 1 to 3, the computation quantity is at most $q_t q_{t+1}$ in each period t. Since $q_t < L_p + 1$ and $L_p = \lceil \frac{\beta^+ + \beta^-}{\gamma - \alpha} \rceil$ is a given constant, the computational complexity of the algorithm IDP is $O(T)$ in the worst case.

5   Managerial Analysis

An example is used to demonstrate characteristics of the optimal solution to the problem MP. In our computations, we set $\alpha = 800$, $\beta^+ = 1700$, $\beta^- = 1200$, $\gamma = 1800$ and $L = 4$. Then $L_v = \lceil \frac{\beta^+ + \beta^-}{\alpha} \rceil = \lceil \frac{29}{8} \rceil = 4$, $L_p = \lceil \frac{\beta^+ + \beta^-}{\gamma - \alpha} \rceil = \lceil \frac{2900}{1000} \rceil = 3$, $L_{end} = \lceil \frac{\beta^-}{\alpha} \rceil = \lceil \frac{12}{8} \rceil = 2$ and $L_p = \lceil \frac{\beta^+}{\gamma - \alpha} \rceil = \lceil \frac{1700}{1000} \rceil = 2$. The demand trajectory is shown in Fig. 1. We computed the optimal state trajectories for the following four cases: 1) neither temporal workers nor the contract period constraint is considered (labeled as 'state1' in Fig. 1); 2) temporal workers are not considered (labeled as 'state2'); 3) the contract period constraint is not considered (labeled as 'state3') and 4) both temporal workers and the contract period constraint are considered simultaneously (labeled as 'state4'). The comparisons among the four results show us the influence of the two factors (i.e., temporal workers and the contract period constraint) on the optimal states. With considering temporal workers, the quantity of regular workers is reduced during the periods over which there are peak demands; With considering the minimal contract period constraint, some regular workers that should be dismissed in periods 5 and 6 are dismissed later in period 7, which makes some unassigned regular workers be reserved from periods 8 to 12.

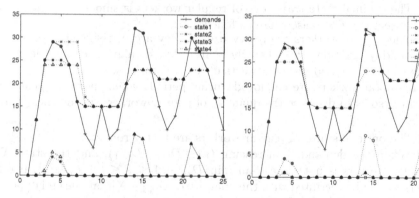

Fig. 1. Comparing solutions   **Fig. 2.** Results when salary of temporal employee changes

Next we discussed the influence of changing parameters L_v, L_p and L on the optimal states. We computed the optimal state trajectories for the following 3 cases: 1) $\gamma = 1800 + 400 * n$, then L_p changes among values $1, 2, 3$; 2) $L = n + 3$, and 3) $\beta^+ = 500 + 400 * n$ and $\beta^- = 400 * n$, then L_v changes among values $2, 3, 4, 5, 6$, for $n = 1, 2, 3, 4, 5$. Other parameters are same as the first example. The demands and the optimal states are illustrated in Fig. 2, Fig. 3 and Fig. 4 respectively, where the optimal states labeled as 'state$i$' are computed when $i = n$.

Finally we showed numerically the influence of the demand change in a period on the optimal states before this period in all cases. The computational results show that the change of demand in a period does not affect the optimal states of regular workers before this period, only changes the state values in and/or after this period.

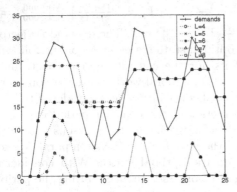 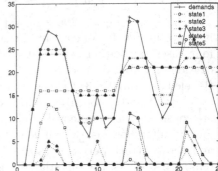

Fig. 3. Results when value L changes   **Fig. 4.** Results when recruitment and dismissal costs change

The above numerical results show the following insights.

1) The optimal state trajectory of regular workers is smoother than the demand trajectory. The reasons are as follows,

- Some regular workers are reserved even they are unassigned (*e.g.*, going on holiday, training, *etc*) when there are valley demands, so as to decrease the recruitment and dismissal costs of regular workers.

- Temporal workers are employed during periods when there are peak demands, so as to decrease the quantity of regular workers during these periods.

2) The optimal states of regular workers are pretty robust.

3) When the demand is increasing (*i.e.*, $D_t < D_{t+1}$), and the state $X[t]$ of regular workers is not more than $\min\{D_s, t+1 \leq s \leq t+L_p+1\}$, regular workers should be recruited with the quantity of $D_{t+1} - X[t]$ at the end of period t; Otherwise, temporal workers should be employed so that the states of regular workers hold a same value of about $\min\{D_s, t+1 \leq s \leq t+L_p\}$ during several future periods, as shown in Fig. 2.

4) When the demand is decreasing (*i.e.*, $D_t > D_{t+1}$), and the quantity $X[t]$ of regular workers is not more than $\max\{D_s, t+1 \leq s \leq t+L_v+1\}$, no regular workers will be dismissed or recruited at the end of period t; Otherwise, some regular workers will be dismissed with the quantity of about $X[t] - \max\{D_s, t+1 \leq s \leq t+L_v+1\}$ at the end of period t, as shown in Fig. 4.

6   Concluding Remarks

We have modeled a dynamic manpower planning problem as a multi-period decision process, and developed an improved dynamic programming algorithm to compute its optimal solution. Our approach is not only computationally efficient, requiring only a time complexity $O(T)$ and independent of the magnitude of the manpower demands, but also capable to reveal some useful insights on the desirable solutions with respect to the data, and therefore allowing management to have better understanding of the impacts and benefits of the solutions.

The model considered in this paper has not been proposed and solved before. However, the model has been simplified to a certain degree to ensure the solution efficiency and it has a gap with the practice. More practical factor (*e.g.*, training requirements, promotion, leave, *etc.*) and/or multiple employee types will be considered in the future study.

Acknowledgment

This research was partly supported in part by NSFC Research Fund No. 60074018, 70321001, the China Postdoctoral Science Foundation (No. 2003034020), SRF for ROCS, SEM (2004) and the Tsinghua-Zhongda Postdoctoral Science Foundation(2003).

References

1. Bartholomew, D.J., Forbes, A.F. and Mclean, S.I.: Statistical techniques for manpower planning (2nd). Wiley, New York, 1991.
2. Bowey, A. M. : Coorporate manpower planning. Manage. Decis. **15**(1977) 421-469
3. Alfares, H.K.: Survey, Categorization, and Comparison of Recent Tour Scheduling Literature. Ann. Opers. Res., **127**(2004) 145-175
4. Cai, X., Li, Y.J., Tu, F.S., *et. al.*: Optimal Manpower Planning with Hierarchical Types of Jobs through dynamic programming approach. Working paper, Dept. of Sys. Eng. & Eng. Manage., The Chinese University of HongKong, HONGKONG, 2003.
5. Li, Y.J., Chen, J., Cai, X.: Optimal Manpower Planning Decision with Single Employee Type Considering Minimal Contract Period. Working paper, Tsinghua University, 2004.
6. Li, X.D., Tu, F.S., Li, Y.J., *et. al.*: Optimal Manpower Recruitment and Dismissal Decision For Single-Type Job, J. of Sys. Sci. & Infor. **2**(2004) 545-555.
7. Verbeek, P.: Decision Support Systems-An application in strategic manpower planning of airline pilots. Eur. J. of Oper. Res. **55**(1991) 368-381.
8. Yu, G., Pachon, J. and Thengvall, B.: Optimization-based Integrated Manpower Management for Airlines. Operations Research in Space & Air, edited by Tito A. Ciriani, Kluwer Academic Publishers, Boston, 2003.
9. Grinold, R.C, Marshall, K.T: Manpower Planning Models. Elsevier Noth-Holland, Inc., New York, NY, 1977.
10. Faalan B., T. Schmitt, Cost-based Scheduling of Workers and Equipment in a Fabrication and Assembly Shop. Oper. Res. **41**(1993) 253-268.

Mechanism Design for Set Cover Games
When Elements Are Agents

Zheng Sun[1,*], Xiang-Yang Li[2,**], WeiZhao Wang[2],
and Xiaowen Chu[1,***]

[1] Hong Kong Baptist University, Hong Kong, China
{sunz, chxw}@comp.hkbu.edu.hk
[2] Illinois Institute of Technology, Chicago, IL, USA
{lixian, wangwei4}@iit.edu

Abstract. In this paper we study the set cover games when the elements
are selfish agents. In this case, each element has a privately known valuation of receiving the service from the sets, *i.e.*, being covered by some
set. Each set is assumed to have a fixed cost. We develop several approximately efficient truthful mechanisms, each of which decides, after
soliciting the declared bids by all elements, which elements will be covered, which sets will provide the coverage to these selected elements, and
how much each element will be charged. For set cover games when both
sets and elements are selfish agents, we show that a cross-monotonic *payment*-sharing scheme does not necessarily induce a truthful mechanism.

1   Introduction

In the past, an indispensable and implicit assumption on algorithm design for interconnected computers has been that all participating computers (called *agents*)
are cooperative; they will behave exactly as instructed. This assumption is being
shattered by the emergence of the Internet, as it provides a platform for distributed computing with agents belonging to self-interested organizations. This
gives rise to a new challenge that demands the study of *algorithmic mechanism
design*, the sub-field of algorithm design under the assumption that all agents
are *selfish* (*i.e.*, they only care about their own benefits) and yet *rational* (*i.e.*,
they will always choose their actions to maximize their benefits).

Assume that there are n agents $\{1, 2, \cdots, i, \cdots, n\}$, and each agent i has
some *private* information t_i, called its *type*. For direct-revelation mechanisms,
the strategy of each agent i is to declare its type, although it may choose to
report a carefully designed lie to influence the outcome of the game to its liking.

---

* The research of the author was supported in part by Grant FRG/03-04/II-21 and
Grant RGC HKBU2107/04E.

** The research of the author was supported in part by NSF under Grant CCR-
0311174.

*** The research of the author was supported in part by Grant RGC HKBU2159/04E.

N. Megiddo, Y. Xu, and B. Zhu (Eds.): AAIM 2005, LNCS 3521, pp. 360–369, 2005.

For any vector $t = (t_1, t_2, \cdots, t_n)$ of reported types, the mechanism computes an output o as well as a payment p_i for each agent i. For each possible output o, agent i's preference is defined by a valuation function $v_i(t_i, o)$. The utility of agent i for the outcome of the game is defined to be $u_i = v_i(t_i, o) + p_i$. An action a_i is called a *dominant strategy* for player i if it maximizes its utility regardless of the actions chosen by other players; a selfish agent will always choose its dominant strategy. A mechanism is *incentive compatible* (IC) if for every agent reporting its type truthfully is a dominant strategy. Another very common requirement in the literature for mechanism design is *individual rationality*: the agent's utility of participating in the outcome of the mechanism is not less than the utility of the agent if it does not participate at all. A mechanism is called *truthful* or *strategyproof* if it satisfies both IC and IR properties.

A classical result in mechanism design is the Vickrey-Clarke-Groves (VCG) mechanism by Vickrey [1], Clarke [2], and Groves [3]. The VCG mechanism applies to maximization problems where the objective function $g(o, t)$ is simply the sum of all agents' valuations. A VCG mechanism is always truthful [3], and is the only truthful implementation, under mild assumptions, to maximize the total valuation [4]. Although the family of VCG mechanisms is powerful, it has its limitations. To use a VCG mechanism, we have to compute the exact solution that maximizes the total valuation of all agents. This makes the mechanism computationally intractable for many optimization problems.

This work focuses on strategic games that can be formulated as the set cover problem. A *set cover game* can be generally defined as the following. Let $\mathcal{S} = \{S_1, S_2, \cdots, S_m\}$ be a collection of multisets (or *sets* for short) of a universal set $U = \{e_1, e_2, \cdots, e_n\}$. Element e_i is specified with an *element coverage requirement* r_i (*i.e.*, it desires to be covered r_i times). The multiplicity of an element e_i in a set S_j is denoted by $k_{j,i}$. Let $d_{\max}$ be the maximum size of the sets in $\mathcal{S}$, *i.e.*, $d_{\max} = \max_j \sum_i k_{j,i}$. Each S_j is associated with a cost c_j. For any $\mathcal{X} \subseteq \mathcal{S}$, let $c(\mathcal{X})$ denote the total cost $\sum_{S_j \in \mathcal{X}} c_j$ of the sets in $\mathcal{X}$. The outcome of the game is a *cover* $\mathcal{C}$, which is a subset of $\mathcal{S}$. Many practical problems can be formulated as a set cover game defined above. For example, consider the following scenario: a business can choose from a set of service providers $\mathcal{S} = \{S_1, S_2, \cdots, S_m\}$ to provide services to a set of service receivers $U = \{e_1, e_2, \cdots, e_n\}$.

* With a fixed cost c_j, each service provider S_j can provide services to a fixed subset of service receivers.
* There may be a limit $k_{j,i}$ on the number of units of service that a service provider S_j can provide to a service receiver e_i.
* Each service receiver e_i may have a limit r_i on the number of units of service that it desires to receive (and is willing to pay for).

A mechanism of the game is to determine an optimal (or approximately optimal) outcome of the game, according to a pre-defined objective function. We design various mechanisms that are aware of the fact that the service receivers and/or the service providers are selfish and rational. In addition to truthfulness, we aim to achieve the following objectives, which are sometimes at odds with each other and thus require proper tradeoffs.

* **Economic Efficiency.** A mechanism is α-efficient if its output is no worse is than α times the optimal solution with respect to the objective function.
* **Budget Balance.** Let $C(S)$ be the total cost incurred by providing services to all agents in S. If $\xi_i(S)$ is the cost charged to each agent $i \in S$, the cost-sharing method is β-budget-balanced if $\sum_{i \in S} \xi_i(S) \geq \beta \cdot C(S)$, for some $0 < \beta < 1$.
* **Fair Cost-Sharing.** We also need to make the cost-sharing method fair so that it encourages agents to participate. Besides the well accepted measures such as *cross-monotonicity* (*i.e.*, the cost share of an agent should not go up if more players require the service), we also consider a less-studied measure, called *fairness under core* (*i.e.*, the cost shares paid by any subset of agents should not exceed the minimum cost of providing the service to them alone), which is derived from game theory concepts [5].
* **No Positive Transfers (NPT).** The cost shares are non-negative.
* **Voluntary Participation (VP).** The utility of each agent is guaranteed to be non-negative if an element reports its bid truthfully.
* **Consumer Sovereignty (CS).** When an agent's bid is large enough, and others' bids are fixed, the agent will get the service.

We first consider the case where the elements to be covered are selfish agents; each e_i has a privately known valuation $b_{i,r}$ of the r-th unit of service to be received. We show that the truthful cost-sharing mechanism designed by a straightforward application of a cross-monotonic cost-sharing scheme is not α-efficient for any $\alpha > 0$. We present another truthful mechanism such that the total valuation of the elements covered is at least $\frac{1}{d_{\max}}$ times that of an optimal solution. This mechanism, however, may have free-riders: some elements do not have to pay at all and are still covered. We then present an alternative truthful mechanism without free-riders and it is at least $\frac{1}{d_{\max} \ln d_{\max}}$-efficient. When the sets are also selfish agents with privately known costs, we show that the cross-monotonic *payment*-sharing scheme does not induce a truthful mechanism; a set could lie about its cost to improve its utility. The positive side is that the mechanism is still truthful for elements.

Previously, Devanur *et al.* [7] studied the truthful cost-sharing mechanisms for set cover games, with elements considered to be selfish agents. In a game of this type, each element will declare its bid indicating its valuation of being covered, and the mechanism uses the greedy algorithm [8] to compute a cover with an approximately minimum total cost. Li *et al.* [6] extended this work by providing a truthful cost-sharing mechanism for multi-cover games. They also designed several cost-sharing schemes to fairly distribute the costs of the selected sets to the elements covered, for the case that both sets and elements are unselfish (*i.e.*, the will declare their costs/bids truthfully). The case of set cover games where sets are considered as selfish agents was also considered. Immorlica *et al.* [9] provided bounds on approximate budget balance for cross-monotone cost-sharing scheme for the set cover games.

2   Preliminaries

Typically, the objective function of a game is defined to be the total valuation of the agents selected by the outcome of the game. In set cover games, when sets are considered to be agents (*e.g.*, [6]), maximizing the total valuation of all selected agents is equivalent to minimizing the total cost of all selected sets. However, if the elements are considered to be agents, the objective becomes to maximize the total valuation of all elements (*i.e.*, the sum of all bids covered). Correspondingly, we need to solve the following optimization problem:

Problem 1. Each element e_i is associated with a *coverage requirement* r_i and a set of bids $B_i = \{b_{i,1}, b_{i,2}, \cdots, b_{i,r_i}\}$ such that $b_{i,1} \geq b_{i,2} \geq \cdots \geq b_{i,r_i}$. An *assignment* $\mathcal{C}$ is defined as the following: i) $\mathcal{C} \subseteq \mathcal{S}$; ii) a bid $b_{i,r}$ can be assigned to at most one set $S_{\pi(i,r)} \in \mathcal{C}$; iii) for any $S_j \in \mathcal{C}$, the *assigned value* $\nu_j(\mathcal{C}) = \sum_{\pi(i,r)=j} b_{i,r}$ is no less than c_j (S_j is "affordable"); iv) $\kappa_{j,i} \leq k_{j,i}$, where $\kappa_{j,i}$ is the number of bids of e_i assigned to S_j; v) if the number γ_i of assigned bids of e_i is less than r_i, then the assigned bids must be the first γ_i bids (with the greatest bid values) of e_i. The *total value* $V(\mathcal{C}) = \sum_{S_j \in \mathcal{C}} \nu_j(\mathcal{C})$ is the sum of all assigned bids in $\mathcal{C}$. The problem is to find an assignment with the maximum total value.

This problem is NP-hard. In fact, the weighted set packing problem, which is NP-complete, can be viewed as a special case of this problem. Therefore, the VCG mechanism cannot be used here if polynomial-time computability is required. In the rest of the paper, we concentrate on designing approximately efficient and polynomial-time computable mechanisms.

All our methods follow a round-based greedy approach: in each round t, we select some set S_{j_t} to cover some elements. After the s-th round, we define the *remaining required coverage* r_i' of an element e_i to be $r_i - \sum_{t'=1}^{s} \kappa_{j_{t'},i}$. For any $S_j \notin \mathcal{C}_{grd}$, the *effective coverage* $k_{j,i}'$ of e_i by S_j is defined to be $\min\{k_{j,i}, r_i'\}$.

The *effective value* (or *value* for short) v_j of S_j is therefore $\sum_{i=1}^{n} \sum_{r=1}^{k_{j,i}'} b_{i,r_i-r_i'+r}$ and it is *affordable* after s-th round if $v_j \geq c_j$.

One scheme is to select a set S_j as long as it is still affordable, and assign all appropriate bids to S_j. However, in this case an element may find it profitable to lie about its bid, as we will show in Section 3. An alternative scheme is to pick a set only if it is *individually affordable*, as defined as the following:

Definition 1. *A set S_j is individually affordable by d bids if it contains at least d bids each with a value no less than $\frac{c_j}{d}$, for some $d > 0$.*

Consequently, only the d largest bids are assigned to S_j, for the maximum d such that S_j is individually affordable by d bids. Notice that here an implicit assumption is that each set S_j can selectively provide coverage to a subset of elements contained by S_j. This is to prevent anybody from taking "free rides." The *modified value* $\tilde{v}_j$ of S_j is defined to be the total value of these bids. The following lemma gives upper bounds on the total value lost by enforcing individually affordable sets:

Lemma 1. *For any set $S_j \in \mathcal{S}$, i) if S_j is individually affordable, the modified value $\tilde{v}_j$ is no less than $\frac{1}{\ln d_{\max}}$ fraction of its value v_j; ii) if S_j is not individually affordable, its value is no more than $\ln d_{\max}$ times the cost c_j of S_j.*

3 Set Cover Games with Selfish Receivers

In this section we first study the case where only elements are selfish.

An obvious solution to designing a truthful mechanism for single-cover set cover games is to use a cross-monotone cost-sharing scheme based on a theorem proved in [10]: a cross-monotone cost-sharing scheme implies a group-strategyproof mechanism when the cost function is submodular, non-negative, and non-decreasing. A cost function C is submodular if $C(T_1) + C(T_2) \geq C(T_1 \cup T_2) + C(T_1 \cap T_2)$. A cost function C is non-decreasing if $C(T_1) \leq C(T_2)$ for any $T_1 \subseteq T_2$. A cost-sharing scheme is group-strategyproof if, for any group of agents who collude in revealing their valuations, if no member is made worse off, then no member is made better off. For set cover games, it is not difficult to show by example that the following cost functions are *not* submodular: the cost $c(\mathcal{C}_{opt})$ defined by the optimal cover $\mathcal{C}_{opt}$ of a set of elements, and the cost defined by the traditional greedy method (*i.e.*, in every round we select the set S_j with the minimum ratio of cost c_j over the number of elements covered by S_j and not covered by sets selected before)[1]. Even if a cost function is submodular, sometimes it may be NP-hard to compute this cost, and thus we cannot use this cost function to design a truthful mechanism. It was shown in [6] that there is a cost function that is indeed submodular: for each element $e_i \in T$, we select the set S_j with the minimum cost that covers e_i. Notice that, if it is a multi-cover set cover game, each set S_j is only eligible to cover an element e_i $k_{j,i}$ times. Let $\mathcal{C}_{lcs}(T)$ be all sets selected to cover a set of elements T. Then $c(\mathcal{C}_{lcs})$ is submodular, non-decreasing, and non-negative.

Given the cost function $c(\mathcal{C}_{lcs})$, it was shown in [6] that the cost-sharing method $\xi_i(T)$, defined as $\xi_i(T) = \sum_{S_j \in \mathcal{C}_{lcs}(T)} \frac{\kappa_{j,i} \cdot c_j}{\sum_a \kappa_{j,a}}$, is budget-balanced, cross-monotone and a $\frac{1}{2n}$-core. Here $\kappa_{j,i}$ is the number of bids of e_i assigned to S_j. For a single-cover set cover game, based on the method described in [10], given the single bid $b_{i,1}$ by each element e_i, we can define a mechanism $M(\xi)$ as follows.

The following theorem is directly implied by the result in [10].

Theorem 1. *The cost-sharing mechanism $M(\xi)$ is group-strategyproof, budget-balanced, and meets NPT, CS, and VP.*

However, this mechanism is not *efficient* at all. We can construct an example to show that it cannot be α-efficient for any $\alpha > 0$. Next, in Algorithm 2, we describe a new greedy algorithm that computes for a single cover game an approximately optimal assignment $\mathcal{C}_{grd}$. Starting with $\mathcal{C}_{grd} = \emptyset$, in each round t' the algorithm adds to $\mathcal{C}_{grd}$ a set $S_{j_{t'}}$ with the maximum effective value.

---

[1] Notice that the greedy method we will present later is different from this traditional greedy set cover method.

Algorithm 1. Mechanism for single cover games via cost-sharing

1: $S^0 = U$; $t = 0$;
2: **repeat**
3:   $S^{t+1} = \{e_i \mid b_{i,1} \geq \xi_i(S^t)\}$; $t = t + 1$;
4: **until** $S^{t-1} = S^t$
5: The output of mechanism $M(\xi)$ is $\tilde{U}(\xi, b) = S^t$,
6: The charge by $M(\xi)$ to an element e_i is $\xi_i(\tilde{U}(\xi, b))$.

Algorithm 2. Greedy algorithm for single cover games

1: $\mathcal{C}_{grd} \leftarrow \emptyset$.
2: For all $S_j \in \mathcal{S}$, x compute effective value v_j.
3: **while** $\mathcal{S} \neq \emptyset$ **do**
4:   pick set S_t in $\mathcal{S}$ with the maximum effective value v_t.
5:   $\mathcal{C}_{grd} \leftarrow \mathcal{C}_{grd} \cup \{S_t\}$, $\mathcal{S} \leftarrow \mathcal{S} \setminus \{S_t\}$.
6:   **for all** $e_i \in S_t$ **do**
7:     $\pi(i, 1) \leftarrow t$; remove e_i from all $S_j \in \mathcal{S}$.
8:   **for all** $S_j \in \mathcal{S}$ **do**
9:     update effective value v_j.
10:    If $v_j < c_j$, then $\mathcal{S} \leftarrow \mathcal{S} \setminus \{S_j\}$.

The following theorem establishes an approximation bound for the algorithm.

Theorem 2. *Algorithm 2 computes an assignment $\mathcal{C}_{grd}$ with a total value* $V(\mathcal{C}_{grd}) \geq \frac{1}{d_{\max}} \cdot V(\mathcal{C}_{opt})$.

Obviously, Algorithm 2 satisfies the monotone property defined in [11]: when an element e_i was selected with a bid $b_{i,1}$, then it will always be selected with a bid $\bar{b}_{i,1} > b_{i,1}$. This monotone property implies that there is always a truthful cost-sharing mechanism using Algorithm 2 to compute its output. Further, Algorithm 2 is a round-based greedy method that satisfies the cross-independence property defined in [11]. Thus, the payment to each element can always be computed in polynomial time. We include the description of this mechanism in the full version of this paper [13].

However, Algorithm 2 and and its induced cost-sharing mechanism together may produce an output such that the payment by a certain element is 0. To avoid this zero payment problem, we use a slightly different algorithm to determine the outcome of the game. Our modified greedy method (described in Algorithm 3) instead only selects individually affordable sets. When a set S_j is added into $\mathcal{C}_{grd}$, the algorithm only assigns to S_j the largest d bids, such that S_j is individually affordable with d bids, for the maximum such d. Using the same argument, we can show that there is a polynomial-time computable and truthful cost-sharing mechanism using Algorithm 3.

On the approximate efficiency of the modified greedy algorithm, we have

Theorem 3. *When only individually affordable sets are allowed to be picked, the assignment $\mathcal{C}_{grd}$ computed by Algorithm 3 has a total value that is: 1) no less*

Algorithm 3. Improved greedy algorithm for single cover games

1: $\mathcal{C}_{grd} \leftarrow \emptyset$.
2: For all $S_j \in \mathcal{S}$, compute the modified value $\tilde{v}_j$.
3: **while** $\mathcal{S} \neq \emptyset$ **do**
4:     pick set S_t in $\mathcal{S}$ with the maximum modified value $\tilde{v}_t$.
5:     $\mathcal{C}_{grd} \leftarrow \mathcal{C}_{grd} \cup \{S_t\}$, $\mathcal{S} \leftarrow \mathcal{S} \setminus \{S_t\}$.
6:     $d_t \leftarrow$ the largest d such that the set S_t is individually affordable by d largest
        unsatisfied bids.
7:     **for all** $e_i \in S_t$ **do**
8:         **if** $b_{i,1}$ is one of the largest d_t unsatisfied bids in S_t **then**
9:             $\pi(i,1) \leftarrow t$; remove e_i from all $S_j \in \mathcal{S}$.
10:     **for all** $S_j \in \mathcal{S}$ **do**
11:         update the modified value $\tilde{v}_j$.
12:         If $\tilde{v}_j < c_j$, then $\mathcal{S} \leftarrow \mathcal{S} \setminus \{S_j\}$.

---

than $\frac{1}{d_{\max}} \cdot V(\mathcal{C}_{opt})$, if the optimal assignment $\mathcal{C}_{opt}$ also allows only individually affordable sets; 2) no less than $\frac{1}{2d_{\max}} \cdot V(\mathcal{C}_{opt})$, if the optimal assignment $\mathcal{C}_{opt}$ allows sets that are not individually affordable, but all sets in $\mathcal{S}$ are individually affordable initially.

Theorem 2 and Theorem 3 can easily be extended to the case of multi-cover. However, when it comes to computing payments, there is a problem: in the multi-cover case, an element can lie in different ways, and it may not be of its best interest if it achieves the maximum utility in the first bid (or the last bid). In that case, how can we compute payments efficiently?

To overcome the computational complexity of computing payments, we design another mechanism using a different greedy algorithm to compute the outcome of the game. This algorithm is the same as Algorithm 3 of [6]. In [6] it is shown that this mechanism produces an outcome with a total cost no more than $\ln d_{\max}$ times the total cost of an optimal outcome. We claim that the outcome is also approximately efficient with respect to the total valuation of the assigned (covered) bids. Further, due to the monotone property, this mechanism is truthful.

Theorem 4. *Algorithm 3 of [6] defines a budget-balanced and truthful mechanism. Further, it is $\frac{1}{d_{\max} H_{d_{\max}}}$-efficient, if all sets are individually affordable initially.*

4   Set Cover Games with Selfish Providers and Receivers

So far, we assume that the cost of each set is publicly known or each set will truthfully declare its cost. In practice, it is possible that each set could also be a selfish agent that will maximize its own benefit, *i.e.*, it will provide the service only if it receives a payment by some elements (not necessarily the elements covered by itself) large enough to cover its cost. In [6], Li *et al.* designed several truthful payment schemes to selfish sets such that each set maximizes its utility when it truthfully declares its cost and the covered elements will pay whatever a charge computed by the mechanism. They also designed a payment sharing scheme that is budget-balanced and in the core.

To complete the study, in this section, we study the scenario when both the sets and the elements are individual selfish agents: each set S_j has a privately known cost c_j, while each element e_i has a privately known bid $b_{i,r}$ for the r-th unit of service it shall receive and is willing to pay for it only if the assigned cost is at most $b_{i,r}$. It is well-known that a cross-monotone *cost* sharing scheme implies a truthful mechanism [10]. Unfortunately, since the sets are selfish agents, it is impossible to design any cost-sharing scheme here, and the best we can do is to design some payment sharing scheme. It was shown in [12] that a cross-monotone payment sharing scheme does *not* necessarily induce a truthful mechanism by using multicast as a running example: a relay node could lie its cost upward or downward to improve its utility.

Given a subset of elements $T \subseteq U$ and their coverage requirement r_i for $e_i \in T$, a collection of multisets $\mathcal{S}$, and each set $S_j \in \mathcal{S}$ with cost c_j, let $M_{\mathcal{S}}$ be a truthful mechanism that will determine which sets from $\mathcal{S}$ will be selected to provide the coverage to *all* elements T, and the payment p_j to each set S_j. We assume that the mechanism is normalized: the payment to an unselected set S_j is always 0. Based on two monotonic output methods, the traditional greedy set cover method (denoted as GRD) and the least cost set method (denoted as LCS) for each element, Li *et al.* [6] designed two truthful mechanisms for set cover games. Let $E(S_j, c, T, M_{\mathcal{S}})$ be the set of elements covered by S_j in the output of $M_{\mathcal{S}}$. In the remaining of the paper, we assume that the mechanism $M_{\mathcal{S}}$ satisfies the property that if a set S_j increases its cost then the set of elements covered by S_j in the output of $M_{\mathcal{S}}$ will *not* increase, i.e., $E(S_j, c|^j d, T, M_{\mathcal{S}}) \subseteq E(S_j, c, T, M_{\mathcal{S}})$ for $d > c_j$. This property is satisfied by all methods currently known for set cover games.

Let $\xi_{i,j}(T)$ be the shared payment by element e_i for its jth copy when the set of elements to be covered is T, given a truthful payment scheme $M_{\mathcal{S}}$ to all sets. Following the method described in [10], given the set U of n elements and their bids $B_1, \cdots, B_n$ we can compute the outcome $\tilde{U}(\xi, B)$ as the limit of the following inclusion monotonic sequence: $S^0 = U$; $S^{t+1} = \{e_i \mid b_{i,j} \geq \xi_{i,j}(S^t)\}$. Notice that here we have to recompute the payments to all sets, and thus the shared payments by all elements, when the set of elements to be covered is changed from S^t to S^{t+1}. In other words, we define a mechanism $M_E(\xi)$ associated with the payment sharing method ξ as follows: the set of elements to be covered is $\tilde{U}(\xi, B)$, the charge to element e_i is $\xi_{i,j}(\tilde{U}(\xi, B))$ if $e_i \in \tilde{U}(\xi, B)$; otherwise its charge is 0. Based on the truthful mechanism using LCS as output for set cover games, Li *et al.* [6] designed a payment sharing mechanism that is budget-balanced, cross-monotone, and in the core.

Hereafter, we assume that for the payment-sharing scheme ξ, the payment p_j to the set S_j is only shared among the elements, i.e., $E(S_j, c, T, M_{\mathcal{S}})$, covered by S_j. This property is satisfied by the payment-sharing methods studied in [6] for set cover games.

Theorem 5. *For set cover games with selfish sets and elements, a truthful mechanism M_S to sets and a cross-monotone payment sharing scheme ξ imply that in mechanism M_E each set S_j cannot improve its utility by lying upward its cost.*

Unfortunately, for set cover games, we show that a truthful mechanism M_S to sets and a cross-monotone payment sharing scheme ξ do *not* induce a truthful mechanism M_E for each element. Figure 1 illustrates such an example when LCS is used as the output, a set s_j can lie its cost downward to improve its utility from 0 to $p_j - c_j$. A similar example can be constructed when the traditional greedy method is used as the output. When set S_2 is truthful, although LCS will select it to cover element e_1 with payment $p_2 = 5$, but the corresponding sharing by e_1 is $\xi_1 = 5$, which is larger then its bid $b_{1,1} = 4$. Consequently, set S_2 will not be selected and element e_1 will not be covered (see Figure 1 (c)). On the other hand, if S_2 lies its cost downward to $\bar{c}_2 = 2$, its payment is still $p_2 = 5$, but now, since it covers elements e_1 and e_2, the shared payments by e_1 and e_2 become $\xi_1 = 3.5$ and $\xi_2 = 1.5$. Thus, the set S_2 becomes affordable by elements e_1 and e_2.

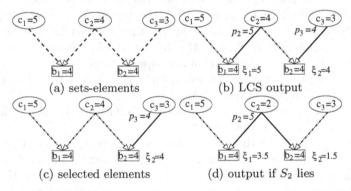

(a) sets-elements   (b) LCS output

(c) selected elements   (d) output if S_2 lies

Fig. 1. An example that a set can lie its cost to improve its utility when LCS is used

We leave it as future work to study whether there exists a truthful mechanism to select selfish sets to cover selfish elements using the combination of a truthful mechanism for sets, and a good payment-sharing method for elements.

5   Conclusion

Strategyproof mechanism design has attracted a significant amount of attentions recently in several research communities. In this paper, we focused the set cover games when the elements are selfish agents with privately known valuations of being covered. We presented several (approximately budget-balanced) truthful mechanisms that are approximately efficient. See [13] for more details about the algorithms and the analysis. Mechanism 1 is based on a cross-monotone cost-sharing scheme and thus is budget-balanced and group-strategyproof. However, in the worse case it cannot be α-efficient for any $\alpha > 0$. The second mechanism is based on Algorithm 2 and its induced cost-sharing mechanism and it produces an output that has a total valuation at least $\frac{1}{d_{\max}}$ of the optimal. However, this mechanism may charge an element 0 payment. The third mechanism, based on Algorithm 3, avoids this zero payment problem, but it is only $\frac{1}{2d_{\max}}$-efficient under some assumptions. We conducted extensive simulations to study the actual total valuations of three mechanisms. In all our simulations, we found that the first mechanism (based on cost-sharing) and the second mechanism have similar efficiencies in practice. As expected, the third mechanism always produces an output that has less total valuations than the other two methods since it only picks sets that are individually affordable.

When the service providers (*i.e.* sets) are also selfish, we show that a cross-monotonic *payment*-sharing scheme does not necessarily induce a truthful mechanism. This is a sharp contrast to the well-known fact [10] that a cross-monotonic *cost*-sharing scheme always implies a truthful mechanism.

References

1. Vickrey, W.: Counterspeculation, auctions and competitive sealed tenders. Journal of Finance **16** (1961) 8–37
2. Clarke, E.H.: Multipart pricing of public goods. Public Choice **11** (1971) 17–33
3. Groves, T.: Incentives in teams. Econometrica **41** (1973) 617–631
4. Green, J., Laffont, J.J.: Characterization of satisfactory mechanisms for the revelation of preferences for public goods. Econometrica **45** (1977) 427–438
5. Osborne, M.J., Rubinstein, A.: A course in game theory. The MIT Press (1994)
6. Li, X.Y., Sun, Z., Wang, W.: Cost sharing and strategyproof mechanisms for set cover games. In: Proceedings of the 22nd Int. Symp. on Theoretical Aspects of Compt. Sci. Volume 3404 of LNCS (2005) 218–230
7. Devanur, N., Mihail, M., Vazirani, V.: Strategyproof cost-sharing mechanisms for set cover and facility location games. In: Proceedings of the 4th ACM EC. (2003) 108–114
8. Chvátal, V.: A greedy heuristic for the set-covering problem. Mathematics of Operation Research **4** (1979) 233–235
9. Immorlica, N., Mahdian, M., Mirrokni, V.S.: Limitations of cross-monotonic cost-sharing schemes. In: Proceedings of the 16th Annual ACM-SIAM SODA. (2005)
10. Moulin, H., Shenker, S.: Strategyproof sharing of submodular costs: budget balance versus efficiency. Economic Theory **18** (2001) 511–533
11. Kao, M.Y., Li, X.Y., Wang, W.: Toward truthful mechanisms for binary demand games: A general framework. In: Proceedings of the 6th ACM EC. (2005).
12. Wang, W., Li, X.Y., Sun, Z., Wang, Y.: Design multicast protocols for non-cooperative networks. In: Proceedings of the 24th IEEE INFOCOM. (2005).
13. Sun, Z., Li, X.Y., Wang, W.,: Mechanism Design For Set Cover Games When Elements Are Agents. Full version of this paper at http://www.cs.iit.edu/~xli/paper/Conf/AAIM_SetCover.pdf.

Graph Bandwidth of Weighted Caterpillars

Zhiyong Lin, Mingen Lin, and Jinhui Xu

Department of Computer Science and Engineering,
University at Buffalo, the State University of New York,
Buffalo, NY 14260, USA
{zlin, mlin6, jinhui}@cse.buffalo.edu

Abstract. Graph bandwidth minimization (GBM) is a classical and challenging problem in graph algorithms and combinatorial optimization. Most of existing researches on this problem have focused on unweighted graphs. In this paper, we study the bandwidth minimization problem of weighted caterpillars, and propose several algorithms for solving various types of caterpillars. More specifically, we show that the GBM problem of caterpillars with hair-length at most 2 and the GBM problem of star-shape caterpillars are NP-complete, and give a lower bound of the graph bandwidth for general weighted graphs. For caterpillars with hair-length at most 1, we present an $O(n \log n \log(n w_{max}))$-time algorithm to compute an optimal bandwidth layout, where n is the total number of vertices in the graph and w_{max} is the maximum edge weight. For caterpillars with hair-length at most k, we give a k-approximation algorithm. For arbitrary caterpillars and general graphs, we give a heuristic algorithm. Experiments show that the solutions obtained by our heuristic algorithm are roughly within a factor of $\log(2n)$ of the lower bound.

1   Introduction

The graph bandwidth minimization (or GBM) problem for a given graph $G = (V, E)$ is to find a linear ordering (layout) of the vertices V such that the maximum difference between the labels (or rankings) of the two endpoints of any edge in E is minimized. The GBM problem is a fundamental problem in the fields of graph algorithms and combinatorial optimization, and finds applications in many other areas. Excellent surveys on existing results can be found in [1, 2].

The GBM problem in general is extremely difficult. It has been shown that the decision version of this problem is NP-complete for general graphs [3]. The problem remains to be NP-complete even for simple graphs such as trees with maximum degree 3 [4] and caterpillars with hair-length no more than 3 [5]. Polynomial time solutions were found only for very special graphs such as caterpillars with hair-length no more than 2 [6] and interval graphs [7]. In 1998, Blache *et al.* showed that there is no PTAS for general graphs and trees [8] unless P=NP. Later, Unger showed that it is NP-hard to approximate this problem within any constant factor [9].

Several approximation algorithms have been obtained for general and special graphs. In [10], Feige presented an $O(\log^3 n \sqrt{\log n \log \log n})$-approximation algo-

N. Megiddo, Y. Xu, and B. Zhu (Eds.): AAIM 2005, LNCS 3521, pp. 370–380, 2005.

rithm using the powerful volume respecting embedding technique. Later, Gupta showed that the approximation ratio for trees can be improved to $O(\log^2 n \sqrt{\log n})$ using similar techniques [11]. For caterpillars, Haralambides *et al.* presented a simple but elegant $O(\log n)$-approximation algorithm [12].

So far most of the researches on the GBM problem have focused on unweighted graphs. Very few results were known for weighted graphs. Part of the reason is that the edge weights could dramatically complicate the optimization task. The objective of the GBM problem on weighted graphs is to minimize the maximum weighted label difference of an edge (i.e., the edge weight times the label difference). In this paper, we study the GBM problem of weighted caterpillars, motivated by interesting applications in VLSI layout.

Comparing to the unweighted counterpart, the GBM problem of weighted caterpillars seems to be much harder to solve. For example, Monien showed that the GBM problem for unweighted caterpillars with hair-length at most 3 is NP-complete [5], and Assmann *et al.* gave an algorithm to find an optimal layout for unweighted caterpillars with hair-length 1 and 2 [6]. For weighted caterpillars, we show that the GBM problem is NP-complete even for caterpillars with hair-length at most 2 and for star-shape caterpillars. To overcome the additional difficulty caused by edge weights, we give a lower bound for general weighted graphs. For caterpillars with hair-length at most 1, we present an $O(n \log n \log(n w_{max}))$-time algorithm to compute an optimal layout, where n is the number of vertices and w_{max} is the maximum edge weight. The algorithm in [6] for unweighted caterpillars cannot guarantee an optimal solution for this case. For caterpillars with hair-length at most k, we present a k-approximation algorithm. For arbitrary caterpillars and general graphs, we give a heuristic algorithm. Experiments show that the solutions obtained by our algorithm are roughly within a factor of $\log(2n)$ of the lower bound.

2   Preliminaries

Let $G = (V, E)$ be a weighted graph with weights $w : E \to Q^+$ and $|V| = n$. A *linear layout*, or simply a *layout* L of G is an ordering of V with a bijective labeling function $L : V \to [n] = \{1, 2, \ldots, n\}$. We say that G has bandwidth B under the layout L, denoted by $b_L(G) \leq B$, if $|L(u) - L(v)| \times w(u, v) \leq B$ for every edge $(u, v) \in E$. The bandwidth $b(G)$ of G is the minimum bandwidth under all possible layouts, i.e., $b(G) = \min_L \{b_L(G) | L \text{ is a layout of } G\}$. We assume that G is connected, since otherwise the bandwidth of G is simply that of the connected component with the largest bandwidth.

Let $e \in E$ be an edge and $w(e)$ be the weight of e. We define *distance* of e as $d(e) = \frac{1}{w(e)}$. The distance $d(P)$ of a path P is the sum of all distances of edges on this path. The *diameter* $d(G)$ of G is defined as the distance of the longest shortest path between any pair of vertices in G.

A *caterpillar* C is a tree which has a simple path, called *backbone*, and various other appendage line graphs attached to the vertices of the backbone. Each line graph is called a *hair*. A hair of the caterpillar is called an h_v subtree if it is attached to a backbone vertex v. The hair-length of h_v is the number of edges of h_v and the hair-length of C is the maximum hair-length of all its hairs. Let u be a vertex in an h_v subtree and x be the length (i.e., the number of edges) of path $v \rightsquigarrow u$. Let u' be another vertex in the same subtree h_v. We say u' is the parent of u if the length of the path $v \rightsquigarrow u'$ is $x - 1$.

3   Lower Bound of General Graphs

In the section, we give a lower bound for the GBM problem of general graphs.

Lemma 1. *For any connected subgraph G' of G, there exists a path $P = v_1 v_2 \ldots v_t$ in G' such that*

$$\sum_{i=1}^{t-1} \left\lfloor \frac{b(G')}{w(v_i v_{i+1})} \right\rfloor \geq |G'| - 1,$$

where P is a shortest path between v_1 and v_t, and $|G'|$ is the number of vertices in G'.

Proof. Omitted in this extended abstract.

Lemma 2. *Let G' be any connected subgraph of G. The bandwidth $b(G) \geq \max_{G'} \frac{|G'|-1}{d(G')}$*

Proof. Omitted in this extended abstract.

4   Algorithm for Caterpillars with Hair-Length At Most 1

Let C_1 be a caterpillar with hair-length no more than 1. We consider the decision version of this problem. That is, given a positive integer b, decide whether there exists a layout L such that $b_L(C_1) \leq b$. Note that although the bandwidth is a rational number, we can always scale up the weights and make it an integer. Below we first present an algorithm for this decision problem. Our algorithm takes a caterpillar C_1 with hair-length no more than 1 and an integer b as inputs and tries to find a layout with bandwidth no more than b. We then show that if the algorithm fails to find such a layout, then there is a subgraph C_1' of C_1 that violates Lemma 1. Thus, $b(C_1) \geq b(C_1') > b$.

We say a number in $[1, |C_1|]$ is *free* if no vertex is permanently labeled with it. Let $L(v)$ be the label of a vertex v. Let u be a hair vertex attached to a backbone vertex v. A possible label for u should be a free number in $[L(v) - \lfloor b \times d(u,v) \rfloor, L(v) + \lfloor b \times d(u,v) \rfloor]$, where $d(u,v)$ is the distance between u and v. Let $D = v_1 v_2 \ldots v_r$ be the backbone of C_1. Our algorithm has the following main steps.

ALGORITHM 1

1. Label permanently the vertices along the backbone D with numbers $0, \lfloor b \times d(v_1, v_2) \rfloor, \lfloor b \times d(v_1, v_2) \rfloor + \lfloor b \times d(v_2, v_3) \rfloor, \ldots, \sum_{i=1}^{r-1} \lfloor b \times d(v_i, v_{i+1}) \rfloor$.
2. For each hair vertex u attached to some backbone vertex v, label it temporarily with the number $L(v) + \lfloor b \times d(u, v) \rfloor$. Note that multiple vertices may be labeled (permanently or temporarily) with a same number.
3. Sort vertices in the non-decreasing order of their labels, and break ties arbitrarily.
4. Scan all hair vertices in the above order and label permanently each hair vertex u with the smallest free number in the range $[L(v) - \lfloor b \times d(u, v) \rfloor, L(v) + \lfloor b \times d(u, v) \rfloor]$, where v is the parent of u. If there is no free number for u, then stop and return false.

Below we show that the above algorithm solves the decision problem of the GBM problem on caterpillars with hair-length at most 1.

Theorem 1. *If the above algorithm fails to find a layout L of C_1 with $b_L(C_1) = b$, then $b(C_1) > b$.*

Proof. Suppose the algorithm returns false. We will show that there exists a subgraph C_1' of C_1 that violates Lemma 1.

In the above algorithm, a failure can occur only in Step 4 when it tries to label some hair vertex, say u, with a temporary label l'. Obviously, there must exist at least another vertex with label l', otherwise u can be labeled permanently with l'.

When the failure occurs, let $L' = [l_1, l_t = l']$ be the longest consecutive sequence of numbers that are occupied (i.e., each $l_i \in L'$ has been assigned as a permanent label to some vertex u_i and $l_1 - 1$ is free). Let $V' = \{u_1, u_2, \ldots, u_t, u\}$. We show that the induced subgraph $C_1' = (V', E')$ violates Lemma 1.

We note that the following properties are true for vertices in V'.

1. Each $u_i, i \in [1, t]$ is either a backbone vertex or a hair vertex attached to some backbone vertex.
2. If $u_i, i \in [1, t]$ is a hair vertex and v_i is its parent, then $v_i \in V'$.

To show that the second property is true, we need to prove that $l_1 \le L(v_i) \le l'$. To see $L(v_i) \le l'$, we note that we always assign to hair vertices temporary labels no less than the permanent labels of their parents and scan hair vertices based on the non-decreasing order of their temporary labels. The temporary label of u_i must be no larger than that of u (i.e., l'), and hence $L(v_i) \le l'$. To show $l_1 \le L(v_i)$, we assume that $l_1 > L(v_i)$. Then in the range $[L(v_i) - \lfloor b \times d(u_i, v_i) \rfloor, L(v_i) + \lfloor b \times d(u_i, v_i) \rfloor]$, we have a free number $l_1 - 1 < L(u_i)$ which can be used to label u_i. This contradicts the fact that we always find the smallest free label for each hair vertex in its range.

The above two properties implies that C_1' is a connected graph since by step 1, backbone vertices in C_1' are always connected .

Let $v_h, v_{h+1}, \ldots, v_k, \ldots, v_g$ be backbone vertices in V'. Suppose u_1 is connected to v_x and u is attached to v_y. Let $P = u_1 v_x v_{x+1} \ldots v_z u_t$ be a path (see Figure 1). Let $d'(v_i, v_j) = \sum_{l=i}^{j-1} \lfloor b \times d(v_l, v_{l+1}) \rfloor$. We have the following claim.

Claim. $\lfloor b \times d(u_1, v_x) \rfloor + d'(v_x, v_z) + \lfloor b \times d(v_z, u_t) \rfloor$ is the maximum over the distances of all paths in C'.

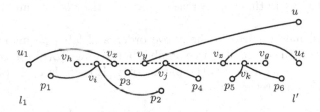

Fig. 1. An ordering of vertices of a subgraph C_1' (vertices are ordered from left to right)

To prove the above claim, we first consider an arbitrary hair h whose both endpoints are in V'. h can be attached to either a backbone vertex v_i between v_h and v_x, a backbone vertex v_j between v_x and v_z, or a backbone vertex v_k between v_z and v_g (see Figure 1). For each of the three cases, we have two subcases depending on whether the hair vertex of h has a smaller or greater label than its parent (i.e., the backbone vertex). Thus we have in total 6 types of hairs, $p_1 v_i$, $p_2 v_i$, $p_3 v_j$, $p_4 v_j$, $p_5 v_k$ and $p_6 v_k$.

For hair $p_1 v_i$, we have $\lfloor b \times d(p_1, v_i) \rfloor + d'(v_i, v_x) \leq \lfloor b \times d(v_x, u_1) \rfloor$, otherwise p_1 would be laid out to the left of u_1 by Step 4 of the algorithm. Also $\lfloor b \times d(p_1, v_i) \rfloor \leq d'(v_i, v_z) + \lfloor b \times d(v_z, u_t) \rfloor$, otherwise p_1 will not be relabeled before u_t. Similarly, we have the following inequality for hairs $p_3 v_j$ and $p_5 v_k$.

$$\lfloor bd(p_3 v_j) \rfloor \leq \lfloor bd(u_1 v_x) \rfloor + d'(v_x v_j)$$
$$\lfloor bd(p_3 v_j) \rfloor \leq d'(v_j v_z) + \lfloor bd(v_z u_t) \rfloor$$
$$\lfloor bd(p_5 v_k) \rfloor \leq \lfloor bd(u_1 v_x) \rfloor + d'(v_x v_k)$$
$$\lfloor bd(p_5 v_k) \rfloor + d'(v_z v_k) \leq \lfloor bd(v_z u_t) \rfloor$$

Similar results hold for hairs $p_2 v_i$, $p_4 v_j$ and $p_6 v_k$.

Putting all the inequalities together, we can show that $\lfloor b \times d(u_1, v_x) \rfloor + d'(v_x, v_k) + \lfloor b \times d(v_z, u_t) \rfloor$ is the maximum over all paths in C_1' by considering all types (in total 15) of paths which start and end with one of the 6 types of hairs. Below we only consider one type of paths $P' = p_2 v_i v_{i+1} \ldots v_j p_3$. The other types of paths can be proved similarly (details are left for the full paper).

$$\lfloor b \times d(p_2, v_i) \rfloor + d'(v_i, v_j) + \lfloor b \times d(v_j, p_3) \rfloor$$
$$\leq \lfloor b \times d(u_1, v_x) \rfloor + d'(v_x, v_j) + \lfloor b \times d(v_j, p_3) \rfloor$$
$$\leq \lfloor b \times d(u_1, v_x) \rfloor + d'(v_x, v_j) + d'(v_j, v_z) + \lfloor b \times d(v_z, u_t) \rfloor$$
$$= \lfloor b \times d(u_1, v_x) \rfloor + d'(v_x, v_z) + \lfloor b \times d(v_z, u_t) \rfloor$$

Since all numbers in $[l_1, l_t]$ are occupied, we have $|C_1'| = l' - l_1 + 2$.

$$\lfloor b \times d(u_1, v_x) \rfloor + d'(v_x, v_k) + \lfloor b \times d(v_z, u_t) \rfloor = l' - l_1 = |C_1'| - 2$$

Thus there is no path satisfies Lemma 1, which implies that the bandwidth of C_1' is larger than b. This concludes the proof of the theorem.   □

Corollary 1. *There is an $O(n \log n \log(n w_{max}))$-time algorithm for finding the bandwidth of C_1, where w_{max} is the maximum edge weight of C_1.*

Proof. Clearly the bandwidth of C_1 is between w_{max} and $n \times w_{max}$. We can use the above algorithm for the decision problem to do a binary search on the interval $[w_{max}, n \times w_{max}]$ to find $b(C_1)$.

The whole algorithm for the decision problem takes $O(n \log n)$ time since we can maintain the free numbers using a balanced binary tree. Thus the total time to find $b(C_1)$ is $O(n \log n \log(n w_{max}))$. Once we have a labeling of C_1, it can be easily changed to a layout by ordering the vertices according to the order of their labels.   □

5   Caterpillars with Hair-Length At Most 2

In this section, we show that the GBM problem of caterpillars with hair-length at most 2 is NP-complete.

Theorem 2. *The graph bandwidth minimization problem of caterpillars with hairs-length at most 2 is NP-complete.*

Clearly the problem is in NP since given a layout and a number b we can verify in polynomial time whether the bandwidth of this layout is no more than b. To prove the NP-hardness of the problem, we follow the approach used in [5] for proving the NP-hardness for unweighted caterpillars and reduce the Multiprocessor Scheduling problem to our GBM problem. Let C_2 be a weighted caterpillar with hair-length at most 2. Given a set $T = \{t_1, t_2, \ldots, t_n\}$ of tasks (t_i is the execution time of the ith task), a deadline D, and m processors, we construct a caterpillar C_2 (see Figure 3) and an integer b such that C_2 has bandwidth b if and only if the tasks in T can be scheduled on the m processors no later than the deadline D. The Multiprocessor Scheduling problem is strongly NP-complete and therefore we can assume that all the t_i are polynomially bounded in n.

The main idea of the reduction is to simulate the scheduling. We use some portions of C_2 to simulate tasks and some portions to simulate processors, called *task portions* and *processor portions* respectively. More specifically, each task t_i is represented by a caterpillar C_1^i with t_i backbone vertices and $p - 1$ hairs (of length 1) on each backbone vertex for some parameter p, and each processor is represented by a caterpillar with $D - 1$ backbone vertices and no hair (see Figure 3). We also use two special caterpillars, called *barrier* and *turning point*

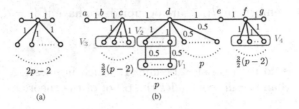

Fig. 2. (a) A barrier of height p. (b) A turning point of height p, $p \equiv 0 \bmod 2$

(see Figure 2), to separate the task portions and processor portions. A barrier is used to separate two processor portions such that a task portion can not be assigned to multiple processor portions. The task portions and processor portions are separated by a turning point. The turning point ensures that in every optimal layout, the task portions and processor portions are always on the same side of the turning point so that vertices representing tasks can be laid out between the backbone vertices representing processors.

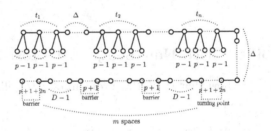

Fig. 3. The caterpillar C_2, $\Delta = 2\{m(D+2)-2\}$, p has to be chosen in an appropriate way

As shown in [5], in order to make the reduction work, a key difficulty is to ensure that the barrier and turning point have the following properties. (1) Each barrier of height p and each turning point of height p (see Figure 2) have bandwidth p; (2) In every optimal layout of the turning point, the first and last backbone vertices (i.e., a and g in Figure 2) are assigned to the same half of the layout.

To ensure the above properties, we modify the structure of the turning point in [5] by assigning different weights to its edges and shortening its hair-length from 3 to 2 (Note that the other part of the constructed caterpillar has hair-length 1). Let $T_p = (V, E)$ denote the turning point of height p. The following lemmas show that the turning point has the expected properties.

Lemma 3. *The bandwidth of the turning point T_p in Figure 2(b) is p.*

Proof. T_p has exactly $6p+1$ vertices. The diameter of T_p is clearly 6. By Lemma 2, we know that the bandwidth of T_p is no less than $\frac{6p+1-1}{6} = p$. Figure 4 gives a layout of T_p with bandwidth exactly p. Thus the lemma follows. $\square$

Lemma 4. *Let L be an optimal layout of the turning point $T_p = (V, E)$ with bijective function $L : V \to \{1, \ldots, 6p + 1\}$ (i.e., $|L(i) - L(j)|w(i, j) \le p$ for all $(i, j) \in E$) and $p \ge 6$. Then either $L(a), L(g) < 3p + 1$ or $L(a), L(g) > 3p + 1$.*

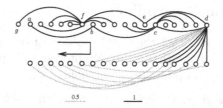

Fig. 4. An optimal layout of the turning point

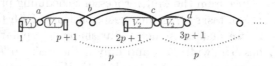

Fig. 5. A layout of a turning point

Proof. Consider the path $P = p_1 \ldots p_k$ that connects the leftmost vertex p_1 and the rightmost vertex p_k in L. The length (in terms of number of edges) of P can be 5 or 6 only.

The only path with length 6 in T_p is $a - b - c - d - e - f - p_k$, $p_k \in \{g\} \cup V_4$ and the weight of each edge on this path is 1. Since $|L(p_k) - L(p_1)| = 6p$ and $|L(u) - L(v)| \le p, \forall (u, v) \in P$, we have $L(d) = 3p + 1$. If the length of P is 5, one of p_1 and p_k must be in V_1. Suppose $p_1 \in V_1$. Since all the edges on P except $p_1 p_2$ have weight 1 and $w(p_1 p_2) = 0.5$, we also have $L(d) = 3p + 1$.

Suppose $L(a) < 3p + 1$ and $L(g) > 3p + 1$, we show below that there will be a contradiction. Since $L(d) = 3p + 1$, the vertices that can have labels $1, \ldots, p$ must be a or in V_1. We will show that $L(a) \in [1, p]$. If not, then $[1, p]$ are all occupied by V_1 and $[2p + 1, 3p]$ are all occupied by V_2. Then $L(a), L(b) \in [p + 1, 2p]$, but c can not be labeled. So $[1, p]$ must be occupied by a and $p - 1$ vertices from V_1 (Figure 5). Then we have $L(b) \in [p + 1, 2p]$ and $L(c) \in [2p + 1, 3p]$. The other $p - 1$ labels in $[2p + 1, 3p]$ must be occupied by $p - 1$ vertices from V_2. Then the vertices in V_3 can have labels within $[L(c) - p, L(c) + p]$. The number of free numbers in $[L(c) - p, L(c) + p]$ is $2p + 1 - 3 - (p - 1) = p - 1$. But we have $\frac{3}{2}(p - 2)$ vertices in V_3 which implies $\frac{3}{2}(p - 2) \le p - 1$. We get $p \le 4$ contradicts to the assumption that $p \ge 6$. $\qquad\square$

With the above lemmas, we can construct the caterpillar C_2 from any instance $\mathcal{Y} = (\{t_1, \ldots, t_n\}, D, m)$ of the Multiprocessor Scheduling problem as in Figure 3 and apply the following two lemmas (proved in [5]) to complete the reduction. This concludes the proof of Theorem 2.

Lemma 5. *If $\mathcal{Y}$ has a solution then C_2 has bandwidth $p + 1 + 2n$.*

Lemma 6. *If $p > 2n(D+4)$ and if C_2 has bandwidth $p+1+2n$, then $\mathcal{Y}$ has a solution.*

6   Star-Shape Caterpillars

In this section, we show that the GBM problem of star-shape caterpillars is NP-complete. It is easy to see that for unweighted star-shape caterpillars, the GBM problem can be solved optimally in polynomial time.

Theorem 3. *The graph bandwidth minimization problem of star-shape caterpillars is NP- complete.*

Proof. Clearly this problem is in NP. Similar to the NP-completeness proof of Theorem 2, we reduce from the Multiprocessor Scheduling problem. Given a set $S = \{t_1, t_2, \ldots, t_n\}$ of tasks (t_i is the execution time of ith task), a deadline D and m processors, we shall construct a star-shape caterpillar C_s and an integer b such that C_s has bandwidth b if and only if the tasks in T can be scheduled on the m processors satisfying the deadline D. We can assume that all the t_is are polynomially bounded in n.

We construct C_s as follows. We first build the backbone. Let $T = \sum_{i=1}^{n} t_i$. We put a set B_1 of $T+1$ nodes to the left. Then we connect a set M of m nodes and a set B_2 of $T+m+1$ nodes to the right. Corresponding to each task i, we construct a hair with t_i nodes. We associate edges with weights as in Figure 6. Each edge with its left endpoint in M represents a processor.

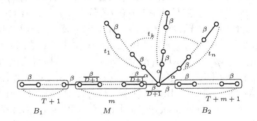

Fig. 6. The caterpillar C_s

Let $b = \beta, \alpha = \frac{\beta}{T+m+1}$. We can prove that there is a feasible scheduling for the Multiprocessor Scheduling problem if and only if we have a layout of C_s with bandwidth b (details are left for the full paper). □

7   Approximation Algorithm for Caterpillars with Hair-Length At Most k

Let C_k be a weighted caterpillar with hair-length at most k. The following simple algorithm yields a layout for C_k.

ALGORITHM 2

1. Let C'_k be the maximal subgraph of C_k such that C'_k is a caterpillar with hair-length at most 1.
2. Apply Algorithm 1 to C'_K and find an optimal layout L'.
3. Put all the descendants of each hair vertex u of C'_k to its left neighboring positions in L' while keeping their orders. Return the resulting layout L.

Theorem 4. *Algorithm 2 is a k-approximation algorithm for the GBM problem of weighted caterpillars with hair-length at most k.*

Proof. Omitted in this extended abstract.

8 Algorithm for Arbitrary Caterpillars and General Graphs

In this section, we present a heuristic algorithm for arbitrary caterpillars and general graphs and show some experimental result. Our algorithm is a generalization of the CutHill-McKcc algorithm [13]. Let $e = (u, v)$ be an edge with weight $w(e)$, $lvl(u)$ be the level of u, d_{max} (d_{min}) be the maximum (minimum) degree of the graph and $d = \frac{d_{max}+d_{min}}{2}$. Our algorithm is as follows.

ALGORITHM 3

1. For each vertex v, set $lvl(v) = \infty$.
2. For each vertex u with degree $\leq d$ do
   (a) Select u as the starting vertex. Set $lvl(v) = 1$.
   (b) For each neighbor v of u, set $lvl(v) = \min(lvl(v), lvl(u) + \frac{w_{max}}{w(u,v)})$.
   (c) The procedure is repeated for each node at levels in the increasing order.
   (d) Label vertices level by level. Store the bandwidth of the layout.
3. Output the minimum bandwidth of all computed layouts.

To examine the performance of our algorithm, we implement it in C++ and test it with a large set of randomly generated graphs. We compare our results with the lower bound given in Lemma 2. (Detailed experimental results are omitted in this extended abstract due to limited space.) Our results suggest that the ratio of our computed bandwidth over the lower bound is roughly a $\log(2n)$ function of the number of vertices.

References

1. Chinn, P., Chvatalova, J., Dewdney, A., Gibbs, N.: The bandwidth problem for graph and matrics - a survey. Journal of Graph Theory **6** (1982) 223–254
2. Diaz, J., Petit, J., Serna, M.: A survey of graph layout problems. ACM Computing Surveys **34** (2002) 313–356

3. Papadimitriou, C.: The np-completeness of the bandwidth minimization problem. Computing (1976) 263–270
4. Garey, M.R., Graham, R.L., Johnson, D.S., Knuth, D.: Complexity results for bandwidth minimization. SIAM Journal on Applied Mathematics **34** (1978) 477–495
5. Monien, B.: The bandwidth minimization problem for caterpillars with hair length 3 is np-complete. SIAM Journal on Algebraic and Discrete Methods **7** (1986) 505–512
6. Assmann, S.F., Peck, G.W., Syslo, M.M., Zak, J.: The bandwidth of caterpillars with hairs of length 1 and 2. SIAM Journal on Algebraic and Discrete Methods **2** (1981) 387–393
7. Sprague, A.P.: An $o(n \log n)$ algorithm for bandwidth of interval graphs. SIAM Journal on Discrete Mathematics **7** (1994) 213–220
8. Glache, G., Karpinski, M., Wirtgen, J.: On approximation intractability of the bandwidth problem. Technical report, TR98-014, Electronic Cooloquium on Computational Complexity (1998)
9. Unger, W.: The complexity of the approximation of the bandwidth problem. In: 37th Annual Symposium on Foundations of Computer Science. (1998) 82–91
10. Feige, U.: Approximating the bandwidth via volume respecting embeddings. Journal of Computer System Science **60** (2000) 510–539
11. Gupta, A.: Improved bandwidth approximation for trees and chordal graphs. Journal of Algorithms **40** (2001) 24–36
12. Haralambides, J., Makedon, F., Monien, B.: Bandwidth minimization: an approximation algorithm for caterpillars. Mathematical Systems Theory (1991)
13. Cuthill, E.H., Mckee, J.: Reducing the bandwidth of sparse symmetric matrices. In: Proceedings of 24th ACM National Conference. (1969) 157–172

An Algorithm for Portfolio's Value at Risk Based on Principal Factor Analysis*

Honggang Xue[1], Chengxian Xu[2], and Chunping Hu[3]

[1] School of Science, Xi'an Jiaotong University, Xi'an,
710049, Shaanxi, P. R. China
xuehg-01@126.com
[2] School of Science, Xi'an Jiaotong University, Xi'an,
710049, Shaanxi, P. R. China
[3] School of Public Policy and Administration, Xi'an Jiaotong University,
Xi'an, 710049, P. R. China
{mxxu, hcp}@mail.xjtu.edu.cn

Abstract. In this paper, we propose principle factor analysis method to reduce the dimensions of a high dimensional random vector in calculating portfolio's Value at Risk. The theoretical foundation, algorithm and numerical example of the method are given. This method outperforms the principle component analysis method. Especially, the advantages of the method are marked, while the factors F's multicollinearity is serious.

Keywords: principal factor analysis, value at risk, principal component analysis, multicollinearity.

1 Introduction

"Market risk" has become one of the most popular buzzwords of the financial markets. Market risk is the uncertainty of future returns due to fluctuation of financial asset quantities such as stock prices, interest rates, exchange rates, and commodity prices. Regulators, commercial and investment banks, and corporate and institution investors are increasingly focusing on more precise measurement of the level of market risks incurred by their institutions.

One of the most widely accepted concepts in market risk management is the value-at-risk (VaR). VaR is defined as the maximum loss with a given confidence level over a given time horizon in a portfolio of financial instruments[1]. Risk management systems based on VaR have been implemented in many financial institutions, asset management institutions, and non-financial corporations. VaR has also been officially accepted and promoted by regulators as sound risk management practice[1].

---

* This work is supported by National Natural Key Product Foundations of China 10231060.

N. Megiddo, Y. Xu, and B. Zhu (Eds.): AAIM 2005, LNCS 3521, pp. 381–391, 2005.
© Springer-Verlag Berlin Heidelberg 2005

To calculate portfolio's VaR there is a need to construct an approximation of the probabilistic distribution of profit and loss($P\&L$). One of the most popular techniques is based on the assumption that the portfolio value can be expressed as a deterministic function of some basic market factors, for example, interest rate, exchange rate, stock market index, and the process is named risk mapping. Having distributions of these market factors one can construct the distribution of the portfolio's value function. The most popular method is delta approach [1-5]. The first order expansion of the value function is used to approximate the distribution at the end of the period. Typically, for short time horizon, the changes in market factors are distributed almost normally and the approximated value of the portfolio is also normally distributed. Suppose all assets contained in a portfolio are linear on the market factors, then the portfolio's VaR calculated by delta-only method is accurate. In general, stock, bond, swap are typical linear instrument.

Suppose the joint distribution of market factors is multivariate normal, and all assets contained in a portfolio are linear, then the portfolio's VaR can be easily calculated.

Consider a portfolio consisting of quantities $X = (x_1, x_2, ..., x_n)^T$ of assets $1, 2, ..., n$ with values $V = (v_1, v_2, ..., v_n)^T$ at time t. The VaR of the portfolio X for some given probability level α, is defined as the level of loss, $\Delta V^*(\alpha)$, such that the probability that the loss $\Delta V \leq \Delta V^*$ is equal to α. Since the joint distribution of the change in the value of market factors is multivariate normal, the calculation of ΔV^* is straightforward. Assume that the value of each asset depends on the time t and the market factors $f_1, f_2, ..., f_K$, or $F = (f_1, f_2, ..., f_K)^T$. Then the portfolio's VaR is given by[5,6]

$$VaR_p = Z_\alpha \sqrt{\delta^T \Sigma_F \delta} \sqrt{\Delta t} \tag{1}$$

where Z_α is the quantile of the standard normal distribution corresponding to confidence level α, and if $\alpha = 0.95$, $Z_\alpha = 1.65$; Δt is a time horizon, usually we set $\Delta t = 1, 5$ or 10 days; δ is the vector of aggregate delta, $\delta = (\delta_1, \delta_2, ..., \delta_K)^T$, $\Sigma_F = (\sigma_{ij})_{K \times K}$ is the variance-covariance matrix of market factors, and $\sigma_{ij} = cov(\Delta f_i, \Delta f_j)$, $\sigma_{ii} = var(\Delta f_i)$,

It follows from (1) that the important thing in calculating portfolio's VaR is to estimate the covariance matrix Σ_F or the coefficient matrix R, of the market factors. For large scale portfolio, with varied asset's classes, the dimension of Σ_F is large, and the calculation of the portfolio's VaR is complicated. When some market factors are strongly linearly related, the multicollinearity of the market factors will be serious, and the matrix Σ_F may not be positive. Principal component analysis(PCA) had been used to overcome these drawback[1,5,7-12]. Because PCA method would twist original message when the multicollinearity of market factors is serious, and the meaning of principal component is not easily explained for financial market risk manages and regulators. The method is difficult to use in practical risk management process. In this paper, the method named principal factor analysis(PFA) is proposed to overcome the drawbacks of the PCA.

In section 2, we introduce portfolio's VaR based on PCA, in section 3, we give PFA's theory and algorithm, in section 4, we use PFA to calculate portfolio's VaR, and in section 5, we compare VaR value based on PCA and PFA methods, at the same time, their advantage and drawback will be given. Conclusion will be given in section 6.

2   VaR Based on PCA

From the previous section we assume that the joint distribution of market factors $f_1, f_2, ..., f_K$ is a multi-normal distribution. Suppose g_1 is the first principal component, g_2 is the second principal component, ..., g_s is the sth principal component($s \leq K$). From the principal component analysis(PCA), $g_1, g_2, ..., g_s$ could be expressed as the equations $g = AF$, where the market factors vectors $F = (f_1, f_2, ..., f_K)^T$, and the coefficients matrix $A = (a_{ij})_{s \times K}$. In PCA, the number s, of the principal components, and the coefficients $a_{j1}, a_{j2}, ..., a_{jK}, j = 1, 2, ..., s$ can be determined as following.

Let $\lambda_1 \geq \lambda_2 \geq ... \geq \lambda_K > 0$ be the eigenvalues of the covariance matrix Σ_F, and $p_1, p_2, ..., p_K$ be the corresponding orthonormal eigenvectors. Then the first s eigenvectors consist of the coefficient matrix A, that is $a_j = (a_{j1}, a_{j2}, ..., a_{jK}) = p_j^T$.

Assume $\bar{\alpha}$ is a threshold value(predetermined maximum contribution rate of the principal components), then the number s, of the principal components is determined by

$$cont(s-1) < \bar{\alpha} \qquad \text{and} \qquad cont(s) \geq \bar{\alpha}$$

where

$$cont(s) = \frac{\sum_{i=1}^{s} \lambda_i}{\sum_{i=1}^{K} \lambda_i}.$$

The portfolio's value function can be rewritten as $V = V(G, t)$, where $G = (g_1, g_2, ..., g_s)^T$ is the principal component vectors. The change of portfolio value is given by

$$\Delta V = \sum_{i=1}^{n} x_i \frac{\partial v_i(G, t)}{\partial t} \Delta t + \sum_{i=1}^{n} x_i \sum_{j=1}^{s} \frac{\partial v_i(G, t)}{\partial g_j} \Delta g_j \equiv \mu_p + \sum_{j=1}^{s} \delta_j \Delta g_j$$

where μ_p is given by (2) and δ_j, the aggregate delta for principal components, is given by

$$\delta_j = \sum_{i=1}^{n} x_i \frac{\partial v_i(G, t)}{\partial g_j}, \qquad j = 1, 2, ..., s.$$

The portfolio's VaR is now calculated as

$$VaR_p = Z_\alpha \sqrt{\delta^T \Sigma_g \delta} \sqrt{\Delta t} \tag{2}$$

where $\delta = (\delta_1, \delta_2, ..., \delta_s)^T$, and Σ_g is the covariance matrix of principal components, which is diagonal matrix with diagonal elements $\sigma_1^2, \sigma_2^2, ..., \sigma_s^2$, where $\sigma_j^2 = var(\Delta g_j), j = 1, 2, .., s$.

From the structure of the covariance matrix Σ_g we see that the principal components are not correlative. This structure could reduce the difficulty of calculating portfolio's VaR. However, the meaning of the principal components is not easily explained since the principal component is the combination of original factors. When the multicollinearity of the market factors is serious, the principal component analysis may distort original messages.

In financial markets, the multicollinearity of market risk factors exists, and in some case, it may be more serious. The U.S treasury market gives such an example. Table 1 presents the monthly VaR of zero-coupon bonds as well as correlations. The maturities of the bonds are from 1 year to 30 years, and the correlation coefficients are from 0.644 to 0.999. The condition number of the

Table 1. Risk and correlations for U.S Bonds (monthly VaR at 95%

Term (year)	VaR %	1y	2y	3y	4y	5y	7y	9y	10y	15y	20y	30y
1	0.470	1										
2	0.987	.897	1									
3	1.484	.886	.991	1								
4	1.971	.886	.976	.994	1							
5	2.426	.855	.966	.988	.998	1						
7	3.192	.825	.936	.965	.982	.990	1					
9	3.913	.796	.909	.942	.964	.975	.996	1				
10	4.250	.788	.903	.937	.959	.971	.994	.999	1			
15	6.234	.740	.853	.891	.915	.930	.961	.976	.981	1		
20	8.146	.679	.791	.832	.860	.878	.919	.942	.951	.991	1	
30	11.119	.644	.761	.801	.831	.853	.902	.931	.943	.975	.986	1

correlation matrix R is $cond(R) = 7.9944 \times 10^4$. In general, when the condition number of the correlation matrix is great than 2×10^3, it is said that the multicollinearity of the random variables is serious[8]. From this example we conclude that the serious multicollinearity exists in financial markets. If the principal component analysis is employed to treat this problem, some important components may be lost. At the same time, since the principal components are combinations of the some market factors, it does not tell which market factor dominates others, and we could not get one hedge strategy which could easily be operated to reduce market risks.

3   Principal Factor Select Algorithm

In this section, we will present the principal factor analysis. Before we give the presentation, the matrix sweep operation, some useful lemmas, and propositions are introduced.

3.1   Matrix Sweep Operation

Suppose the matrix $V = (v_{ij})_{p \times p}$ with $v_{ii} \neq 0$, $i = 1, 2, ..., p$. We define a new matrix $B = (b_{kl})_{p \times p}$, with

$$b_{ii} = \frac{1}{v_{ii}}, \quad b_{il} = \frac{v_{il}}{v_{ii}}, \quad b_{li} = -\frac{v_{li}}{v_{ii}}, \ l \neq i, \quad b_{kl} = v_{kl} - \frac{v_{ki}v_{il}}{v_{ii}}, \ k \neq i, \ l \neq i, \ (3)$$

where i is given. This operation is defined as a matrix sweep operation pivoted on v_{ii} from V to B, and is denoted $S_i V = B$.

Lemma 1. $S_i S_i = V$,   and $S_i S_j V = S_j S_i V$.

Let the matrix V be decomposed into four submatrices

$$V = \begin{bmatrix} V_{11} & V_{12} \\ V_{21} & V_{22} \end{bmatrix}$$

where the dimensions of V_{11} and V_{22} are r and $(p - r)$.

Lemma 2.   If $S_{ir}...S_{i2}S_{i1}V = B$, and let

$$B = \begin{bmatrix} B_{11} & B_{12} \\ B_{21} & B_{22} \end{bmatrix}$$

with $B_{11} \in R^{r \times r}$, $B_{22} \in R^{(p-r) \times (p-r)}$, then

$$B_{11} = V_{11}^{-1}, \quad B_{12} = V_{11}^{-1}V_{12}, \quad V_{21} = -V_{21}V_{11}^{-1}, \quad B_{22} = V_{22} - V_{21}V_{11}^{-1}V_{12}. \tag{4}$$

Suppose $E(X) = U = (\mu_1, \mu_2, ..., \mu_p)^T$ is the mean vector of the random vector $X = (x_1, x_2, .., x_p)^T$, the covariance matrix of X is $D(X) = V = (v_{ij})_{p \times p}$. Let the matrix V and the vectors x and U be similarly decomposed as

$$V = \begin{bmatrix} V_{11} & V_{12} \\ V_{21} & V_{22} \end{bmatrix}, \quad X = \begin{pmatrix} X^1 \\ X^2 \end{pmatrix}, \quad U = \begin{pmatrix} U^1 \\ U^2 \end{pmatrix}$$

where
$X^1 = (x_{i1}, x_{i2}, .., x_{ir})^T$,
$X^2 = (x_{j1}, x_{j2}, ..., x_{j(p-r)})^T$,
$U^1 = E(X^1) = (\mu_{i1}, \mu_{i2}, ..., \mu_{ir})^T$,
$U^2 = E(X^2) = (\mu_{j1}, \mu_{j2}, ..., \mu_{j(p-r)})^T$.

Lemma 3.   If the inverse of the matrix V_{11} exists, and we set

$$Z = \begin{bmatrix} Z^1 \\ Z^2 \end{bmatrix} = \begin{bmatrix} X^1 \\ X^2 - V_{21}V_{11}^{-1}X^1 \end{bmatrix},$$

then the covariance matrix of Z is given by

$$D \begin{bmatrix} Z^1 \\ Z^2 \end{bmatrix} = \begin{bmatrix} V_{11} & 0 \\ 0 & V_{22} - V_{21}V_{11}^{-1}V_{12} \end{bmatrix}.$$

From Lemma 3, we can get the following propositions.

Proposition 1. $V_{22} - V_{21}V_{11}^{-1}V_{12} \geq 0$, i.e, the matrix is positive semidefinite.

Proposition 2.   If the diagonal elements of the matrix $V_{22} - V_{21}V_{11}^{-1}V_{12}$ are zero, then

$$V_{22} - V_{21}V_{11}^{-1}V_{12} = 0$$
$$X^2 = V_{21}V_{11}^{-1}(X^1 - U^1) + U^2, \qquad a.s$$

The conclusions are obvious, and we omit the proof.

3.2   Principal Factor Select Strategy

Let V be the covariance matrix of the random vector X. In the principal component analysis, the principal components are the linear combinations of the random variables $x_1, x_2, ..., x_p$, and the variance of the ith principal component is the ith eigenvalue of V, and the sum of all principal component's variance is the trace of the matrix V. The sum of all the principal component's variances is the trace of the matrix V, denoted by $tr(V)$, and is also the total variance of the random vector X. As we mentioned above, when the multicollinearity of X is serious, $tr(V)$ will contain much more similar messages, that will magnify some principal component's effect.

In this subsection, we consider total variance, $tr(V)$, of the random vector X, according to the sizes of variances. In other words, the variable with large variance would have strong ability to interpret X. We make sweep operations on the matrix V pivoted on the maximum diagonal elements step by step. We will get r variables $x_{i1}, x_{i2}, ..., x_{ir}$ as the principal factors of the vectors X, with the kth variable corresponding to the kth pivot element. Now, the question is how to determine the value of r?

The random vector $X^2 - V_{21}V_{11}^{-1}X^1$ is the remainder of X^2 deporting the linear section of X^1. Since $X^2 - V_{21}V_{11}^{-1}X^1$ is not correlative with X^1(see Lemma 3.3), and $D(X^2 - V_{21}V_{11}^{-1}X^1) = V_{22} - V_{21}V_{11}^{-1}V_{12}$, the total variance, $tr(V_{22} - V_{21}V_{11}^{-1}V_{12})$ represents the ability of $X^2 - V_{21}V_{11}^{-1}X^1$ to interpret X. With the principal factor selection proceeding, the variance of the matrix $V_{22} - V_{21}V_{11}^{-1}V_{12}$, i.e, $tr(V_{22} - V_{21}V_{11}^{-1}V_{12})$ will become small and small. Let d_i be the pivot elements, $i = 1, 2, ..., r$, from lemma 2 and 3, $\sum_{i=1}^{r} d_i$ is similar the total variance $tr(V)$, which represent the ability of the principal factors vector X^1 to interpret original random vector X, and it deduct the repeated messages existing in the principal factor vector X^1. Thus, to determine the principal factors, we set

$$\tilde{\delta} = tr(V_{22} - V_{21}V_{11}^{-1}V_{12}) \tag{5}$$

and

$$\delta = \frac{\sum_{i=1}^{r} d_i}{\sum_{i=1}^{r} d_i + \tilde{\delta}} \tag{6}$$

A threshold value(maximum contribution rate) α^1 is predetermined, and a value of r is determined such that $\delta \geq \alpha^1$. The principal factor selection process is complicated and r principal factors $X^1 = (x_{i1}, x_{i2}, ..., x_{ir})^T$ are obtained. For the vectors X^2, we have the following formula

$$X^2 \approx U^2 + V_{21}V_{11}^{-1}(X^1 - U^1). \tag{7}$$

(Note that if $\delta \geq \alpha^1$, we regard $\tilde{\delta} \approx 0$, and it follows from proposition 2, that the above formula hold). The following is the algorithm for principal factor selection.

Algorithm 1. Principal Factor Selection

Step 1. Give the threshold value α^1, and set $V^0 = V$, $I^{(0)} = \emptyset$, $J^{(0)} = \{1, 2, ..., p\}$, and find $l_1 \in J^{(0)}$, such that $v_{l1,l1}^0 = maxmax\{v_{l,l}^0 : l \in J^{(0)}\}$ hold. $v_{l1,l1}^0 \Rightarrow \delta_1$, $v_{l1,l1}^0 \Rightarrow d_1$, go to step 2;

Step 2. In the rth $(r = 1, 2, ..., p-1)$ step, we make the sweep operation S_{ir} to $V^{(r-1)}$ to obtain $V^{(r)} = S_{ir}V^{(r-1)}$, and reset $I^{(r)} = I^{(r-1)}\bigcup\{l_r\}$, $J^{(r)} = J^{(r-1)}\backslash\{l_r\}$, $\sum_{j \in J^{(r)}} v_{j,j}^r \Rightarrow \tilde{\delta}$, $\frac{\delta_1}{\delta_1 + \tilde{\delta}} \Rightarrow \delta$. If $\delta \geq \alpha^1$, then terminate the algorithm, and $I^{(r)}$ is the principal factor index set. If $\delta < \alpha^1$, we find $l_{r+1} \in J^{(r)}$, such that $v_{l_{r+1},l_{r+1}}^r = max\{v_{l,l}^r : l \in J^{(r)}\}$. $\delta_1 + v_{l_{r+1},l_{r+1}}^r \Rightarrow \delta_1$, $v_{l_{r+1},l_{r+1}}^r \Rightarrow d_{r+1}$, and go to the $(r+1)$th step.

Remark: $I^{(i)}$ is the principal factors index set. When the algorithm terminates, the index set $I^{(r)} = \{l_1, l_2, ..., l_r\}$, $J^{(r)} = \{j_1, j_2, ..., j_{p-1}\}$. and $I^{(r)} \bigcup J^{(r)} = \{1, 2, ..., p\}$ are available. The principal factor vector is X^1 corresponding to the set $I^{(r)}$. d_1 is the variance of the first principal factor x_{l_1}, d_2 is the variance of x_{l_2} deporting the linear section of x_{l_1}, ..., and d_r is the variance of x_{l_r} deporting the linear section of $x_{l_1}, x_{l_2}, ..., x_{l_{r-1}}$ with $d_1 \geq d_2 \geq ... \geq d_r > 0$.

4   Portfolio's VaR Based on *PFA*

Suppose a portfolio's value function is $V = V(f_1, f_2, .., f_K, t)$. It follows from section 1 that the change of the portfolio's value is given by

$$\Delta V = \mu_p + \sum_{j=1}^{K} \delta_j \Delta f_j = \mu_p + \delta^T \Delta F$$

where $\Delta F = (\Delta f_1, \Delta f_2, ..., \Delta f_K)^T$ is the vector of the market factor value's change, $\delta = (\delta_1, \delta_2, ..., \delta_K)^T$ is the aggregate delta vector.

Assume that we have get the principal factor vector $F^1 = (f_1, f_2, ..., f_r)^T$, and the remained factor vector $F^2 = (f_{r+1}, f_{r+2}, ..., f_K)^T$. If the predetermined threshold value α^1 approaches 1(usually $\alpha^1 \geq 85\%$), then the following formula holds

$$F^2 = U^2 + V_{21}V_{11}^{-1}(F^1 - U^1), \qquad a.s. \tag{8}$$

Thus we can get

$$\Delta F^2 = V_{21}V_{11}^{-1}\Delta F^1, \qquad a.s. \tag{9}$$

We rewrite ΔV as

$$\Delta V = \mu_p + {\delta^1}^T \Delta F^1 + {\delta^2}^T \Delta F^2 \tag{10}$$

where $\delta^1 = (\delta_1, \delta_2, ..., \delta_r)^T$, $\delta^2 = (\delta_{r+1}, \delta_{r+2}, ..., \delta_K)^T$. Substituting (12) into (13) generates

$$\Delta V = \mu_p + ({\delta^1}^T + {\delta^2}^T V_{21}V_{11}^{-1})\Delta F^1. \tag{11}$$

So the variance of ΔV is

$$var(\Delta V) = \beta^T \Sigma_{PF} \beta \tag{12}$$

where $\beta = \delta^1 + V_{11}^{-1}V_{12}\delta^2$, and Σ_{PF} is the covariance matrix of the principal factors. The portfolio's VaR is then calculated as the following

$$VaR_p = Z_\alpha \sqrt{\beta^T \Sigma_{PF} \beta}\sqrt{\Delta t} \tag{13}$$

where β_j is the jth element of β.

5   The Comparison of the PFA VaR with the PCA VaR

In this section, we will compare the portfolio's VaR based on PFA and PCA by examples, and the advantages and the disadvantages are discussed.

5.1   The Numerical Example

We consider the example of Table 1. Four threshold values of α^1(from 85% to 99%) are used and the principal factors are found. In table 2, f_i is the principal factor, d_i is the pivoted element.

From Table 2 we see that when the predetermined threshold value $\alpha^1 = 90\%$, there are four principal factors, two short time zero-coupon bonds of 2-years and 1-year, two long time zero-coupon bonds of 30-years and 7-years, their ability to interpret origin message reach 96.56%. When the portfolio's VaR is calculated, we only evaluate one four dimension covariance matrix consisted with these four principal factors.

Next we consider the portfolio in which the return is expressed by

$$y = 0.0416 + 0.2968f_1 - 0.9734f_2 + 2.5014f_3 - 1.1958f_4 + 0.0485f_5 \tag{14}$$
$$-2.2396f_6 + 2.0527f_8 + 1.3745f_9 - 0.8752f_{10} - 0.1902f_{11} \tag{15}$$

where f_1 is the 1 year zero-coupon bond's return, ..., f_{11} is the 30 years zero-coupon bond's return. It's VaR is calculated when time horizon choose one day,

Table 2. The order of the principal factors and pivoted elements

threshold value(α^1 %)	principal factor order	1	2	3	4	5	6	7
85	f_i	2y	30y	1y				
	d_i	1	0.4209	0.1918				
90	f_i	2y	30y	1y	7y			
	d_i	1	0.4209	0.1918	0.0386			
95	f_i	2y	30y	1y	7y			
	d_i	1	0.4209	0.1918	0.0386			
99	f_i	2y	30y	1y	7y	20y	4y	
	d_i	1	0.4209	0.1918	0.0386	0.0230	0.0070	

Table 3. The portfolio's VaR with 11 factors

time horizon (day)	1	5	10
portfolio's VaR	0.9467	2.1169	2.9938

Table 4. The portfolio's VaR based on PFA

threshold value(α^1 %)	85	90	95	99
1 day	0.8946	0.9012	0.9012	0.9063
5 days	2.0004	2.0151	2.0151	2.0266
10 days	2.8290	2.8497	2.8497	2.8661

five days and ten days, respectively, with some threshold values. The confidence level is $\alpha = 95\%$, and the initial value of the portfolio is 100 unit.

When the portfolio's VaR is calculated using the origin 11 market factors, the results are given in table 3.

When the portfolio's VaR is calculated based on the principal factor analysis, the results are given in table 4.

For the portfolio's VaR based on the principal component analysis, the results are given in table 5. When the parameters are the same as the calculations portfolio's VaR based on PFA.

Comparing these tables, we can find that the portfolio's VaR based on PFA is better than the portfolio's VaR based on PCA, because the VaR based on PFA is closer to the portfolio's VaR calculated by origin 11 market factor than the VaR based on PCA is.

5.2   The Advantages and Disadvantages of the PFA VaR and PCA VaR

The principal factor analysis has more advantages than the principal component analysis has. Firstly, PFA needs few computer time than PCA, because in every sweep operation, at most p^2 multiplications and divisions are operated.

Table 5. The portfolio's VaR based on PCA

threshold value(α^1 %)	85	90	95	99
1 day	0.8862	0.8862	0.9004	0.9037
5 days	1.9817	1.9817	2.0133	2.0208
10 days	2.8025	2.8025	2.8473	2.8579

If r principal factors are carried out, there are at most $r \times p^2$ multiplications and divisions operations. At the same time, because in every sweep operation, we select the maximum element in $V_{22} - V_{21}V_{11}^{-1}V_{12}$, it is a stable algorithm. Secondly, we use the principal factor vector F^1(with dimension r) to replace the origin vector F(with dimension p). If the multicollinearity is serious, then $r \ll p$, so we only calculate covariance matrix Σ_F(with dimension $r \times r$), the aim to reduce the dimensions is realized. Thirdly, the meaning of the principal factors is clear, this is a very important point for financial risk managers. Lastly, the PFA overcomes the serious multicollinearity (existing in origin random vector F) better than PCA dose. From the previous example we could understand this point. From these advantages we see that the VaR based on PFA is more accurate than VaR based on PCA. However, the disadvantages of the PFA exist. If the multicollinearity is not serious, in other words, the correlation of the random factors is not high, then we will have many principal factors, i.e, r will be close to the origin factor number p, and we could not reduce the high dimension of the random vector obviously. But this disadvantage also exists for the principal component analysis.

6   Conclusion

In this paper, we discussed the principal component analysis for calculating the portfolio' value at risk, it is a good method to reduce the dimensions of a high dimension matrix, but it's performance will reduce when the multicollinearity of market risk factors is serious, so we propose principle factor analysis method to reduce the dimensions of a high dimensional random vector in calculates portfolio's Value at Risk. The theoretical foundation, algorithm and numerical example of the method are given. This method outperforms the principle component analysis method. Especially, the advantages of the method are marked, while the factors F's multicollinearity is serious.

References

[1] Jorion,P.. Value at Risk: The New Benchmark for Controlling Market Risk. McGraw-Hill, New York, 1997.
[2] Duffie, D. and J. Pan, An Overview of Value at Risk, The Journal of Derivatives, 4(3):9-49,1997.

[3] Davé, R., and S. Gerhard, On the Accuracy of VaR Estimates Based on the Variance-Covariance Approach, Olsen Associates Research Institute, Zurich, 1998.
[4] J.P.Morgan Bank. RiskMetrics Technical Manual, New York: J.P.Bank, 1995.
[5] Longerstaey,R., Spencer,M., RiskMetrix^{TM}-Technical Document, fourth ed. Morgan Guaranty Trust Company, New York, 1996.
[6] Mark Britten-Jones, and Stephen M. Schaefer. Non-Linear Value-at-Risk, European Finance Review, 2: 161-187, 1999.
[7] Cindy I. N., Robin D. H., and Richard B. F., A Synthetic Factor Approach to the Estimation of Value-at-Risk of a Portfolio of Interest Rate Swaps, Journal of Banking Finance, 24:1903-1932, 2000.
[8] Fang, K.-T., and Zhang, R.-T., Generalized Multivariate Analysis, Beijing:Science Press, 1992.
[9] Knez, P., Litterman,R., Scheinkman,J., Explorations Into Factors Explaining money market returns, Journal of Finance, 49(5):1861-1882, 1994.
[10] Longstaff, F., and E. Schwartz, Interest Rate Volatility and the Term Structure:A Two-Factor General Equilibrium Model, Journal of Finance, 47:1259-1283, 1992.
[11] Lewis-Beck, M.S.(Ed.). Factor Analysis and Related Thchniques. Sage, London, 1994.
[12] Litterman, R., Scheinkman, J.. Common factors affecting bond returns. Financial Strategies Group Discussion Paper. Goldman Sachs, New York, Stemper, 1988.

An Approximation Algorithm
for Embedding a Directed Hypergraph on a Ring

Kang Li[1] and Lusheng Wang[2]

[1] School of Information Science and Engineering,
Shandong University, Jinan, Shandong Province, P. R. China
kangli@sdu.edu.cn
[2] Department of Computer Science, City University of Hong Kong,
Kowloon, Hong Kong
lwang@cs.cityu.edu.hk

Abstract. We study the problem of embedding a directed hypergraph on a ring that has applications in optical network communications. The undirected version (MCHEC) has been extensively studied. It was shown that the undirected version was NP-complete. A polynomial time approximation scheme (PTAS) for the undirected version has been developed. In this paper, we design a polynomial time approximation scheme for the directed version.

1 Introduction

Embedding hyperedges of a hypergraph as paths on a ring is an important problem with applications in various areas such as optical network communications, parallel computation and electronic design automation.

Optical fiber networks are increasingly substituting traditional communication networks in modern communications. The technique for design multiple channels simultaneously across an optical link using a different wavelength for each channel is called wavelength-division multiplexing (WDM)[3]. In a WDM network, to set up a connection for a request, a path between the two nodes of the request is selected and a wavelength is assigned to every edge in the path. In some cases, different wavelengths must be used if two paths share a common edge. This requirement is known as the *wavelength-continuity* constraint. Current optical technologies impose limitations on the number of available wavelengths per fiber [6, 7, 8]. Typically, the number is between 20-100 per fiber.

Routing on WDM networks is an important problem in optical fiber communication. A typical topology in network design is a ring. A lot of work has been done for routing on rings. Raghavan and Upfal studied routing techniques for different kinds of optical fiber networks including rings, trees and meshes [3]. The objective is to route the network to serve all the requests such that the ring congestion, i.e., the maximum number of times that an edge can be used, is minimized.

N. Megiddo, Y. Xu, and B. Zhu (Eds.): AAIM 2005, LNCS 3521, pp. 392–399, 2005.
© Springer-Verlag Berlin Heidelberg 2005

The problem of embedding hyperedges on a ring was originally proposed for electronic design automation, where the objective is to route within a minimum-area rectangle [1, 2]. The problem of embedding undirected hyperedges on a ring with minimum congestion (MCHEC) has applications in parallel computing as well as multicast routing. Here we study the problem of embedding directed hyperedges on a ring. It models the case where the links in the network is directed.

For the undirected case (MCHEC), Ganley and Cohoon in [5] proved that the problem was NP-hard and gave a ratio-3 approximation algorithm. They also gave an algorithm that can solve the case where the congestion is a constant. Several ratio-2 approximation algorithms were given in [9, 10]. Gu and Wang presented a ratio-1.8 approximation algorithm [12]. Recently, Deng and Li proposed a polynomial time approximation scheme (PTAS) for the problem. Their approach comes from the techniques for string problems [11].

In this paper, we study the problem of embedding directed hyperedges on a ring. We extend the method in [13] to get a polynomial time approximation scheme for the directed version. We have developed a new technique for the proofs of some key lemmas. This technique can also be applied to the undirected case. The new proofs allow us to reduce the time complexity of the algorithms in [13] by a factor of $O(m)$, where m is the total number of hyperedges.

The paper is organized as follows: We first give some definitions in Section 2. Section 3 deals with a special case, where the number of hyperedges m is $O(\log n)$. Section 4 gives an algorithm that solves the case where the number of hyperedges m is of $O(C_{opt})$, and c_{opt} is the minimum congestion cost of an optimal embedding. The general case is solved in Section 5.

2   Preliminaries

A *ring* of n nodes is a directed graph $R = (V, E_R)$, where $V = \{1, 2, \ldots, n\}$ is the set of n vertices on the ring and $E_R = \{e_i^+ = (i, i+1), e_i^- = (i+1, i) | i = 1, 2, \ldots, n\}$ is the set of $2n$ directed edges on the ring, where $n+1$ is treated as 1. Consider the same set of vertices $V = \{1, 2, \ldots, n\}$. A directed hyperedge $h = (u, S)$ is a pair, where $u \in V$ is the source of the hyperedge and $S \subseteq V - \{s\}$ is the set of sinks. In communication applications, each hyperedge h represents a request that asks to send a message from u to every vertex in S. Let $H = (V, E_H)$ be a directed hypergraph with the same set of vertices V and a set of m directed hyperedges $E_H = \{h_1, h_2, \ldots, h_m\}$.

Let $h_j = (u_j, S_j)$ be a hyperedge in E_H, where $S_j = \{i_1^j, i_2^j, \ldots, i_{k_j}^j\}$. $k_j = |S_j|$ denotes the total number of sink vertices in the hyperedge. For convenience, we use i_0^j to denote u_j. Assume that the $k_j + 1$ vertices $i_0{}^j, i_1{}^j, \ldots, i_{k_j}^j$ follow the clockwise order on the ring. P_k^j denotes the segment of vertices on the ring from vertex i_k^j to vertex i_{k+1}^j for $k = 0, 1, \ldots, k_j - 1$ and $P_{k_j}^j$ denotes the segment of vertices on the ring from vertex $i_{k_j}^j$ to vertex i_0^j. In order to realize the request h_j on the ring, one can cut one of the paths P_k for $k = 0, 1, \ldots k_j$

and obtain two directed paths on the ring both starting from i_0^j. This forms an embedding of h_j on the ring. For a hyperedge h_j, there are $k_j + 1$ different embeddings, one for each P_k^j (by cutting P_k^j). An *embedding* of h_j is a P_k^j *embedding* if P_k^j is cut.

Given an embedding x of all the hyperedges E_H, the *congestion* $e_i^+(x)$ or $e_i^-(x)$ of a directed edge e_i^+ or e_i^- is the number of times that the edge e_i^+ or e_i^- is used in the embedding. When x is clear, we also use $c(e_i^+)$ and $c(e_i^-)$ to denote the congestion. The problem here is to find an embedding for each $h_j \in E_H$ such that every edge e_i^+ and e_i^- on the ring is used at most c times and c is minimized. We refer the problem as the *embedding directed hyperedges on the ring* problem (EDHR for short).

3   Enumerating Method for $m = O(\log n)$

In this section, we give an algorithm with ratio $1 + \frac{1}{r}$ for the case where there are $O(\log n)$ hyperedges in E_H. The basic idea of our algorithm is to choose $2r$ edges on the ring, where r is a constant related to the ratio, and for each hyperedge h_j we only have to cut a P_k^j that contains one of the $2r$ selected edges on the ring. By doing this, for each h_j we only have to consider $2r$ choices (instead of $k_j + 1$ choices for an optimal solution). Since there are at most $O(\log n)$ hyperedges, the time complexity is $O((2r)^{O(\log n)})$, that is polynomial.

Let $i_1, i_2, \ldots, i_{2r}$ be $2r$ indices representing the $2r$ edges $(i_1, i_1 + 1), (i_2, i_2 + 1), \ldots, (i_{2r}, i_{2r} + 1)$ on the ring R. Let $x = (x_1, x_2, \ldots, x_m)$ be an embedding of H, where x_j indicates the choice P_k^j that is cut for the embedding of h_j. We use $E_j(x)$ to denote the segment P_k^j that is cut for the embedding x of h_j.

Let $Q_{i_1, i_2, \ldots, i_{2r}}(x)$ be a set of indices of hyperedges such that j is in $Q_{i_1, i_2, \ldots, i_{2r}}(x)$ if and only if $E_j(x)$ contains at least one of the $2r$ edges $(i_1, i_1 + 1), (i_2, i_2 + 1), \ldots, (i_{2r}, i_{2r} + 1)$.

Lemma 1. *Let x be any embedding of H. For any fixed index $1 \le i_1 \le n$, there exist $2r - 1$ indices $i_2, i_3, \ldots, i_{2r}$ such that for any embedding x' satisfying $x_j' = x_j$ for every $j \in Q_{i_1, i_2, \ldots, i_{2r}}(x)$, we have*

$$e_i^+(x') - e_i^+(x) \le \frac{1}{r} e_i^+(x) \text{ and } e_i^-(x') - e_i^-(x) \le \frac{1}{r} e_i^-(x)$$

for any directed edge e_i^+ and e_i^- in E_R on the ring.

Proof. We prove the lemma by giving a way to find the $2r - 1$ indices. Let c be the congestion for the embedding x that we want to approximate. First, we select an arbitrary edge, say, $e_1^+ = (1, 2) = (i_1, i_1 + 1)$ on the ring. In the embedding of x, there are at least $m - 2c$ h_j's with $e_1^+ \in E_j(x)$. That is, there are at most $2c$ h_j's with $e_1^+ \notin E_j(x)$. Let H_r be the set of the (at most $2c$) h_j's with $e_1^+ \in E_j(x)$. We use the following method to select the remaining (at most) $2r - 1$ indices.

1. **for** a remaining edge e_g^+ on the ring R **do**
2. **if** there are more than $\frac{c}{r}$ hyperedges $h_j \in H_r$ with $e_g^+ \in E_j(x)$, **then** we select the index q and set $H_r = H_r - \{j | e_g^+ \in E_j(x)\}$.
3. **if** the size of H_r is more than $\frac{c}{r}$ **then** goto Step 1 **else** stop.

The above procedure will stop after at most $2r - 1$ iterations since each time the size of H_r is reduced by at least $\frac{c}{r}$ and the original size of H_r is at most $2c$.

Now, consider any edge e_i^+ or e_i^- with index i ($1 \le i \le n$) not selected in the above procedure. For the embedding x, the number of h_j's that are not cut at edge e_i^+ or e_i^- in x is at most the size of H_r, that is upper bounded by $\frac{c}{r}$. Thus, the lemma holds.   □

Note that, in the proof of Lemma 1, we assume that x is known when selecting the $2r - 1$ indices. In fact, x will be the optimal solution that we do not known. However, we can go through all possible sets of $2r - 1$ indices in polynomial time. Based on Lemma 1, we can solve the problem as follows:

1. try all possible choices of $i_2, i_3, \ldots, i_{2r}$.
2. for each $h_j \in H$, try the $2r - 1$ choices for cutting the path P_k^j containing $e_{i_1}^+, e_{i_2}^+, \ldots,$ or $e_{i_{2r}}^+$.

Step 1 takes $O(n^{2r-1})$ time and Step 2 needs $O((2r)^m) = O((2r)^{O(\log n)}) = n^{O(\log 2r)}$ time.

Theorem 1. *There is a PTAS with ratio $1 + \frac{1}{r}$ that runs in $O(n^{2r-1} \times n^{O(\log 2r)})$ time when $m = O(\log n)$.*

4   The Algorithm for $c \geq O(\log n)$ and $c = O(m)$

In this section, we consider the case where $c = O(m)$. We use linear programming and randomized rounding approach. Let $h_j = (u_j, S_j)$ be a hyperedge. We define $k_j + 1$ variables, $x_{j,1}, x_{j,2}, \ldots, x_{j,k_j+1}$. $x_{j,l} = 1$ indicates that P_l^j is cut for the embedding of h_j. For each segment P_q^j of h_j and an edge e_i^+ on the ring, we have a constant $\mu_{i,q,j}$. $\mu_{i,q,j} = 1$ if edge e_i^+ is in the segment P_q^j of h_j. Otherwise, $\mu_{i,q,j} = 0$. We have the following LP formulation.

$$\min c;$$

$$\sum_{l=1}^{k_j+1} x_{j,l} = 1;$$

$$c(e_i^+) = \sum_{j=1}^{m} \sum_{q=1}^{k_j+1} \mu_{i,q,j}(x_{j,q+1} + x_{j,q+2} + \cdots + x_{j,k_j}) \le c; \qquad (1)$$

$$c(e_i^-) = \sum_{j=1}^{m} \sum_{q=1}^{k_j+1} \mu_{i,q,j}(x_{j,0} + x_{j,1} + \cdots + x_{j,q-1}) \le c; \qquad (2)$$

For a fixed hyperedge h_j and a directed edge e_i^+ (e_i^-) on the ring, there is only one $\mu_{i,q,j}$ for $q = 0, 1, \ldots k_j$ with value 1. Consider such a segment P_q^j, where e_i^+ and e_i^- are in the segment. e_i^+ is used in the embedding of h_j if one of the segments $P_{q+1}^j, P_{q+2}^j, \ldots, P_{k_j}^j$ is cut for the embedding of h_j. Likewise, e_i^- is used in the embedding of h_j if one of the segments $P_0^j, P_1^j, \ldots, P_{q-1}^j$ is cut for the embedding of h_j. Therefore, we have (1) and (2).

The fractional version of the linear programming problem can be solved in polynomial time. After we get a fractional solution $x_{j,l}$, independent, with probability $x'_{j,l}$, we set $\hat{x}_{j,l} = 1$ and $\hat{x}_{j,h} = 0$ for the rest of h. Thus, we obtain an integer solution for the LP problem. Let c_{opt} be the optimal congestion of the LP formulation. Similar to Lemma 3 in [13], we can prove that

Theorem 2. *Assume that $m \geq c_1 \log n$, and $c_{opt} = c_2 \times m$. Let $\hat{x}$ be the 0-1 solution obtained by randomized rounding. With probability at least $1 - n^{1 - \frac{1}{3}\epsilon^2 c_2^2 c_1}$, for any e_i^+ and e_i^- in E_R,*

$$e_i^+(\hat{x}) \leq (1 + \epsilon)c_{opt},$$

and

$$e_i^-(\hat{x}) \leq (1 + \epsilon)c_{opt}.$$

Proof. To prove the theorem, we need the following Lemma originally from [4].

Lemma 2. *Let $X_1, X_2, \ldots, X_n$ be n independent random $0 - 1$ variables, where x_i takes 1 with probability p_i, $0 < p_i < 1$. Let $X = \sum_{i=1}^n X_i$, and $\mu = E[X]$. Then for $\delta > 0$, $\mathbf{Pr}(X > \mu + \delta n) < exp(-\frac{1}{3}n\delta^2)$.*

For a fixed i and a fixed j, only one $\mu_{i,q,j}$ is 1 and the rest are 0. For $l = 0, 1, \ldots, k_j$, consider such an l with $\mu_{i,q,l} = 1$. For a fixed j, only one $x_{j,l}$ is rounded to 1. Thus, $\mu_{i,q,j}(x_{j,q+1} + x_{j,q+2} + \cdots + x_{j,k_j})$ and $\mu_{i,q,j}(x_{j,0} + x_{j,1} + \cdots + x_{j,q-1})$ are also randomly rounded to either 1 or 0 and are independently for different j's. Therefore, both $c(e_i^+) = \sum_{j=1}^m \sum_{q=1}^{k_j+1} \mu_{i,q,j}(x_{j,q+1} + x_{j,q+2} + \cdots + x_{j,k_j})$ and $c(e_i^-) = \sum_{j=1}^m \sum_{q=1}^{k_j+1} \mu_{i,q,j}(x_{j,0} + x_{j,1} + \cdots + x_{j,q-1})$ are sums of m independent $0 - 1$ random variables. Set

$$E[c(e_i^+)] = \sum_{j=1}^m \sum_{q=1}^{k_j+1} \mu_{i,q,j} E[x_{j,q+1} + x_{j,q+2} + \cdots + x_{j,k_j}] = \mu_i^+ \leq c_{opt},$$

and

$$E[c(e_i^-)] = \sum_{j=1}^m \sum_{q=1}^{k_j+1} \mu_{i,q,j} E[x_{j,0} + x_{j,1} + \cdots + x_{j,q-1}] = \mu_i^- \leq c_{opt}.$$

From Lemma 2, for any fixed δ,

$$\mathbf{Pr}(c(e_i^+) > \mu_i^+ + \delta m) \leq exp(-\frac{1}{3}\delta^2 m).$$

Consider the set of all clockwise edges $\{e_1^+, e_2^+, \ldots, e_n^+\}$,

$$\mathbf{Pr}(c(e_i^+) > \mu_i^+ + \delta m \text{ for at least one } e_i^+ \in E_R) \leq n \times exp(-\frac{1}{3}\delta^2 m).$$

Similarly, we can show that

$$\mathbf{Pr}(c(e_i^-) > \mu_i^- + \delta m \text{ for at least one } e_i^- \in E_R) \leq n \times exp(-\frac{1}{3}\delta^2 m).$$

By assumption, $m \geq C \log n$. Thus, we have

$$n \times exp(-\frac{1}{3}\delta^2 m) \leq n^{1-\delta^2 C/3}.$$

Therefore, we get a randomized algorithm to find a solution x for the problem with probability at lease $1 - 2n^{1-\delta^2 C/3}$ such that for any $e_i^+ \in E_R$ and $e_i^- \in E_R$, $c(e_i^+) \leq \mu_i + \delta m \leq c_{opt} + \epsilon c_{opt}$, and $c(e_i^-) \leq c_{opt} + \epsilon c_{opt}$, where $\epsilon = \frac{\delta}{c}$.   □

Using the standard de-randomization method for packing integer programs [4], we can have a polynomial deterministic algorithm.

Theorem 3. *There is a PTAS for EDHR when $m \geq O(\log n)$ and $c_{opt} \geq O(m)$.*

5   The General Algorithm

The linear programming and randomized rounding approach in Section 4 does not work for the case where c_{opt} is small comparing with m. Here we propose a method that decomposes the set of all hyperedges into two groups so that we can give approximate embeddings using different methods for the two groups.

Consider $2r$ indices i_1, i_2, ..., i_{2r} of edges in E_R. Let $e_{i_1}^+$, $e_{i_2}^+$, ..., $e_{i_{2r}}^+$ be $2r$ edges on the ring. We define

$$R_{i_1, i_2, \ldots, i_{2r}} = \{1 \leq j \leq m| \text{ there exist an } l \text{ such that } e_{i_k}^+ \in P_l^j \text{ for any } k \in \{1, 2, \ldots, 2r\}\}.$$

Let $U_{i_1, i_2, \ldots, i_{2r}} = \{1, 2, \ldots, m\} - R_{i_1, i_2, \ldots, i_{2r}}$. Let x_{opt} be an optimal embedding. $x_{opt}|R_{i_1, i_2, \ldots, i_r}$ and $x_{opt}|U_{i_1, i_2, \ldots, i_r}$ denote the reduced embeddings of x_{opt} on the sets of hyperedges $R_{i_1, i_2, \ldots, i_r}$ and $U_{i_1, i_2, \ldots, i_r}$, respectively.

Lemma 3. $|U_{i_1, i_2, \ldots, i_{2r}}| \leq 4rc_{opt}$ and $|R_{i_1, i_2, \ldots, i_{2r}}| \geq m - 4rc_{opt}$.

Proof. Consider an $E_j(x_{opt})$ containing all the $2r$ edges $e_{i_1}^+$, $e_{i_2}^+$, ..., $e_{i_{2r}}^+$. For each edge $e_{i_k}^+$, there are at most $2c_{opt}$ H_j's such that $e_{i_k}^+ \notin E_j(x_{opt})$. Thus, there are at most $4rc_{opt}$ h_j's in total with $e_{i_k}^+ \notin E_j(x_{opt})$ for some i_k. Therefore, $|R_{i_1, i_2, \ldots, i_r}| \geq m - 4rc_{opt}$.

By definition, $|U_{i_1, i_2, \ldots, i_r}| \leq 4rc_{opt}$.   □

Let x^{i_1} be an embedding of h_j's in $R_{i_1, i_2, \ldots, i_{2r}}$ such that every h_j is cut at edge $e_{i_1}^+$. Now, we want to show that

Lemma 4. *For any fixed index* $1 \leq i_1 \leq n$, *there exist* $2r - 1$ *indices* $1 \leq i_2, i_3, \ldots, i_{2r} \leq n$ *such that for every edge* e_i^+ *and* e_i^- *in* E_R,

$$e_i^+(x^{i_1}) - e_i^+(x_{opt}|R_{i_1,i_2,\ldots,i_r}) \leq \frac{1}{r} c_{opt}, \text{ and } e_i^-(x^{i_1}) - e_i^-(x_{opt}|R_{i_1,i_2,\ldots,i_r}) \leq \frac{1}{r} c_{opt}.$$

Proof. To show the existence of the $2r - 1$ indices, we give a way to find the $2r - 1$ indices assuming that x_{opt} is known. First, we select an arbitrary edge, say, $e_1^+ = (1, 2) = (i_1, i_1 + 1)$ on the ring. In the embedding of x_{opt}, there are at least $m - 2c_{opt}$ h_j's with $e_1^+ \in E_j(x_{opt})$. That is, there are at most $2c_{opt}$ h_j's with $e_1^+ \notin E_j(x_{opt})$. (If we cut all the m h_j's in H at edge e_1^+, there are at most $2c_{opt}$ h_j's that are embedded in a way different from that of x_{opt}.) Let H_r be the set of the (at most $2c_{opt}$) h_j's with $e_1^+ \in E_j(x_{opt})$.

For every edge e_g^+ on the ring, if there are more than $\frac{c_{opt}}{r}$ hyperedges in H_r with $e_g^+ \in E_j(x_{opt})$, then we select the index q. Consider the set $R_{i_1,g}$ of indices. $j \in R_{i_1,g}$ if $e_g^+ \in E_j(x_{opt})$ and $e_1^+ \in E_j(x_{opt})$. If we cut all the (at most $m - 2c_{opt}$) h_j's with j in $R_{i_1,g}$ on edge e_1^+, there are at most $2c_{opt} - \frac{c_{opt}}{r}$ h_j's that are embedded in a way different from that of x_{opt}. Set $H_r = H_r - \{j | e_g^+ \in E_j(x_{opt})\}$ (the set of at most $2c_{opt} - \frac{c_{opt}}{r}$ h_j's that are embedded in a way different from that of x_{opt}). If the size of H_r is more than $\frac{c_{opt}}{r}$ then we can repeat the process and find another edge e_g (index). The process continues until the size of H_r is less than $\frac{c_{opt}}{r}$. The above procedure will stop after at most $2r - 1$ iterations since each time the size of H_r is reduced by at least $\frac{c_{opt}}{r}$ and the original size of H_r is at most $2c_{opt}$.

Now, consider any edge e_i^+ with index i ($1 \leq i \leq n$) not selected in the above procedure. The number of h_j's in $R_{i_1,i_2,\ldots,i_{2r}}$ that are not cut correctly at edge e_i^+ in x^{i_1} is at most the size of H_r, that is upper bounded by $\frac{c_{opt}}{r}$. Thus, the lemma holds. □

Theorem 4. *There is a PTAS for the EDHR problem.*

Proof. We first compute $U_{i_1,i_2,\ldots,i_r}$ and $R_{i_1,i_2,\ldots,i_{2r}}$.
Case 1: $|U_{i_1,i_2,\ldots,i_{2r}}| \leq C \log n$: We use the enumerating approach in Section 3 to compute an embedding for the set of hyperedges in $U_{i_1,i_2,\ldots,i_r}$. For the hyperedges in $R_{i_1,i_2,\ldots,i_r}$, we simply cut the ring at edge e_1^+. From Lemma 4 and Theorem 1, the ratio is $\frac{1}{r} + \frac{1}{r}$.
Case 2: $|U_{i_1,i_2,\ldots,i_r}| > C \log n$: We use the LP and randomized rounding approach in Section 4 to compute an embedding for the set of hyperedges in $U_{i_1,i_2,\ldots,i_{2r}}$. For the hyperedges in $R_{i_1,i_2,\ldots,i_r}$, we simply cut the ring at edge e_1^+. The LP formulation is as follows:

$$\min \ c;$$
$$\sum_{l=1}^{k_j+1} x_{j,l} = 1 \text{ for } j = 1, 2, \ldots, |U_{1_1,i_2,\ldots,i_{2r}}|;$$
$$\sum_{j=1}^{|U_{i_1,i_2,\ldots,i_{2r}}|} \sum_{q=1}^{k_j+1} \mu_{i,q,j}(x_{j,q+1} + x_{j,q+2} + \ldots + x_{j,k_j}) \leq c - c(e_i^+|R);$$

$$\sum_{j=1}^{|U_{i_1,i_2,\ldots,i_{2r}}|} \sum_{q=1}^{k_j+1} \mu_{i,q,j}(x_{j,0} + x_{j,1} + \ldots + x_{j,q-1}) \le c - c(e_i^- | R),$$

where $c(e_i^+ | R)$ and $c(e_i^- | R)$ are the number of times that e_i^+ and e_i^- are used for the embedding of h_j's in $R_{i_1,i_2,\ldots,i_{2r}}$.

Theorem 3 and Lemma 6 ensure that the ratio is $1 + \epsilon$ for any ϵ. The standard de-randomization approach gives a deterministic algorithm. □

Remarks. The NP-hardness of the directed version is still open.

Acknowledgements. The work is fully supported by a grant from the Research Grants Council of the Hong Kong Special Administrative Region, China [Project NO. City U 1196/03E].

References

1. B.S. Baker and R.Y. Pinter: An algorithm for the optimal placement and routing of a circuit within a ring of pads. Proc. 24th Symp. Foundations of Computer Science. (1983) 360-370
2. A. Frank, T. Nishizeki, N. Saito, H. Suzuki, E. Tardos: Algorithms for routing around a rectangle. Discrete Applied Mathematics. **40**, (1992) 363-378
3. P. Raghavan and E. Upfal: Efficient routing in all-optical networks. Proc. of the 26th Annual ACM Symposium on the Theory of Computing. (1994) 134-143
4. R. Motwani and P. Raghavan: Randomized algorithms. (1995) Cambridge Univ. Press
5. J. L. Ganley and J. P. Cohoon: Minimum-congestion hypergraph embedding in a cycle. IEEE Trans. on Computers. **46**, No.5, (1997) 600-602
6. S. Khanna: A polynomial time approximation scheme for the SONET ring loading problem. Bell Labs Tech. J.. **2**, (1997) 36-41
7. A. Schrijver, P. Seymour and P. Winkler: The ring loading problem. Siam Journal on Discrete Mathematics. **11**, No.1, (1998) 1-14
8. G. Wilfong and P. Winkler: Ring routing and wavelength translation. Proc. of the ninth annual ACM-SIAM symposium on Discrete Algorithms (SODA'98), San Francisco, California.(1998) 333-341
9. T. Gonzalez: Improved approximation algorithm for embedding hyperedges in a cycle. Information Processing Letters. **67**, (1998) 267-271
10. S. L. Lee, H. J. Ho: Algorithms and complexity for weighted hypergraph embedding in a cycle. Proc. of the 1st International Symposium on Cyber World. (CW2002)
11. M. Li, B. Ma and L. Wang: On the closest string and substring problems. J.ACM. **49**,(2002)157-171
12. Q.P. Gu and Y. Wang: Efficient algorithm for embedding hypergraph in a cycle. Proceedings of the 10th International Conference on High Performance Computing. (2003) 85-94, Hyderabad, India
13. X. Deng and G. Li: A PTAS for Embedding Hypergraph in a Cycle (Extended Abstract). the 31st International Colloquium on Automata, Languages and Programming (ICALP 2004), (2004) 433-444

On Product Covering in Supply Chain Models: Natural Complete Problems for $W[3]$ and $W[4]^\star$

Jianer Chen[1,2] and Fenghui Zhang[1]

[1] Department of Computer Science, Texas A&M University,
College Station, TX 77843-3112, USA
{chen, fhzhang}@cs.tamu.edu
[2] College of Information Science and Engineering,
Central-South University, ChangSha, Hunan 410083 P.R. China

Abstract. The field of supply chain management has been growing at a rapid pace in recent years, both as a research area and as a practical discipline. In this paper, we study the computational complexity of product covering problems in 3-tier supply chain models, and present natural complete problems for the classes $W[3]$ and $W[4]$ in parameterized complexity theory. This seems the first group of natural complete problems for higher levels in the parameterized intractability hierarchy (i.e., the W-hierarchy), and the first precise complexity characterizations of certain optimization problems in the research of supply chain management. Our results also derive strong computational lower bounds and inapproximability for these optimization problems.

1 Introduction

Parameterized complexity theory [9] is a recently proposed and promising approach to the central issue of how to cope with intractable problems – as is so frequently the case in the natural world of computing. An example is the NP-complete problem VERTEX COVER (determining whether a given graph has a vertex cover of size k), which now is solvable in time $O(1.285^k + kn)$ [5] and becomes quite practical for various applications. The other direction of the research is the study of *parameterized intractability*, based on a parameterized intractability hierarchy, the W-hierarchy $\bigcup_{t \geq 1} W[t]$. Under a parameterized reduction, the *fpt-reduction*, a large number of well-known computational problems have been proved to be complete for certain levels of the W-hierarchy [9]. For example, CLIQUE, INDEPENDENT SET, SET PACKING, V-C DIMENSION, and WEIGHTED 3-SAT are complete for the class $W[1]$, and DOMINATING SET, HITTING SET, SET COVER, and WEIGHTED SAT are complete for the class $W[2]$. The completeness of a problem in a level of the W-hierarchy characterizes precisely the parameterized complexity of the problem.

---

* This research is supported in part by US NSF under Grants CCR-0311590 and CCF-0430683 and by China NNSF Grants No. 60373083 and No. 60433020.

N. Megiddo, Y. Xu, and B. Zhu (Eds.): AAIM 2005, LNCS 3521, pp. 400–410, 2005.

However, no complete problem is known for any level $W[t]$ for $t > 2$, except the generic problems based on weighted satisfiability on bounded depth circuits and their variations [2, 9][1]. Therefore, it is interesting to know whether high levels of the W-hierarchy, which are defined in terms of formal mathematics, catch the complexity of certain natural computational problems.

In this paper, we present natural complete problems for the classes $W[3]$ and $W[4]$, based on computational problems studied in the areas of supply chain management. The study of supply chain management has been growing at a rapid pace in recent years, as a research area and as a practical discipline [10, 14]. It has provided extremely rich contexts for the definition of new large-scale optimization problems. Efforts to improve supply chain management have gained the attention of academic researchers, along with the enthusiastic support of government and industry. Therefore, our completeness results in the W-hierarchy for computational problems in the study of supply chains will also contribute to the understanding of this new computation model. Moreover, based on the recent research on parameterized intractability and inapproximability [4], our results also imply directly inapproximability for these problems.

We give a quick review on the related background (see [9] for more details).

A *parameterized problem* consists of instances of the form (x, k), where x is the *problem description* and k is an integer called the *parameter*. A parameterized problem Q is *fixed parameter tractable* if it can be solved by an algorithm of running time $O(f(k)n^{O(1)})$, where f is a function independent of $n = |x|$. Denote by FPT the class of all fixed parameter tractable problems.

A Π_t-*circuit* of n input variables $x_1, \ldots, x_n$ is a $(t+1)$-levelled circuit in which (1) the 0-th level is a single output gate that is an AND-gate; (2) each level-t gate is an input gate labelled by either x_i (a positive literal) or $\overline{x}_i$ (a negative literal), $1 \le i \le n$; (3) the outputs of a level-j gate can only be connected to the inputs of level-$(j-1)$ gates; and (4) AND-gates and OR-gates are organized into t alternating levels. A circuit is *monotone* (resp. *antimonotone*) if all its input gates are labelled by positive literals (resp. negative literals). A circuit represents naturally a boolean function. A truth assignment α to the variables of a circuit C *satisfies* C if α makes the output gate of C have value 1. The *weight* of an assignment α is the number of variables assigned value 1 by α.

The problem *weighted satisfiability on Π_t-circuits*, briefly WCS[t], consists of instances of the form (C, k), where C is a Π_t-circuit that is satisfied by an assignment of weight k. The W-*hierarchy*, $\bigcup_{t \ge 1} W[t]$, in parameterized complexity theory is defined based on WCS[t] via a new reduction, the *fpt-reduction*. We say that a parameterized problem Q is *fpt-reducible* to another parameterized problem Q' if there are two recursive functions f and g, and an algorithm A of running time bounded by $f(k)|x|^{O(1)}$, such that on an input (x, k), the algorithm A pro-

---

[1] We note that a similar situation has occurred in the study of the popular polynomial time hierarchy, for which complete problems for the first level Σ_1^p =NP have been extensively studied while the research on natural complete problems for higher level Σ_t^p for $t > 1$ has just started recently [15, 16].

duces a pair (x', k'), where $k' \leq g(k)$, and (x, k) is a yes-instance of Q if and only if (x', k') is a yes-instance of Q'. It is easy to verify that the fpt-reducibility is transitive [9]. For an integer $t \geq 2$, a parameterized problem Q_1 is in the class $W[t]$ if Q_1 is fpt-reducible to the problem $\text{WCS}[t]$, a parameterized problem Q_2 is $W[t]$-*hard* if the problem $\text{WCS}[t]$ is fpt-reducible to Q_2 (or equivalently, if all problems in $W[t]$ are fpt-reducible to Q_2), and a parameterized problem Q_3 is $W[t]$-*complete* if Q_3 is in $W[t]$ and is $W[t]$-hard. In particular, the problem $\text{WCS}[t]$ is a generic $W[t]$-complete problem for $t \geq 2$.[2]

We briefly review the related concepts in supply chain management (see [7, 8, 18] for detailed and systematic discussions, and see [10, 11, 14] for more recent progresses). The underlying structure of a supply chain model is a network consisting of various functional units (such as material suppliers, manufactures, storages, marketing/sales and retailers, and customers) and connections between different units (in the means of both material and information). A supply chain may have numerous tiers in the case of that substructure of manufactures forms a lengthy network itself [14]. *Supply chain management* involves the management of flows between and among the units in a supply chain to maximize total profitability [10]. The research in supply chain management includes the studies in *strategic-*, *tactical-*, and *operational-*level decisions [10]. In particular, tactical-level decisions, which is the subarea directly related to our current paper, are concerned with medium-range planning efforts, such as production and distribution quantity planning among multiple existing facilities, system-wide inventory policies, and distribution frequency decisions between facilities.

2   Single Product Cover and $W[3]$-Completeness

We follow the supply chain model studied in [17], which is a slight generalization of the model studied in [12]. The model is a 3-tier supply chain that consists of three kinds of units: (material) *suppliers*, (product) *manufacturers*, and *retailers*, such that:

(1) A supplier can be linked to a manufacturer, and a manufacturer can be linked to a retailer, standing for transportations/transactions between the units (link capacity is assumed unlimited); (2) A supplier can *provide* certain materials; (3). A manufacturer can *produce* a product if all needed materials for the product are provided by suppliers linked to the manufacturer; (4) A retailer has supply of a product if a manufacturer linked to the retailer produces the product.

Such a supply chain can be modelled by a directed graph $G = (S \cup M \cup R, E)$, where each unit is represented as a vertex in G and each directed edge in E represents a link between the corresponding units, here S is the set of all suppliers, M is the set of all manufacturers, and R is the set of all retailers. The objective of optimization studied in the current paper on this model is to

---

[2] The corresponding definitions for the class $W[1]$ are somehow special and not directly related to our discussion, thus are omitted. The readers are referred to [9] for details.

maximize the *channel profit* [7, 18], that is, to study the strategies that ensure that all retailers have supply of certain products they want to carry. In particular, we say a product *covers* all retailers if all retailers have supply of the product.

Now suppose that we want to test a new product in market at the widest range of customers, using as little experimental resource (i.e., suppliers) as possible and without overloading any supplier. For this, we assign at most one kind of material needed for the new product to each supplier and would like that the new product covers all retailers. Obviously, the problem is directly related to the complexity of the product, i.e., the number k of different kinds of materials needed for the product. Formally, the problem can be formulated as the following parameterized problem:

> 3-SCM SINGLE-PRODUCT COVER: Let $G = (S \cup M \cup R, E)$ be a supply chain model, k an integer, and suppose that we are going to produce a new product that requires k different kinds of materials. Is it possible to pick k suppliers, each provides a different kind of material for the new product, such that the product covers all retailers?

Before proving our main result in this section, we first define the problem WCS⁻[3]. The problem WCS⁻[3] is a subproblem of the problem WCS[3] that requires that in the input pair (C, k) the Π_3-circuit C be antimonotone (i.e., all input gates of C be labelled by negative input literals). It is known that the problem WCS⁻[3] is also $W[3]$-complete [9]. Thus, to prove the $W[3]$-completeness for the problem 3-SCM SINGLE-PRODUCT COVER, it suffices to derive fpt-reductions between WCS⁻[3] and 3-SCM SINGLE-PRODUCT COVER.

Theorem 1. 3-SCM SINGLE-PRODUCT COVER *is* $W[3]$-*complete.*

Proof. We first present an fpt-reduction from WCS⁻[3] to 3-SCM SINGLE-PRODUCT COVER. Let (C, k) be an instance of WCS⁻[3], where C is an antimonotone Π_3-circuit. Let g_0 be the output AND-gate of C (which is at level 0), L_1 be the set of OR-gates at level 1 in C (whose outputs are inputs to g_0), L_2 be the set of AND-gates at level 2 in C (whose outputs are inputs to gates in L_1), and L_3 be the set of input gates in C (which are inputs to gates in L_2 and are labelled by negative input literals).

Construct a 3-tier supply chain model $G = (S \cup M \cup R, E)$ as follows: (1) each retailer ρ_i in R corresponds to an OR-gate u_i in L_1; (2) each manufacturer μ_i in M corresponds to an AND-gate v_i in L_2; (3) each supplier σ_i in S corresponds to an input gate $\overline{x}_i$ in L_3. The vertices in G are connected in the following way: (1) there is a link from a manufacturer μ_i to a retailer ρ_j if and only if the corresponding AND-gate v_i is an input to the corresponding OR-gate u_j; and (2) there is a link from a supplier σ_i to a manufacturer μ_j if and only if the corresponding input gate $\overline{x}_i$ is *not* an input to the corresponding AND-gate v_j (note that C is an antimonotone circuit). This completes the description of the 3-tier supply chain model G. We prove that the circuit C has a satisfying assignment α of weight k if and only if we can pick k suppliers in the supply chain G, each for a different kind of material for a new product that needs k kinds of materials, so that the new product covers all retailers.

Suppose that the circuit C has a satisfying assignment α of weight k. Let X_k be the set of k variables in C that are assigned value 1 by α. Let S_k be the k suppliers corresponding to the k input variables in X_k. We show that we can pick the k suppliers in S_k, each for a different kind of material for the new product that needs k kinds of materials, such that the product covers all retailers in G. Consider any manufacturer μ_i in M. If μ_i has the supply for all k kinds of materials for the new product, i.e., if μ_i has links from all the k suppliers in S_k, then by the construction of the supply chain model G, the corresponding AND-gate v_i in C has no input from any input gate $\overline{x}_j$ where x_j is an input variable in X_k. Therefore, under the assignment α, all inputs to the gate v_i have value 1 and the output of v_i has value 1. On the other hand, if the manufacturer μ_i does not receive supply from a supplier σ_j in S_k, then the input gate $\overline{x}_j$ is an input to the AND-gate v_i, and under the assignment α, the output of gate v_i has value 0. In summary, the AND-gate v_i outputs value 1 if and only if the corresponding manufacturer μ_i has supply from all k suppliers in S_k and is able to produce the new product. Now, a retailer ρ_i has supply of the new product if and only if it has a link from a manufacturer μ_j that can produce the new product, which by the above analysis if and only if the corresponding AND-gate v_j in L_2 outputs value 1 under the assignment α. Since the retailer ρ_i has a link from a manufacturer μ_j if and only if the corresponding OR-gate u_i in C has input from the corresponding AND-gate v_j, we conclude that the retailer ρ_i has supply of the new product if and only if the corresponding OR-gate u_i in C outputs value 1 under the assignment α. Finally, since the output AND-gate g_0 of C is connected to all OR-gates in L_1, we conclude that the circuit C has value 1 if and only if all retailers have supply of the new product. In consequence, if α is a satisfying assignment for the circuit C, then picking the k suppliers in S_k results in the new product that covers all retailers in G.

Conversely, suppose there is a set S_k of k suppliers, each for a different kind of material for the new product such that the new product covers all retailers. We let X_k be the k input variables in the circuit C corresponding to the k suppliers in S_k. Let α be a weight-k assignment to C that assigns value 1 to the k variables in X_k and value 0 to all other input variables. Then following exactly the same reasoning as above, we can verify that the assignment α satisfies the circuit C.

This completes the analysis of the reduction from WCS$^-$[3] to 3-SCM SINGLE-PRODUCT COVER. The reduction is obviously an fpt-reduction. In conclusion, we have proved that the problem 3-SCM SINGLE-PRODUCT COVER is $W[3]$-hard.

To show that the problem 3-SCM SINGLE-PRODUCT COVER is in $W[3]$, it suffices to show that 3-SCM SINGLE-PRODUCT COVER is fpt-reducible to WCS$^-$[3]. The construction is very similar to the one described above: for an instance (G, k) of 3-SCM SINGLE-PRODUCT COVER, where $G = (S \cup M \cup R, E)$ is a supply chain model and k is an integer, we construct an instance (C, k) of WCS$^-$[3], where each level-1 OR-gate in C corresponds to a retailer in R, each level-2 AND-gate in C corresponds to a manufacturer in M, and each input gate in C (labelled by a negative literal) corresponds to a supplier in S. A level-1 OR-gate has an input from a level-2 AND-gate if and only if the corresponding retailer has a link from the corresponding manufacturer in G, and a level-2 AND-gate has an input from an input gate if and only if the corresponding manufacturer has *no* link from the corresponding supplier. Now by exactly the same method, we can verify that the circuit C has a satisfying assignment of weight k if and only if there are k suppliers, each for a different kind of material for the new product, such that the new product covers all retailers. In consequence, the problem 3-SCM SINGLE-PRODUCT COVER is in the class $W[3]$.

This proves that 3-SCM SINGLE-PRODUCT COVER is $W[3]$-complete.   □

3   Multiple Product Cover and $W[4]$-Completeness

To describe $W[4]$-complete problems, we consider a general model of 3-tier supply chains by allowing a supplier to provide multiple kinds of materials, a manufacturer to produce multiple kinds of products and a retailer to carry multiple kinds of products. The proofs of the theorems in this section can be found in [6].

We first consider a covering problem by a line of homogeneous (i.e., similar) products. Formally, let P be a given line of homogeneous products and let T be a set of materials, where each product π in P is associated with a set of materials in T that are needed for producing this product. In a 3-tier supply chain $G = (S \cup M \cup R, E)$, each supplier σ in S is associated with a list of materials in T that the supplier σ can provide, each manufacturer μ in M is associated with a list of products in P that the manufacturer μ can produce when necessary materials are provided by suppliers linked to μ, and each retailer ρ in R is associated with a suggested list of products in P that the retailer ρ is interested in carrying when the products are produced by the manufacturers linked to ρ. We are interested in the following problem in supply chain management: for a new line P of homogeneous products, we want to use limited amount of resource (i.e., a small number of suppliers) to test the product market in the widest range of customers (i.e., make all retailers have supply of some of the new products). This is formulated as the following parameterized problem.

> GENERAL 3-SCM H-PRODUCT-LINE COVER: Given a line P of homogeneous products, a general supply chain model $G = (S \cup M \cup R, E)$, and an integer k, is it possible to pick k suppliers providing materials for the products in P so that each retailer has supply of some products in its associated product list?

Theorem 2. GENERAL 3-SCM H-PRODUCT-LINE COVER *is $W[4]$-complete.*

The $W[4]$-completeness also provides precise complexity characterization for other computational problems in supply chain management. For example, suppose now that a firm is interested in investigating the market for a set P of *non-homogeneous* products. The 3-tier supply chain is again given as a network of suppliers, manufacturers, and retailers, where each supplier is given as before and associated with a set of materials that can be provided by the supplier.

Each manufacturer μ is associated with a set T_μ of materials and a set P_μ of products such that when all materials in T_μ are provided by suppliers linked to μ, the manufacturer μ can produce all products in P_μ. Finally, each retailer ρ is associated with a *requested* list of products that *must* be carried by the retailer ρ. This supply chain model gives a parameterized problem as follows:

> GENERAL 3-SCM PRODUCT-SET COVER: Given a product set P, a general supply chain model $G = (S \cup M \cup R, E)$ as described above, and an integer k, is it possible to pick k suppliers in S providing materials for the products in P so that every retailer in R has supply of *all* products in its associated product list?

The main difference between GENERAL 3-SCM H-PRODUCT-LINE COVER and GENERAL 3-SCM PRODUCT-SET COVER is that in the former model each retailer only needs to carry *some* of the products in its associated list while in the latter model each retailer must carry *all* products in its associated list.

Theorem 3. GENERAL 3-SCM PRODUCT-SET COVER *is $W[4]$-complete.*

4   Computational Lower Bounds and Inapproximability

Theorem 1, Theorem 2 and Theorem 3 provide strong lower bounds for the complexity of the problems 3-SCM SINGLE-PRODUCT COVER, GENERAL 3-SCM H-PRODUCT-LINE COVER, and GENERAL 3-SCM PRODUCT-SET COVER.

Theorem 4. *For any recursive function f, 3-SCM SINGLE-PRODUCT COVER cannot be solved in time $f(k)m^{O(1)}n^{o(k)}$ unless $W[2] = FPT$, and GENERAL 3-SCM H-PRODUCT-LINE COVER and GENERAL 3-SCM PRODUCT-SET COVER cannot solved in time $f(k)m^{O(1)}n^{o(k)}$ unless $W[3] = FPT$, where n is the number of suppliers and m is the size of the instance of the problems.*

Proof. Suppose that the problem 3-SCM SINGLE-PRODUCT COVER could be solved in time $f(k)m^{O(1)}n^{o(k)}$, then by the fpt-reduction from WCS$^-$[3] to 3-SCM SINGLE-PRODUCT COVER given in Theorem 1, it is easy to see that the problem WCS$^-$[3] can also be solved in time $f(k)m^{O(1)}n^{o(k)}$, where m is the instance size and n is the number of input variables in the circuit. By Theorem 4.2 in [4], it would imply $W[2] = FPT$. The lower bounds for GENERAL 3-SCM H-PRODUCT-LINE COVER and GENERAL 3-SCM PRODUCT-SET COVER can be proved in the same way using the same theorem in [4].   □

Since it is generally believed that $W[t] \neq FPT$ for all $t > 0$, Theorem 4 provides a computational lower bound $f(k)m^{O(1)}n^{\Omega(k)}$ for the problems 3-SCM SINGLE-PRODUCT COVER, GENERAL 3-SCM H-PRODUCT-LINE COVER, and GENERAL 3-SCM PRODUCT-SET COVER. Note that this is an asymptotically tight lower bound for the problems as the algorithm that exhaustively enumerates and examines all subsets of k suppliers in a problem instance solves the problems in time $O(m^2 n^k)$ trivially.

Theorem 4 further implies inapproximability results for certain optimization problems in 3-tier supply chain management. For this, we need to first review some related terminologies in approximation algorithms. The readers are referred to [1] for more detailed definitions and more comprehensive discussions.

An *optimization problem* Q consists of a set of *instances*, where each instance x is associated with a set of *solutions*. Each solution y of an instance x of Q is assigned an integral *value* $f_Q(x, y)$. The problem Q is a *maximization* (resp. *minimization*) problem if for each instance x of Q, we are looking for a solution of maximum (resp. minimum) value. Such a solution is called an *optimal solution* for the instance, whose value is denoted by $opt_Q(x)$.

An algorithm A is an *approximation algorithm* for an optimization problem Q if, for each instance x of Q, A returns a solution $y_A(x)$ for x. The solution $y_A(x)$ has an *approximation ratio* r if it satisfies the following condition:

$$\max\{opt_Q(x)/f_Q(x, y_A(x)), f_Q(x, y_A(x))/opt_Q(x)\} \leq r$$

The approximation algorithm A has an *approximation ratio* r if for any instance x of Q, the solution $y_A(x)$ constructed by the algorithm A has an approximation ratio bounded by r. A *polynomial time approximation scheme* (PTAS) for Q is an algorithm A' that on an instance x of Q and a real number $\epsilon > 0$, constructs a solution for x whose approximation ratio is bounded by $1 + \epsilon$, and the running time of A' is bounded by a polynomial of $|x|$ for each fixed ϵ [1].

Consider the following optimization problems in supply chain management:

3-SCM MOST-COMPLICATED PRODUCT COVER: given a 3-tier supply chain G, select the largest number k of suppliers in G, each for a different kind of material, such that a new product that needs the k materials can be produced and all retailers in G have supply of the new product.

GENERAL 3-SCM MIN-RESOURCE H-PRODUCT-LINE COVER: given a line P of homogeneous products and a general 3-tier supply chain G (as defined in GENERAL 3-SCM H-PRODUCT-LINE COVER), select the minimum number of suppliers in G for the product line P, such that each retailer in G has supply of some products in its associated product list.

GENERAL 3-SCM MIN-RESOURCE PRODUCT-SET COVER: given a set P of non-homogeneous products and a general 3-tier supply chain G (as defined in GENERAL 3-SCM PRODUCT-SET COVER), select the minimum number of suppliers in G for the product set P, such that each retailer in G has supply of all products in its product list.

Theorem 5. *For any recursive function f, 3-SCM MOST-COMPLICATED PRODUCT COVER has no PTAS of running time $f(1/\epsilon)m^{O(1)}n^{o(1/\epsilon)}$ unless $W[2] = FPT$, and GENERAL 3-SCM MIN-RESOURCE H-PRODUCT-LINE COVER and GENERAL 3-SCM MIN-RESOURCE PRODUCT-SET COVER have no PTAS of running time $f(1/\epsilon)m^{O(1)}n^{o(1/\epsilon)}$ unless $W[3] = FPT$, where n is the number of suppliers and m is the instance size of the problems.*

Proof. According to the definition given in [3], the parameterized problem 3-SCM SINGLE-PRODUCT COVER is the parameterized version of the optimization problem 3-SCM MOST-COMPLICATED PRODUCT COVER. Suppose that 3-SCM MOST-COMPLICATED PRODUCT COVER has a PTAS of running time $f(1/\epsilon)m^{O(1)}n^{o(1/\epsilon)}$, then by Theorem 5.1 in [3], its parameterized version, i.e., the problem 3-SCM SINGLE-PRODUCT COVER can be solved in time $f(2k)m^{O(1)}n^{o(k)}$, which, by Theorem 4, would imply $W[2] = FPT$. The inapproximability for GENERAL 3-SCM MIN-RESOURCE H-PRODUCT-LINE COVER and GENERAL 3-SCM MIN-RESOURCE PRODUCT-SET COVER can be proved using the same logic.   □

Since it is commonly believed in parameterized complexity theory that $W[t] \neq FPT$ for all $t \geq 1$, Theorem 5 implies that even for a moderate error bound $\epsilon > 0$, any PTAS for the problems, if exists, will become impractical.

5   Final Remarks

This paper studies the complexity issues for certain computational problems arising from the research in supply chain management, and characterizes these problems in terms of parameterized completeness in higher levels in the W-hierarchy. The research contributes to both parameterized complexity theory and to the study of supply chain management. For parameterized complexity theory, we presented the first group of natural complete problems for the classes $W[3]$ and $W[4]$, which had no known natural complete problems except the generic complete problems WCS[3] and WCS[4] and their variations. For the study of supply chain management, to the authors' knowledge, our results provide first group of precise complexity characterizations for certain computational problems in the area, which derive directly strong computational lower bounds and inapproximability results for the problems. The hardness results of these problems will provide useful information in the study of supply chain management.

A supply chain model has its units classified into different types, which makes it natural to map the computation in the supply chain model to that of bounded depth circuits. However, the mapping is not always straightforward and in many cases must be designed carefully. As we have seen in the current paper, problems on 3-tier supply chains can either correspond to the class $W[3]$, which is associated with the satisfiability problem on Π_3-circuits of 3 levels, or correspond to the class $W[4]$, which is associated with the satisfiability problem on Π_4-circuits of 4 levels. Our more recent research studied a computational problem, HARMFUL WASTE SOURCES, on the recycling model proposed in [13]. The problem is concerned with whether there are k waste sources who can pollute all markets. This recycling system is a 4-tier supply chain model, consisting of waster sources, recycling centers, processors, and markets. However, our study shows that the problem HARMFUL WASTE SOURCES is actually $W[2]$-complete (i.e., corresponding to the satisfiability problem on Π_2-circuits of 2 levels). Therefore, the computational complexity of the problems in supply chain management does not directly depend on the number of tiers in the model but is more closely related to the actual applications. In particular, the research in supply chain management has opened an area in computational complexity and optimization, and provided very rich contexts for new large-scale optimization problems that are both theoretically interesting and practically important.

References

1. G. AUSIELLO, P. CRESCENZI, G. GAMBOSI, V. KANN, A. MARCHETTI-SPACCAMELA, AND M. PROTASI, *Complexity and Approximation: Combinatorial Optimization Problems and Their Approximability Properties*, Springer-Verlag, Berlin Heidelberg, 1999.
2. M. CESATI, Compendium of parameterized problems, Department of Computer Science, Systems, and Industrial Engineering, University of Rome "Tor Vergata", Available at http://bravo.ce.uniroma2.it/home/cesati/research/compendium.pdf.
3. J. CHEN, Parameterized computation and complexity: a new approach dealing with NP-hardness, *Journal of Computer Science & Technology* **20**, (2005), pp.18-37.
4. J. CHEN, X. HUANG, I. KANJ, AND G. XIA, Linear FPT reductions and computational lower bounds, *Proc. 36th ACM Symp. on Theory of Computing* (STOC'04), (2004), pp.212-221.
5. J. CHEN, I. KANJ, AND W. JIA, Vertex cover: further observations and further improvements, *Journal of Algorithms* **41**, (2001), pp.280-301.
6. J. CHEN AND F. ZHANG, On product Covering in Supply Chain Models: Natural Complete Problems for W[3] and W[4], *Technical Report No 2005-3-2 (http://www.cs.tamu.edu/academics/tr/tamu-cs-tr-2005-3-2)*, Dept. of Computer Science, Texas A&M University, 2005.
7. S. CHOPRA AND P. MEINDL, *Supply Chain Management: Strategy, Planning, and Operations*, Prentice-Hall, Upper Saddle River, NJ, 2001.
8. M. C. COOPER, D. M. LAMBERT AND J. D. PAGH, Supply Chain Management: More Than a New Name for Logistics, *The International Journal of Logistics Management* **8-1**, (1997), pp.1-13
9. R. DOWNEY AND M. FELLOWS, *Parameterized Complexity*, Springer-Verlag, New York, 1999.
10. J. GENUES AND P. PARDALOS, Network optimization in supply chain management and financial engineering: an annotated bibliography, *Networks* **42-2**, (2003), pp.66-84.
11. J. GENUES, P. PARDALOS, AND H. ROMEIJN (Editors), Supply Chain Management: Models, Applications, and Research Directions, *Kluwer Series in Applied Optimization* **62**, Kluwer, Dordrecht, The Netherlands, 2002.
12. M. KHOUJA, Optimizating inventory decisions in multi-stage multi-customer supply chain, *Transportation Research Part E* **39**, (2003), pp. 193-208.
13. A. NAGURNEY AND F. TOYASAKI, Reverse supply chain management and electronic waste recycling: a multitiered network equilibrium framework for e-cycling, *Transportation Research D, (2003)*,in press.
14. H. MIN, G.ZHOU, Supply chain modeling: past, present and future, *Computers & Industrial Engineering* **43**, (2002), pp.231-249.
15. M. SCHÄFER AND C. UMANS, Completeness in the polynomial-time hierarchy: part I: a compendium, *SIGACT News* **33-3**, (2002), pp.32-49.

16. M. SCHÄFER AND C. UMANS, Completeness in the polynomial-time hierarchy: part II, *SIGACT News* **33-4**, (2002), pp.22-36.
17. J. SHEU, Locating manufacturing and distribution centers: an integrated supply chai-based spatial interaction approach, *Transportation Research Part E* **39**, (2003), pp. 381-397.
18. D. SIMCHI-LEVI, P. KAMINSKY, AND E. SIMCHI-LEVI, *Designing and Managing the Supply Chain: Concepts, Strategies, and Case Studies*, Irwin McGraw-Hill, Boston, MA, 2000.

Assign Ranges in General Ad-Hoc Networks

Janka Chlebíková[1,*], Deshi Ye[2,**], and Hu Zhang[3,***]

[1] Faculty of Mathematics, Physics and Informatics, Comenius University,
Mlynská dolina, 842 48 Bratislava, Slovakia
chlebikova@fmph.uniba.sk
[2] Department of Mathematics, Zhejiang University,
Hangzhou 310027, China
dye@math.zju.edu.cn
[3] Department of Computing and Software, McMaster University,
1280 Main Street West, Hamilton, ON L8S 4K1, Canada
zhanghu@mcmaster.ca

Abstract. In this paper we study the MINIMUM RANGE ASSIGNMENT problem in static ad-hoc networks with arbitrary structure, where the transmission distances can violate triangle inequality. We consider two versions of the MINIMUM RANGE ASSIGNMENT problem, where the communication graph has to fulfill either the h-strong connectivity condition (MIN-RANGE(h-SC)) or the h-broadcast condition (MIN-RANGE(h-B)). Both homogeneous and non-homogeneous cases are studied. By approximating arbitrary edge-weighted graphs by paths, we present probabilistic $O(\log n)$-approximation algorithms for MIN-RANGE(h-SC) and MIN-RANGE(h-B), which improves the previous best ratios $O(\log n \log \log n)$ and $O(n^2 \log n \log \log n)$, respectively [21]. The result for MIN-RANGE(h-B) matches the lower bound [20] for the case that triangle inequality for transmission distance holds (which is a special case of our model). Furthermore, we show that if the network fulfils certain property and the distance power gradient α is sufficiently small, the approximation ratio is improved to $O((\log \log n)^\alpha)$.

---

* This work was performed in part when this author was working at the University of Kiel, Germany. Research supported in part by APVT grant APVT-20-018902, and by EU-Project ARACNE, Research Training Network, Approximation and Randomized Algorithms in Communication Networks, HPRN-CT-1999-00112.
** This work was performed in part when this author was studying at the University of Kiel, Germany. Research supported in part by a DAAD Sandwich Project, and by NSFC(10231060).
*** This work was performed in part when this author was studying at the University of Kiel, Germany. Research supported in part by the DFG Graduiertenkolleg 357, Effiziente Algorithmen und Mehrskalenmethoden, by EU Thematic Network APPOL II, Approximation and Online Algorithms for Optimization Problems, IST-2001-32007, by EU Project CRESCCO, Critical Resource Sharing for Cooperation in Complex Systems, IST-2001-33135, by an MITACS grant of Canada, by the NSERC Discovery Grant DG 5-48923, and by the Canada Research Chair program.

N. Megiddo, Y. Xu, and B. Zhu (Eds.): AAIM 2005, LNCS 3521, pp. 411–421, 2005.
© Springer-Verlag Berlin Heidelberg 2005

1   Introduction

Nowadays *wireless communication network* plays an important role in the daily life due to the significant drop in the prices of equipments and the progress in new technology. In wireless communication networks there is no infrastructure backbone and for each device the radio signal transmission is conducted in a finite range around it. The locations of devices and the ranges can be adjusted dynamically in order to fulfil certain communication quality requirement and to extend the lifetime of the networks. In general the wireless devices are portable with only limited power resources (e.g., batteries). High quality of communication usually consumes more energy and reduce the network lifetime, and vice verse [3]. Hence, a crucial issue of wireless communication networks is to minimize the energy consumption as well as to keep required communication quality. Among the models of new generation wireless communication networks, the *ad-hoc wireless network* based on *multi-hop* plays a very promising role [15].

The (static) RANGE ASSIGNMENT problem on d-dimensional Euclidean space ($d \geq 1$) is defined as follows. We are given a set S of stations (radio transmitter/receivers) on d-dimensional Euclidean space and the distances between all pairs are known according to their coordinates. The stations can communicate with each other by sending/receiving radio signals. The message communication happens via *multi-hop transmission*, i.e., a message is delivered from the source to the destination through some intermediate stations and each station in this transmission chain other than the source station is in the coverage range of its predecessor. A range assignment for a set of station S is a function $r : S \rightarrow \mathbb{R}^+$, which indicates the radii that stations can cover. For a station $v \in S$ associated with a range $r(v)$ in a network-wise range assignment, its energy consumption (power consumption) is $cost(r(v)) = c(v)(r(v))^\alpha$, where $c(v)$ is a parameter depending on the individual device. The *distance-power gradient* α is a positive real number, usually in the interval $[1, 6]$ in practice. When $c(v)$ is a constant for all $v \in S$, we call the model *homogeneous*, otherwise it is *non-homogeneous* [1]. It is worth noting that the non-homogeneous model can be asymmetric, as the energy consumption for a device on u to cover v may differ from the energy consumption of another device to conduct the same transmission. The overall energy consumption of a range assignment r is defined as $cost(r) = \sum_{v \in S} cost(r(v)) = \sum_{v \in S} c(v)(r(v))^\alpha$.

A range assignment r yields a directed communication graph $G_r = (S, E_r)$, such that for each pair of stations u and v there exists a directed edge $(u, v) \in E_r$ if and only if v is at an (Euclidean) distance at most $r(u)$ from u. For the purpose of a variety of communication requirements, the communication graph G_r must fulfil one of following two properties Π_h for any fixed h, $h \in \{1, \dots, n-1\}$, where n is the number of stations:

- *h-strong connectivity*: from every station to any other station, G_r must contain a directed path of at most h hops (edges),
- *h-broadcast*: G_r must contain a directed source spanning tree rooted at a source station with depth at most h.

The goal of the MINIMUM RANGE ASSIGNMENT problem (shortly, MIN-RANGE) is to find a range assignment r for a given S such that G_r fulfils a given property Π_h and the overall energy consumption $cost(r)$ is minimized. We use notations MIN-RANGE(h-SC) (respectively, MIN-RANGE(h-B)) for the corresponding MINIMUM RANGE ASSIGNMENT problems, when Π_h is the property of h-strong connectivity (respectively, h-broadcast). The detailed description of the problem can be also found in [20].

Known Results: Previously attention was mainly paid to the MINIMUM RANGE ASSIGNMENT problems defined on one dimensional (Euclidean) space, which is equivalent to the case that a set S of stations are placed along a line (or a path). Polynomial time algorithms by dynamic programming were addressed for both homogeneous and non-homogeneous cases for MIN-RANGE(h-B) on one dimensional (Euclidean) space in [14, 9, 1]. In the homogeneous case the MIN-RANGE(h-SC) problem is polynomial time solvable for $h = 2$ (respectively, $h = n - 1$) within a running time $O(n^3)$ (respectively, $O(n^4)$) as presented in [7] (respectively [14]). However, for any other h it is still open whether the MIN-RANGE(h-SC) problem on one dimensional space can be solved in polynomial time. Clementi et al. [7] proposed a 2-approximation algorithm for any $h \in \{2, \ldots, n - 1\}$ for the homogeneous case with a complexity $O(hn^3)$. Furthermore, the algorithm can be extended to the non-homogeneous case. There are only few results on d-dimensional Euclidean space for $d \geq 2$ (see [8, 20]). In case of $h = n - 1$, algorithms based on the minimum spanning tree technique can deliver a solution with a constant approximation ratio for the MIN-RANGE(h-B) problem [6], where the constant ratio depends on the dimension d and the distance-power gradient α. For the MIN-RANGE(h-SC) problem and $h = n - 1$, a 2-approximation algorithm was addressed in [14]. Recently, Calinescu et al. [4] developed an $O(\log^\alpha n)$-approximation algorithm for the MIN-RANGE(h-B) problem on d dimensional Euclidean space. They also presented $(O(\log n), O(\log n))$ bicriteria approximation algorithms for both MIN-RANGE(h-B) and MIN-RANGE(h-SC) problems. In [10] the MIN-RANGE(h-SC) problem was proved in $\mathcal{Av}\text{-}\mathcal{APX}$ for any fixed $h \geq 1$ and the problem is $\mathcal{APX}$-hard on d-dimensional Euclidean space for $d \geq 3$. $O(\min\{\log n \log \log n, (\log n)^\alpha\})$- and $O(n^2 \min\{\log n \log \log n, (\log n)^\alpha\})$-approximation algorithms for MIN-RANGE(h-B) and MIN-RANGE(h-SC) on metric space were proposed in [21], respectively. This was the first work to explore MINIMUM RANGE ASSIGNMENT problem on general spaces. In their model, the triangle inequality is still required for the transmission distance.

Our Contributions: In this paper, we first propose a new model of the MINIMUM RANGE ASSIGNMENT problem. We show that our model is a generalization of the previous models and is realistic. We notice that the transmission cost for the same device is not homogeneous on space, i.e., the costs from different locations to cover the same Euclidean distance can be different due to environmental factors. In this case it is invalid to measure the cost by Euclidean distance. Thus we consider the problem with a station set S on a space with transmission dis-

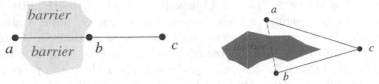

Fig. 1. Example of difference between the transmission distance and the geometric distance

Fig. 2. Example that the triangle inequality is violated

tance (see Subsection 2.1) instead of the original Euclidean distance. In such an instance, the transmission distance can even violate the triangle inequality, and no previous study remains valid in this case. Our main ideas are as follows. We first present a probabilistic algorithm to approximate any edge-weighted graph by a collection of paths, such that for any pair of nodes the expected distortion of shortest path distance is at most $O(\log n)$, where n is the number of nodes in the graph. The paths in the collection and the corresponding probability distribution are given by solving a packing problem [13], and an approximate solver of the MINIMUM LINEAR ARRANGEMENT problem [19] is employed as an oracle. With this algorithm we are able to approximate the general static ad-hoc networks to paths and run known algorithms in [1, 7, 9, 14] for the MINIMUM RANGE ASSIGNMENT problem on one dimensional Euclidean space (lines or paths). Therefore this strategy leads to probabilistic $O(\log n)$-approximation algorithms for the MINIMUM RANGE ASSIGNMENT problem (both MIN-RANGE(h-B) and MIN-RANGE(h-SC)) for general static ad-hoc networks. The ratio for the MIN-RANGE(h-B) problem reaches the lower bound for networks that triangle inequality is valid for transmission distance [20]. It is worth noting that the case in [20] is only a special case of our model as here we allow violation of the triangle inequality. Furthermore, if the input graph of station set fulfils certain property, we show that the approximation ratio can be further reduced to $O((\log \log n)^{\alpha})$.

2   Preliminaries

2.1   Our Model

In most of previous works for the MINIMUM RANGE ASSIGNMENT problem either h is set as $n - 1$ [14, 6], or the station set is on one dimensional (Euclidean) space [14, 7, 9, 1]. For studies of MINIMUM RANGE ASSIGNMENT problem on multidimensional space, the transmission cost is measured by the (Euclidean) geometric distance [14, 6, 4]. Even in [21], the triangle inequality must hold for the transmission cost.

The one dimensional model has already been extensively studied as a good approximation of the real instances. However, demands on models on multidimensional space are increasing, as they are more precise to characterize the real ad-hoc networks. Furthermore, from the engineering point of view, the model with fixed $h = n - 1$ is not practical and hard to control by current protocols [5]. Finally, an important issue is the measurement of the transmission cost in the ad-hoc networks. In almost all of previous works the transmission cost is assumed to be characterized by the (Euclidean) geometric distance. We notice that due to environmental condition, that assumption is not always true. For instance (Figure 1), three stations are along a line, and $d_E(a, b) = d_E(b, c)$, where d_E is the Euclidean geometric distance. However, there are some barriers (forests, buildings, etc.) between a and b, while there is no barrier between b and c. In such an instance, it costs more energy to send signals from b to a than from b to c. Thus the geometric distance is not sufficient to measure the real energy cost in transmission though it is a good approximation. We further notice that on metric spaces the triangle inequality is assumed valid, and this assumption could be violated in real wireless communication networks. For instance (See Figure 2), three stations are on a plane. There is a solid barrier (e.g. mountains, large buildings, etc.) between stations a and b, while there is no such solid barrier between station pairs a, c and b, c. In this example $d_E(a, b) \leq d_E(a, c)$ and $d_E(a, b) \leq d_E(b, c)$. However, it costs much more to launch a signal transmission between a and b due to the solid barrier. Therefore the energy cost for direct transmission between a and b can be greater than the sum of costs of transmissions between a, c and b, c. Thus the triangle inequality does not hold. When the transmission cost between a and b is unbounded (which does happen in real world), it is impossible to build an ad-hoc network for this instance with only $h = 1$ hop and a bounded overall energy cost.

In this paper, we propose a new model of the MINIMUM RANGE ASSIGNMENT problem in ad-hoc communication networks. In this model, h can be any integer number in $\{1, \ldots, n - 1\}$. Furthermore, the stations are on arbitrary spaces, and the transmission cost between each pair can be arbitrary. We propose a concept *transmission distance*, which is a scalable quantity. Given a station set S and a distance power gradient α, we can measure the *minimum* energy cost $cost(u, v)$ of *directly* sending signals from any station $u \in S$ to any other $v \in S \setminus \{u\}$ with a standard wireless device $c(u) = 1$. The transmission distance between the station pair (u, v) is defined as $d(u, v) = (cost(u, v))^{-\alpha}$. In an instance of MINIMUM RANGE ASSIGNMENT problem $(G(S, E, l_G), \alpha, \Pi_h)$ of our model, we are given a complete edge weighted graph G, a distance power gradient $\alpha \geq 1$ and a required property Π_h of the communication graph (either h-strong connectivity or h-broadcast), for $h \in \{1, \ldots, n - 1\}$. In the weighted graph G, the vertex set S is the station set and the weight $l(u, v)$ of any edge (u, v) is the transmission distance (which can violate the triangle inequality) between the two endpoints u and v. The edge weight can be infinity if the transmission cost between the corresponding two endpoints is unbounded. Same as the previous models, the goal is to find a network-wise arrangement r such that the property Π_h holds in the resulting communication graph and the overall energy cost $cost(r)$ is minimized. We also notice that this model of the MINIMUM RANGE ASSIGNMENT problem generalizes the previous models.

2.2   Notations and Definitions

Given a graph, we can embed it in a simpler graph such that the distance between each pair of vertices are approximately preserved. This technique can be
employed to solve some hard problems on arbitrary graphs, as an arbitrary graph
may have a very complicated structure. We will propose the idea of *probabilistic approximation* of weighted graphs by a collection of simpler graphs (e.g.,
paths). This is a generalization of the concept of probabilistic approximation of
metric spaces addressed in [2], because here edge weights can violate triangle
inequality.

Given two graphs $G_1 = (V, E_1, l_1)$ and $G_2 = (V, E_2, l_2)$ with the same node
set V, G_1 *dominates* G_2 if and only if $d_{G_1}(u, v) \geq d_{G_2}(u, v)$ for all pair u,
$v \in V$, where $d_{G_i}(u, v)$ is the shortest path distance between node pair u and
v in G_i. G_1 is also called a *non-contracting* embedding of G_2. The *distortion* is
defined as $\max_{u,v \in V} d_{G_1(u,v)}/d_{G_2(u,v)}$. Suppose that $\mathcal{H}$ is a collection of graphs
that have the same node set V as another graph G. Assuming that each graph
in $\mathcal{H}$ dominates G, $\mathcal{H}$ is defined to ρ-*probabilistically approximate* G if there is
a probability distribution μ over $\mathcal{H}$ such that for each pair of nodes in V the
expected distance between them in a graph $H \in \mathcal{H}$ chosen according to μ is at
most ρ times the distance between the pair in G, i.e., $E[d_H(u, v)] \leq \rho d_G(u, v)$.

In this paper, we will develop an algorithm to $O(\log n)$-probabilistically approximate any graph by a collection of paths (see Section 3). Based on this
algorithm, we are able to generalize the existing algorithms for the MINIMUM
RANGE ASSIGNMENT problems in ad-hoc networks on lines (paths) to arbitrary
networks. For the deterministic version of the problem, Matoušek [16] shows
that any metric can be embedded into the real line with a distortion $O(n)$. This
result is existentially tight as the n-cycle can not be embedded into a line with
distortion $o(n)$ [12, 18].

3   Approximate a Graph by a Collection of Paths

We are given an edge-weighted graph $G(V, E, l_G)$, where $|V| = n$, $|E| = m$,
and a weight function $l_G : E \rightarrow \mathbb{R}_0^+$. The weight function l_G can violate the
triangle inequality. Without loss of generality, we assume that the diameter of G
is bounded by one. Otherwise a simple scaling method can be employed with a
running time bounded by $O(n^2)$. Let $\mathcal{P} = \{P_1, \ldots, P_N\}$ be a collection of paths,
each connecting all nodes in V. Each path in $\mathcal{P}$ dominates the graph G, i.e.,
$d_{P_i}(u, v) \geq d_G(u, v)$ for any pair $u, v \in V$ and $i \in \{1, \ldots, N\}$. Here the distance
functions d_{P_i} and d_G are shortest path distance in the path P_i and the graph
G, respectively. In addition, we assign a real number $x_i \in [0, 1]$ to every path
$P_i \in \mathcal{P}$, which represents the probability distribution μ over the path collection
$\mathcal{P}$, and the sum of x_i is 1. Denote by λ the distortion of each edge and $l_G(e)$
the edge length of $e \in E$. The following linear program is to find the probability
distribution μ that minimizes the expected edge distortion in $P \in \mathcal{P}$:

$$\min \lambda$$
$$\text{s.t. } \sum_{i=1}^{N} d_{P_i}(e)x_i \leq \lambda l_G(e), \text{for all edge } e \in E; \tag{1}$$
$$\sum_{i=1}^{N} x_i = 1;$$
$$x_i \geq 0.$$

Here the first set of constraints indicates that the expected distortion of every edge $e \in E$ in all paths in $\mathcal{P}$ is bounded by λ. The other constraints are directly from the definition of probability distribution. Notice that in (1) the number of variables (i.e., the number of paths in $\mathcal{P}$) can be exponentially large. We notice that (1) can be formulated as a packing problem described in [13] and the packing constraints $f_e(x) = \sum_{i=1}^{N} d_{P_i}(e)x_i/l_G(e)$ are nonnegative linear functions, and the set $B = \{x = (x_1, \ldots, x_N)^T | \sum_{i=1}^{N} x_i = 1, x_i \geq 0\}$ is indeed a simplex. We will apply the approximation algorithms in [13] to solve this packing problem. In order to develop an algorithm for (1), we need to consider the block problem in advance, which is related to the dual problem and the structure of the set B. As showed in [13], given a price vector $y \in Y = \{(y_1, \ldots, y_m) | \sum_{e \in E} y_e = 1, y_e \geq 0\}$, the block problem is to find an $\hat{x} \in B$ such that $y^T f(\hat{x}) = \min_{x \in B} y^T f(x)$. With the formulation of the packing constraints in (1), the block problem can be simplified as follows:

$$\min_{x \in B} \sum_{e \in E} \left(y_e \sum_{i=1}^{N} \frac{d_{P_i}(e)}{l_G(e)} x_i \right) = \min_{x \in B} \sum_{i=1}^{N} \left(x_i \sum_{e \in E} y_e \frac{d_{P_i}(e)}{l_G(e)} \right) = \min_{P_i \in \mathcal{P}} \sum_{e \in E} \frac{y_e}{l_G(e)} d_{P_i}(e).$$

The last equality holds because we can choose one path P_i minimizing the sum $\sum_{e \in E} y_e d_{P_i}(e)/l_G(e)$, and set its corresponding probability $x_i = 1$ (and the probabilities of other paths are set as 0) to achieve the optimum. Denote by $w(e) = y_e/l_G(e)$ the weight on edge e. Therefore the goal of the block problem is to find a path P connecting all nodes in G such that the value $\sum_{e \in E} w(e) d_P(e)$ is minimized with the given weight function w, for all edge $e \in E$.

The block problem actually is equivalent to the MINIMUM LINEAR ARRANGE-MENT problem (MLA). The problem is defined as follows: Given a graph $G(V, E)$ and nonnegative edge weights $w(e)$ for all $e \in E$, where $|V| = n$ and $|E| = m$. The goal is to find a linear arrangement of the nodes $\sigma : V \to \{1, \ldots, n\}$ that minimizes the sum of the weighted edge lengths $|\sigma(u) - \sigma(v)|$, over all $(u, v) \in E$. If we define the overall cost as $c = \sum_{(u,v) \in E} w(u, v) |\sigma(u) - \sigma(v)|$, then the goal of the MINIMUM LINEAR ARRANGEMENT problem is to minimize the total cost c. Then we can place all vertices $u \in V$ on a path P (i.e., a one dimensional Euclidean space) and the coordinates are their arrangements $\sigma(u)$. It is obvious that the weight in the MINIMUM LINEAR ARRANGEMENT problem corresponds to the weight function in our block problem and the length $|\sigma(u) - \sigma(v)|$ corresponds to the distance in the path $d_P(u, v)$. Therefore we can directly apply the algorithms for the MINIMUM LINEAR ARRANGEMENT problem to solve our block problem to generate a path. However, the MINIMUM LINEAR ARRANGEMENT problem is $\mathcal{NP}$-hard [11]. The best known algorithm for the MINIMUM LINEAR ARRANGEMENT problem is proposed by Rao et al. [19] and the approximation

Table 1. Approximation Algorithm $\mathcal{RA}$ for MINIMUM RANGE ASSIGNMENT

Algorithm $\mathcal{RA}(G(S, E, l_G), \alpha, \Pi_h)$:

1. Construct a complete graph M_G for the given network G, such that for any pair $u, v \in S$, the weight of the edge (u, v) in M_G is $l_{M_G}(u, v) = (l_G(u, v))^{\alpha}$.
2. Run **Algorithm** $\mathcal{AG}(M_G(S, E_{M_G}, l_{M_G}), \mathcal{P}, x)$ in Section 3 to generate a collection $\mathcal{P}$ of paths with a probability distribution μ over it; the generated paths can be represented in one-dimensional (Euclidean) lines such that neighbours are in (Euclidean) distance one.
3. Run algorithms for the MINIMUM RANGE ASSIGNMENT problems (MIN-RANGE(h-SC) and MIN-RANGE(h-B)) for static ad-hoc networks on one dimensional (Euclidean) space for the reduced instances $(P(S_P, E_P, 1), 1, \Pi_h)$, for any $P \in \mathcal{P}$ chosen according to the probability distribution μ.

---

ratio is $O(\log n)$. Thus we are able to construct an $O(\log n)$-approximation algorithm for the linear program (1).

We denote by $\mathcal{AG}(G(V, E, l_G), \mathcal{P}, x)$ the $O(\log n)$-approximation algorithm based on [13]. Due to the space limit, we do not give the details here. According to [13], we have the following result:

Theorem 1. *Given a graph $G = (V, E, l_G)$, $|V| = n$, $|E| = m$, and an edge weight function $l_G : E \to \mathbb{R}_0^+$, there exists an algorithm that generates a collection $\mathcal{P}$ of $O(m \log m)$ paths and a probability distribution $\mu(x)$ over the collection $\mathcal{P}$, such that for any edge $e \in E$, the expected distortion of e in $\mathcal{P}$ is bounded by $O(\log n)$. The running time of the algorithm is $O(m \log m(\beta + m \log \log m))$ time, where β is the running time of the approximate* MINIMUM LINEAR ARRANGEMENT *solver.*

4   Approximation Algorithms for the Minimum Range Assignment Problem in Static Ad-Hoc Networks

For any fixed h, let $(G(S, E, l_G), \alpha, \Pi_h)$ be an instance for the MINIMUM RANGE ASSIGNMENT problem (MIN-RANGE(h-SC) or MIN-RANGE(h-B)) in static ad-hoc networks with arbitrary structure, where S is the station set, l_G is the transmission distance in the complete graph G, α is the distance-power gradient, and Π_h is the property of the communication graph (h-strong connectivity or h-broadcast). Our approximation algorithm is in Table 1. Hence, we obtain the following theorem for the approximation algorithm $\mathcal{RA}$:

Theorem 2. *There exists a probabilistic $O(\log n)$-approximation algorithm for* MIN-RANGE(h-SC) *and* MIN-RANGE(h-B) *in general static ad-hoc networks running in at most $O(n^2 \log n(\beta + n^2 \log \log n) + h n^4)$ time, where β is the running time of the approximate* MINIMUM LINEAR ARRANGEMENT *solver.*

Table 2. Approximation Algorithm $\mathcal{RA}'$ for MINIMUM RANGE ASSIGNMENT in a special case

---

Algorithm $\mathcal{RA}'(G(S,E,l_G),\alpha,\Pi_h,\Omega)$:

1. Run **Algorithm** $\mathcal{AG}(G(S,E,l_G),\mathcal{P},x)$ in Section 3 to generate a collection $\mathcal{P}$ of paths with a probability distribution μ over it;
2. Run algorithms for the MINIMUM RANGE ASSIGNMENT problems (MIN-RANGE(h-SC) and MIN-RANGE(h-B)) for static ad-hoc networks on one dimensional (Euclidean) space for the reduced instances $(P(S_P,E_P,1),1,\Pi_h)$, for any $P \in \mathcal{P}$ chosen according to the probability distribution.

Due to the limit of space we do not give the proof in this version. It is worth noting that the approximation ratio of Algorithm $\mathcal{RA}$ for the MIN-RANGE(h-B) problem is $O(\log n)$, while the lower bound for the MIN-RANGE(h-B) problem on metric spaces, where triangle inequality holds, is also $O(\log n)$ [20].

5   Improved Approximation Ratio for a Special Case

A graph H is a *minor* of another graph G if H can be obtained from G by deleting and contracting some edges of G. Denote by $K_{r,r}$ the $r \times r$ complete bipartite graph. We define a property Ω as follows: An instance $(G(S,E,l),\alpha,\Pi_h,\Omega)$ of the MINIMUM RANGE ASSIGNMENT problem in general ad-hoc networks has the property Ω if and only if (1) the graph G does not contain $K_{r,r}$-minors for any $r \geq 3$; (2) $\alpha \leq O(\log\log n/\log\log\log n)$. For any instance $(G(S,E,l_G),\alpha,\Pi_h,\Omega)$ of the MINIMUM RANGE ASSIGNMENT problem we have the two-step algorithm in Table 2. Then we have the following theorem:

Theorem 3. *There exists a probabilistic $O((\log\log n)^\alpha)$-approximation algorithm for* MIN-RANGE(h-SC) *and* MIN-RANGE(h-B) *in general static ad-hoc networks when property Ω holds.*

We claim that indeed the instances with property Ω are not rare. It is obvious that for a fixed value of α (which is usually in the interval $[1,6]$ in practice), a large station set S can result in the second assumption of the property Ω. In fact, an instance on a planar graph with $\alpha = 2$, a set of $n = 16$ stations is sufficient for the property Ω. We believe that many real applications belong to this category.

Remark: In general, a mobile ad-hoc network consists of mobile nodes that are connected via wireless links. By the DSDV (Destination-Sequence Distance-Vector) protocol [17], it needs to compute a temporary ad-hoc networks based on the routing table. Thus, DSDV is the base to study a static ad-hoc networks and apply the results in static networks to the mobile one. Here we can also generalize our algorithms to the model of mobile ad-hoc networks (dynamical model) according to the DSDV protocol. Assuming that the time for radio signals to reach any node in the network is no more than the time step for updating the routing table, our algorithm $\mathcal{RA}$ or $\mathcal{RA}'$ lead to an $O(\min\{\log n, (\log\log n)^\alpha\})$-approximation algorithm for any instance of the MINIMUM RANGE ASSIGNMENT problemin general mobile ad-hoc networks.

References

1. C. Ambuehl, A. E. F. Clementi, M. D. Ianni, A. Monti, G. Rossi and R. Silvestri, The range assignment problem in non-homogeneous static ad-hoc networks, *Proc. of 4th International Workshop on Algorithms for Wireless, Mobile, Ad Hoc and Sensor Networks*, WMAN 2004.
2. Y. Bartal, Probabilistic approximation of metric spaces and its algorithmic applications, *Proc. of the 37th IEEE FOCS*, 1996, 184–193.
3. G. Calinescu, S. Kapoor, A. Olshevsky, A. Zelikovsky, Network lifetime and power assignment in ad hoc wireless networks, *Proc. of the 11th ESA*, 2003, 114–126.
4. G. Calinescu, S. Kapoor, and M. Sarwat, Bounded hops power assignment in ad-hoc wireless networks, *Proc. of the IEEE Wireless Communications and Networking Conference*, WCNC 2004.
5. I. Chlamtac, M. Conti, J. J.-N. Liu, Mobile ad hoc networking: imperactives and challenges, *Ad Hoc Networks*, 1 (2003) 13-64.
6. A. E. F. Clementi, P. Crescenzi, P. Penna, G. Rossi and P. Vocca, On the complexity of computing minimum energy consumption broadcast subgraph. *Proc. of 18th STACS*, 2001, LNCS 2010, 121–131.
7. A. E. F. Clementi, A. Ferreira, P. Penna, S. Perennes, R. Silvestri, The minimum range assignment problem on linear radio networks, *Algorithmica*, 35(2) 2003, 95–110.
8. A. E. F. Clementi, G. Huiban, P. Penna, G. Rossi and Y. C. Verhoeven, Some recent theoretical advances and open questions on energy consumption in ad-hoc wireless networks. *ARACNE 2002*, 23–38.
9. A. E. F. Clementi, M. D. Ianni and R. Silvestri, The minimum broadcast range assignment problem on linear multi-hop wireless networks, *Theoretical Computer Science*, 1-3 (299) 2003, 751–761.
10. A. E. F. Clementi, P. Penna, and R. Silvestri, On the power assignment problem in radio networks, *Technical Report ECCC TR00-054*, 2000.
11. M. Garey and D. Johnson, Computer and Intractability: A Guide to the Theory of NP-Completeness, W. H. Freeman and Company, NY, 1979.
12. A. Gupta, Steiner points in tree metrics don't (really) help, *Proc. of the 12th ACM/SIAM SODA*, 2001, 220–227.
13. K. Jansen and H. Zhang, Approximation algorithms for general packing problems with modified logarithmic potential function, *Proc. of 2nd IFIP International Conference on Theoretical Computer Science*, TCS 2002.
14. L. M. Kirousis, E. Kranakis, D. Krizanc and A. Pelc, Power consumption in packet radio networks, *Theoretical Computer Science*, (243) 2000, 289-305.
15. G. S. Lauer, Packet Radio Routing, (chap. 11), Printice-Hall, Englewood Cliffs, NJ, 1995.
16. J. Matoušek, Bi-lipschitz embeddings into low dimensional Euclidean spaces, *Comment. Math. Univ. Carolinae*, 31(3) (1990), 589-600.

17. C. E. Perkins, P. Bhagwat, Highly dynamic destination-sequence distence-vector routing (DSDV) for mobile computers, *Computer Communications Review* (October 1994) 234-244.
18. Y. Rabinovich and R. Raz, Lower bounds on the distortion of embedding finite metric spaces in graphs, *GEOMETRY: Discrete and Computational Geometry*, 19, 1998.
19. S. Rao and A. W. Richa, New approximation techniques for some ordering problems, *Proc. of the 9th ACM-SIAM SODA*, 1998, 211–218.
20. G. Rossi, The range assignment problem in static ad-hoc wireless networks. *Ph.D. Thesis*, 2003.
21. D. Ye and H. Zhang, The range assignment problem in static ad-hoc networks on metric spaces, *Proc. of the 11th Colloquium on Structural Information and Communication Complexity* (Sirocco 2004), LNCS 3104, 291-302.

Inverse Problems
of Some NP-Complete Problems

Siming Huang

Institute of Policy and Management, Chinese Academy of Sciences,
Beijing 100080, P.R. China

Abstract. The Knapsack problem and integer programming are NP-complete problems. In this paper we show that the inverse problem of Knapsack problem can be solved with a pseudo-polynomial algorithm. We also show that the inverse problem of integer programming with fixed number of constraints is pseudo-polynomial.

Keywords: Inverse problem, Knapsack problem, integer programming, pseudo-polynomial algorithm.

1 Introduction

Given an optimization problem:

$$\min\{f(c, x) | x \in D\}, \tag{1}$$

where $c \in R^n$ is a parameter vector, D is the feasible region of x, $f(c, x)$ is the objective function. Given a feasible solution x^0 of (1), is there $\bar{c} \in R^n$ such that x^0 is the optimal solution of (1) with $\bar{c}$ as parameter vector? Formally, let

$$F(x^0) = \{\bar{c} \in R^n | \min\{f(\bar{c}, x) | x \in D\} = f(\bar{c}, x^0)\},$$

if $F(x^0) \neq \emptyset$, define

$$\min\{\|c - \bar{c}\| | \bar{c} \in F(x^0)\}, \tag{2}$$

where $\|.\|$ denotes the norm of the vector, the popular choices for the norms are l_1, l_2 and l_∞. we call (2) the inverse problem of (1).

The research on inverse optimization problems have attracted some attention over the last decade. For example, Burton and Toint [3],[4] and Burton, Pulleyblank and Toint [5], Ahuja and Orlin [1], Dial [6] have studied inverse shortest path problem, Huang and Liu [7],[8] have considered inverse linear programming and applied it to inverse matching problem and inverse minimum cost flow problem respectively; Zhang, Liu and Ma [12], Sokkalingam, Ahuja and Orlin [1], and Ahuja and Orlin [2] studied inverse minimum spanning tree problem. For a survey on inverse combinatorial optimization problems we refer the reader to Ahuja and Orlin [1], and Hueberger [9]. Most of the inverse problems studied so far are polynomial problems, and it has been shown that most of their inverse problems

N. Megiddo, Y. Xu, and B. Zhu (Eds.): AAIM 2005, LNCS 3521, pp. 422–426, 2005.

are polynomial problems too [9]. The results on the inverse problems of NP-complete problems are rare due to their difficulties. In this paper, we first show that the inverse problem of Knapsack is pseudo-polynomial; we also show that the inverse problem of integer programming with fixed number of constraints is pseudo-polynomial.

The paper is organized as follows: in section 2 we introduce the Knapsack problem and show that its inverse problem is pseudo-polynomial. In section 3, we show that the inverse problem of integer programming with fixed number of constraints is also pseudo-polynomial. We will give some concluding remarks in section 4.

2   The Inverse Knapsack Problem

The Knapsack problem can be stated as the following:

$$\text{Min}\{c^T x | a^T x \geq B; x \in \{0, 1\} - vector\} \tag{3}$$

where $c = (c_1, \ldots, c_n) \in R^n$ is an rational vector, $a = (a_1, \ldots, a_n) \in Z^n$ and B are integer vector and integer number respectively.

Using the definition of inverse problem given in section 1, the inverse Knapsack problem can be stated as follows: given a vector $x^0 \in \{0, 1\}$ such that $a^T x^0 \geq B$, we want to perturb the vector c to $c + \theta$ such that x^0 is an optimal solution of above Knapsack problem with cost vector $c + \theta$ under the condition that $\|\theta\|$ is minimal.

To show that the inverse Knapsack problem is pseudo-polynomial, we construct a directed graph first as in [10]. Let $S = \text{Max}\{|a_1|, \ldots, |a_n|\}$, construct the directed graph $G = (V, A)$ with vertex set

$$V := \{0, 1, \ldots, n\} \times \{-nS, \ldots, +nS\}$$

and arc set

$$A := \{((j, k), (i, k'))|j = i - 1, \text{and } k' - k = 0, \text{or } a_i\}.$$

Define the length of arc $((i - 1, k), (i, k')) \in A$ as follows:

$$l((i - 1, k), (i, k')) = \begin{cases} c_i, & \text{if } k' = k + a_i \\ 0, & \text{if } k' = k. \end{cases}$$

It is easy to check that the directed graph has no directed cycle.

To show the relationship between the feasible solutions of the Knapsack problem (3) and the directed paths of graph G, we have the following lemma.

Lemma 2.1. x is a feasible solution of (3) if and only if there exists a directed path P of G from $(0, 0)$ to (n, B') for some $B' \geq B$ such that the length of P is $c^T x$.

Proof.   Let P be any directed path in G from $(0, 0)$ to (n, B') for some $B' \geq B$, define

$$x_i = \begin{cases} 1, & \text{if } ((i - 1, k), (i, k + a_i)) \in P \text{ for some } k; \\ 0, & \text{if } ((i - 1, k), (i, k)) \in P \text{ for some } k. \end{cases}$$

It is easy to see that $x = (x_1, \ldots, x_n)$ is a feasible solution of the Knapsack problem, and the length of P is equal to $c^T x$. Therefore the Knapsack problem of (3) can be solved by finding a shortest path from $(0, 0)$ to (n, B') for some $B' \geq B$.

Conversely, let x^0 be any feasible solution of the above Knapsack problem, then by mathematical induction one can easily construct a path P of G from $(0, 0)$ to (n, B') by same method, where $B' = a^T x^0 = \sum_{x_i^0 = 1} a_i$, such that the length of the path P is $c^T x^0$. For example, if $x_n^0 = 1$, put arc $((n - 1, B' - a_n), (n, B')) \in P$, otherwise put arc $((n - 1, B'), (n, B')) \in P$, and so on. Eventually we can construct a directed path P of G from $(0, 0)$ to (n, B') such that the length of the path P is $c^T x^0$. Hence the inverse Knapsack problem can be solved by inverse shortest path problem of G.

Since the directed graph G has no cycle, therefore the inverse shortest path problem of such graph can be solved in $O(m)$ time according to Dial [6], where $m = |A| = O(n^3 S^2)$. Hence we can obtain the following result immediately.

Theorem 2.2. The inverse problem of Knapsack problem (3) can be solved by a pseudo-polynomial algorithm with complexity $O(n^3 S^2)$.

3   Inverse Problem of Integer Programming

In this section we consider the inverse problem of integer programming with fixed number of constraints. Consider the integer linear programming problem:

$$\text{Max}\{c^T x | Ax = b, x \geq 0; x \in Z^n\}, \tag{4}$$

where Z^n denotes the n-dimensional integral vectors, $A = (a_{ij}) \in Z^{m \times n}$ is an integral $m \times n$ matrix, and $b = (b_1, \ldots, b_m) \in Z^m$ and $c = (c_1, \ldots, c_n) \in Z^n$ are integral vectors. We assume m is fixed, and n can vary.

The inverse integer programming problem can be stated as follows: given a feasible solution x^0 of (4), we want to perturb the vector c to $c + \theta$ such that x^0 is an optimal solution of the integer programming with cost vector $c + \theta$ under the condition that $\|\theta\|$ is minimal.

To show that the inverse integer programming is pseudo-polynomial, we construct a directed graph similar to the last section, and show that an feasible solution of (4) corresponding to a directed path in the graph D.

Let $S = \text{Max}\{|a_{ij}|, |b_i| | i = 1, \ldots, m; j = 1, \ldots, n\}$. It is well known (Theorem 17.1 of [10]) that if (4) is finite, than it has an optimal solution with components at most $(n + 1)(mS)^m$. Let $U := (n + 1)S(mS)^m$. Construct a directed graph $D = (V, E)$ with vertex set, whose elements are $(m + 1)$-dimensional vectors, as follows:

$$V := \{0, \ldots, n\} \times \{-U, \ldots, +U\}^m$$

and arc set E given by

$$((j, u''), (i, u')) \in E \Leftrightarrow j = i - 1, u'' - u' = kA_i, k \in Z_+$$

where A_i denotes the ith column of A.

Define the length of arc $((i-1, u''), (i, u')) \in E$ as follows:

$$l((i-1, u''), (i, u')) = -kc_i.$$

Similar to the lemma 2.1 of last section, we have the following lemma.

Lemma 3.1 x is a feasible solution of (4) if and only if there exists a path P of D from $(0, 0)$ to (n, b) such that the length of P is $-c^T x$.

Proof.   Let P be any directed path from $(0, 0)$ to (n, b), define

$$x_i = k, \text{if } ((i-1, u'), (i, u' + kA_i)) \in P \text{ for some } k.$$

It is easy to see that $x = (x_1, \ldots, x_n)$ is a feasible solution of the integer programming (4), and the length of P is $-c^T x$.

On the other hand, let $x = (x_1, \ldots, x_n)$ be any feasible solution of the integer programming (4), we can construct a path P of D from $(0, 0)$ to (n, b) as follows: put arc $((n-1, b - x_n A_n), (n, b)) \in P$ and so on, then P is a path from $(0, 0)$ to (n, b) in D. Furthermore, the length of the P is $-c^T x$.

Lemma 3.1 states that a feasible solution x of the integer programming (4) corresponding to a directed path P from $(0, 0)$ to (n, b) in D. If there is no directed path from $(0, 0)$ to (n, b) exists, than (4) is infeasible, since if (4) is feasible, it has a feasible solution x with all components at most $(n+1)(mS)^m$. If such a path does exist, and if the 'LP-relaxation' $\text{Max}\{c^T x | Ax = b, x \geq 0\}$ is unbounded, (4) is also unbounded by Theorem 16.1 of [10]. If the 'LP-relaxation' $\text{Max}\{c^T x | Ax = b, x \geq 0\}$ is finite, than the shortest path from $(0, 0)$ to (n, b) gives the optimal solution for (4). Therefore the inverse problem of integer programming (4) can be solved by solving the inverse shortest path problem of D.

It is easy to check that the directed graph D also has no cycle, therefore the inverse shortest path problem of such graph can also be solved in $O(m) = O(|E|) = O(nU^2) = O(n^3 S^2 (mS)^{2m})$ time by Dial [6]. So we have the following theorem.

Theorem 3.2.   The inverse problem of integer programming (4) can be solved by a pseudo-polynomial algorithm with complexity $O(n^3 S^2 (mS)^{2m})$.

4   Conclusions

In this paper we have shown that the inverse problem of Knapsack problem is pseudo-polynomial. We have also shown that the inverse problem of integer programming with fixed number of constraints is pseudo-polynomial. The idea of the proof is simple. We first construct a directed graph, then we show that a feasible solution of the Knapsack problem of (3) and integer programming of (4) corresponding to a path of the graph from one vertex to another one. Therefore the inverse problems can be solved by solving the shortest path problems of the directed graphs G and D respectively. To the best of our knowledge this is the first result for the inverse problem of NP-complete problems.

An interesting and open question is: are there NP-complete problems whose inverse problems are polynomial? This will be the future research topic.

Acknowledgement

This research was partially supported by National Science Foundation of China Under No. 19731010, No. 70171023, No. 79970052.

References

[1] Ahuja, R.K. and J.B. Orlin, Combinatorial algorithms for inverse network flow problems, Networks, **40**(4): 181-187(2002).

[2] Ahuja, R.K. and J.B. Orlin, A fast algorithm for the inverse spanning tree problem, J. of Algorithms, **34**, 177-193(2000).

[3] Burton, D. and Ph.L. Toint, On an instance of the inverse shortest paths problem, Mathematical Programming, **53**, 45-61(1992).

[4] Burton, D. and Ph.L. Toint, On the use of an inverse shortest paths algorithm for recovering linearly correlated costs, Mathematical Programming, **63**, 1-22(1994).

[5] Burton, D., B. Pulleyblank and Ph. L. Toint, The inverse shortest paths problem with upper bounds on shortest paths costs, In **Network Optimization**, edited by P. Pardalos, D.W. Hearn and W.H. Hager, Lecture notes in Economics and Mathematical Systems, Volume 450, pp. 156-171(1997).

[6] Dial, B., Minimum-revenue congestion pricing, Part 1: A fast algorithm for the single-origin case. Technical Report, The Volpe National Transportation Systems Center, Kendall Square, Cambridge, MA 02142.

[7] Huang, S. and Z. Liu, On the inverse problem of linear programming and its application to minimum weight perfect k-matching, European Journal of Operational Research, **112**, 421-426(1999).

[8] Huang, S. and Z. Liu, On the inverse minimum cost flow problem, Advances in Operations Research and Systems Engineering, World Publishing, Co., 30-37(1998).

[9] Heuberger, C., Inverse combinatorial optimization: A survey on problems, methods, and results, J. of Combinatorial Optimization,

[10] Schrijver, A., **Theory of Linear and Integer Programming**, John Wiley & Sons, 1986.

[11] Sokkalingam, P.T., R.K. Ahuja and J.B Orlin, Solving inverse spanning tree problems through network flow techniques, Operations Research, **47**, 291-298(1999).

[12] Zhang, J., Z., Liu and Z., Ma, On the inverse problem of minimum spanning tree with partition constraints, Mathematical Methods of Operations Research, **44**, 171-188(1996).

Level of Repair Analysis and Minimum Cost Homomorphisms of Graphs

Gregory Gutin[1], Arash Rafiey[1], Anders Yeo[1], and Michael Tso[2]

[1] Department of Computer Science,
Royal Holloway University of London,
Egham, Surrey TW20 OEX, UK
{gutin, arash, anders}@cs.rhul.ac.uk
[2] School of Mathematics, University of Manchester,
P.O. Box 88, Manchester M60 1QD, UK
mike.tso@manchester.ac.uk

This paper is dedicated to the memory of Lillian Barros

Abstract. Level of Repair Analysis (LORA) is a prescribed procedure for defence logistics support planning. For a complex engineering system containing perhaps thousands of assemblies, sub-assemblies, components, etc. organized into several levels of indenture and with a number of possible repair decisions, LORA seeks to determine an optimal provision of repair and maintenance facilities to minimize overall life-cycle costs. For a LORA problem with two levels of indenture with three possible repair decisions, which is of interest in UK and US military and which we call LORA-BR, Barros (1998) and Barros and Riley (2001) developed certain branch-and-bound heuristics. The surprising result of this paper is that LORA-BR is, in fact, polynomial-time solvable. To obtain this result, we formulate the general LORA problem as an optimization homomorphism problem on bipartite graphs, and reduce a generalization of LORA-BR, LORA-M, to the maximum weight independent set problem on a bipartite graph. We prove that the general LORA problem is NP-hard by using an important result on list homomorphisms of graphs. We introduce the minimum cost graph homomorphism problem and provide partial results. Finally, we show that our result for LORA-BR can be applied to prove that an extension of the maximum weight independent set problem on bipartite graphs is polynomial time solvable.

1 Introduction

Level of Repair Analysis (LORA) is a prescribed procedure for defence logistics support planning (see, e.g., Crabtree and Sandel [9] and the website of the UK MoD Acquisition Management System at www.ams.mod.uk/ams). For a complex engineering system containing perhaps thousands of assemblies, sub-assemblies, components etc. organized into $\ell \geq 2$ levels of *indenture* and with $r \geq 2$ possible repair decisions, LORA seeks to determine an optimal provision of repair and maintenance facilities to minimize overall life-cycle costs.

N. Megiddo, Y. Xu, and B. Zhu (Eds.): AAIM 2005, LNCS 3521, pp. 427–439, 2005.
© Springer-Verlag Berlin Heidelberg 2005

Barros [4] and Barros and Riley [6] provide a generic integer programming formulation of the LORA optimization problem for systems with ℓ levels of indenture and r possible repair decisions (including the non-repair option). A special case with $\ell = 2$ and $r = 3$, which we call LORA-BR, is of particular importance because it corresponds to recommendations in certain UK and US military standard handbooks, see Barros and Riley [6]. In French military standards, $\ell = 2$ and $r = 5$. Notice that the actual research of Barros and Riley was only for LORA-BR [5] for which the corresponding software have been developed.

While Barros [4] solves LORA-BR using a general purpose IP solver, Barros and Riley [6] outline a specialized branch-and-bound heuristic, which appears to be more efficient in computational experiments. Their heuristic is based on a relaxation of LORA-BR into a pair of uncapacitated facility location (UFLP) problems. A branch-and-bound procedure then employs local search heuristics to satisfy additional side constraints ensuring consistency between repair decisions for pairs of items nested on adjacent indenture levels. Since UFLP is NP–hard, it could be expected that LORA-BR would also be intractable. However, the surprising result of this paper is that LORA-BR is polynomially solvable and this is achieved by reducing its generalization, LORA-M (defined in Section 3), to the maximum weight independent set problem on a bipartite graph.

As it was pointed out above, the case of two levels of indenture is of particular interest (e.g., in UK, USA and French military). For clarity of exposition, in the rest of this paper apart from Section 4, we restrict ourselves to *two levels* of indenture, $\ell = 2$, but our approach can be extended to arbitrary ℓ as demonstrated in Section 4.

For a pair of graphs $H = (V(H), E(H))$ and $B = (V(B), E(B))$, a mapping $k : V(B) \rightarrow V(H)$ such that if $xy \in E(B)$ then $k(x)k(y) \in E(H)$ is called a *homomorphism* of B to H. To study the LORA problem, we show how to formulate it as a problem of finding a homomorphism of minimum cost belonging to a certain class of homomorphisms of a bipartite graph to a fixed bipartite graph. This allows us to use a nontrivial result on the list H-homomorphism problem from [10] to easily show that the general LORA problem with $\ell = 2$ is NP–hard. We also prove that LORA-M is polynomial time solvable.

The formulation of the LORA problem in terms of special homomorphisms leads us to the introduction of the minimum cost H-homomorphism problem (MCHP): For a fixed graph H and an input graph G given together with costs $c_z(u)$, the cost of mapping a vertex $u \in V(G)$ to $z \in V(H)$, verify whether there is a homomorphism of G to H, and if one exists, find such a homomorphism k that minimizes $\sum_{u \in V(G)} c_{k(u)}(u)$. MCHP extends the well-studied list H-homomorphism problem [13]. We use our results for the LORA problem to obtain the corresponding results for MCHP. In particular, we show that if H is a bipartite graph with the complement being an interval graph, then MCHP is polynomial time solvable. In contrast, if H is not bipartite with the complement being a circular arc graph, then MCHP is NP–hard.

We also use our results to show that the bipartite case of the critical independent set problem (defined in Section 6), which generalizes the maximum weight independent set problem, is polynomial time solvable.

In this paper, all graphs are finite, undirected, and simple (i.e., without loops or multiple edges). For standard graph-theoretical terminology and notation, see, e.g., Asratian, Denley and Haggkvist [3] or West [15]. For terminology and results on homomorphisms, see Hell and Nesetril [13].

The rest of the paper is organized as follows. In Section 2, we provide formulations of LORA-BR and the general LORA problem with $\ell = 2$ in terms of graph homomorphisms. We prove that the general LORA problem with $\ell = 2$ is *NP*–hard. In Section 3, we show how to solve a generalization of LORA-BR, LORA-M with $\ell = 2$, in polynomial time. In Section 4, we extend the general LORA problem with $\ell = 2$ to the general LORA problem with arbitrary $\ell \geq 2$. In Section 5, we introduce the minimum cost H-homomorphism problem and show that the results of Sections 2 and 3 can be easily extended to it. Finally, in Section 6 we apply a result from Section 3 to solve the bipartite case of the critical independent set problem in polynomial time.

2   LORA-BR and General LORA with $\ell = 2$

Consider first a special case of LORA with $\ell = 2$ and $r = 3$ following Barros [4] and Barros and Riley [6] (we will call this special case *LORA-BR*). We refer to the first level of indenture in LORA-BR as *subsystems* $s \in S$ and the second level of indenture as *modules* $m \in M$. The distribution of modules in subsystems can be given by a bipartite graph $G = (V_1, V_2; E)$ with partite sets $V_1 = S$ and $V_2 = M$. For arbitrary $s \in V_1$ and $m \in V_2$, $sm \in E$ if and only if module m is in subsystem s. We consider G to be an arbitrary bipartite graph and denote its vertex set V ($V = V_1 \cup V_2$).

There are $r = 3$ available repair decisions for each level of indenture: "discard", "local repair" and "central repair", labelled respectively D, L, C (subsystems) and d, l, c (modules). To be able to use a decision $z \in \{D, L, C, d, l, c\}$, we have to pay a fixed cost c_z. Assume also known additive costs (over a system life-cycle) $c_z(u)$ of prescribing repair decision z for subsystem or module u.

We wish to minimize the total cost of choosing a subset of the six repair decisions and assigning available repair options to the subsystems and modules subject to the following constraints:

If a module m occurs in subsystem s (i.e., $sm \in E$) we impose the following logical restrictions on the repair decisions for the pair (s, m) motivated through practical considerations:

$$R_1 : D_s \Rightarrow d_m, R_2 : l_m \Rightarrow L_s,$$

where D_s, d_m denote the decisions to discard subsystem s, module m, respectively, etc. Notice that even though module m may be common to several subsystems we are required to prescribe a unique repair decision for that module.

R_1 has the interpretation that a decision to discard subsystem s necessarily entails discarding all enclosed modules. R_2 is a consequence of R_1 and a policy of "no backshipment" which rules out the local repair option for any module enclosed in a subsystem which is sent for central repair [6].

To get a complete graph-theoretical formulation of LORA-BR, we will use the notion of a homomorphism of graphs that generalizes the notion of coloring (see, e.g., Hell and Nesetril [13]). For a pair of graphs $H = (V(H), E(H))$ and $B = (V(B), E(B)$, a mapping $k : V(B) \rightarrow V(H)$ such that if $xy \in E(B)$ then $k(x)k(y) \in E(H)$ is called a *homomorphism* of B to H.

Let $F_{BR} = (Z_1, Z_2; T)$ be a bipartite graph with partite sets $Z_1 = \{D, C, L\}$ (subsystem repair options) and $Z_2 = \{d, c, l\}$ (module repair options) and with edges $T = \{Dd, Cd, Cc, Ld, Lc, Ll\}$. Let $Z = Z_1 \cup Z_2$. Observe that any homomorphism k of G to F_{BR} such that $k(V_1) \subseteq Z_1$ and $k(V_2) \subseteq Z_2$ satisfies the rules R_1 and R_2. Indeed, let $u \in V_1$, $v \in V_2$, $uv \in E$. If $k(u) = D$ then $k(v) = d$, and if $k(v) = l$ then $k(u) = L$.

Let $L_i \subseteq Z_i$, $i = 1, 2$. We call a homomorphism k of G to F_{BR} an (L_1, L_2)-*homomorphism* of G to F_{BR} if $k(u) \in L_i$ for each $u \in V_i$, $i = 1, 2$. Now LORA-BR can be formulated as the following graph-theoretical problem: We are given a bipartite graph $G = (V_1, V_2; E)$, $V = V_1 \cup V_2$, and we consider homomorphisms k of G to F_{BR}. Mapping of $u \in V$ to $z \in Z$ (i.e., $k(u) = z$) incurs a real cost $c_z(u)$. The use of a vertex $z \in Z$ in a homomorphism k (i.e., $k^{-1}(z) \neq \emptyset$) incurs a real cost c_z. We wish to choose subsets $L_i \subseteq Z_i$, $i = 1, 2$, and find an (L_1, L_2)-homomorphism k of G to F_{BR} that minimize

$$\sum_{u \in V} c_{k(u)}(u) + \sum_{z \in L_1 \cup L_2} c_z. \tag{1}$$

We call the expression in (1) the *cost* of k.

The graph-theoretical formulation of LORA-BR can be naturally extended as follows: The above problem with F_{BR} replaced by an arbitrary *fixed* bipartite graph $F = (Z_1, Z_2; T)$ is called the *general LORA problem with $\ell = 2$*. Let $Z = Z_1 \cup Z_2$. Notice that the general LORA problem with $\ell = 2$ extends the generic formulation of the LORA problem with $\ell = 2$ given in [6]. The formulation of the general LORA problem (with arbitrary ℓ) provided in Section 4 extends the generic formulation of the LORA problem (with arbitrary ℓ) given in [6].

To prove that the general LORA problem with $\ell = 2$ is *NP*–hard, we will use an important result on the list H-homomorphism problem defined below. Suppose that we are given a pair of graphs H and B and a list $\Lambda(v) \subseteq V(H)$ for each $v \in V(B)$. A homomorphism $f : V(B) \rightarrow V(H)$ such that $f(v) \in \Lambda(v)$ for each $v \in V(B)$ is called a Λ-*homomorphism*. For a fixed H, the *list H-homomorphism problem* asks whether there exists a Λ-homomorphism f of B to H for an input graph B with lists Λ.

A graph $P = (V(P), E(P))$ is a *circular arc graph* if there is a family of arcs A_v, $v \in V(P)$, on a fixed circle, such that $xy \in E(P)$ if and only if A_x and A_y intersect. Feder, Hell and Huang [10] obtained the following important result.

Theorem 1. *If H is a bipartite graph with the complement being a circular arc graph, then the list H-homomorphism problem is polynomial time solvable. Otherwise, the problem is* NP*-complete.*

Observe that, if H is bipartite, we may restrict inputs B of the list H-homomorphism problem to bipartite graphs since there is no homomorphism of a non-bipartite graph to H. Brightwell [7] found the first proof that the general LORA problem with $\ell = 2$ is *NP*–hard. Since his proof does not use Theorem 1, our proof turns out to be shorter and it gives a stronger result.

Theorem 2. *The general LORA problem with $\ell = 2$ is* NP*–hard provided the complement of F is not a circular arc graph.*

Proof: Let F be a bipartite graph and assume that the complement of F is not a circular arc graph (see Theorem 1). Let a bipartite graph G and lists Λ be an input of the list F-homomorphism problem. Define costs $c_z(u)$ for each $z \in V(F)$ and $u \in V(G)$ as follows: $c_z(u) = 0$ if $z \in \Lambda(u)$ and $c_z(u) = 1$, otherwise. We put $c_z = 0$ for each $z \in V(F)$. In other words, the use of each vertex $z \in V(F)$ in homomorphisms of G to H is free. In this case, in the general LORA problem with $\ell = 2$, we can always put $L_1 \cup L_2 = V(F)$.

Let $G_1, G_2, \ldots, G_g$ be components of G and let $F_1, F_2, \ldots, F_f$ be components of F. Let Z_1^j, Z_2^j be partite sets of F_j for every $j = 1, 2, \ldots, f$. Observe that there exists a Λ-homomorphism of G to F if and only if for each $i = 1, 2, \ldots, g$ there is a $j(i) \in \{1, 2, \ldots, f\}$ such that there exists a Λ-homomorphism of G_i to $F_{j(i)}$. However, there is a Λ-homomorphism of G_i to $F_{j(i)}$ if and only if the minimum cost of either a $(Z_1^{j(i)}, Z_2^{j(i)})$-homomorphism of G_i to $F_{j(i)}$ or a $(Z_2^{j(i)}, Z_1^{j(i)})$-homomorphism of G_i to $F_{j(i)}$ is equal to 0 (with the costs defined above). Thus, we have a polynomial time Turing-reduction [12] from the *NP*–complete list H-homomorphism problem to the general LORA problem with $\ell = 2$. Hence, by the definition of the *NP*–hardness (see Section 5.1 in [12]), the general LORA problem with $\ell = 2$ is *NP*–hard.   □

3   LORA-M with $\ell = 2$

Let $B = (W_1, W_2; E)$ be a bipartite graph. For a vertex $z \in W_1 \cup W_2$, let $N(z)$ be the set of vertices adjacent to z. Orderings $x_1, x_2, \ldots, x_{|W_1|}$ and $y_1, y_2, \ldots, y_{|W_2|}$ of vertices of W_1 and W_2, respectively, are called *monotone* if $N(x_i) \subseteq N(x_{i+1})$ and $N(y_j) \subseteq N(y_{j+1})$ for each $i = 1, 2, \ldots, |W_1| - 1$ and $j = 1, 2, \ldots, |W_2| - 1$. A bipartite graph B is called *monotone* if it has monotone orderings of its partite sets. Observe that if $x_1, x_2, \ldots, x_{|W_1|}$ and $y_1, y_2, \ldots, y_{|W_2|}$ are monotone orderings, then $x_p y_q \in E$ implies that $x_s y_t \in E$ for each $s \geq p$ and $t \geq q$.

Notice that the bipartite graph F_{BR} corresponding to the rules R_1 and R_2 of LORA-BR is monotone (consider orderings D, C, L and l, c, d), so are the bipartite graphs corresponding to R_1 and R_2 separately (there might be a situation when one of the rules is not used). Interestingly, monotone bipartite graphs form a family of so-called convex bipartite graphs; several families of convex bipartite graphs have been found useful in various applications, see [3].

Let $B = (W_1, W_2; E)$ be a bipartite graph, let $n = |W_1| + |W_2|$ and let $m = |E|$. One can test whether B is monotone in time $O(m+n)$ as follows. Order vertices of W_1 and W_2 separately according to their degrees $\deg(z)$, $x_1, x_2, \ldots, x_{|W_1|}$ and $y_1, y_2, \ldots, y_{|W_2|}$, such that $\deg(x_i) \leq \deg(x_{i+1})$ and $\deg(y_j) \leq \deg(y_{j+1})$ for each $i = 1, 2, \ldots, |W_1|-1$ and $j = 1, 2, \ldots, |W_2|-1$. Observe that these orderings are monotone if and only if $N(x_i) \subseteq N(x_{i+1})$ and $N(y_j) \subseteq N(y_{j+1})$ for each $i = 1, 2, \ldots, |W_1| - 1$ and $j = 1, 2, \ldots, |W_2| - 1$. We can use counting sort (see Chapter 9 of [8]) to get the orderings according to degrees in time $O(n)$. The remaining computations can be carried out in time $O(m)$.

The general LORA problem restricted to fixed monotone bipartite graphs $F = (Z_1, Z_2; T)$ is called *LORA-M*. We assume that we have monotone orderings $x_1, x_2, \ldots, x_{|Z_1|}$ and $y_1, y_2, \ldots, y_{|Z_2|}$ of Z_1 and Z_2, respectively. We reduce LORA-M to the maximal weight independent set problem on bipartite graphs. Recall that a vertex set I of a graph is *independent* if there is no edge between vertices of I.

In the next theorem, we will consider a bipartite graph B with partite sets W_1, W_2 and nonnegative vertex weights $p(u)$, $u \in V(B)$, and the following (s,t)-network $\mathcal{N}(B)$: add new vertices s and t to B, append all arcs su of capacity $p(u)$, vt of capacity $p(v)$ for all $u \in W_1$ and $v \in W_2$, and orient every edge xy of B, where $x \in W_1$, from x to y (these arcs are of capacity ∞). For results on flows and cuts in networks see [8].

Theorem 3. *If (S,T) is a minimum cut in $\mathcal{N}(B)$, $s \in S$, then $(S \cap W_1) \cup (T \cap W_2)$ is a maximum weight independent set in B. One can find a maximum weight independent set in B in time $O(n_1^2\sqrt{m} + n_1 m)$, where $n_1 = |U_1|$ and $m = |E(B)|$.*

The structural part of Theorem 3 is well-known, cf. Frahling and Faigle [11] (a similar result is described in [14]). The complexity claim follows from the fact that one can find a minimum cut in $\mathcal{N}(B)$ in time $O(n_1^2\sqrt{m} + n_1 m)$ by first finding a maximum flow by the bipartite preflow-push algorithm of Ahuja et al. [2] and then finding a minimum cut (e.g., by finding vertices reachable from s in the residual network using depth-first search).

Let us return to LORA-M and formulate it as a maximization problem. Choose sets $L_i \subseteq Z_i$, $i = 1, 2$. Let $u \in V_i$ and set lists $\Lambda(u) = L_i$, $i = 1, 2$. Recall that $x_1, x_2, \ldots, x_{|Z_1|}$ and $y_1, y_2, \ldots, y_{|Z_2|}$ are monotone orderings of Z_1 and Z_2. Assume that $u \in V_1$, $x_p, x_q \in \Lambda(u)$, $p < q$ and $c_{x_p}(u) > c_{x_q}(u)$. Observe that since $c_{x_p}(u) > c_{x_q}(u)$ and F is monotone, an optimal (L_1, L_2)-homomorphism k will not map u to x_p. Thus, we may reduce the list $\Lambda(u)$ of possible images of u by deleting x_p. Certainly, we may *reduce* all $\Lambda(v)$, $v \in V_1$, such that if $x_r, x_s \in \Lambda(v)$ and $r < s$, then $c_{x_r}(v) \leq c_{x_s}(v)$. We call such a list $\Lambda(v)$ *reduced*. Similarly, one defines the reduced list of a vertex in V_2.

For a vertex $u \in V$, we can get the reduced list $\Lambda(u)$ in time $O(1)$ by the following simple procedure (the running time is constant since F is fixed). To simplicity the description, assume that $u \in V_1$. The input is $\Lambda(u) := L_1 =$

$\{x_{p(1)}, x_{p(2)}, \ldots, x_{p(t)}\}$, $p(1) < p(2) < \cdots < p(t)$. We start from $x_{p(t)}$. We compare $c_{x_{p(t)}}(u)$ with $c_{x_{p(t-1)}}(u)$, $c_{x_{p(t-2)}}(u), \ldots$ and find the maximal i such that $c_{x_{p(i)}}(u) \leq c_{x_{p(t)}}(u)$. We delete from $\Lambda(u)$ all $x_{p(i+1)}, x_{p(i+2)}, \ldots, x_{p(t-1)}$. We compare $c_{x_{p(i)}}(u)$ with $c_{x_{p(i-1)}}(u)$, $c_{x_{p(i-2)}}(u), \ldots$ and continue as above. Thus, we can obtain the reduced lists $\Lambda(v)$, $v \in V$, in time $O(|V|)$.

In the reminder of this section, we will use the following notation for the reduced lists: $\Lambda(u) = \{z_{p(1)}, z_{p(2)}, \ldots, z_{p(|\Lambda(u)|)}\}$, where $p(1) < p(2) < \cdots < p(|\Lambda(u)|)$ and $z = x$ if $u \in V_1$ and $z = y$, otherwise.

Recall that a homomorphism k of G to F is a Λ-homomorphism if $k(u) \in \Lambda(u)$ for each $u \in V$. Observe that LORA-M is equivalent to the problem of choosing sets $L_i \subseteq Z_i$, $i = 1, 2$ and finding a Λ-homomorphism k of G to F that minimize the cost of k, where $\Lambda(u)$ is the reduced list for $u \in V$.

Now we replace the costs by *weights*. Let M be the maximum of all costs in LORA-M (i.e., $c_z(u)$'s and c_z's). For each pair of vertices $z \in Z_i$ and $u \in V_i$, $i = 1, 2$, let $w_z(u) = M - c_z(u)$ and for each vertex $z \in Z$ let $w_z = M - c_z$. Notice that, by the definition, all the weights are nonnegative. Let k be a Λ-homomorphism of G to F. The *weight* of k is defined as

$$\sum_{u \in V} w_{k(u)}(u) + \sum_{z \in L_1 \cup L_2} w_z. \tag{2}$$

Observe that LORA-M is equivalent to the problem of choosing sets $L_i \subseteq Z_i$, $i = 1, 2$ and finding a Λ-homomorphism k of G to F that maximize the weight of k, where $\Lambda(u)$ is the reduced list for $u \in V$.

We now prove the following main result of the paper.

Theorem 4. *For fixed subsets L_i, $i = 1, 2$, LORA-M with $\ell = 2$ can be solved in time $O(n_1^2 \sqrt{m} + n_1 m + n)$, where $n_1 = |V_1|$, $n = |V|$ and $m = |E|$.*

Proof: Recall that all our graphs have no loops. If F is edgeless, then there is no homomorphism of G to F. Thus, we may assume that $x_{|U_1|}y_{|U_2|} \in T$. Since L_i, $i = 1, 2$, are fixed, for simplicity, we will assume that all weights $w_{ij} = 0$ in (2). Let $\Lambda(u)$ be the reduced list for each $u \in V$ (we have shown how to find these lists in time $O(n)$).

Let W be a constant larger than $\max\{w_j(u) : u \in V, j \in \Lambda(u)\}$. Construct a new graph H with $\sum_{u \in V} |\Lambda(u)|$ vertices:

$$V(H) = \{u_z : u \in V, z \in \Lambda(u)\}.$$

Let an edge $u_x v_y$ be in H if $uv \in E$ and $xy \notin T$. Let $u \in V$. For every $j \in \{1, 2, \ldots, |\Lambda(u)|\}$, let the weight $w(u_{z_{p(j)}})$ be equal to $w_{z_{p(j)}}(u) + W$, if $j = |\Lambda(u)|$, and equal to $w_{z_{p(j)}}(u) - w_{z_{p(j+1)}}(u)$, otherwise. Since each list $\Lambda(u)$ is reduced, the weights of the vertices of H are nonnegative.

Clearly, if we replace, in G, a vertex $u \in V$ by $|\Lambda(u)|$ independent copies such that there is an edge between a copy of u and a copy of v if and only if $uv \in E$, then we obtain a supergraph G^* of H. Since G is bipartite, so is G^* and, thus, H.

Observe that, by monotonicity of F, if $u_{x_{p(i)}}, u_{x_{p(j)}}, v_{y_{p(f)}}, v_{y_{p(g)}}$ are vertices of H, $j \geq i$, $g \geq f$ and $u_{x_{p(i)}} v_{y_{p(f)}} \notin E(H)$, then $u_{x_{p(j)}} v_{y_{p(g)}} \notin E(H)$ as well. We call this property of H *index-antimonotonicity*.

Assume that there exists a Λ-homomorphism k of G to F. Let $k(u) = z_{p(i_u)}$. Then the set $\{u_{z_{p(i_u)}} : u \in V\}$ is independent in H. Moreover, by index-antimonotonicity of H,

$$S = \cup_{u \in V}\{u_{z_{p(j)}} : i_u \leq j \leq |\Lambda(u)|\} \tag{3}$$

is an independent set in H. Observe that S contains $S' = \{u_{z_{p(\Lambda(u))}} : u \in V\}$ and the weight of S is equal to that of the homomorphism plus $W \times |V|$ (we use telescopic sums).

Assume that a maximum weight independent set S in H contains S'. Then map each $u \in V$ to $k(u) = z_{p(i_u)}$ such that $i_u = \min\{j : u_{z_{p(j)}} \in S\}$. By maximality, S is of the form (3) or, due to index-antimonotonicity of H, S may be extended to (3) by adding some vertices of zero weight. Observe that the weight of S is equal to that of the homomorphism plus $W \times |V|$. If a maximum weight independent set S in H does not contain S', then S' is not an independent set in H (since the weight of S' is larger than the weight of S) and, thus, there is no Λ-homomorphism of G to F.

Thus, there is an Λ-homomorphism of G to F if and only if a maximum weight independent set in H contains S'. If there is an Λ-homomorphism of G to F, then this homomorphism corresponds to a maximum weight independent set S in H. It remains to observe that we may apply Theorem 3 to find a maximum weight independent set of H. $\square$

There are less than $a = 2^{|Z_1|+|Z_2|}$ choices of nonempty L_1 and L_2. Since F is fixed, a is a constant. Thus, we obtain the following:

Theorem 5. *LORA-M with $\ell = 2$ can be solved in time* $O(n_1^2\sqrt{m} + n_1 m + n)$, *where n_1, n and m are defined in Theorem 4.*

4   General LORA Problem and LORA-M

Let $\ell \geq 2$ be a constant. An ℓ-partition $X_1, X_2, \ldots, X_\ell$ of a set X is a collection of subsets of X such that $X_i \cap X_j = \emptyset$ for each $i \neq j$ and $X_1 \cup X_2 \cup \cdots \cup X_\ell = X$. An ℓ-partition $X_1, X_2, \ldots, X_\ell$ of the vertex set X of a graph H is called *layered* if, for each edge xy of H, there exists an index i such that one vertex of xy is in X_i and the other is in X_{i+1}. Observe that a graph H with a layered ℓ-partition is bipartite with partite sets $\cup\{X_i : 1 \leq i \leq \ell, \ i \equiv 1 \pmod 2\}$ and $\cup\{X_i : 1 \leq i \leq \ell, \ i \equiv 0 \pmod 2\}$.

Let $G = (V, E)$ be a graph with a layered ℓ-partition $V_1, V_2, \ldots, V_\ell$ of V. Let $F = (U, T)$ be a fixed graph with a layered ℓ-partition $U_1, U_2, \ldots, U_\ell$ of U. Let $L_i \subseteq U_i, i = 1, 2, \ldots, \ell$. We call a homomorphism k of G to F an $(L_1, L_2, \ldots, L_\ell)$-*homomorphism* of G to H if $k(u) \in L_i$ for each $u \in V_i, i = 1, 2, \ldots, \ell$.

We formulate the *general LORA problem* as follows: We are given a graph G as above and we consider homomorphisms k of G to F. Mapping $u \in V$ to $z \in U$ (i.e., $k(u) = z$) incurs a real cost $c_z(u)$. The use of a vertex $z \in U$ in a homomorphism k (i.e., $k^{-1}(z) \neq \emptyset$) incurs a real cost c_z. We wish to choose subsets $L_i \subseteq U_i$, $i = 1, 2, \ldots, \ell$, and find an $(L_1, L_2, \ldots, L_\ell)$-homomorphism k of G to F that minimizes

$$\sum_{u \in V} c_{k(u)}(u) + \sum_{z \in L} c_z, \qquad (4)$$

where $L = \cup_{i=1}^{\ell} L_i$. Notice that the graph F is fixed and is not part of the input.

By Theorem 2, the general LORA problem is NP–hard (even the general LORA problem in which all costs $c_z(u) = 0$ for $u \in V_i$, $i \geq 3$, is NP–hard). To define (the general) LORA-M for $\ell \geq 2$, let us define ℓ-monotone graphs. Let $F = (U, T)$ be a fixed graph with a layered ℓ-partition $U_1, U_2, \ldots, U_\ell$; F is called ℓ-*monotone* if there is an ordering $z_1^i, z_2^i, \ldots, z_{|U_i|}^i$ of vertices of U_i for each $i = 1, 2, \ldots, \ell$ such that the subgraph $F[U_j \cup U_{j+1}]$ of F induced by $U_j \cup U_{j+1}$ is monotone with $z_1^j, z_2^j, \ldots, z_{|U_j|}^j$ and $z_1^{j+1}, z_2^{j+1}, \ldots, z_{|U_{j+1}|}^{j+1}$ being monotone orderings for each $j = 1, 2, \ldots, \ell - 1$. LORA-M is the general LORA problem with F being ℓ-monotone. Similarly to Theorem 5, one can prove the following:

Theorem 6. *LORA-M with fixed $\ell \geq 2$ can be solved in time $O(n_1^2 \sqrt{m} + n_1 m + n)$, where n_1 is the number of vertices in the smaller partite set of input graph G, $n = |V(G)|$ and $m = |E(G)|$.*

5   Minimum Cost H-Homomorphism Problem

This paper provides a motivation to study the following *minimum cost H-homomorphism problem* (*MCHP*): For a fixed graph H and an input graph G given together with costs $c_z(u)$, the cost of mapping a vertex $u \in V(G)$ to $z \in V(H)$, verify whether there is a homomorphism of G to H, and if one exists, find such a homomorphism k that minimizes $\sum_{u \in V(G)} c_{k(u)}(u)$.

An argument similar to that in the proof of Theorem 2 shows that MCHP problem generalizes the list H-homomorphism problem and that if H is not bipartite with the compliment being circular arc graph, then MCHP is NP–hard.

Theorem 7. *If $H = (U_1, U_2, ; T)$ is a monotone bipartite graph, then MCHP can be solved in time $O(n^2 \sqrt{m} + nm + n)$, where n is the number of vertices in the input graph G and m is the number of edges in G.*

Proof: Let $t(n, m) = O(n^2 \sqrt{m} + nm + n)$. Since H is bipartite (and loopless), if there is a homomorphism of G to H, then G is bipartite. We can check whether G is bipartite in time $O(m+n)$ using the breadth-first search. So we may assume that $G = (V_1, V_2; E)$ is bipartite.

Assume that G and H are connected. Then for each homomorphism k of G to H, we have either $k(V_i) \subseteq U_i$ or $k(V_i) \subseteq U_{3-i}$ for every $i = 1, 2$. Thus, to find an optimal homomorphism of G to H, it suffices to compute an optimal (U_1, U_2)-homomorphism and optimal (U_2, U_1)-homomorphism and compare their costs. By Theorem 4, the total running time for finding the two optimal homomorphisms is $t(n, m)$.

If H is disconnected, then by the definition of monotonicity, H consists of isolated vertices and at most one component H', which is not an isolated vertex. The case when all components of H are isolated vertices is trivial, so we may assume that H' does exist.

Assume that G consists of components $G_1, G_2, \ldots, G_b$. Observe that every homomorphism k of G to H consists of b 'independent' homomorphisms k_i : $G_i \rightarrow H$. In fact, if G_i has more than one vertex that k_i maps G_i into H' and, by the above, we can find an optimal homomorphism of G_i to H' in time $t(n_i, m_i)$, where $n_i = |V(G_i)|$ and $m_i = |E(G_i)|$. If G_i is a vertex v, k_i may map it to any vertex of H and, in an optimal k_i it maps G_i into z with minimum $c_z(u)$, $z \in U_1 \cup U_2$. The running time to find such a vertex z is $t(1,0) = O(1)$. To complete our proof, it suffices to observe that $\sum_{i=1}^{b} t(n_i, m_i) = t(n, m)$.   □

The following theorem allows us to relate the NP-hardness and polynomial solvable cases above. Recall that a graph $P = (V(P), E(P))$ is an *interval graph* if there is a family of intervals I_v, $v \in V(P)$, of the real line, such that $xy \in E(P)$ if and only if I_x and I_y intersect. The *clique covering number* of a graph B is the minimum number of complete subgraphs of B covering $V(B)$.

Theorem 8. *A bipartite graph H is monotone if and only if its complement $\bar{H}$ is an interval graph with clique covering number two.*

Proof: First assume that H is a monotone bipartite graph with partite sets $\{v_1, v_2, \ldots, v_k\}$ and $\{w_1, w_2, \ldots, w_l\}$. By the definition of a bipartite monotone graph we may assume that $v_i w_j \in E(H)$ implies that $v_{i'} w_{j'} \in E(H)$ for all $i' \geq i$ and $j' \geq j$. Let $m(j)$ be defined as the least index such that $v_{m(j)} w_j \in E(H)$. Now consider the following intervals:

$$
\begin{array}{ll}
s_i = [i, k+1] & \text{for all } i = 1, 2, \ldots, k \\
t_j = [0, m(j) - \frac{1}{2}] & \text{for all } j = 1, 2, \ldots, l
\end{array}
$$

Let B be the interval graph obtained from the above intervals, such that $V(B) = S \cup T$, where $S = \{s_1, s_2, \ldots, s_k\}$ and $T = \{t_1, t_2, \ldots, t_l\}$ and there is an edge between two vertices if and only if the corresponding intervals intersect. Note that both S and T form a clique in B. Furthermore $s_i t_j \in E(B)$ if and only if $i < m(j)$, which happens if and only if $v_i w_j \notin E(H)$. Therefore $B = \bar{H}$, and we have completed one direction.

So assume that $\bar{H}$ is an interval graph with clique covering number two. Let $[s_i, t_i]$, $i = 1, 2, \ldots, k$, denote the intervals corresponding to one of the cliques in the clique cover of size two and let $[s_i', t_i']$, $i = 1, 2, \ldots, l$, denote the intervals corresponding to the other clique in the clique cover. Let T denote the minimum value of all t_i and let T' denote the minimum value of all t_i'. Without loss of generality we may assume that $T \leq T'$. Again without loss of generality we may assume that $t_1 \geq t_2 \geq \ldots \geq t_k$ and $s_1' \leq s_2' \leq \ldots \leq s_l'$.

Assume that $[s_i, t_i]$ and $[s_j', t_j']$ do not intersect. Suppose that $t_j' < s_i$, which implies that $t_k < s_i$ contradicting the fact that $[s_k, t_k]$ and $[s_i, t_i]$ intersect. Therefore we must have $t_i < s_j'$, which implies that $[s_a, t_a]$ and $[s_b', t_b']$ do not intersect for any $a \geq i$ and $b \geq j$. Therefore $\bar{H}$ is the compliment of a monotone bipartite graph. □

The last two theorems imply the following:

Theorem 9. *If H is a bipartite graph and its compliment is an interval graph, then MCHP can be solved in time $O(n^2\sqrt{m} + nm + n)$, where n is the number of vertices in an input graph G and m is the number of edges in G.*

Let P_5 be the path with 5 vertices. The graph P_5 is not a monotone bipartite graph, but its complement is a circular arc graph. Thus, there remains a gap between the set of graphs H for which we showed that the problem is NP-hard and for which we proved that it is tractable. It would be interesting to close the gap. We considered some directed extension of the 2-SAT approach of [10], but they did not appear to be useful.

6   LORA-BR and Critical Independent Set Problem

Let Q be an arbitrary graph. For a set $X \subseteq V(Q)$, let $N(X) = \cup_{x \in X}\{y \in V(Q) : xy \in E(Q)\}$. Let p, q be a pair of functions from $V(Q)$ to the set of nonnegative reals. In the *critical independent set problem (CISP)* we seek

$$\text{argmax}\{\sum_{a \in A} p(a) - \sum_{c \in N(A)} q(c) : \ A \text{ is an independent vertex set in } Q\}.$$

Clearly, CISP is NP–hard as the maximum weight independent set problem on arbitrary graphs is CISP with $q(u) = 0$ for each $u \in V(Q)$. Ageev [1] proved that CISP is polynomial time solvable if $p(u) = q(u)$ for each $u \in V(Q)$. This generalized the corresponding result of Zhang [16] for $p(u) = q(u) = 1$ for each $u \in V(Q)$. We will show that CISP can be solved in polynomial time on bipartite graphs for arbitrary functions p and q.

Theorem 10. *CISP on a bipartite graph $G = (V_1, V_2; E)$, $V = V_1 \cup V_2$, can be solved in time $O(n_1^2\sqrt{m} + n_1 m + n)$, where $n_1 = |V_1|$, $n = |V|$ and $m = |E|$.*

Proof: Observe that LORA-BR with fixed lists $L_1 = V_1$, $L_2 = V_2$ may be reformulated as follows: Given a bipartite graph $G = (V_1, V_2, E)$ and three weights $w_i(v)$, $i = 1, 2, 3$, for each vertex $v \in V$, we color every vertex of G in one of the colors 1,2,3 such that if a vertex is colored 1, then all its neighbors must be colored 3. Assigning a color i to a vertex v contributes weight $w_i(v)$ to the total weight of the coloring. We seek a coloring of maximum total weight.

Observe that if $w_1(u) < w_2(u)$ for some $u \in V$, then there is an optimal coloring for which u is not colored 1. Thus, we may set $w_1(u) := w_2(u)$ and keep a record, say $(u, 1, 2)$, that indicates that if, in an optimal coloring that we found u is colored 1, we recolor it 2. Similar arguments allow us to assume that $w_1(u) \geq w_2(u) \geq w_3(u)$ for each $u \in V$.

Consider an optimal coloring, in which A is the set of vertices assigned color 1. Then A is independent, all vertices of $N(A)$ must have color 3 and all vertices of $B = V(G) - A - N(A)$ may have color 2. The total weight of the coloring is

$$\sum_{a\in A} w_1(a) + \sum_{c\in N(A)} w_3(c) + \sum_{b\in B} w_2(b) = \sum_{d\in V} w_2(d) - \sum_{c\in N(A)} w_{2,3}(c) + \sum_{a\in A} w_{1,2}(a),$$

where $w_{2,3}(c) = w_2(c) - w_3(c)$, $w_{1,2}(a) = w_1(a) - w_2(a)$.

Choose weight functions w_1, w_2, w_3 as follows: $w_1(u) = p(u) + q(u)$, $w_2(u) = q(u)$, $w_3(u) = 0$ for each $u \in V(G)$. Since $\sum_{d\in V} w_2(d)$ is a constant, we observe that CISP on G (and functions p and q) can be reduced to LORA-BR with fixed $L_1 = V_1$, $L_2 = V_2$. It remains to apply Theorem 4. $\square$

Acknowledgements

We'd like to thank Graham Brightwell, David Cohen and Martin Green for valuable discussions on the topic of the paper. Research of the first three authors was partially supported by the Leverhulme Trust. Research of Gutin and Rafiey was supported in part by the IST Programme of the European Community, under the PASCAL Network of Excellence, IST-2002-506778.

References

1. Ageev, A.A.: On finding critical independent and vertex sets. SIAM J. Discrete Math. 7 (1994) 293–295
2. Ahuja, R.K., Orlin, J.B., Stein, C., Tarjan, R.E.: Improved algorithms for bipartite network flows. SIAM J. Comput. 23 (1994) 906–933
3. Asratian, A.S., Denley, T.M.J., Haggkvist, R.: Bipartite Graphs and Their Applications. Cambridge University Press, Cambridge (1998)
4. Barros, L.L.: The optimisation of repair decisions using life-cycle cost parameters. IMA J. Management Math. 9 (1988) 403–413
5. Barros, L.L., Private communications with M. Tso (2001)
6. Barros, L.L., Riley, M.: A combinatorial approach to level of repair analysis. Europ. J. Oper. Res. 129 (2001) 242–251
7. Brightwell, G. Private communications with G. Gutin (Jan., 2005)
8. Cormen, T.H., Leiserson, C.E., Rivest, R.L.: Introduction to Algorithms. MIT Press, Cambridge, MA (1990)
9. Crabtree J.W., Sandel, B.C.: 1989 Army level of repair analysis (LORA). Logistics Spectrum 1989 (Summer) 27–31
10. Feder, T., Hell, P., Huang, J.: List homomorphisms and circular arc graphs. Combinatorica 19 (1999) 487–505
11. Frahling, G., Faigle, U.: Combinatorial algorithm for weighted stable sets in bipartite graphs. To appear in Discrete Appl. Math.
12. Garey, M.R., Johnson, D.S.: Computers and Intractability. Freeman and Co., San Francisco (1979)
13. Hell, P., Nesetril, J.: Graphs and Homomorphisms. Oxford University Press, Oxford (2004)

14. Hochbaum, D.: Provisioning, Shared Fixed Costs, Maximum Closure, and Implications on Algorithmic Methods Today. Management Sci. 50 (2004) 709–723
15. West, D.: Introduction to Graph Theory. Prentice Hall, Upper Saddle River, N.J. (1996)
16. Zhang, C.-Q.: Finding critical independent sets and critical vertex subsets are polynomial problems. SIAM J. Discrete Math. 3 (1990) 431–438

A Schedule Algebra Based Approach to Determine the K-Best Solutions of a Knapsack Problem with a Single Constraint

Subhash C. Sarin*, Yuqiang Wang, and Dae B. Chang

Virginia Polytechnic Institute and State University, Blacksburg VA 24061, USA
sarins@vt.edu

Abstract. In this paper, we develop a new and effective schedule algebra based algorithm to determine the K-best solutions of a knapsack problem with a single constraint. Computational experience with this algorithm is also reported and it is shown to dominate both the dynamic programming and branch and bound based procedures when applied to this problem.

1 Introduction

While solving a problem, it is sometimes desirable to have knowledge about the K-best solutions instead of knowing just the optimal solution. As also mentioned by Eppstein [3], one need for this arises when one is unable to capture all the constraints to start with. Thus, as the new constraints intervene or conditions change, the calculated optimal solution may become infeasible or unacceptable. Having knowledge about the K-best solutions in that case may be better than trying to solve the problem from scratch. This is particularly relevant if the optimization procedure is implemented on-line in an interactive mode.

The types of problems which encounter instances in which one may not be able to capture all the constraints to start with are transportation problems, communication network and allocation problems. We consider here a special type of allocation problem, namely, the knapsack problem with single constraint. This problem is well known in the literature and has been the subject of extensive research. Numerous papers have been published on the investigation of its generalizations, applications and optimization. Surveys and reviews on this topic can be found in [2],[5] and [6]. This problem also arises as a subproblem in the solution of single or higher-dimensional packing problems as discussed in Sarin [7] and Sarin and Ahn [8]. In the solution of two or higher dimensional packing problems, not all feasibility constraints can be represented mathematically, and hence, the solution obtained by the knapsack problem may become unacceptable when tested for feasibility, thereby, requiring the need to generate K-best solutions.

---

* Corresponding author.

N. Megiddo, Y. Xu, and B. Zhu (Eds.): AAIM 2005, LNCS 3521, pp. 440–449, 2005.

The problem to find the K-best solutions can also be viewed as a generalization of finding the shortest path in a network graph. The latter has received extensive treatment in the literature during the last three decades. A complete bibliographical list in this respect is maintained at http://www.ira.uka.de/bibliography/Theory/k-path.html, which includes about 500 papers.

In this paper, we present an approach that relies on schedule algebra to find the K-best solutions for the standard knapsack problem. The computational experience with this algorithm indicates that it is quite efficient. Its effectiveness is further illustrated by comparing its performance with those of the dynamic programming and branch and bound approaches.

2 The Knapsack Problem

The knapsack problem that we consider has the following mathematical expression:

$$\text{Max} \qquad c_1x_1 + c_2x_2 + \cdots + c_nx_n \qquad (1)$$
$$\text{subject to :} \quad a_1x_1 + a_2x_2 + \cdots + a_nx_n \leqslant b$$
$$x_i \geq 0 \text{ and Integer, } \forall i = 1, \cdots, n.$$

Without loss of generality, we assume all the coefficients in (1) to be non-negative integers.

A network approach, which converts this problem to a network graph, was first used by Shapiro [9]. The network corresponding to (1) consists of $(b + 1)$ nodes, which represents all possible values of the left side of the constraint, i.e., $0,1,\ldots$ and b. An arc exists between two nodes s and t if $s + a_i = t$, for some i. The value of that arc is the coefficient of the corresponding variable in the objective function i.e., c_i. This network can be further represented as a matrix. The number of rows and columns of this matrix are equal to $(b + 1)$ and the entries represent the values of the arcs between nodes. Let us designate this network matrix by A. To demonstrate the construction of the network and the corresponding matrix, consider the following example.

Example 1:

$$\text{Max} \qquad 3x_1 + x_2 + 2x_3$$
$$\text{subject to :} \quad 2x_1 + 3x_2 + 4x_3 \leqslant 5$$
$$x_i \geq 0 \text{ and Integer, } \forall i = 1, 2, 3.$$

The network for this knapsack problem is shown in Fig.1(a) and the matrix to represent this network is shown in Fig.1(b). Now, the problem of finding the K-best solutions of the knapsack problem is equivalent to finding the K-best paths of this network, and matrix A plays a central role in this determination. To manipulate matrix A, we use a schedule algebra, which was first presented by Giffler [4].

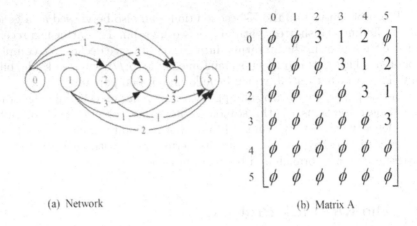

(a) Network

$$\begin{array}{c} \\ 0 \\ 1 \\ 2 \\ 3 \\ 4 \\ 5 \end{array}
\begin{array}{cccccc}
0 & 1 & 2 & 3 & 4 & 5 \\
\phi & \phi & 3 & 1 & 2 & \phi \\
\phi & \phi & \phi & 3 & 1 & 2 \\
\phi & \phi & \phi & \phi & 3 & 1 \\
\phi & \phi & \phi & \phi & \phi & 3 \\
\phi & \phi & \phi & \phi & \phi & \phi \\
\phi & \phi & \phi & \phi & \phi & \phi
\end{array}$$

(b) Matrix A

Fig. 1. The network of Example 1 and its corresponding matrix A

3   Schedule Algebra

Our schedule algebra treats a matrix as a rectangular array of sets. A set which is not empty will contain a finite number of entries. An entry in a non-empty set may consist of positive or negative integers. A "0" entry represents numerical zero or zero magnitude. However, when being added or multiplied, it is regarded as one unit. If a set is empty, it contains the simple entry "ϕ", which is referred to as "zero" or "logical zero". (The "ϕ" in Fig.1(b) can be interpreted accordingly). Therefore, the unit matrix I under schedule algebra, contains "0" along the diagonal and "ϕ" elsewhere. Likewise, the unit vector $\underline{e}_m$ only has "0" in its mth position and "ϕ" elsewhere. In general, we represent a vector by $\underline{a}$ with its ith element by a_i.

The addition and multiplication operators in our schedule algebra also differ from the ones in the traditional algebra. They are defined as follows:

Addition operator "$\oplus$":

$$a_i \oplus b_i = \begin{cases} \{a_i\}, & \text{if } a_i = b_i; \\ \{\phi\}, & \text{else if } a_i = -b_i; \\ \{a_i, b_i\}, & \text{else.} \end{cases}$$

For the purpose of illustration, consider the following example. If $\underline{a} = \{3, 4, 7\}$ and $\underline{b} = \{-3, 0, 5\}$, then $\underline{a} \oplus \underline{b} = \{0, 4, 5, 7\}$. The "$\oplus$" operator is commutative, associative and distributive as in the traditional algebra.

Multiplication operator "$\odot$":

$$a_i \odot b_i = \begin{cases} \{\phi\}, & \text{if } a_i = \{\phi\} \text{ and/or } b_i = \{\phi\}. \\ |a_i| \oplus |b_i|, & \text{if } a_i \text{ and } b_i \neq 0, \text{ and have the same sign.} \\ -(|a_i| \oplus |b_i|), & \text{if } a_i \text{ and } b_i \neq 0, \text{ and have the opposite sign.} \\ \pm b_i, & \text{if } a_i = \pm\{0\} \text{ and } b_i \neq \{\phi\}. \end{cases}$$

For example, if $\underline{a} = \{-2, 4\}$ and $\underline{b} = \{0, -3\}$, then $\underline{a} \odot \underline{b} = \{-2, 2, 3, -3, 4, -4\}$. Like the "$\oplus$" operator, the "$\odot$" operator also has the properties of communitation, association and distribution.

Let E, F and G be matrices, and e_{ij}, f_{ij} and g_{ij} be their elements in the ith row and j^{th} column, respectively. Then, we have the following properties.

(i) If $G = E \oplus F$, then $g_{ij} = e_{ij} \oplus f_{ij}$, $\forall i, j$.

(ii) If $G = E \odot F$, then $g_{ij} = \sum_k (e_{ik} \odot f_{kj})$, $\forall i, j$.

(iii) $E^0 = I$ and $E^j = E^{j-1} \odot E$, for integer $j \geq 1$.

In complying with the definitions of schedule algebra, the matrix A in Fig.1(b) has the following properties:

(i) It only contains non-negative elements since there are no loops or backward arcs in the network.

(ii) The diagonal entries and those below the diagonal are "ϕ".

(iii) Its Sth power follows the properties of Markov chains.

(iv) Every entry a_{ij}^S of matrix A^S is the length of path from node i to node j after exactly S arc steps.

From property (iv), it can be seen that if $S > b + 1$, then all entries of A^S are empty because there is no way to go from node i to node j after S steps. Furthermore, if we define the summation of the powers of A by A^*, that is,

$$A^* = A^1 \oplus A^2 \oplus A^3 \oplus \cdots \oplus A^S \oplus \cdots,$$

then it can be inferred that A^* must converge within a finite number of steps $S \geq b + 1$. A^* represents a matrix of all possible path lengths between any two nodes. Since we are interested in determining K-best path values starting from node "0", we need to find only the first row of A^*.

4   Algorithm KBS

Let Z denote A^*. Shier [10] presented an iterative scheme called the generalized Jacobi method to determine the K-best paths by computing Z explicitly. But, this scheme requires high storage space and computation time, and thus, is very inefficient. An implicit way to calculate Z was presented by Wongseelashote[11], which is to solve the following equation:

$$Z \odot (I \ominus A) = I, \quad \text{where } I \ominus A \text{ can be viewed as } I \oplus (-A).$$

The matrix Z captures all the path values between any two nodes i and j. In our case, since we are interested in determining the paths from node 0, we need to determine only the first row of Z. In other words, we need to solve the following equation:

$$[z_{00} z_{01} \cdots z_{0b}] \odot (I \ominus A) = [0\phi \ldots \phi]$$

or more explicitly,

$$[z_{00}z_{01}\cdots z_{0b}] \odot \begin{bmatrix} 0 & -a_{01} & -a_{02} & \cdots & -a_{0b} \\ \phi & 0 & -a_{01} & \cdots & -a_{0b-1} \\ \vdots & \vdots & \vdots & \ddots & \vdots \\ \phi & \phi & \phi & \cdots & 0 \end{bmatrix} = [0\phi \cdots \phi]$$

Multiplying $[z_{00}z_{01}\cdots z_{0b}]$ by the first column of $(I \ominus A)$, $z_{00} \odot \{0\} = 0 \Rightarrow z_{00} = 0$.
Multiplying $[z_{00}z_{01}\cdots z_{0b}]$ by the second column of $(I \ominus A)$,
$z_{00} \odot \{-a_{01}\} \oplus z_{01} \odot \{0\} = \phi \Rightarrow z_{01} = z_{00} \odot \{a_{01}\} = a_{01}$,
and so on. For the last multiplication,
$z_{00} \odot \{-a_{0b}\} \oplus z_{01} \odot \{-a_{0b-1}\} \oplus \cdots \oplus z_{0b} \odot \{0\} = \phi$.
from which z_{0b} can be computed. From these computations one can see that to
compute a z value, one needs knowledge about all previous z values. This method
still requires excessive storage and computation time. The following observation
helps us in making improvement in this regard.
*Observation: To determine the K-best solutions, one needs to determine only the
K-best paths to all nodes.*

By this observation, we can truncate a z entry, say z_{0i}, to contain up to K-best
values while previously z_{0i} contains all the possible path values between node
0 and node i. This would help in reducing both the storage and computation
time while guaranteeing the determination of the K-best solutions. For further
improvement, consider the following modification of the above observation: determine K-best paths for only a subset of nodes while for the remaining nodes,
determine the best paths. In particular, divide the set of nodes into two groups,
namely, G1 and G2. G1 contains nodes from 0 to $\bar{n}$ for some $0 \leqslant \bar{n} \leqslant b-1$,
and G2 contains the remaining nodes. The best paths are computed for nodes
in G1 and using these best path values, K-best path values are computed for
the nodes in G2. This logic leads to the following network algorithm to find the
K-best solutions for the knapsack problem introduced in Sect.2.

Algorithm KBS:

Initialization:
   (i) Set matrix A in accordance with the arc values of the network.
   (ii) Construct matrix $(I \ominus A)$ and fix $\bar{n}$.
   (iii) Set $j = 0$.

Part One: (Determination of K-best path values)

S1: Set $j = j + 1$. Compute z_{oj}. If $j \leqslant \bar{n}$, go to S2(a); otherwise, go to S2(b).
S2: a) Find the maximum value among the elements of z_{oj}; go to S1.
       b) Find the K-best path values among the elements of z_{oj}. If j=b, go to S3;
       otherwise, go to S1.
S3: Find the K-best path values among the elements of $z_{0j}, \forall j = 1, \cdots, b$, and
       sort them in descending order. Denote these by $\zeta_1, \cdots, \zeta_k$.

Part Two: (Retrieval of the paths corresponding to $\zeta_j, \forall j = 1, \cdots, k$.)

S4: Set $j = 1$ and $i = 0$.

S5: Set $i = i + 1$, let $\zeta_t = \zeta_j - a_{ib}$. If $\zeta_t \in z_{0i}$, go to S6; otherwise, go to S7.

S6: If $\zeta_t - a_{0i} = 0$, go to S8; else, let $b = i, \zeta_j = \zeta_t$, $i = 0$, go to S5.

S7: If $i = b - 1$, make $b = b - 1$, $i = 0$, go to S5; else, go to S5.

S8: Count the number of times a coefficient of the objective function is used in the path. That gives the value of the corresponding variable. If $j = j + 1$, stop; otherwise, set $k = k + 1$, $i = 0$ and go to S5.

Next, we illustrate this algorithm by using an example problem.

Example 2:

Consider the following knapsack problem:

$$\text{Max} \qquad 3x_1 + 5x_2 + 4x_3 + 6x_4$$
$$\text{subject to}: \quad x_1 + 2x_2 + 3x_3 + 4x_4 \leqslant 5$$
$$x_i \geq 0 \text{ and Integer}, \quad \forall i = 1, 2, 3, 4.$$

Let $K = 3$. The matrices A and $(I \ominus A)$ are depicted as follows:

$$A = \begin{bmatrix} \phi & 3 & 5 & 4 & 6 & \phi \\ \phi & \phi & 3 & 5 & 4 & 6 \\ \phi & \phi & \phi & 3 & 5 & 4 \\ \phi & \phi & \phi & \phi & 3 & 5 \\ \phi & \phi & \phi & \phi & \phi & 3 \\ \phi & \phi & \phi & \phi & \phi & \phi \end{bmatrix} \quad ; \quad I \ominus A = \begin{bmatrix} 0 & -3 & -5 & -4 & -6 & \phi \\ \phi & 0 & -3 & -5 & -4 & -6 \\ \phi & \phi & 0 & -3 & -5 & -4 \\ \phi & \phi & \phi & 0 & -3 & -5 \\ \phi & \phi & \phi & \phi & 0 & -3 \\ \phi & \phi & \phi & \phi & \phi & 0 \end{bmatrix}$$

Let $\bar{n} = 5-1 = 4$. By solving $Z \odot (I \ominus A) = I$, $z_{00} \odot \{0\} = \{0\}$ implies $z_{00} = \{0\}$; go to S2(a). Since $z_{00} = \{0\}$; go to S1. $z_{00} \odot \{-3\} \oplus z_{01} \odot \{0\} = \{\phi\} \Rightarrow z_{01} = z_{00} \odot \{3\} = \{3\}$; go to S2(a). Since $z_{01} = \{3\}$; go to S1. $z_{02} \odot \{0\} = 5 \odot z_{00} \oplus 3 \odot z_{01} = \{5, 6\}$; go to S2(a). $z_{02} = \{6\}$; go to S1. $z_{03} \odot \{0\} = 4 \odot z_{00} \oplus 5 \odot z_{01} \oplus 3 \odot z_{02} = \{4, 8, 9\}$; go to S2(a). $z_{03} = \{9\}$; go to S1. $z_{04} \odot \{0\} = 6 \odot z_{00} \oplus 4 \odot z_{01} \oplus 5 \odot z_{02} \oplus 3 \odot z_{03} = \{6, 7, 11, 12\}$; go to S2(a). $z_{04} = \{12\}$; go to S1. $z_{05} \odot \{0\} = 6 \odot z_{01} \oplus 4 \odot z_{02} \oplus 5 \odot z_{03} \oplus 3 \odot z_{04} = \{15, 14, 13, 10, 9\}$; go to S2(b). $z_{05} = \{15, 14, 13\}$; go to S3. The three largest path values are $\zeta_1 = 15$, $\zeta_2 = 14, \zeta_3 = 13$. Next, trace an elementary path corresponding to the Kth largest path value. For $\zeta_k = 15$, test $a_{15} = 6$, $\zeta_t = \zeta_k - a_{15} = 15 - 6 = 9$; since $9 \notin z_{01}$, discard. Test $a_{25} = 4$, $\zeta_t = 15 - 4 = 11$; since $11 \notin z_{02}$, discard. Test $a_{35} = 5$, $\zeta_t = 15 - 5 = 10$; since $10 \notin z_{03}$, discard. Test $a_{45} = 3$, $\zeta_t = 15 - 3 = 12$; since $12 \in z_{04}$, accept $a_{45} = 3$; $\zeta_k = \zeta_t = 12$ and go to S5. Following the same procedure, the elementary path corresponding to ζ_1 is $X_1 = 5$ and $X_2 = X_3 = X_4 = 0$. Other paths can be found similarly. Note that for the selected value of $\bar{n} = 4$, the procedure does not generate three best solutions. This fact is discussed further in the next section.

5 Computational Experience and Results

The computational experience is reported from two viewpoints, (i) determination of the value of $\bar{n}$ for the use in Algorithm KBS; (ii) comparison of CPU times of Algorithm KBS with two other approaches viz dynamic programming and branch and bound based algorithms for obtaining the K-best solutions.

The dynamic programming (DP) algorithm determines the K-best solutions for problem (1) introduced in Sect.2 by utilizing the following recursive equation:

$$F_i^K(\xi) = \text{KMax}\{c_i x_i + F_{i-1}^K(\xi - a_i x_i)\}; \forall i = 1, 2, \cdots, n. \qquad (2)$$

$$x_i = 0, 1, \cdots, \left\lfloor \frac{\xi}{a_i} \right\rfloor.$$

The notation $\lfloor y \rfloor$ designates the greatest integer less than y. $F_i^K(\xi)$ represents the K-best values for a particular ξ, where $\xi = 0, 1, \cdots, b$, and hence, is a vector of size K. KMax represents the K-best values obtained from among all values of x_i for a particular ξ. Initializing $F_0^K = 0$, the K-best solution can be found by solving (2) recursively.

The branch and bound (BB) method first converts problem (1) to a (0-1) integer problem by replacing x_i in (1) by $\sum_{j=0}^{m} 2^j y_j$, where $y_j = 0, 1$, and m is an integer and is determined by $2^m \leqslant U_i \leqslant 2^{m+1} - 1$, where $U_i = \left\lfloor \frac{b}{a_i} \right\rfloor$, is the integer upper bound of x_i. By this substitution, problem (1) takes the following form:

$$\begin{aligned}
\text{Max} \qquad & q_1 y_1 + q_2 y_2 + \cdots + q_p y_p \qquad (3) \\
\text{subject to:} \qquad & d_1 y_1 + d_2 y_2 + \cdots + d_p y_p \leqslant b \\
& y_i = 0, 1, \forall i = 1, \cdots, p.
\end{aligned}$$

Also, let the y's in (3) be arranged in the non-increasing order of (q_i/d_i), for $i = 1, \ldots, p$. The enumeration is carried out by systematically constructing a tree while fixing values of variables to 0 to 1 in a sequential fashion. Each node of the enumeration tree represents a solution to (3). During the implementation of the enumeration scheme, K-best solutions are stored and constantly updated. The smallest of these K-best solutions is the lower bound. A node is fathomed if any of the following conditions hold.

 (i) The solution satisfies the constraint of (3) as an equality and is integer.
 (ii) The solution is infeasible.
(iii) The value of the solution or its integer part is less than that of the current lower bound.

Following Balas and Zemel [1], the variables are divided into two groups to cut down on the size of the enumeration tree. The variables in one group can take values 0 to 1 and it is called the free variable group while the variables in the other group are set to zeros and it is called the zero variable group. The following criterion was found to be very effective in dividing the variables into these two groups. Let the variable corresponding to the smallest coefficient in the constraint be $\bar{y}$. If $\bar{y} \geq 2K$, then the free variable group contains variables from y_1 to $\bar{y}$ and the remaining variables belong to the zero variable group. However, if $\bar{y} < 2K$, then $\bar{y}$ is the $2K_{th}$ variable from the start, that is, $\bar{y} = y_{2k}$, and the two groups are defined as before.

In our experimentation, all the algorithms were programmed in C language and were run on an IBM Thinkpad T23 computer. The problem data were generated to obtain various (c_i/a_i) ratios, the coefficients of the variables in the objective function and constraints. The values of a_i, $i = 1, \ldots, b-1$, were chosen as follows: $a_1 = 1, a_2 = 2, \cdots, a_{b-1} = b - 1$; while those of c_i were generated uniformly between 1 and $10N$, where N is the total number of variables. An experimentation was conducted to determine an effective value of $\bar{n}$ for use in the network algorithm. The values of K that we considered for the experimentation were 10,15, 30 and 40. For $K=10$ and 15, the problem sizes considered consisted of 50,60,70,80,90 and 100 number of variables, while for $K=30$ and 40, the problem sizes consisted of 200, 300, 400 and 500 number of variables. Three values of $\bar{n}$ were considered, namely, $n-2, n-K, n-2K$. Ten observations regarding the number of K-best solutions missed by the algorithm were taken for each problem size, K and $\bar{n}$ combination. Thus, in all, 600 problems were run. We recorded the number of K-best solutions missed for each value of $\bar{n}$ used. These were determined by comparing the K-best solutions obtained using the specified value of $\bar{n}$ and the $\bar{n}$ value close to zero. Clearly, if $\bar{n} = b - 1$, Algorithm KBS generates the best path value while for $\bar{n}=0$, it can generate all possible path values. For $0 < \bar{n} < b - 1$, the algorithm will generate K-best solutions only for certain values of K. However, the smaller the value of $\bar{n}$, the greater is the computational time. Ideally, from the viewpoint of saving computation time and storage, one needs to determine the largest number $\bar{n}$ which can generate the desired K-best solutions. Our results indicated that when $\bar{n} = n - 2$, KBS misses a significant number of best solutions. And, as the problem size and the value of K increase, so does the number of solutions it misses. However, when $\bar{n} = n - K$, KBS was found to perform almost as well as the case when $\bar{n} = n - 2K$, with regard to the number of solutions it misses. The average number of K-best solutions missed for $\bar{n} = n - K$ were found to be in the range from 1.1 to 2.9, and this value was not related to the problem size or the value of K. In other words, when $\bar{n} = n - K$, the performance of KBS does not deteriorate as the problem size and the value of K increase. In fact, even for the case when forty best solutions are required and the problem contains 400 variables, it was found to miss only one solution on average. Hence, $\bar{n} = n - K$ was found to be a reasonable compromise value of $\bar{n}$ to use.

Next, we performed experimentation on large size problems to compare the performance of KBS with those of dynamic programming and branch and bound based procedures. The results are shown in Table 1. The problem size was varied from 50 to 20,000 variables, while three values of K were chosen, namely, 1, 10 and 20. Note that the branch and bound based approach requires more CPU time than that required by either Algorithm KBS or the procedure based on DP. Moreover, as the problem size increases, its CPU time increases exponentially. In fact, it cannot solve a problem having more than 100 variables in realistic CPU time. Hence, it is the least attractive among the three procedures. For the DP based procedure, it doesn't perform significantly worse than Algorithm KBS from CPU point of view, when the problem contains less than 100 variables.

However, as the problem size exceeds 100, its CPU time rises more quickly than that of Algorithm KBS. Besides, the DP based procedure suffers from excessive storage requirements. In fact, it cannot be run on the Thinkpad T23 computer when the problem size exceeds 6000 variables for $K=1$, 3000 variables for $K=10$, and 2000 variables for $K=20$.

Table 1. CPU time(in secs.)to obtain K-best solutions by Algorithm KBS, Dynamic Programming and Branch and Bound based approaches

| Problem Size (n) | CPU Time (in secs.) to obtain the K-best solutions when | | | | | | | | |
| | $K=1$(Optimal) | | | $K=10$ | | | $K=20$ | | |
	KBS	DP	BB	KBS	DP	BB	KBS	DP	BB
50	0	0	0.791	0.01	0.07	0.761	0.15	0.08	0.771
60	0	0.01	3.995	0.03	0.03	3.865	0.15	0.11	4.296
70	0	0	19.778	0.02	0.08	20.19	0.14	0.11	19.288
80	0.02	0	85.322	0.06	0.04	85.61	0.13	0.08	85.913
90	0	0.01	322.824	0.05	0.06	325.3	0.15	0.09	324.216
100	0	0	1172.02	0.05	0.09	1175	0.13	0.1	1178.262
200	0	0.02	* [1]	0.11	0.2	*	0.21	0.29	*
300	0	0.04	*	0.24	0.28	*	0.23	0.661	*
400	0.01	0.05	*	0.21	0.541	*	0.39	1.152	*
500	0	0.09	*	0.18	0.771	*	0.56	1.863	*
600	0.04	0.16	*	0.31	1.122	*	0.831	2.694	*
800	0.06	0.25	*	0.68	2.053	*	1.462	4.917	*
900	0.05	0.29	*	0.761	2.623	*	1.942	6.199	*
1000	0.03	0.38	*	0.66	3.265	*	1.972	7.701	*
2000	0.29	1.482	*	3.184	45.385	*	8.462	95.97	*
3000	0.3	3.485	*	5.467	123.438	*	14.41	- [2]	*
4000	0.09	6.489	*	10.945	-	*	37.523	-	*
5000	0.57	10.354	*	14.28	-	*	38.465	-	*
6000	1.522	23.073	*	28.561	-	*	70.921	-	*
9000	1.442	-	*	54.969	-	*	155.032	-	*
15000	5.177	-	*	73.575	-	*	206.687	-	*
20000	7.04	-	*	201.99	-	*	784.127	-	*

On the other hand, when $K = 1$, Algorithm KBS only takes less than 10 seconds to solve a problem with up to 20000 variables. When $K = 10$ and $K = 20$, it still takes less than 10 seconds to solve a problem with up to 2000 variables and a few minutes to solve a problem with tens of thousands of variables. Consequently, it is clear that Algorithm KBS is far superior to the DP and BB-based procedures with respect to both the computational time and storage requirement, while generating close to true K-best solutions.

---

[1] '*' indicates that CPU time required is greater than 10^5 seconds.

[2] '-' indicates that the algorithm requires an image size beyond the limit of the C programming language.

6   Conclusions

In this paper, we have presented a new and effective schedule algebra based algorithm to generate the K-best solutions of the knapsack problem with a single constraint. An empirical formula is used to reduce the CPU time required by the algorithm to solve large size problems. Our computational results show that the proposed algorithm dominates the procedures based on dynamic programming and branch and bound approaches with respect to the computational time and storage requirement,while generating close to true K-best solutions.

References

[1] Balas, E. and Zemel, E., "An Algorithm for Large Zero-One Knapsack Problem," *Operations Research*, 1980, Vol.28, No.5, 1130-1154.
[2] Bretthauer K.M. and Shetty B., " The nonlinear knapsack problem - algorithms and applications," *European Journal of Operational Research*, 2002, Vol.138, Issue 3, 459-472.
[3] Eppstein D. "Finding the k shortest paths," *SIAM Journal of Computing*, 1998, Vol. 28, 652-73.
[4] Giffler, B., "Schedule Algebra: A Progress Report," *Naval Research Logistics Quarterly*, 1968, Vol.15, 255-280.
[5] Lin, E., "A bibliographical survey on some well-known non-standard knapsack problems," *INFOR*, 1998, Vol. 36, Issue 4, 274-317.
[6] Salkin, H.M. and Kluyver, C., "Knapsack problem – Survey," *Naval Research Logistics*, 1975, Vol. 22, Issue 1, 127-144.
[7] Sarin, S.C., "Mixed Disc Packing Problem: Part I," *IIE Transactions*, 1983, Vol.15, No.1, 37-45.
[8] Sarin, S.C. and Ahn, S., "Mixed Disc Packing Problem: Part II," *IIE Transactions*,1983, Vol. 15, No.2, 91-98.
[9] Shapiro, J.F., "Dynamic Programming Algorithm for Integer Programming Problem I: The Integer Programming Problem Viewed as a Knapsack Type Problem," *Operations Research*, 1968, Vol.16, No.1, 103-121.
[10] Shier,D.R., "Interactive Methods for Determining the K Shortest Paths in a Network," *Networks*, 1976, Vol.6, 205-229.
[11] Wongseelashote, A., "An Algebra for Determining all Path-Values in a Network with Application to K-Shortest-Paths Problems," *Networks*, 1976, Vol.6, 307-334.

Point Sets and Frame Algorithms
in Management

José H. Dulá

Virginia Commonwealth University, Richmond, VA 23284, USA

Abstract. Consider a finite point set $\mathcal{A}$ in m-dimensional space and the polyhedral hulls it generates from constrained linear combinations of its elements. There are several interesting management problems that are modelled using these point sets and the resulting polyhedral objects. Examples include efficiency/performance evaluation, ranking and ordering schemes, stochastic scenario generation, mining for the detection of fraud, etc. These applications require the identification of frames; that is, the extreme elements of the polyhedral sets, a computationally intensive task. Traditional approaches require the solution of an LP for each point in the point set. We discuss this approach as well as a new generation of faster, output-sensitive, algorithms.

1 Notation and Definitions

The finite point set $\mathcal{A} = \{a^1, \ldots, a^n\}$, composed of n points in $\Re^m$ can be linearly combined to produce a *polyhedral hull*, pol($\mathcal{A}$). The polyhedral hull is a generalization of the familiar convex and conical hulls of the data. A more formal representation of these sets in terms of the point set $\mathcal{A}$ is:

- The convex hull: $\mathrm{con}(\mathcal{A}) = \{y \in \Re^m | y = \sum_j a^j \lambda_j, \; \sum_j \lambda_j = 1, \; \lambda_j \geq 0, \; \forall j\}$.
- The conical hull: $\mathrm{pos}(\mathcal{A}) = \{y \in \Re^m | y = \sum_j a^j \lambda_j, \; \lambda_j \geq 0, \; \forall j\}$.
- The polyhedral hull: $\mathrm{pol}(\mathcal{A}) = \{y \in \Re^m | y = \sum_j a^j \lambda_j + \sum_k v^k \mu_k, \; \sum_j \lambda_j = 1, \lambda_j \geq 0, \mu_k \geq 0 \; \forall j, k\}$.

These polyhedral sets are *externally* defined. Figure 1 depicts examples of these sets in two dimensions. We use the same point set with eight points to generate a convex hull, a conical hull, and a polyhedral hull obtained combining a convex hull and two rays, $v^1 = (1,0)$ and $v^2 = (0,-1)$ which define a recession cone along which the set is unbounded. This example illustrates properties that are true in general:

- The hulls are defined by subsets of the data set: these points are extreme elements (extreme points or extreme rays) of the final hull.
- The set of extreme points of the convex hull is a superset for extreme elements of all other hulls.
- The extreme points (rays) of the polyhedral hull are a minimal subset of the generators required to describe the same set.

N. Megiddo, Y. Xu, and B. Zhu (Eds.): AAIM 2005, LNCS 3521, pp. 450–459, 2005.
© Springer-Verlag Berlin Heidelberg 2005

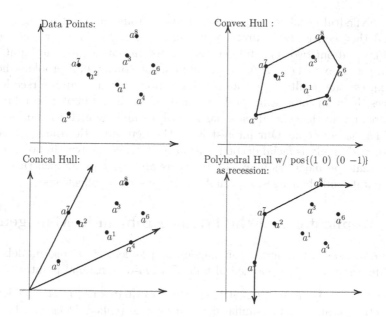

Fig. 1. A point set combined into three different hulls

The extreme elements of the hull of a point set, i.e., extreme rays for a pointed cone, extreme points for a convex hull and an unbounded polyhedral hull (provided the recession cone is already minimally defined and its extreme rays known), are called the *frame*. The following definition formalizes this.

Definition 1. *The frame, $\mathcal{F}$, of a polyhedral hull is a minimal subset of the point set, $\mathcal{A}$, such that* $\mathrm{pol}(\mathcal{A}) = \mathrm{pol}(\mathcal{F})$.

The following observations apply:

- Frame elements correspond to extreme points (rays) of the polyhedral set (assumptions needed; e.g., cones must be pointed, no duplication, no two points are multiples for cones, recession cone is minimally defined and known for general polyhedral hulls).
- Non-frame generators are nonessential or "redundant".
- All inferences about the geometry of the polyhedral hull can be made using the frame.
- In large data sets where $n \gg m$, the cardinality of the frame is typically a small fraction of n.

The original problem of identifying the extreme points of a convex hull of a finite set of points in multidimensional space is known by various names. In [1] Fukuda refers to it as the "redundancy-removal" problem for a point set, in [2] Edelsbrunner describes it without naming it, and in [3] it is a "convex-hull" problem. We have opted for Gerstenhaber's term in [4] and refer to it as the "frame" problem. It is important to stress from the outset that the frame problem is not the polyhedral facial decomposition problem from computational geometry, also called the "convex hull" problem by many, e.g., [2]. This terminology creates confusion about two very different problems. In both cases, the input is a finite point set. The goal of the facial decomposition-convex hull problem, however, is to express the resultant multidimensional polyhedron as an intersection of half-spaces. It is an interesting problem in its own right but it is combinatorial in nature and, in the general case, has no polynomial algorithms. That convex-hull problem is our topic. Our interest is on the deterministic frame problem which is not as difficult as facial decomposition. In fact, it is solved in polynomial time. The frame problem has many applications and presents its own theoretical and computational challenges, especially in large-scale applications.

2   Applications of the Frame Problem in Management

There are several important management problems that use models based on finding the frame of a polyhedral hull of an m-dimensional point set.

Data Envelopment Analysis (DEA). DEA evaluates the relative efficiency of a collection of functionally similar firms or entities (called 'Decision Making Units' or DMUs) in transforming multiple inputs into multiple outputs. The study proceeds to identify the points from the data set on a subset of the boundary of a polyhedral hull known as the *efficient frontier*. The final result of a DEA study is a classification of DMUs as either efficient or inefficient. The data points in the efficient frontier correspond to the efficient DMUs; the rest are inefficient. DEA provides information about inefficient DMUs that can be used to assess how inputs can be decreased or outputs increased to attain efficiency. The frame of the hull is sufficient for the efficiency classification and scoring. The fourth panel in Figure 1 depicts the efficient frontier of a DEA problem in $\Re^2$ where the first component corresponds to an input and the second to an output. This frontier is defined by the points a^5, a^7 and a^8. DEA problems with several thousand DMUs in eight to ten dimensions occur in the banking industry.

Ranking and Ordering. Barnett states that "There is no natural basis for ordering multivariate data" [13]. Nevertheless there is a demand from the part of decision makers for meaningful ranking schemes applied to complex entities described by multiple attributes. One approach is "hull peeling". This is a purely geometric approach based on defining a layer of the data as the points on the boundary of the polyhedral hull. If these points are removed, the remaining data define a new hull within the previous one and hence a second layer or depth. The process can be repeated to produce a sequence of nested layers that may be used as a partial ordering of the data. The process requires the repeated solutions to frame problems.

Outlier Detection; e.g. Security, Fraud, etc. Without actually attaining extreme values in any single dimension, interesting entities may be identified by having extreme values when dimensions are combined. For examples, the tax return with the largest sum of the charitable contributions and employee deductions may reveal an interesting tax return; or the individual whose money transfer events and total monetary value of the transfers is the largest when added up may merit further review. The record that emerges as a geometric outlier is just one of, possibly, many using this simple criterion if the combination of the two values are weighted differently. All such points are, in the same sense, geometric outliers and all would be interesting for different reasons.

In general, a data point, a^{j^*}, is a geometric outlier if it is a support for a polyhedral set generated by the data. The shape of this polyhedral set is based on assessments as to whether only larger or smaller magnitudes (or both) for the attribute's value are the priority. Each of the m relations in the system $\sum_j a^j \lambda_j$ can have three possibilities: $\leq, \geq$, and $=$ leading to 3^m different polyhedral sets.

The frame problem described here has clear data mining applications. The scale of these problems is potential massive. Research needs to be performed on the performance of the frame algorithms at these scales. Also, the multitude of shapes and properties that emerge need to be studied for their properties and impact on algorithmic performance. Modeling the type of situations and applications that lead to some of these new shapes is interesting as well.

Stochastic Programming. It is not necessary to expound here on the importance of mathematical programming under uncertainty in management and decision making. The frame problem has been identified as critical in the stochastic programming area by Wallace and Wets [14], where it appears in the process of determining if the second stage of a two-stage stochastic program has relatively complete recourse. Wallace and Wets state that "there is a lot to be gained by a more efficient implementation (of an algorithm to find the frame of the convex hull, than one based on solving linear programs)".

3   Deterministic Procedures to Find Frames of Polyhedral Hulls

In [1], Fukuda notes that solving the frame problem "might end up in a very time consuming job for large n (say > 1000)". However, applications with many more than 1000 points are not uncommon. Therefore, there is an interest in studying the algorithmic and computational aspects of the frame problem.

The first procedure for identifying the frame is a result of a direct application of the external definition of pol($\mathcal{A}$). A data point, a^{j^*}, is an extreme element of a polyhedral hull if and only if it does not belong to the hull of the rest of the data set, $\mathcal{A} \setminus a^{j^*}$. Therefore, to test whether $a^{j^*} \in \mathcal{F}$, check the feasibility of the system

$$\mathbf{S}(\mathcal{A}, j^*) = \begin{cases} \sum_{j \neq j^*} a^j \lambda_j + \sum_k v^k \mu_k = a^{j^*}, \\ \sum_{j \neq j^*} \lambda_j = 1, \\ \lambda_j \geq 0, \mu_k \geq 0 \forall j, k. \end{cases}$$

Each point can be tested individually using the system, hence the following, "naive", deterministic procedure.

"Naive" Procedure to Identify Frame Elements.

For $j = 1$ to n do:
   Step 1. $j^* \leftarrow j$.
   Step 2. Verify feasibility of System $\mathbf{S}(\mathcal{A}, j^*)$.
   Step 3. Classify DMU j^*:
     If $\mathbf{S}(\mathcal{A}, j^*)$ is infeasible
      Then: $a^{j^*} \in \mathcal{F}$,
      Else: $a^{j^*} \in \mathcal{J}$.

Testing the feasibility of a linear system such as $\mathbf{S}(\mathcal{A}, j^*)$ can be done a number of ways in polynomial time. The most practical is linear programming. Therefore, the naive procedure requires the solution to n LPs each roughly dimension $m \times n$. We obtained predictable results given what is known about solving LPs in an implementation using IMSL's subroutine 'ddlprs' [5]. Figure 2.a shows how the time required to solve individual LPs for a typical implementation using a point set in $\Re^5$ behaves almost linearly with respect to cardinality. Since a complete implementation requires n LPs, the total time behaves closely as a factor of the square of the number of LPs solved; this is clearly evident in Figure 2b.

The reports in Figure 2 represent a direct basic implementation. Preprocessing; e.g., sorting by component for convex hulls, etc. can can reveal the status of some points without paying a full LP price. The impact of such opportunistic schemes especially on large data sets is limited.

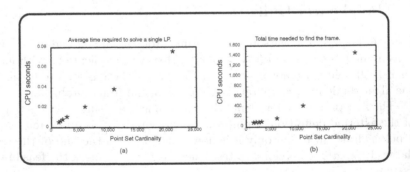

Fig. 2. Naive Procedure Performance as a Function of Data Set Cardinality: (a) Single LP. (b) Full Implementation (n Iterations)

Substantial performance enhancements occur with what is known about the mechanics of solving LPs and the geometry of finitely generated hulls. The LPs solved from one iteration to the next in the naive procedure differ only in their right-hand side. Indeed, implementing advanced basis reoptimization, i.e., "hot starts", yields substantial reductions in time. It turns out that only frame elements can be part of an LP basis. Therefore, if a^{j^*} is identified as nonessential, it can be removed from the coefficient matrix in the LP in subsequent iterations. This is known as "restricted basis entry" (RBE) and it can provide further important gains.

Output-sensitive algorithms for the frame problem of a convex hull that do not rely on naive feasibility verification were independently proposed in [6], [7], [8], and [9]. The rough sketch of an idea for an algorithm that is essentially the same as in [6] is provided in [7] . Separate and independently developed output-sensitive frame algorithms appear in [8] and [9] . This algorithm differs from [6] and [7] in an important way. What is done in [6] and [7] using inner products, [8] and [9] do with full-data LPs. It is not immediately clear how this approach compares to the inner product way since the full-data LPs actually have the potential of uncovering more than one new frame element every time they are invoked. The algorithm in [8] was cleaned up, formalized, and corrected in [10]. There are no computational results in [7], [8], [9], and [10] so it is difficult to predict or compare performances.

The complexity and performance of output-sensitive algorithms depend on the cardinality of the frame. They are based on the general principles that if $\hat{\mathcal{A}} \subseteq \mathcal{A}$ then $\mathrm{pol}(\hat{\mathcal{A}}) \subseteq \mathrm{pol}(\mathcal{A})$; and if $\tilde{a}$ is interior to $\mathrm{pol}(\hat{\mathcal{A}})$ then $\tilde{a}$ is interior to $\mathrm{pol}(\mathcal{A})$. Finally if $\tilde{a} \notin \mathrm{pol}(\hat{\mathcal{A}})$, then there exists at least one new frame element. This information can be used to update $\hat{\mathcal{A}}$.

A formal output-sensitive algorithm for finding general frames is presented next:

Algorithm BuildHull.

Initialization: Select any subset of the frame: $\hat{\mathcal{F}}$.
For $j = 1$ to n do:
    Step 1. $j^* \leftarrow j$.
    Step 2. $a^{j^*} \in \mathrm{pol}(\hat{\mathcal{F}})$? (*Solve special LP*)
            Yes: Classify a^{j^*} as 'interior'. Next j.
            No:
                    1. Find separation and support of $\mathrm{pol}(\hat{\mathcal{F}})$.
                       (*From LP solution.*)
                    2. Find new extreme point (*Sort inner products*).
                    3. Update $\hat{\mathcal{F}}$ by including new extreme point.
                    4. New extreme point is a^{j^*}?
                       Yes. Next j.
                       No. Go To Step 2.

456   J.H. Dulá

The initialization can be any nonempty subset of the frame, including a single point. Step 2 of `BuildHull` requires the solution of a special LP. This LP needs to satisfy several properties:

- It must be feasible and bounded for any point in $\Re^m$. This way, a meaningful solution is always available for use in the following two tasks.
- Its solution must conclusively resolve whether $a^{j^*} \in \text{pol}(\hat{\mathcal{F}})$ and, if the test point is external then ...
- Its solution must provide a separating and supporting hyperplane between a^{j^*} and $\text{pol}(\hat{\mathcal{F}})$.

Specialized LP formulations are available for different polyhedral hulls in [6], [11], and [12].

The algorithm is best understood with an illustration. The six panels in Figure 3 depict two full iterations using a two-dimensional point set. The first iteration tests an interior point and therefore does not execute the substeps in Step 2. The second iteration tests an exterior point and therefore a separating and supporting hyperplane is generated and a new frame element is identified. Consider the point set from Figure 1 and suppose we apply `BuildHull` to find its convex hull. Here is a panel by panel explanation:

- **Panel a.** The initializing frame elements are a^4, a^5 and a^8. This becomes the "current partial hull", $\hat{\mathcal{F}}$, until a new frame element is identified and $\hat{\mathcal{F}}$ is updated.
- **Panel b.** Iteration 1: $j^* = 1$, Steps 1 and 2. Point a^1 is interior to $\hat{\mathcal{F}}$. (In general, an LP solution is used to make this determination). It gets classified as interior. The first iteration concludes by incrementing j.
- **Panel c.** Second iteration begins: $j^* = 2$. An LP solution reveals that the second test point, a^2, is exterior to the current partial hull. Therefore, there exist a separating and supporting hyperplane between a^2 and $\hat{\mathcal{F}}$. We proceed to execute instructions in the substeps of Step 2.
- **Panel d.** The separating and supporting hyperplane is the line through a^5 and a^8. This hyperplane is obtained from the LP solution.
- **Panel e.** The separating and supporting hyperplane is used to find a new frame element by translating it away from the current partial hull. On the way, we meet the points a^2 and a^7. Since a^7 is farthest, and there are no ties, it must necessarily be an extreme point and gets classified as a frame element. Note that this translation is equivalent to evaluating the hyperplane at each point on the same side of the separation. These operations require simple inner products. The list of known frame elements is updated with the point a^7.
- **Panel f.** The new partial hull, $\hat{\mathcal{F}}$, is now composed of a^4, a^5, a^7 and a^8. The next iteration does not increment the index j. This way, the point a^2 will be tested again.
- **Remaining Iterations.** The point a^2 will be identified as interior in Iteration 3. Iteration 4 identifies the point a^3 as interior. Iteration 5, the last iteration, will uncover a^6 as a frame element.

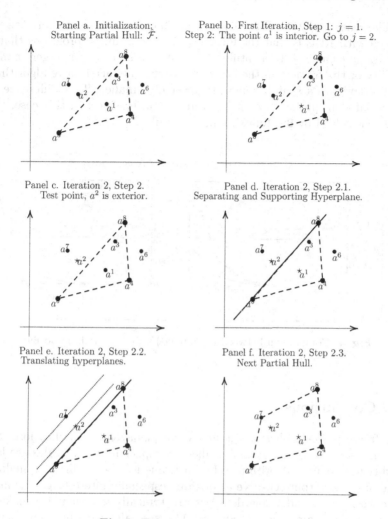

Panel a. Initialization;
Starting Partial Hull: $\mathcal{F}$.

Panel b. First Iteration, Step 1: $j = 1$.
Step 2: The point a^1 is interior. Go to $j = 2$.

Panel c. Iteration 2, Step 2.
Test point, a^2 is exterior.

Panel d. Iteration 2, Step 2.1.
Separating and Supporting Hyperplane.

Panel e. Iteration 2, Step 2.2.
Translating hyperplanes.

Panel f. Iteration 2, Step 2.3.
Next Partial Hull.

Fig. 3. Two iterations of `BuildHull`

Remark 1. Note that any implementation of procedure `BuildHull` should check whether a point has already been classified before proceeding to Step 2.

A comparison between a naive implementation enhanced with RBE and `BuildHull` appears in Figure 4. (Hot starts benefit both procedures). The results reported are typical. The "frame density" is the percentage of points that are elements of the frame. We say a point set has low density when its extreme elements are up to 15% of the cardinality of the set and we classify point sets with more than 35% as high density. The two panels in the figure show that the performance of `BuildHull` is clearly output-sensitive with respect to the frame density. In either case, `BuildHull` dominates and this domination becomes increasingly dramatic as the the cardinality of the point set increases and the frame density decreases.

The reason why BuildHull is better than a naive algorithm even when enhanced with RBE is that the former solves small linear programs that grow as new frame elements are identified but these LPs are never larger than the cardinality of the frame. On the other hand, the enhanced naive algorithm begins solving large LPs that become progressively smaller. This difference becomes accentuated in problems with low frame density since, in this case, the linear programs solved by BuildHull remain small.

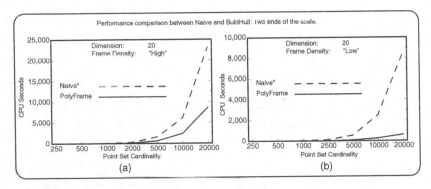

Fig. 4. Comparison between (enhanced) naive algorithm and BuildHull

4   Conclusions

The frame problem plays a role in several management applications. Solving it presents serious computational challenges especially when scales are large. The traditional "naive" algorithm to find a frame is severely limiting making a new generation of output-sensitive algorithms especially interesting. The fact that in practice $n \gg m$ and frame densities are typically less than 1% makes output-sensitive algorithms such as BuildHull especially valuable. The new output-sensitive algorithms will allow the solution of much larger frame management problems.

References

1. K. Fukuda, *Frequently asked questions in polyhedral computation*, http://www.cs.mcgill.ca/~fukuda/soft/polyfaq/, 2000.
2. H. Edelsbrunner, *Algorithms in combinatorial geometry*, Monographs in Theoretical Computer Science, 10, 1987, Springer-Verlag, New York.
3. J. B. Rosen, G.L. Xue, A.T. Phillips, *Efficient computation of extreme points of convex hulls in $\Re^d$*, Advances in Optimization and Parallel Computing, P.M. Pardalos, ed., North Holland, Amsterdam, 1992, pp. 267-292.
4. M. Gerstenhaber, *Theory of convex polyhedral cones*, Activity Analysis of Production and Allocation, T. C. Koopmans, ed., Chapman and Hall, London, 1951, pp. 298-316.

5. Application Development Tools., *IMSL Stat Library*. Visual Numerics, Inc. Houston, TX, 1994.
6. J. H. Dulá and R.V. Helgason *A new procedure for identifying the frame of the convex hull of a finite collection of points in multidimensional space.*, Eur. J. Oper. Res., 92, 1996, pp. 352-367.
7. K.L. Clarkson, *More output-sensitive geometric algorithms*, Proc. 35th IEEE Sympos. Found. of Comput. Sci., Santa Fe, NM, 1996, pp. 695-702.
8. T. Ottmann, S. Schuierer, and S. Soundaralakshimi, *Enumerating extreme points in higher dimension*, Proc. 12th Annual Sympos. on Theoret. Aspects of Comput. Sci., Lecture Notes in Computer Science, 900, Springer, Berlin, 1995, pp. 562-570.
9. T. M. Chan, *Output-sensitive results on convex hulls, extreme points, and related problems.*, Discrete & Comput. Geometry, 16, 1996, pp. 369-387.
10. T. Ottmann, S. Schuierer, and S. Soundaralakshimi, *Enumerating extreme points in higher dimension*, Nordic J. Comput. 8, 2001, pp. 179-192
11. J. H. Dulá, R.V. Helgason, and N. Venugopal *An algorithm for identifying the frame of a pointed finite conical hull.*, INFORMS J. Computing, 10, 1997, pp. 323-330.
12. J. H. Dulá and F.J. Lopéz, *Algorithms for the frame of a finitely generated unbounded polyhedron*, INFORMS J. Computing, 2005, to appear.
13. V. Barnett, *The ordering of multivariate data*, J. Roy. Statist. Soc. A, 1976, pp. 318-344.
14. S. W. Wallace, and R.J.B. Wets, *Preprocessing in stochastic programming: the case of linear programs*, ORSA Journal on Computing, 4, 1992, pp. 45-59.

Mining a Class of Complex Episodes
in Event Sequences

H.K. Dai and G. Wang

Computer Science Department, Oklahoma State University,
Stillwater, Oklahoma 74078, USA
{dai, guohuan}@cs.okstate.edu

Abstract. This work extends the existing parallel- and serial-episode data mining algorithms to that for parallel connection of serial (PoS) episodes. The PoS-episodes can model more general situations and preserve the sequence information as well. The PoS-episode mining algorithm can provide episode-mining users a powerful mining tool and make the episode mining more flexible. To use the PoS-episode mining algorithm, users need to decide reasonable parameters like window width and minimum frequency ratio. Concepts and methods are provided by using Web log mining as example to illustrate the applicability of the PoS-episode mining and show how to decide reasonable parameters as well as evaluate the mining process.

1 Preliminaries

Data mining is a task that extracts or "mines" knowledge from large amounts of data. Sequence data mining is one of the branches of data mining where the data can be viewed as a sequence of events and each event has an associated time of occurrence. Examples of such data are telecommunication network alarms, occurrences of recurrent illnesses, Web site traversal actions, etc [1].

Sequence data can be modeled using episodes [5]. An episode is an acyclic directed graph representing partial ordering of a group of events. Episodes can be classified into serial episodes, parallel episodes, and complex episodes, according to the nature of partial ordering among the events. An episode is serial (parallel) if the underlying partial ordering is total (trivial, respectively). An episode is complex if it is not serial or parallel.

The Apriori algorithm [2] [3] [6] can be used in mining frequent episodes. It employs a bottom-up strategy and utilizes the known/prior information to reduce the search space. There are two main steps in each iteration of the algorithm, candidate generation and frequent-episode recognition. The Apriori algorithm terminates when there is no more candidate generated.

When working on sequence data mining, it is always desirable to maintain the relative sequence information in the output of date mining in terms of input data sequence. Serial episode is capable of keeping sequence information. However, the episode configuration is rigidly restricted and has to be totally ordered.

N. Megiddo, Y. Xu, and B. Zhu (Eds.): AAIM 2005, LNCS 3521, pp. 460–471, 2005.

The capability of expression of the serial episode therefore is limited to certain real-world scenarios. Parallel-episode mining does not take the events sequence information into account and is less useful in sequence data mining area.

Complex episode is ideal for modeling the real world. However, mining complex episodes is not a trivial task, due to the topological complexity of episode generation and the computational complexity of episode recognition. This work takes a step out toward a compromise direction by providing algorithms for mining parallel connection of serial (PoS) episodes, which is an interesting subclass of complex episodes. An experiment that illustrates the applicability of the PoS-episode mining is performed by studying Web page traversal pattern mining using Web server log data set.

Given a set E of event types, an event is an ordered pair (A,t) where $A \in E$ is an event type and t is a nonnegative integer, the occurrence time of the event. We assume that all events last one time unit. An event sequence on E is a triple (s, T_s, T_e), where $s = \{(A_i, t_i)\}_{i=1}^n$ for some positive integer n, is an ordered sequence of events such that $A_i \in E$ for all $i \in \{1, 2, \ldots, n\}$, and $t_i < t_{i+1}$ for all $i \in \{1, 2, \ldots, n-1\}$. Furthermore, T_s and T_e are positive integers: T_s and T_e are the starting time and ending time of the event sequence, respectively, and $T_s \leq t_i < T_e$ for all $i \in \{1, 2, \ldots, n\}$.

A window on an event sequence (s, T_s, T_e) is an event sequence (w, t_s, t_e), where $t_s < T_e$ and $t_e > T_s$, and w consists of those events/pairs $(A,t) \in s$ where $t_s \leq t < t_e$. We abbreviate an event sequence (s, T_s, T_e) and a window (w, t_s, t_e) by s and w, respectively. The time span $t_e - t_s$ is called the width of the window w, and it is denoted by $width(w)$. Given an event sequence s and an integer w_width, we denote by $W(s, w\_width)$ the set of all windows w on s such that $width(w) = w\_width$.

An episode is an acyclic directed graph with node set V and edge set $\preceq$. An episode α is a triple $(V, \preceq, g)$, where V is a set of nodes, $\preceq$ is a partial order on V, and $g : V \to E$ is a function associating each node with an event type in E of an event sequence such that the temporal order of the event types of $g(V)$ is given by the partial order $\preceq$ on V. We define the size of episode α, $|\alpha|$ as $|V|$. Depending on the underlying graph structure, we can divide episodes into three categories: serial episodes, parallel episodes, and complex episodes. An episode $\alpha = (V, \preceq, g)$ is serial if its partial order $\preceq$ on V is a total order; and α is parallel if its partial order $\preceq$ on V is trivial/empty. An episode is complex if it is neither serial nor parallel. A complex episode can be reduced to the recognition of a hierarchical combination of serial and parallel episodes. PoS-episodes form a subclass of complex episodes. A PoS-episode consists of two serial episodes that share the common start and the end event types.

For two episodes $\alpha = (V, \preceq, g)$ and $\alpha' = (V', \preceq', g')$, α' is a subepisode of α if the underlying graph structure of α' is a subgraph of that of α; that is, there exists an injection $f : V' \to V$ such that $g'(v) = (g \circ f)(v)$ for all $v \in V'$, and for all $v, w \in V'$ with $v \preceq' w$, we have $f(v) \preceq f(w)$.

An episode $\alpha = (V, \preceq, g)$ occurs in an event sequence $s = ((A_i, t_i)_{i=1}^n, T_s, T_e)$ if there exists an injection $h : V \to \{1, 2, \ldots, n\}$ from the node set of α to the set of all event-indices of s such that $g(v) = A_{h(v)}$ for all $v \in V$, and for all $v, w \in V$ with $v \preceq w$ and $v \neq w$, we have $t_{h(v)} < t_{h(w)}$.

The frequency of an episode is the fraction of windows in which the episode occurs out of all possible windows. Given an event sequence $s = ((A_i, t_i)_{i=1}^n, T_s, T_e)$ and a window width w_width, the frequency of an episode α in s is:

$$fr(\alpha, s, w\_width) = \frac{|\{w \in W(s, w\_width) \mid \alpha \text{ occurs in } w\}|}{|W(s, w\_width)|}.$$

We say that α is frequent in s if $fr(\alpha, s, w\_width) \geq min\_fr$, where min_fr is a prescribed threshold of minimum frequency ratio given by the user. The collection of all frequent episodes is denoted by $F(s, w\_width, min\_fr)$ with respect to the given s, w_width, and min_fr.

Once we find all frequent episodes, we can use them to generate rules that describe the relationship between episodes. An episode rule is an expression $\beta \Rightarrow \gamma$ where β is a subepisode of γ. The confidence of the episode rule is:

$$confidence(\beta \Rightarrow \gamma) = \frac{fr(\gamma, s, w\_width)}{fr(\beta, s, w\_width)}.$$

The candidate-generation algorithms for serial episodes and parallel episodes are slightly different, but both of them follow an Apriori-like algorithm that performs a level-wise search. From the episode set with only one event, that is, set of all size-1 episodes, the search algorithm first computes a collection of candidate episodes, then checks the frequencies by scanning the event sequence. The episodes with frequency of at least a prescribed threshold form the frequent-episode set of current episode size. The algorithm then computes the collection of candidate episodes for next episode size (from the frequent-episode set of current episode size) and generates the frequent-episode set of next episode size accordingly, until no frequent episodes are generated at certain episode size. In the candidate-generation algorithm, a parallel episode is represented as a lexicographically sorted array of event types. Parallel and serial episodes in their episode collections are also sorted by lexicographical order. Candidates can be generated by arranging appropriate combinations of two episodes of size l that share the first $l - 1$ common event types.

Comparing with the candidate-generation algorithms, the episode-recognizing algorithms for parallel episodes and serial episodes are in different approaches. For parallel-episode recognition, each candidate parallel episode α is associated with two identifiers: an event counter $\alpha.event\_count$ indicating the number of events of α in the current window, and a frequency counter $\alpha.freq\_count$ cumulating the total number of windows in which α occurs.

Serial episodes can be recognized by using automata [5]. The algorithm constructs an automaton for each serial episode α every time the first event A_{first} of α comes into the window. The active state of an automaton reflects a prefix of α in the window. When the same state A_{first} of α leaves the window, the corresponding automaton is removed. When an automaton reaches its accepting state at time t, it means that the corresponding episode is entirely in the window. Since there could be multiple instances of the same automaton existing in a window, the starting time t is saved in $\alpha.inwindow$ if no other automata for α are in the accepting state. When an automaton for α in the accepting state is removed and no other automata for α in the accepting state, the counter $\alpha.freq\_count$ is increased by the number of windows that α has been remained in the window.

2   Supporting Algorithms

The PoS-episode mining algorithms consist of two parts, generation of candidate PoS-episodes and recognition of frequent PoS-episodes.

2.1   Generation of Candidate PoS-Episodes

The output of the serial-episode mining algorithm is the frequent-serial episode array, FSE, which is used as the input of the PoS-episode candidate-generation algorithm.

The frequent serial-episode array FSE is the array of all frequent serial-episode sets; $FSE[i]$ is the set of all frequent serial episodes that have the same size i, that is , have the same number of events. The number of frequent serial-episode sets in FSE and the number of episodes in $FSE[i]$ are denoted by $|FSE|$ and $|FSE[i]|$, respectively. The serial episodes of size i in $FSE[i]$ are sorted in the lexicographical order, and $FSE[i][j]$ gives the jth serial episode of size i in $FSE[i]$.

A PoS-episode α is a parallel connection of two serial episodes that connects two events together. We label these two serial subepisodes as α_1 and α_2. Since we employ state-transition automata to recognize serial episodes, we use the term "state" interchangeable with the term "event" when the context is clear. The sizes of α_1 and α_2 can be different but each is at least 3. Since α_1 and α_2 share the same starting state and ending state, the size of α is $\alpha.size = \alpha_1.size + \alpha_2.size - 2$, and the size of a PoS-episode is at least 4. A PoS-episode α remains the same if we interchange its serial subepisodes α_1 and α_2. If we have $\alpha_1 = \alpha_2$, then α becomes a serial episode.

Algorithm 1 generates all possible candidate PoS-episodes α consisting of two serial episodes α_1 and α_2 with $\alpha_1 < \alpha_2$ (with respect to the lexicographic order). We assume that all the event types have been mapped into contiguous integer numbers beginning from 1. Thus, we use the event type as the index of arrays in our algorithm.

According to the Apriori property that all non-empty subsets of a frequent set must also be frequent, the two subepisodes α_1 and α_2 must also be frequent serial episodes for any given α that is frequent. Since FSE is the complete set of frequent serial episodes, Algorithm 1 can explore all possible combinations of the serial-episode pairs and obtain the complete candidate PoS-episode set. Also, because FSE consists of only frequent serial episodes, the candidate PoS-episode set generated by Algorithm 1 is a small set since it utilizes the known frequent serial-episode information. This can greatly reduce the search space when detecting the frequent PoS-episodes.

Algorithm 1. Generating all the possible PoS-episode using the frequent serial-episode set.
Input: An array FSE of frequent serial-episode set.
Output: An array C of all candidate PoS-episodes.
begin
  $C := \emptyset$;
  for $i := 3$ to $|FSE|$ do
    for $j := 1$ to $|FSE[i]|$ do
      $\alpha := FSE[i][j]$;
      for $m := i$ to $|FSE|$ do   // since $\alpha < \beta$ (in lexicographical order)
        for $n := 1$ to $|FSE[m]|$ do
          $\beta := FSE[m][n]$;
          if $(\alpha \geq \beta)$ then
            continue;
          if $(\alpha[1] = \beta[1]$ and $\alpha[\alpha.size] = \beta[\beta.size])$ then
            construct candidate PoS-episode γ based on α and β;
            $C := C \cup \{\gamma\}$;
  end Algorithm 1;

2.2   Recognition of Frequent PoS-Episodes

Frequent PoS-episodes can be recognized by constructing two deterministic finite
automata M_1 and M_2 for each candidate PoS-episode α as shown in Figure 2:
M_1 and M_2 correspond to the two serial subepisodes α_1 and α_2 of the candidate
PoS-episode α.

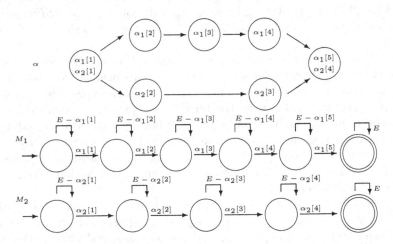

Fig. 1. Two deterministic finite automata M_1 and M_2 for an example PoS-episode α

When symbols, that is, event types are fed in, both M_1 and M_2 will do state
transitions according to the input event types. A PoS-episode α is recognized
when both M_1 and M_2 are in the accepting states. During the recognition, a
sliding window with width w_width is advancing along the time axis. For the
event sequence $\{(A_i, t_i)\}_{i=1}^{n}$, the timestamp of the first sliding window is t_1 and
the timestamp of the last sliding window is $t_n + w\_width - 1$. Each move represents
the passing of one time unit and will bring in zero or one event (A, t), where t
is the timestamp of current window.

Algorithm 2 recognizes all of the frequent PoS-episodes from the given event sequence, window size, and the minimum frequency threshold. This algorithm initializes a pair of instances of automata (M_1, M_2) for each PoS-episode α whose first event type is the same as the event that comes into the sliding window. The automata M_1 and M_2 are equivalent to the subepisodes α_1 and α_2, respectively, of PoS-episode α. The instances of automata pair (M_1, M_2) will be removed when the same event falls out of the sliding window. A PoS-episode α is completely enclosed in the window if both M_1 and M_2 of α have reached their accepting states. We call the earliest time that α is enclosed in the sliding window the entrance time of α. The entrance time is saved into the identifier $\alpha.inwindow$. When an automata pair in the accepting states is removed, we calculate the window number that the PoS-episode α stayed in the sliding window and save it into $\alpha.freq\_count$.

There may be multiple instances of automata of PoS-episode α that exist in the sliding window. Some of them may be in the accepting state. However, according to the definition of frequency of an episode, only the number of windows that contain the PoS-episode will be counted. In other words, for a certain window, it is important that PoS-episode α is present or not, while the number of α is not important. Thus the entrance time of α is defined as the time that the instance of automata of α reaches the accepting states and no other instances of α in the window are in the accepting states. The removal time of automata of α in the accepting states is defined as the time that the automata of α are removed and no other instance of the automata in the accepting states.

We use a two-dimensional array $waits[i][j]$ to keep track of the automata pair (M_1, M_2) that accept event type i in M_1 and j in M_2. In the algorithm, symbols α_1 and α_2 are used interchangeably with M_1 and M_2. The automata pair (M_1, M_2) is preserved in the linked list. The cell $waits[i][j]$ is the head of the linked list. The automata pair (M_1, M_2) is represented in the form (α, i, j, T, w) where α is the PoS-episode and the data structure of α preserves the states and transition information, i and j are the next states in M_1 and M_2, T is the time when the first state of α enters the window, and w is the description of the last transitions of the automata pair (M_1, M_2) as shown below:

w	description of the last transitions of automata pair (M_1, M_2)
0	both M_1 and M_2 moved forward one state
$i \in \{1, 2\}$	M_i moved forward one state and the other automaton remained in the same state
3	both M_1 and M_2 remained in the same states

All the automata pairs that are initialized at time T are linked at $beginset[T]$. If the automata pair is removed, then it is removed from this linked list. Unlike the automata pair stored in the $waits[i][j]$ that i and j are the next states for transitions, the automata pair (α, i, j, T, w) stored in $beginset[T]$ give the states i and j on which transitions just happened.

For each input symbol, all the transitions are organized in a linked list $transitions$. Automata stored in $transitions$ are in the form of (α, i, j, T, A) where

α, i, j, and T have the same semantics as $beginset[T]$ and A is the input symbol, that is, current processing event.

3   PoS-Episode Mining on a Large Web Log Data Set

The PoS-episode mining algorithms are studied by carrying out Web page traversal pattern mining using Web server log data set. When a user is visiting a Web site, each visit of a Web page can be viewed as an event, in which the Web page name is the event type and the visiting time is the event time. Thus, a Web site access event has the form of (Web page name, visiting time). All the events in a Web server log file constitute of an event sequence. The experiment data set is the Web site (cybermath.okstate.edu) access log data. This Web site is a remote education Web site that belongs to College of Arts and Sciences of Oklahoma State University. It provides college-level calculus courses to high-school students, who use this Web site to view lessons, take quizzes, and communicate with other users. The basic information of the Web site and log data set is:

Web site name	cybermath.okstate.edu
individual Web pages number	304
log starting time – ending time	08/Jan/2003:09:09:03 – 28/Feb/2003:19:18:22
total log entries	44955
effective Web pages entries	19710
average visiting length: Web site / Web page	493 seconds / 44 seconds

In this experiment, the purpose of data mining is to discover the Web page traversal patterns. In the Web site, each Web page has more than one links that point to other Web pages and may even have links point back to itself. If we treat each Web page as a node and each link as an edge, then we can view the Web pages in a Web site as a complex graph. From one node to another node, there could be many ways to move along. To improve the Web site design or to efficiently put advertisements on the right Web pages, the frequent Web page traversal patterns are needed.

One of the characters of the Web log data is the sparseness of visit events compared to the total available timestamps. During the night, there are long gaps of silence without users' visit. This leads to a very small minimum frequency ratio to be set. On the other hand, the algorithms iterate on every sliding window. In the Web server log, there could be an event in any second. So, the step for the sliding windows is one second. The sparseness of the event can slow down the computation time dramatically. Thus, condensing of Web log data is needed. The data condensing process should not introduce new relationship among any events. This means that any two adjacent events with a gap width $w_{\mathrm{gap}} > w\_width + 1$ should be moved together into a gap width of $w\_width + 1$ between them. All other relative distances between events are preserved during the condensing process.

Based on the statistics collected for the average visit lengths for Web site and Web page (493 seconds and 44 seconds, respectively), the experimental window sizes are chosen. We tabulate below the frequent PoS-episodes number in different combination of window width and minimum frequency ratio obtained

Algorithm 2. Recognize all frequent PoS-episodes from the given event sequence, window size, and the minimum frequency threshold.

Input: Given the number of event type ET, the candidate PoS-episode array C, an event sequence $s = (s, T_s, T_e)$, a window width w_width, and a frequency threshold min_fr.

Output: The set $FPoS$ of frequent PoS-episodes, that is, $F(s, w\_width, min\_fr)$.

begin

// initialization

for $i := 1$ to ET do

  for $j := 1$ to ET do

    $waits[i][j] := \emptyset$;

for each $\alpha \in C$ do

  $waits[\alpha_1[1]][\alpha_2[1]] := waits[\alpha_1[1]][\alpha_2[1]] \cup \{\alpha, 1, 1, -1, -1)\}$;

  $\alpha.freq\_count := 0$; $\alpha.inwindow := -\infty$;

// recognition

for $start := T_s - w\_width + 1$ to T_e do

  $t := start + w\_width - 1$;

  $beginset[t] := \emptyset$; $transitions := \emptyset$;

  for all events $(A, t') \in s$ such that $t' = t$ do

    for all $(\alpha, i, j, T, w) \in waits[A][*] \cup waits[*][A]$ do

      if $i = |\alpha_1|$ and $j = |\alpha_2|$ and $\alpha.inwindow = -\infty$ then

        $\alpha.inwindow = start$;

      if $i = j = 1$ then

        $transitions := transitions \cup \{(\alpha, 1, 1, t, A)\}$;

      else

        $transitions := transitions \cup \{(\alpha, i, j, T, A)\}$;

        if $w = 0$ then

          $beginset[T] := beginset[T] - \{(\alpha, i - 1, j - 1)\}$;

        else if $w = 1$ then

          $beginset[T] := beginset[T] - \{(\alpha, i - 1, j)\}$;

        else if $w = 2$ then

          $beginset[T] := beginset[T] - \{(\alpha, i, j - 1)\}$;

        else if $w = 3$ then

          $beginset[T] := beginset[T] - \{(\alpha, i, j)\}$;

        $waits[\alpha_1[i]][\alpha_2[j]] := waits[\alpha_1[i]][\alpha_2[j]] - \{(\alpha, i, j, T, w)\}$;

  for all $(\alpha, i, j, T, w) \in transitions$ do

    if $w = \alpha_1[i]$ and $w = \alpha_2[j]$ then

      if $i < \alpha_1.size$ and $j < \alpha_2.size$ then

        $waits[\alpha_1[i + 1]][\alpha_2[j + 1]] := waits[\alpha_1[i + 1]][\alpha_2[j + 1]] \cup \{(\alpha, i + 1, j + 1, T, 0)\}$; $w' := 0$;

      else if $i < \alpha_1.size$ then

        $waits[\alpha_1[i + 1]][\alpha_2[j]] := waits[\alpha_1[i + 1]][\alpha_2[j]] \cup \{(\alpha, i + 1, j, T, 1)\}$; $w' := 1$;

      else if $j < \alpha_2.size$ then

        $waits[\alpha_1[i]][\alpha_2[j + 1]] := waits[\alpha_1[i]][\alpha_2[j + 1]] \cup \{(\alpha, i, j + 1, T, 2)\}$; $w' := 2$;

    else if $w = \alpha_1[i]$ then

      if $i < \alpha_1.size$ then

        $waits[\alpha_1[i + 1]][\alpha_2[j]] := waits[\alpha_1[i + 1]][\alpha_2[j]] \cup \{(\alpha, i + 1, j, T, 1)\}$; $w' := 1$;

      else

        $waits[\alpha_1[i]][\alpha_2[j]] := waits[\alpha_1[i]][\alpha_2[j]] \cup \{(\alpha, i, j, T, 3)\}$; $w' := 3$;

    else if $w = \alpha_2[j]$ then

      if $j < \alpha_2.size$ then

        $waits[\alpha_1[i]][\alpha_2[j + 1]] := waits[\alpha_1[i]][\alpha_2[j + 1]] \cup \{(\alpha, i, j + 1, T, 2)\}$; $w' := 2$;

      else

        $waits[\alpha_1[i]][\alpha_2[j]] := waits[\alpha_1[i]][\alpha_2[j]] \cup \{(\alpha, i, j, T, 3)\}$; $w' := 3$;

    $beginset[T] := beginset[T] \cup \{(\alpha, i, j, T, w')\}$;

  for all $(\alpha, i, j, T, w) \in beginset[start - 1]$ do

    if $i = \alpha_1.size$ and $j = \alpha_2.size$ then

      if no other α in current sliding window in the accepting state then

        $\alpha.freq\_count := \alpha.freq\_count - \alpha.inwindow + start$; $\alpha.inwindow := -\infty$;

      else

        continue;

    else if $w = 0$ then $waits[\alpha_1[i + 1]][\alpha_2[j + 1]] := waits[\alpha_1[i + 1]][\alpha_2[j + 1]] - \{(\alpha, i + 1, j + 1, T)\}$;

    else if $w = 1$ then $waits[\alpha_1[i + 1]][\alpha_2[j]] := waits[\alpha_1[i + 1]][\alpha_2[j]] - \{(\alpha, i + 1, j, T)\}$;

    else if $w = 2$ then $waits[\alpha_1[i]][\alpha_2[j + 1]] := waits[\alpha_1[i]][\alpha_2[j + 1]] - \{(\alpha, i, j + 1, T)\}$;

    else if $w = 3$ then $waits[\alpha_1[i]][\alpha_2[j]] := waits[\alpha_1[i]][\alpha_2[j]] - \{(\alpha, i, j, T)\}$;

$FPoS := \emptyset$;

for all PoS-episodes $\alpha \in C$ do

  if $\alpha.freq\_count/(T_e - T_s + w\_width) \geq min\_fr$ then $FPoS := FPoS \cup \{\alpha\}$;

end Algorithm 2;

from the experiment. The threshold of minimum frequency ratio for each window size is decided by making the algorithms generate suitable number of output frequent PoS-episodes since too many or too few output of frequent PoS-episodes are not desirable. In this work, we use the output frequent PoS-episodes number range of 10 − 1000 to decide the minimum frequency ratios. In the experiment, we first guessed a minimum frequency ratio and used it to get the frequent PoS-episodes number under this ratio. Then we gradually increased or decreased this ratio to let the output frequent PoS-episodes number fall into the desired range.

frequent PoS-episodes number in different (window width, minimum frequency ratio)-combination

window width (seconds)	0.8	1	1.5	2	3	4	5	6	7	8	9	10	12	15	18	20
30	887	639	280	171	45	4										
50				812	311	144	38	13	6							
100						813	532	214	136	13	12	10				
200								870	668	534	259	143	36	8		
300										877		552	170	30		3
400												717	544	133	22	8
500												1059	638	179		44	19

By running the PoS algorithms on the parameters (window width and minimum frequency ratio) described above, we collect all the candidate-episode data and the frequent-episode data for both serial episodes and PoS-episodes, and tabulate below the detailed result of PoS-episode mining. For each category, the candidate episodes are generated first and then the frequent episodes are recognized by the database passes.

PoS-episode mining						serial-episode mining			
	minimum	number of					minimum	number of	
window	frequency	candidate	frequent	fresh		window	frequency	candidate	frequent
size	ratio	PoS-episodes			F_{fresh} (%)	size	ratio	serial episodes	
30	0.008	1004	887	27	3.0	30	0.008	2222	373
30	0.01	723	639	73	11.4	30	0.01	1896	302
30	0.015	301	280	4	1.4	30	0.015	1383	201
30	0.02	186	171	20	11.7	30	0.02	1321	144
30	0.03	49	45	3	6.7	30	0.03	992	91
30	0.05	5	4	1		30	0.05	462	31
50	0.02	847	812	10	1.2	50	0.02	1440	260
50	0.03	354	311	4	1.3	50	0.03	1232	171
50	0.04	165	144	25	17.4	50	0.04	964	110
50	0.05	45	38	3	7.9	50	0.05	759	76
50	0.06	16	13	1	7.7	50	0.06	483	44
50	0.07	10	6	1	16.7	50	0.07	449	35
100	0.04	832	813	5	0.6	100	0.04	1281	218
100	0.05	559	532	61	11.5	100	0.05	1020	172
100	0.06	236	214	21	9.8	100	0.06	890	131
100	0.07	159	136	20	14.7	100	0.07	521	88
100	0.08	48	42	0	0.0	100	0.08	478	66
100	0.09	25	25	0	0.0	100	0.09	442	49
100	0.1	14	10	1	10.0	100	0.1	433	38
200	0.06	952	870	22	2.5	200	0.06	1078	219
200	0.07	726	668	54	8.1	200	0.07	991	183
200	0.08	559	534	63	11.8	200	0.08	658	152
200	0.09	291	259	33	12.7	200	0.09	557	126
200	0.1	168	143	23	16.1	200	0.1	511	95
200	0.12	42	36	3	8.3	200	0.12	456	61
200	0.15	12	8	1	12.5	200	0.15	395	34
300	0.08	967	877	27	3.1	300	0.08	892	204
300	0.1	577	552	68	12.3	300	0.1	608	152
300	0.12	201	170	28	16.5	300	0.12	496	103
300	0.15	33	30	3	10.0	300	0.15	449	55
300	0.2	4	3	0	0.0	300	0.2	370	24

400	0.1	848	717	71	9.9	400	0.1	705	186
400	0.12	567	544	65	11.9	400	0.12	562	150
400	0.15	166	133	23	17.3	400	0.15	480	86
400	0.18	23	22	2	9.1	400	0.18	405	45
400	0.2	12	8	1	12.5	400	0.2	393	33
500	0.1	1212	1059	53	5.0	500	0.1	797	219
500	0.12	774	638	69	10.8	500	0.12	634	169
500	0.15	213	179	26	14.5	500	0.15	503	108
500	0.18	58	44	6	13.6	500	0.18	443	65
500	0.2	20	19	6	31.6	500	0.2	405	44

The ratio of the frequent-episode number to the candidate-episode number can provide how efficient the candidate-generation algorithms are. The larger the ratio is, the more efficient the algorithms are. From the table above, we can see that the PoS candidate-generation algorithm has a high ratio and the efficiency is high. Only about 10% of the candidates turn out to be not frequent.

Among the discovered frequent PoS-episodes, there are two categories of PoS-episodes. We name one the fresh PoS-episode and the other the associating PoS-episode. A PoS-episode α is an associating PoS-episode if there is at least one input string on event-type set that makes the automata pair (M1, M2) for α to reach their accepting states and also makes at least one frequent serial-episode based automaton reach the accepting state. We say that these frequent serial episodes are associated with the PoS-episode α. Apparently, there could be one or more serial episodes associated with a PoS-episode. A PoS-episode is a fresh PoS-episode if it is not an associating PoS-episode. An associating PoS-episode α represents a class of serial episodes that accept the same inputs with α. A fresh PoS-episode is a new discovery of the PoS-episode mining algorithm and no frequent serial episode is associated with it. All the associated serial episodes of a fresh PoS-episode are eliminated from the output of serial-episode mining because their frequencies are lower than the minimum frequency threshold.

We can use the fraction of fresh PoS-episode, F_{fresh}, to evaluate the performance of the parameter settings PoS-episode mining:

$$F_{\text{fresh}} = \frac{\text{number of fresh PoS-episodes}}{\text{number of all frequent PoS-episodes}}.$$

A high fraction F_{fresh} of fresh PoS-episode indicates that the PoS-episode mining algorithm picks up a large number of frequent PoS-episodes that have been missed by the serial-episode mining algorithm. The table above lists the numbers of fresh PoS-episode and the values of F_{fresh}. We can see from the tabulated statistics that for a specific window size, the number of frequent PoS-episodes decreases while the minimum frequency ratio increases. However, the F_{fresh} increases first, then reaches a maximum value, and then begins to decrease. We can easily discover that when F_{fresh} reaches its maximum value, the number of fresh PoS-episodes and the number of frequent serial episodes reach reasonable values for data mining users who will eventually examine the result manually. If we plot out these window sizes and minimum frequency ratios corresponding to the maximum values of F_{fresh}, we can find that they are almost in a linear relation. This can give the PoS-episode mining a guide on how to choose the mining parameters.

H.K. Dai and G. Wang

For an associating PoS-episode α, its sliding window count $C_{\text{associating}}$ has such relationship with the count of the associated serial episodes $C_{\text{associated}}(\beta)$, where β is an associated serial episode of α, as the following equation:

$$\sum_{\text{all } \beta} C_{\text{associated}}(\beta) \geq C_{\text{associating}} \geq \max_{\text{all } \beta} C_{\text{associated}}(\beta).$$

We define the variable $R_{\text{associating}}$ as the ratio of the maximum value of the $C_{\text{associated}}$ to $C_{\text{associating}}$:

$$R_{\text{associating}} = \frac{\max_{\text{all } \beta} C_{\text{associated}}(\beta)}{C_{\text{associating}}},$$

which can be used to evaluate the discreteness of the associated serial episodes of an associating PoS-episode. A value of $R_{\text{associating}} = 1.0$ indicates that there is only one serial episode associated with the PoS-episode. A low value of $R_{\text{associating}}$ gives a hint that there could be several associated serial episodes and each of them has a low frequency.

4 Conclusion

This work extends the existing parallel- and serial-episode mining algorithms by developing the PoS-episode mining algorithms, which provide episode-mining users a powerful mining tool and make the episode mining more flexible. As examples of how to analyze the real PoS-episode mining case and how to decide reasonable parameters, an experiment is performed in this work by studying the Web server log data set and mining the Web page traversal patterns. Concepts and methods are provided for this specific mining case to evaluate the mining process. Our future work focuses on the topological and algorithmic aspects of episode mining, possibly without candidate generation [4]: the consideration of more complex episode topologies of higher-order serial and parallel hierarchical structures, which also lends itself to parallelization for algorithm development.

References

1. R. Agrawal, T. Imielinski, and A. Swami. Mining association rules between sets of items in large databases. In *Proceedings of the 1993 Association for Computing Machinery Special Interest Group on Management Of Data International Conference on Management of Data*, pages 207–216. Association for Computing Machinery, May 1993.
2. R. Agrawal and R. Srikant. Fast algorithms for mining association rules. In *Proceedings of 20th International Conference on Very Large Data Bases*, pages 487–499, September 1994.
3. R. Agrawal and R. Srikant. Mining sequential patterns. In *Proceedings of the Eleventh International Conference on Data Engineering*, pages 3–14, March 1995.
4. J. Han, J. Pei, Y. Yin, , and R. Mao. Mining frequent patterns without candidate generation: a frequent-pattern tree approach. *Data Mining and Knowledge Discovery*, 8(1):53–87, 2004.

5. H. Mannila, H. Toivonen, and A. I. Verkamo. Discovery of frequent episodes in event sequences. *Data Mining and Knowledge Discovery*, 1(3):259–289, 1997.
6. J. Pei and J. Han. Can we push more constraints into frequent pattern mining? In *Proceedings of the Sixth Association for Computing Machinery Special Interest Group on Knowledge Discovery and Data mining International Conference on Knowledge Discovery and Data Mining*, pages 350–354. Association for Computing Machinery, August 2000.

Locating Performance Monitoring Mobile Agents in Scalable Active Networks

Amir Hossein Hadad, Mehdi Dehghan, and Hossein Pedram

Amirkabir University, Computer Science Faculty, Tehran, Iran
a_haddad@itrc.ac.ir, {dehghan, pedram}@ce.aut.ac.ir

Abstract. The idea of active networks has been emerged in recent years to increase the processing power inside the network. The intermediate nodes such as routers will be able to host mobile agents and many management tasks can be handled using autonomous mobile agents inside the network. One of the important limitations, which should be considered in active networks, is the restricted processing power of active nodes. In this paper, we define an optimal location problem for monitoring mobile agents in a scalable active network as a p-median problem, which is indeed a kind of facility location problem. The agents are responsible to monitor and manage the performance of all of the network nodes such that the total monitoring traffic overhead is minimized. Then we proposed two methods of finding an appropriate sub set of intermediate nodes for hosting mobile agents. In our first method, we have not considered the limited processing power of active nodes, which host mobile agents. In our second method, we have solved the problem so that the processing loads of host nodes do not exceed a predefined threshold. Since p-median problems are NP-complete and the search space of these problems is very large, our methods are based on genetic algorithms. We have tested our two methods for finding mobile agents optimal locations on four network topologies with different number of nodes and compared the obtained location. By this comparison, we have shown the importance of considering processing load limitation for active nodes as a parameter in choosing them as hosts of mobile agents in a scalable active network. The proposed locations in our second method eliminates the probability of CPU overload in the active nodes hosting the mobile agents and reduces the processing time required for finding the optimal locations of mobile agents.

Keywords: Active Networks, mobile agents, Performance Monitoring, P-Median Problem, Genetic Algorithm.

1 Introduction

Due to increasing need for data processing power in computer networks, end nodes' processing power does not seem to be enough. In addition, increasing number of nodes in large-scale networks has made it difficult to update the communication pro-

N. Megiddo, Y. Xu, and B. Zhu (Eds.): AAIM 2005, LNCS 3521, pp. 472–482, 2005.

tocols and handling the complex management processes. For such reasons idea of Active networks has been proposed in 1997   [1]. Active network is a network in which intermediate nodes such as routers have processing power to run applications such as end nodes.

On the other hand, complexities in management tasks demand autonomy in management applications. One of the new management technologies to address this problem is Mobile agents. Mobile Agents are autonomous software applications, which are able to migrate to different nodes in a heterogeneous network. Using mobile agents in network management has many advantages. Some of these advantages are effective resource usage, traffic reduction, and real time interaction ([2]- [4]). Additionally a management system, which uses mobile agent for its management tasks, has distributed structure. Such a management system would be highly scalable and flexible ([3], [4]).

Introducing Active Networks was a milestone for effective usage of mobile agents in network management. Variety of researches for using mobile agents in different management tasks is an evidence for this ([5]- [9]). However, there are some limitations for using mobile agents in Active Networks. One of these restrictions is limited processing power of active nodes ([10]-[12]). As the result, active nodes have limited power for hosting mobile agents and performing their management tasks.

In reference   [4] mobile agents have been used for performance monitoring. Goal of this research was to locate monitoring agents' locations in end nodes of a large-scale network. Since the network is not active, no processing power capacity restriction has been assumed for end nodes. In this paper, we extend the work that has been done in   [4] for Active Networks and we have assumed active nodes processing power capacity limitation in addition to other problem limitations.

In next section, we would present the mobile agents Location problem as a p-median problem. Then we would explain in detail our solution, which is based on genetic algorithms in part 3. In 4th part, simulation results are presented. The final part is conclusion, which expresses achieved results and further works.

2 Mobile Agent Location Problem

Suppose there is a central management workstation, which is going to send mobile agents for performance monitoring on the network nodes. A sub set of active nodes is selected for hosting the mobile agents by management workstation. In this selection, host nodes are chosen so that:

- Performance monitoring traffic of the mobile agent is near minimal
- Their Processing usage after hosting the mobile agents wouldn't be beyond than a predefined threshold

Solving the problem for active nodes with processing power limitation, make it a kind of modified p-median problem. This new kind of p-median problem is capacitated p-median problem [13]. We present a formulation of p-median problem in an integer

programming proposed in [14]. In this presentation, it is possible to have each vertex
of graph as both demand and facility. In our case, this is useful, because mobile agent
host nodes (facilities) and the nodes, which are going to be monitored (demands), are
the same in network topology. In other word, each of the active nodes can be a mobile
agent host.

p-median problem:

$$Min \sum_{i=1}^{n} \sum_{j=1}^{n} a_i d_{ij} x_{ij} \tag{1}$$

Subject to the restriction:

$$\sum_{j=1}^{n} x_{ij} = 1, \quad i = 1,2,...,n \tag{2}$$

$$x_{ij} \leq y_j, \quad i = 1,2,...,n \tag{3}$$

$$\sum_{j=1}^{n} y_j = p, \tag{4}$$

$$x_{ij}, y_j \in \{0,1\}, \quad i, j = 1,2,...,n \tag{5}$$

Where:
n = total number of vertices in the graph,
a_i = demand of vertex i,
d_{ii} = distance from vertex i to vertex j,
p = number of facilities used as medians,
a_p, d_{ii} are positive real numbers,.

$$x_{ij} = \begin{cases} 1 \ if \ vertex \ i \ is \ assigned \ to \ facility \ j \\ 0 \ otherwise \end{cases} \tag{6}$$

$$y_j = \begin{cases} 1 \ if \ vertex \ j \ chosen \ as \ facility \\ 0 \ otherwise \end{cases} \tag{7}$$

The objective function (1) minimizes the sum of the (weighted) distances between the
demand vertices and the median set. The constraint set (2) guarantees that all demand
vertices are assigned to exactly one median. The constraint set (3) prevents that a
demand vertex be assigned to a facility that was not selected as a median. The total
number of median vertices is defined by (4) as equal to p. Constraint (5) ensures that
the values of the decision variables x and y are binary (0 or 1).

The main difference between a capacitated p-median problem and p-median problem is two constraints [13]:

1. each facility can satisfy only a limited number of demands (capacity restrictions)
2. all demand points must be satisfied by respecting the capacities of the facilities
   selected as medians

Since in our work active host nodes have capacitated processing power, this version of the problem is a good candidate for our case. As it is mentioned, goal of this work is to minimize monitoring traffic, which is sent from mobile agents to the management workstation. Monitoring traffic is divided into three types:

- Traffic sent from monitored nodes to the mobile agent, which is responsible to monitor them. It is represented by "Traff".
- Traffic sent from mobile agents to the root. It is represented by "rtTraff".
- Traffic of sending the mobile agents to host nodes. It is represented by "rtTraff".

Therefore, the integer programming formulation of our problem would be:

$$Min\,(\sum_{i=1}^{n} \sum_{j=1}^{n} Traff_{ij} d_{ij} x_{ij} + \sum_{j=1}^{p} d_{j} (rtTraff_{j} + MaTraff_{j})) \tag{8}$$

Subject the following constrains:

$$Load_i < Threshold \quad i = 1,2,..< n \tag{9}$$

$$Traff_{ij} > rtTraff_j > MaTraff_j \tag{10}$$

Where:

$Traff_{ij}$: j is index of monitored node, and i is index of active node which is hosting a mobile agent,

$rtTraff_i$: i is index of active node which is hosting a mobile agent,

$MaTraff_i$: i is index of active node which is hosting a mobile agent,

$Load_i$: processing load of i'th node after starting performance monitoring,

$Threshold$: Threshold defined for processing load of active nodes.

Constraint (9) is equivalent to the constraints (1) and (2) of capacitated p-median problem. Constraint (10) is added to make the performance monitoring processes beneficial than the case of monitoring the performance of network without using mobile agents. In simulation, we use a performance-monitoring task, which satisfies constraint (10) regarding monitoring traffic.

3 Proposed Method for Optimally Locating Mobile Agents Hosts

In our proposed method, we use a genetic algorithm for finding near optimal location of mobile agent hosts. In this algorithm, solution is a bit string chromosome, which shows location of the hosts. Length of this bit string is equal to number of network nodes. Bits of the chromosome equal to 1 are location of the mobile agent hosts. We assume that mobile agents are only able to monitor their one-hob distance nodes and their host. Structure of chromosome coding is the same as [17]. In figure 1 a sample chromosome and its meaning in the Active Network is shown.

Proposed solution has been presented for two cases: Case in which goal is to minimize monitoring traffic and case in which goal is to minimize and satisfy processing load constraint.

Assumed conditions for this problem are as follows:

- Each mobile agent is only able to monitor one-hob distance nodes in the Active Network.
- Mobile Agents should monitor all the nodes.

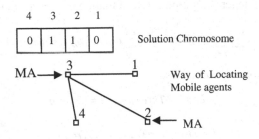

Fig. 1. way of locating mobile agents by a chromosome

The Genetic Algorithm Parameters

Following adjustment is used for the genetic algorithm parameters:

Mutation
Several simulations with different mutation rates have been performed and the best results belong to 0.03 mutation rate. Simulations results are generated using this mutation rate.

Crossover
Different methods of crossover are used in our simulations. Best results are obtained for two-point crossover.

Using Migration
Migration is used in solving this problem. Using migration would increase performance of algorithm for searching problem space in our case, which result in better solutions.

Fitness Function
In this paper, there are defined two different fitness functions. The aim of the genetic algorithm here is to minimize these functions. One of them is defined without regarding the (9) constrain (we refer to this fitness function type 1). The other one is defined regarding the (9) constrain (we refer to this fitness function type 2).

Fitness function type 1

$$Ft1 = Ma2RootTraff \ / \ MaxTraff +$$
$$Nodes2MaTraff \ / \ MaxTraff + OverlapRate \qquad (11)$$

Fitness function type 2

$$Ft2 = Ma2RootTraff \ / \ MaxTraff +$$
$$Nodes2MaTraff \ / \ MaxTraff + \qquad (12)$$
$$OverlapRate + OverloadedNodes \ / \ N$$

Where:

Ma2RootTraff : Traffic sent from mobile agents to the root,
Nodes2MaTraff : Traffic sent from monitored nodes to the mobile agent which is responsible to monitor them,
OverloadedNodes: Number of overloaded nodes,
MaxTraff: Traffic sent from nodes to management workstation in case there is no mobile agent,
OverlapRate: Value of this parameter shows the goodness of mobile agents' locations regarding the monitored nodes. If a part of these nodes have not been monitored or have been monitored more than 1 times, the value of this parameter would be increased. Calculation of the value of it is as follows:

$$OverlapRate = \exp(-1 \ / \ | \ 1 - \sum_{i=1}^{N} (visitedNodes_i \ / \ N) \ |) \qquad (13)$$

visitedNodes$_i$: Number of nodes, which are assigned as facility for monitoring the *i*'th node,
N: Number of active nodes.

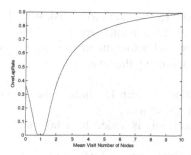

Fig. 2. Different values of *OverlapRate* for different mean of assigning mobile agents for monitoring network nodes. As it could be seen in the figure when this mean is equal to 1 (each node of the network is only is monitored by one mobile agent), the value of Overlap rate is minimum

Genetic algorithm has been run for both these fitness functions and the simulation results are presented in the next part of the paper.

4   Simulation Results

In this part of the paper, genetic algorithm simulation results for four networks have been presented. These networks have 15, 25, 35, and 50 nodes. CPU of the computer, which the simulation has been run on, is Centrino 1.5 GHz with 512Mbytes ram. We define a performance monitoring process in which the mean of 100 parameters of all active nodes of the network should be collected. The duration of this process is assumed 1 hour and 40 minutes, and performance information is collect each 1 minute from each node. Size of each parameter is supposed 16 bytes. This way each node sends 160 Kbytes of traffic data to mobile agent, which is responsible to monitor it (in whole of the monitoring task duration). Mobile agents send 1.6 Kbytes to root for each node they are monitoring. The reason for reduction in size of traffic sent by mobile agents is that they calculate the mean of 100 parameters and then send the result to the root. Size of mobile agents has been chosen two Kbytes, based on Grasshopper framework mobile agent's size[18]. In a more formal way:

$Traff_{ij}$= 160 Kbytes for all i and j,
$rtTraff_i$ = 1.6 Kbytes for all i,
$maTraff_i$ = 2 Kbytes for all i.

The processing load of active nodes is supposed to be a percentage between 20 and 65 percent. The processing load of simple and host nodes are calculated using the following simple formula:

$$sLoad = CLoad + p * ratio \qquad (14)$$

$$hLoad = CLoad + n * p * ratio + maLoad \qquad (15)$$

Where:
$sLoad$: simple node load after the monitoring task started,
$hLoad$: host node load after the monitoring task started,
$CLoad$: Node-processing load before the monitoring task started,
n: number of nodes monitored by the host,
p: parameter number,
$ratio$: it is a constant value less than 1, which is the ratio of processing load for monitoring one parameter of active node,
$maLoad$: processing load, this is added to a host node for running a mobile agent on it.

Figure 3 shows the processing load of active nodes after starting the monitoring process for 15 and 25 node networks for fitness functions type one and two. It can be seen in this figure that not considering the processing load can leads the active nodes to overload.

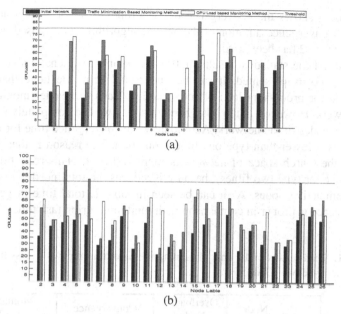

(a)

(b)

Fig. 3. (a) network with 15 nodes. There is one overload in this network using type one fitness function. (b) Network with 25 nodes. There are 2 overloads in this network using type one fitness function

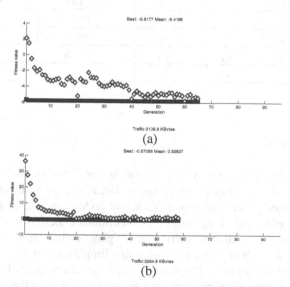

(a)

(b)

Fig. 4.(a) Simulation results using fitness type 1 (b) Simulation results using fitness type 2. Diamonds are fitness mean for the whole population of each generation and solid squares are fitness of best chromosome in each population

Figure 4 shows the convergence of the proposed method for a network with 10 nodes. Convergence duration of the genetic algorithm is decreased, when the fitness function type 2 has been used.

In table 1 the statistical results of 10 times running of genetic algorithm for different networks is presented. It can be understand form the table that considering a threshold for processing load of active nodes, leads to heavier monitoring traffic in the network. However, in the other hand this way we can completely prevent overload in active nodes of the network. Additionally the convergence time for type two fitness functions is lower than type one fitness function. The reason is that, in the type two fitness the search space of network is more restricted. This reduction in time is an advantage for type two fitness, because it reduces the total needed to gather information from active nodes. As it can be seen in table 1, total time of genetic algorithm convergence is proper in comparison with total monitoring task time.

Table 1. Mean results of 10 time of running the algorithm for two fitness functions type one (GA1) and type two (GA2)

	Node Number	Overloaded Nodes percentage	Convergence time	Monitoring Traffic Kbytes
GA1	15	12	5.79	2962.1
GA2	15	0	4.71	3284.8
GA1	25	16	23.64	4959.4
GA2	25	0	20.48	5445.1
GA1	35	14	45.02	8140
GA2	35	0	37.4	9267
GA1	50	13	72.06	13263
GA2	50	0	57.23	11342

5   Conclusion

In deploying mobile agents systems in the Active Networks, we should consider the processing load constrains of active nodes. Because as it is shown in this work, supposing unlimited processing power for active nodes can lead to active several nodes overloads in using mobile agents. We could eliminate these overloads by adding overload parameter to our fitness function in our genetic algorithm. In the other hand, our simulation results show that considering this limitation can result in finding the mobile agents locations faster. The reason is that the search space is more restricted this way. The total time of monitoring task in comparison with the time of finding locations by genetic algorithm was proper in our results.

In further work, we would solve the problem for mobile agents, which are able to monitor the whole network from their location. We have to change the structure of chromosome to solve the problem in this case. In one hand, we can achieve more optimal solution for location problem, but in the other hand, this might cause to increase the complexity of finding locations of mobile agents.

References

[1]   D. L. Tennenhouse, J.M. Smith, W. D. Sincoskie, D. J. Wetherall, and G.J. Minden. "A survey of Active Network research", IEEE Communications Magazine, pages 80 - 86, January 1997.

[2]   S. Green, L. Hurst, B. Nangle, P. Cunningham, F. Sommers and R. Evans, "Software Agents: A review", Technical Report, Department of Computer Science, Triniy College, Deblein, Irland, 1998.

[3]   D. Chess, C. G. Harrison and A. Kershenbaum, "mobile agents: Are They a Good Idea?" G. Vigna (ed.), mobile agents and Security, LNCS 1419, Springer Verlag, 1998.

[4]   Antonio Liotta, "Towards Flexible and Scalable Distributed Monitoring with mobile agents", Doctor of Philosophy Dissertation, University of London, July 2001.

[5]   D. Gavalas, D. Greenwood, M. Ghanbari, and M. O'Mahony, "An infrastructure for distributed and dynamic network management based on mobile agent technology." In Proceedings of the IEEE International Conference on Communications (ICC99), 1362-1366, 1999.

[6]   Daniel Rossier, Rudolf Scheurer, "An Ecosystem-inspired mobile agent Middleware for Active Network Management", Swisscom Corporate Technology, University of Fribourg, Switzerland, 2003.

[7]   Breugst, M. Magedanz, "mobile agents – Enabling Technology for Active Intelligent Network Implementation", IEEE Network, Vol. 12, No. 3, May 1998. R. Kazi, P. Morreale, "mobile agents for Active Network Management", Stevens Institute of Technology, IEEE, 1999.

[8]   A. Galis, D. Griffin, W. Eaves, G. Pavlou, S. Covaci, R. Broos — "Mobile Intelligent Agents in Active Virtual Pipes" — in "Intelligence in Services and Networks" — Springer Verlag Berlin Heildelberg, April 2000.

[9]   Rumeel Kazi, Patricia Moreale, "mobile agents for Active Network Management", Stevens Institute of Technology, IEEE communications surveys, 1999.

[10]   Celestine Brou et al. , " Future Active IP Networks (FAIN) ", GMD company, Initial Specification of Case Study Systems, May 2001.

[11]   Florian Baumgartner, Torsten Braun, and Bharat Bhargava, "Design and Implementation of a Python-Based Active Network Platform for Network Management and Control", Design and Implementation of a Python-Based Active Network Platform for Network Management and Control, pages 177-190, Springer-Verlag, 2002.

[12]   L. Kencl, J. L. Boudec, "Adaptive Load Sharing for Network Processors", In Proceedings of INFOCOM '02, IEEE, June 2002.

[13]   E. S. Correa, M. T. A. Steiner, A. A. Freitas, C. Carnieri, "A genetic algorithm for solving a capacitated p-median problem", Numerical Algorithms 35, pages 373–388, Kluwer Academic Publishers, 2004.

[14]   C. Revelle, R. Swain, Central facilities location, Geographical Analysis 2 (1970) 30–42.

[15]   S. Shephered, A. Sumali, "A Genetic Algorithm Based Approach to Optimal Toll Level and Location Problems", Networks and Spatial Economics (4), pages 161-179, Kluwer Academic Publishers, 2004.

[16]   O. Alp, E. Erkut, Z. Drezner, "An Efficient Genetic Algorithm for the p-Median Problem", Annals of Operations Research 122, pages 21–42, Kluwer Academic Publishers, 2003.

[17]   J. H. Jaramillo, J. Bhadury, R. Batta, "On the use of genetic algorithms to solve location problems", Computers & Operations Research 29, pages 761-779, Elsevier Science, 2002.

[18]   Grasshopper Development System, Light Edition v2.2.1, Programmer's Guide, IKV++ GmbH, [URL:http://www.grasshopper.de], Berlin, 2001.

Author Index

Lecture Notes in Computer Science

For information about Vols. 1–3427

please contact your bookseller or Springer